CBS | PROBLEMS and
SOLUTIONS *Series*

Problems and Solutions in
NETWORK
ANALYSIS

- 📖 This book is suitable for examinations of all Indian universities.
- 📖 Solutions of **460 problems** are given in most simplified manner.
- 📖 All types of problems are included and the solutions given in detail.
- 📖 To explain the basics lucidly, computerised figures have been liberally added in the solutions. This volume contains more than **675 line diagrams**.

CBS Problems and Solutions *Series*

- 📖 Engineering Thermodynamics
- 📖 Strength of Materials
- 📖 Integrated Electronics
- 📖 Communication Systems
- 📖 Signals and Systems
- 📖 Power Systems
- 📖 Electrical Machines and Transformers
- 📖 Network Analysis
- 📖 Power Electronics
- 📖 Control Systems
- 📖 Analog Systems

 *plus many ... more new topics*

CBS PROBLEMS and SOLUTIONS Series

Problems and Solutions in NETWORK ANALYSIS

R. GOPAL
Engineer

CBS

CBS Publishers & Distributors Pvt. Ltd.

New Delhi • Bengaluru • Chennai • Kochi • Kolkata • Mumbai
Hyderabad • Nagpur • Patna • Pune • Vijayawada

ISBN: 81-239-1196-3

First Edition: 2005
Reprint: 2008, 2012, 2016

Published by:
Satish Kumar Jain for CBS Publishers & Distributors Pvt. Ltd.,
4819/XI Prahlad Street, 24 Ansari Road, Daryaganj, New Delhi - 110002
delhi@cbspd.com, cbspubs@airtelmail.in • www.cbspd.com
Ph.: 23289259, 23266861, 23266867 • Fax: 011-23243014

Corporate Office: 204 FIE, Industrial Area, Patparganj, Delhi - 110 092
Ph: 49344934 • Fax: 011-49344935
E-mail: publishing@cbspd.com • publicity@cbspd.com

Branches:
• *Bengaluru:* 2975, 17th Cross, K.R. Road, Bansankari 2nd Stage,
 Bengaluru - 70 • Ph: +91-80-26771678/79 • Fax: +91-80-26771680
 E-mail: cbsbng@gmail.com, bangalore@cbspd.com
• *Chennai:* No. 7, Subbaraya Street, Shenoy Nagar, Chennai - 600030
 Ph: +91-44-26681266, 26680620 • Fax: +91-44-42032115
 E-mail: chennai@cbspd.com
• *Kochi:* Ashana House, 39/1904, A.M. Thomas Road, Valanjambalam,
 Ernakulum, Kochi • Ph: +91-484-4059061-65
 Fax: +91-484-4059065 • E-mail: cochin@cbspd.com
• *Kolkata:* 6-B, Ground Floor, Rameshwar Shaw Road, Kolkata - 700014
 Ph: +91-33-22891126/7/8 • E-mail: kolkata@cbspd.com
• *Mumbai:* 83-C, Dr. E. Moses Road, Worli, Mumbai - 400018
 Ph: +91-9833017933, 022-24902340/41 • E-mail: mumbai@cbspd.com

Representatives:
• Hyderabad: 0-9885175004   • Nagpur: 0-9021734563
• Patna: 0-9334159340   • Pune: 0-9623451994
• Vijayawada: 0-9000660880

Printed at:
J.S. Offset Printers, Delhi

Preface

Network Analysis is one of the fundamental subjects for the students of electronics, electronics and telecommunication and electrical engineering. The author feels it a great pleasure in presenting this very compex subject through typical problems and their solutions.

This book covers all important and major topics taught at the undergraduate level in an engineering college. All the **460 problems** have been discussed at length, and more than **675 line diagrams** given extensively to help the reader understand the basics and grasp the logic. This book is suitable for preparing for all university examinations and also to face the competitions. This feature is expected to help the reader in visualising the patterns of problems and the extent and level of difficulty in solving the questions likely to be encountered in the examinations, and thus tackling the problems with confidence and success.

Although every care has been taken to ensure accuracy, yet some errors might have crept in. The author will be grateful if these are brought to his notice so that these could be rectified in the subsequent printings and editions of the book. Readers are requested to send their suggestions for further improvement and revision to me through the publishers at the following e-mail address: cbspubs@del3.vsnl.net.in.

R. GOPAL

Contents

① Introduction and Basic Concepts

Problem 1.1. In a certain copper conductor, the current density is 250 amp/cm², and there are 5×10^{22} free electrons in 1 cm³ of copper. What is the mean electron drift in centimetres per second?

Solution :

Current density $J = 250$ amp / cm²

No. of free electrons $N = 5 \times 10^{22}$ per cm³.

Electron charge $e = 1.6021 \times 10^{-19}$ coulomb.

∴ Drift velocity $v = J/Ne$ or $v = J/N.e$

$$= 250/5 \times 10^{22} \times 1.621 \times 10^{-19} \text{ cm / sec}$$

$$v = 3.12 \times 10^{-2} \text{ cm / sec } \textbf{Ans.}$$

Problem 1.2. The current in a circuit varies according to the equation $i = 2e^{-t}$ amperes for t greater than zero, and is zero for t less than zero. Find the total charge that passes through the circuit in coulombs.

Solution :

$$i\,(t) = \begin{cases} 0, & t < 0 \\ 2e^{-t}, & t > 0 \end{cases}$$

So, $\quad q\,(t) = \int\limits_{-\infty}^{0} i\,dt + \int\limits_{0}^{\infty} i\,dt = 0 + \int\limits_{0}^{\infty} 2e^{-t}\,dt$

$$= 2\left[\frac{e^{-t}}{-1}\right]_0^\infty = -2\left[e^{-\infty} - e^0\right]$$

$$= -2[0-1] = 2 \text{ coulomb } \mathbf{Ans.}$$

Problem 1.3. A simple capacitor is constructed from two parallel plates of metal separated by a dielectric material. Assuming no fringing of the electric field at the edges of the plates, show that the capacitance of this capacitor is

$$C = \frac{\varepsilon A}{d} F$$

Where A is the area of either plate in meters^2, d is the distance of separation of the plates in meters, and

$$\varepsilon = K\varepsilon_0 = 8.854 \times 10^{-12} K$$

Where K is the relative dielectric constant (which is 1 for air).

Solution:

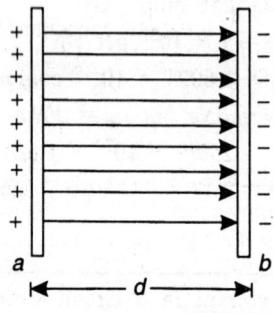

Fig. P. 1.3

Given, Area of plate $= A$ (m^2)

Distance between plates $= d$ (m)

Dielectric constant $\varepsilon = K\varepsilon_0 = 8.854 \times 10^{-12}$ K.

Let the charge on the top plate be q. So charge density on the plate will be $q / A = D$.

From Gauss's law $q = \int_S D \cos\theta \cdot ds$.

where D is flux density, ds is an increment of surface and θ is the angle between D and ds. Flux density and dielectric field are related by the equation $D = \varepsilon E$ [where ε = Dielectric constant of the medium and E = electric field].

$$\therefore \qquad E = \frac{D}{\varepsilon} = \frac{q}{\varepsilon A}$$

Now voltage can be expressed in terms of electric field as:

$$v_{ab} = \int_{a}^{b} E \cos \theta \cdot dl.$$

For parallel plate capacitor $\cos \theta = 1$.

So,

$$v = E \int_{a}^{b} dl = E \cdot [L] \frac{b}{a} = E [b - a] = E \cdot d.$$

so,

$$v = E.d$$

or,

$$v = \left(\frac{q}{\varepsilon A} \right) \cdot d$$

or,

$$v = \left(\frac{d}{\varepsilon A} \right) q = \frac{q}{c}$$

where,

$$C = \frac{\varepsilon A}{d} \text{ is called capacitance.}$$

$$\therefore \qquad C = \frac{\varepsilon A}{d} \text{ Ans.}$$

Problem 1.4. The tuning capacitor that is located in radio receivers is represented in Fig. P. 1.4. The plates are separated by air by a

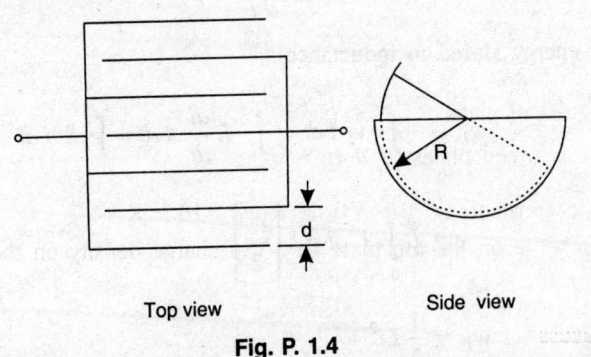

Top view   Side view

Fig. P. 1.4

distance d; the movable plates and the fixed plates are each connected together as shown in the figure. Neglecting fringing at the edges, determine the maximum capacitance of the tuning capacitor.

Solution:

From the side view, it appears that the maximum capacitance would occur when the area of the plate is maximum and maximum area occurs

when two half plate coincided, i.e. Area $= \dfrac{\pi R^2}{2}$.

From top view it appears that there are six capacitance in parallel separated by a distance d. So the maximum capacitance is given as:

$$(C)_{max} = \frac{6 \times \varepsilon A_{max}}{d} = \frac{6 \times \dfrac{\pi R^2}{2}}{d}$$

[Where ε = dielectric constant]

$$(C)_{max} = 3 \varepsilon \pi R^2 / d \text{ Farad } \textbf{Ans.}$$

Problem 1.5. From the defining equation for energy, $w = \displaystyle\int_{-\infty}^{t} vi \, dt$,

show that, for the inductance, $w_L = \dfrac{1}{2} Li^2$ and $w_L = \dfrac{1}{2} \psi^2 / L$.

Solution:

For an inductance, voltage is given as:

$$v_L = L \frac{d i}{d t}.$$

Hence energy stored in inductance,

$$w_L = \int_{-\infty}^{t} v_L \, i \, dt = \int_{-\infty}^{t} L \frac{di}{dt} . i \, dt = \int_{0}^{t} L . i \, di$$

$$= L \int_{0}^{i} i . di = L \left[\frac{i^2}{2} \right]$$

$$w_L = \frac{1}{2} Li^2$$

$\therefore \qquad \psi = i.L \quad$ or, $\quad i = \dfrac{\psi}{L}$

$\therefore \qquad w_L = \dfrac{1}{2} L. \left(\dfrac{\psi}{L}\right)^2, \quad w_L = \dfrac{1}{2}\dfrac{\psi^2}{L} \quad$ **Ans.**

Problem 1.6. From the equation for energy in Prob. 1.5, show that for the capacitance, $w_c = \dfrac{1}{2} Cv^2$ and $w_c = \dfrac{1}{2} Dq^2$.

Solution:

Current through capacitance $\quad i = C\dfrac{dv}{dt}$

Energy stored in capacitance

$$w_c = \int\limits_{-\infty}^{t} v.i.dt = \int\limits_{o}^{v} v.c\dfrac{dv}{dt}\,dt = c\int\limits_{o}^{v} v.dv = c\left[\dfrac{v^2}{2}\right]_0^v$$

$$w_c = \dfrac{1}{2}cv^2$$

$\therefore \qquad D = \dfrac{1}{c} = \dfrac{v}{q}, \; \therefore \; w_c = \dfrac{1}{2}Dq^2 \quad$ **Ans.**

Problem 1.7. Assume that the inductance parameter is defined as the constant relating stored energy and the current squared by the equation $w_L = \dfrac{1}{2}Li^2$. Making use of the relationship $p = vi$, show that for constant inductance the voltage across the inductor is $v_L = L\left(di/dt\right)$.

Solution:

Given $\qquad\qquad w_L = \dfrac{1}{2}Li^2$

$\therefore \quad$ Power $\qquad p = \dfrac{dw_L}{dt} = \dfrac{d}{dt}\left(\dfrac{1}{2}Li^2\right)$

When L is constant.
$$p = \frac{1}{2} L . 2i \frac{di}{dt}$$

$$p = L . i \frac{di}{dt} = v . i$$

or
$$v = L . \frac{di}{dt} \text{ Ans.}$$

Problem 1.8. A number of devices and systems make use of banks of capacitors, such as power transmission systems, nuclear particle accelerators, and electronic flash units used in photography. Capacitors for these devices are designed using a dielectric material selected to prevent breakdown at the operating voltage. For a given dielectric constant, show that the energy stored is directly proportional to the volume of capacitors. From this, show that for a given dielectric material the cost per joule storage of capacitors is approximately constant and independent of voltage rating or capacitance value per individual capacitor.

Solution:

We know that capacitor stored energy is equal to

$$w_c = \frac{1}{2} c v^2$$

$\because$ Voltage across capacitor, where d is the distance between two plates.

so,
$$v = k d, \text{ [where } k = \text{constant]}$$

$\therefore$
$$w_c = \frac{1}{2} \left(\frac{\varepsilon A}{d} \right) k^2 d^2$$

$$w_c = \frac{1}{2} \varepsilon A . d k^2 = \frac{1}{2} \varepsilon k^2 (A.d)$$

or,
$$w_c \propto (A.d) \pi = \text{Volume of capacitors.}$$

So, $w_c \propto$ Volume of capacitors

Since the energy stored is also directly proportional to the area of plate, if we increase the area of plates, the capacitor cost will also increase but the stored energy will also increase. That is why the cost per Joule storage of capacitiors is approximately constant. And it is independent of voltage rating or capacitance value per individual capacitor. **Ans.**

Problem 1.9. The voltage and charge in a certain nonlinear capacitor are related by the equation $q = Kv^{1/3}$. For this capacitor, compute the energy stored as a function of capacitor charge.

Solution :

Given

$$q = k v^{1/3}, \text{ or } v = \frac{q^3}{k^3}$$

or,

$$i = \frac{d q}{d t}$$

and capacitor stored energy

$$w_c = \int_{-\infty}^{t} v . i \, d t$$

So,

$$w_c = \int_{-\infty}^{t} \frac{q^3}{k^3} \cdot \frac{dq}{dt} \cdot dt = \int_{o}^{q} \frac{q^3}{k^3} . d q$$

or,

$$w_c = \frac{1}{k^3} \left[\frac{q^4}{4} \right]_0^q = \frac{1}{4 k^3} \left[q^4 \right] \quad \textbf{Ans.}$$

Problem 1.10. A capacitor of capacitance 1-μF is charged to 200 V. If the stored energy is used with 100 per cent efficiency to lift a 100-lb boy, through what distance will he be lifted?

Solution:

Total energy stored in the capacitor

$$w_c = \frac{1}{2} C v^2 = \frac{1}{2} 1 \times 10^{-6} (200)^2$$

$$w_c = 2 \times 10^{-2} \text{ J}$$

Since we know that 1 ft-lb = 1.356 Joules. and here the efficiency of energy used is 100 per cent. So the total energy stored in capacitor will fully utilized to lift a 100-lb boy. So from energy balance:

$$x.100 (1.356) \text{ J} = 2 \times 10^{-2} \text{ J}$$

where x = distance by which boy is lifted (in ft)

So, $x = \dfrac{2 \times 10^{-2}}{100 \times 1.356}$ ft $= 1.475 \times 10^{-4}$ ft

So, $x = 1.475 \times 10^{-4}$ ft **Ans.**

Problem 1.11. A current of 1 amp is supplied by a source to an inductor of value $\dfrac{1}{2}$ H. Compute the energy stored in the inductor. What happens to this energy if the source is replaced by a short circuit?

Solution:

Given $L = \dfrac{1}{2}$ H, current $i = 1$ amp

So, energy stored in inductor is equal to

$$w_L = \frac{1}{2} L i^2 = \frac{1}{2}\left(\frac{1}{2}\right)(1)^2$$

$$w_L = \frac{1}{4} \text{ Joules}$$

If the source is replaced by a short circuit, then the energy stored in the coil would be converted into power loss through the resistance of the coil.

Problem 1.12. A time-variable inductor changes with time as shown in Fig. P. 1.12 with C replaced by L in henrys and C_m replaced by L_m. This inductor is connected to a current source of constant value I_0 amperes. Determine V_L, the voltage across the inductor.

Solution:

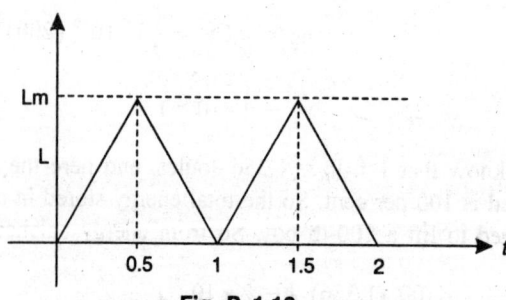

Fig. P. 1.12

when current I_0 through inductor is constant and inductance is variable, than voltage across inductor is given as:

$$V_L = I_0 \frac{dL}{dt}$$

$$L = \begin{cases} \dfrac{L_m}{0.5} t, & 0 < t < 0.5 \text{ and } 1 < t < 1.5 \\[2mm] \dfrac{-L_m}{0.5} t, & 0.5 < t < 1 \text{ and } 1.5 < t < 2 \end{cases}$$

So, $\quad V_L = \begin{cases} I_0 \dfrac{d}{dt}\left(\dfrac{L_m}{0.5} t\right) = 2\, I_0\, L_m, & 0 < t < 0.5 \text{ and } 1 < t < 1.5 \\[3mm] I_0 \dfrac{d}{dt}\left(\dfrac{-L_m}{0.5} t\right) = -2\, I_0\, L_m, & 0.5 < t < 1 \text{ and } 1.5 < t < 2 \end{cases}$

$\therefore \quad V_L = \begin{cases} 2 I_0\, L_m, & 0 < t < 0.5 \text{ and } 1 < t < 1.5 \\[2mm] -2\, I_0\, L_m, & 0.5 < t < 1 \text{ and } 1.5 < t < 2 \end{cases} \qquad \textbf{Ans.}$

Problem 1.13. In the circuit shown the switch K is closed at $t = 0$ (the reference time). The current flowing in the circuit is given by the equation $i(t) = (1 - e^{-t})$ amp, $t > 0$. At a certain time the current has a value of 0.63 amp.

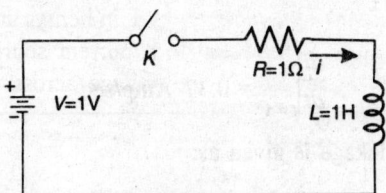

Fig. P. 1.13

(a) At what rate is the current changing?
(b) What is the value of the total flux linkages?
(c) What is the rate of change of flux linkages?
(d) What is the voltage across the inductor?
(e) How much energy is stored in the magnetic field of the inductor?

(f)   What is the voltage across the resistor?

(g)   At what rate is energy being stored in the magnetic field of the inductor?

(h)   At what rate is energy being dissipated as heat?

(i)   At what rate is the battery supplying energy?

Solution:

KVL equation when the switch K is closed:

$$iR + L\frac{di}{dt} = V$$

or, $$i \cdot 1 + 1 \cdot \frac{di}{dt} = 1$$

or, $$i(t) = (1 - e^{-t}) \text{ amp}$$

At time $t = \tau$ (time constant), current reaches to 63% of the maximum value. So,

$$0.63 = \left(1 - e^{-\tau}\right)$$

or, $$e^{-\tau} = 1 - 0.63 = 0.37$$

or, $$-\tau = -0.994 \text{ sec.}$$

or, $$\tau = -0.994 \text{ sec.}$$

(a)   The rate of charging current is given as: $\dfrac{di}{dt} = e^{-t}$

So, $$\left.\frac{di}{dt}\right|_{t=\tau} = e^{-\tau} = e^{-0.994} = 0.37 \text{ Amp/sec.}$$

$\therefore$ $$\left.\frac{di}{dt}\right|_{t=\tau} = 0.37 \text{ Amp/sec}$$

(b)   Total flux linkage is given as:

$$\Psi = L \cdot i = (1 - e^{-\tau}) = 1 \times 0.63 = 0.63 \text{ weber}$$

$\therefore$ $$\Psi = 0.63 \text{ weber}$$

(c)   Rate of change of flux linkage is given as:

$$\frac{d\Psi}{dt} = L \cdot \frac{di}{dt} = 1 \cdot e^{-t} = e^{-\tau}$$

$$\frac{d\Psi}{dt} = 0.37 \text{ weber/sec}$$

(d) Voltage across inductor is given as:

$$V_L = L\frac{di}{dt} = 1 \cdot e^{-t} = e^{-\tau}$$

or,　　　　　$V_L = 0.37$ volts

(e) Energy stored in inductor is given as:

$$w_L = \frac{1}{2}Li^2 = \frac{1}{2} \times 1 \cdot (1 - e^{-\tau})^2 = \frac{1}{2}(0.63)^2$$

$$w_L = 0.19845 \text{ joules.}$$

(f) Voltage across resistor is given as:

$$V_R = i \cdot R = (1 - e^{-t}) \cdot R = (1 - e^{-\tau}) \cdot R$$

$$= (0.63) \cdot R = 0.63 \text{ volts}$$

$\therefore$　　　　　$V_R = 0.63$ volts

(g) Rate at which energy stored in inductor is given as:

$$\frac{dw_L}{dt} = \frac{d}{dt}\left(\frac{1}{2}Li^2\right) = \frac{1}{2}L\left(2 \cdot i\frac{di}{dt}\right) = L \cdot i\frac{di}{dt}$$

$$\frac{dw_L}{dt} = 1 \cdot (1 - e^{-t})(e^{-t})$$

or,　　$\left.\frac{dw_L}{dt}\right|_{t=\tau} = (1 - e^{-\tau}) \cdot e^{-\tau} = (0.63) \times 0.37$

$$\left.\frac{dw_L}{dt}\right|_{t=\tau} = 0.233 \text{ watts}$$

(h) The rate at which heat is dissipated is:

$$\frac{dP_R}{dt} = \frac{d}{dt}(i^2 R) = 2 i \cdot R\frac{di}{dt}$$

$$\frac{dP_R}{dt} = 2(1 - e^{-t}) \cdot 1 \cdot e^{-t}$$

or,　　$\left.\frac{dP_R}{dt}\right|_{t=\tau} = 2(1 - e^{-\tau}) \cdot e^{-\tau} = 2 \times 0.63 \times 0.37$

$$\left.\frac{dP_R}{dt}\right|_{t=\tau} = 0.466 \text{ W}$$

(i) Battery supply energy at the rate of

$$\frac{d}{dt}(v \cdot i) = v \cdot \frac{di}{dt} = 1 \cdot e^{-t}$$

or, $\left.\dfrac{d}{dt}(v \cdot i)\right|_{t=\tau} = e^{-\tau} = 0.37$ W

∴ $\dfrac{d}{dt}(v \cdot i) = 0.37$ W   **Ans.**

Problem 1.14. In the circuit shown the capacitor is charged to a voltage of 1 V, and at $t = 0$ the switch K is closed. The current in the circuit is known to be of the form $i(t) = e^{-t}$ amp, $t > 0$. At a certain time the current has a value of 0.37 amp.

(a) At what rate is the voltage across the capacitor changing?

(b) What is the value of the charge on the capacitor?

(c) What is the time rate of chance of the product Cv?

(d) What is the voltage across the capacitor?

(e) How much energy is stored in the electric field of the capacitor?

(f) What is the voltage across the resistor?

(g) At what rate is energy being taken from the electric field of the capacitor?

(h) At what rate is energy being dissipated as heat?

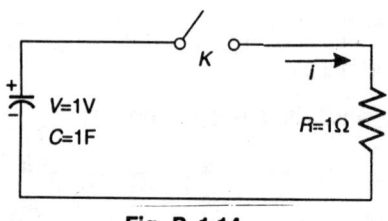

Fig. P. 1.14

Solution:

Given $C = 1F$, $R = 1\Omega$ and current in the circuit

$i(t) = e^{-t}$ amp, $t > 0$.

At time $t = \tau, i(t) = 0.37$ amp

So, $0.37 = e^{-\tau}$ or, $\tau = 0.994$ sec

(a) $$\because i(t) = C\frac{dV_c}{dt}$$

or, $$\frac{dV_c}{dt} = \frac{1}{C}i(t) = \frac{e^{-t}}{C}$$

$$\left.\frac{dV_c}{dt}\right|_{t=\tau} = \frac{e^{-\tau}}{C} = \frac{0.37}{1} \text{ volts/sec.}$$

So, $$\left.\frac{dV_c}{dt}\right|_{t=\tau} = 0.37 \text{ volts/sec.}$$

(b) Charge on the capacitor is given as:

$$q_c = CV_c = 1.V_c = i \cdot R = i \cdot 1 = i$$

So, $$q_c = i = e^{-t}$$

$$\left.q_c\right|_{t=\tau} = e^{-\tau} = 0.37 \text{ coulombs}$$

(c) Rate of CV_c is given as:

$$\frac{d}{dt}(CV_c) = C\frac{dV_c}{dt} = i(t) = e^{-t}$$

$$\left.C\frac{dV_c}{dt}\right|_{t=\tau} = e^{-\tau} = 0.37 \text{ volts/sec}$$

(d) Voltage across capacitor

$$V_c = i.R = i.1 = i$$

So, $$V_c = e^{-t} \text{ or, } \left.V_c\right|_{t=\tau} = e^{-\tau} = 0.37 \text{ volts}$$

(e) Energy stored in capacitor

$$w_c = \frac{1}{2}CV_c^2 = \frac{1}{2}\times 1 \cdot (0.37)^2$$

$$w_c = 0.06845 \text{ joules}$$

(f) Voltage across resistor in given as:

$$V_R = i \cdot R = e^{-t} \cdot 1$$

or, $$\left.V_R\right|_{t=\tau} = e^{-\tau} = 0.37 \text{ volts}$$

$\therefore$ $\qquad$ $V_R\big|_{t=\tau} = 0.37$ volts

(g) Rate of energy from capacitor is given as:

$$\frac{d}{dt}\left(\frac{1}{2}CV^2\right) = \frac{1}{2}C\,2V\frac{dV}{dt} = V\frac{dV}{dt}$$

$$= 0.37 \times 0.37 = (0.37)^2 = 0.1369 \text{ watts/sec}$$

$\therefore$ $\qquad$ $\frac{d}{dt}\left(\frac{1}{2}CV^2\right) = 0.1369$ watts/sec

(h) Energy being dissipated in the form of heat at the rate of:

$$\frac{d}{dt}(i^2R) = R\,2\cdot i\cdot\frac{di}{dt}$$

$$= 1\cdot 2\cdot e^{-t}\cdot(-1)\cdot e^{-t} = -2\cdot(e^{-t})^2$$

or, $\qquad$ $\frac{d}{dt}(i^2R)\bigg|_{t=\tau} = -2\cdot(e^{-\tau})^2 = -2\cdot(0.37)^2$

$$\frac{d}{dt}P_R\bigg|_{t=\tau} = -0.4107 \text{ watts/sec } \textbf{Ans.}$$

Problem 1.15. Show that the following quantities all have the dimension of time: (a) RC; (b) L/R; (c) $\sqrt{LC}$. Show that (d) R^2C has the dimension of inductance, (e) $\sqrt{LC}$ has the dimension of resistance, (f) L/R^2 has the dimension of capacitance.

Solution:

(a) For RC $\qquad$ $R = \dfrac{V}{i}$ and $C = \dfrac{i\cdot t}{V}$ $\left[i.e.,\ i = C\dfrac{dV}{dt}\right]$

So, $\qquad$ $RC = \dfrac{V}{i}\cdot\dfrac{i\cdot t}{V} = t$ (sec)

So, RC has dimension of time.

(b) For L/R $\qquad$ $R = \dfrac{V}{i}$ and $L = \dfrac{V\cdot t}{i}$ $\left[\because V = L\dfrac{di}{dt}\right]$

So, $\dfrac{L}{R} = \dfrac{V \cdot t}{i \cdot V / i} = t \text{ (sec)}$

So, $\dfrac{L}{R}$ has also the dimension of time.

(c) For $\sqrt{LC}$ $\qquad C = \dfrac{i \cdot t}{V}$ and $L = \dfrac{V \cdot t}{i}$

so, $\qquad \sqrt{LC} = \sqrt{\dfrac{i \cdot t}{V} \cdot \dfrac{V \cdot t}{i}} = \sqrt{t^2} = t \left(\text{sec}\right)$

so, $\sqrt{LC}$ has also the dimension of time.

(d) For $R^2 C$ dimension of inductance (in term of V, i, t) is given as:

$$L = \dfrac{V \cdot t}{i} \quad \left[\because V = L \dfrac{di}{dt} \right] \tag{1}$$

$$\text{Dimension of } R^2 C = \left(\dfrac{V}{i}\right)^2 \cdot \dfrac{i \cdot t}{V} = \dfrac{V \cdot t}{i} \tag{2}$$

equation (1) = equation (2)

So, dimension of $R^2 C$ and L are equal

(e) Dimension of $\sqrt{L/C}$:

$$\sqrt{\dfrac{V \cdot t \cdot V}{i \cdot i \cdot t}} = \sqrt{\dfrac{V^2}{i^2}} = \dfrac{V}{i} = R$$

So, dimension of $\sqrt{L/C}$ is equal to the dimension of resistance.

(f) Dimension of L/R^2

$$\dfrac{L}{R^2} = \dfrac{V \cdot t}{i \cdot \dfrac{V^2}{(i)^2}} = \dfrac{V \cdot t \cdot i^2}{i \cdot V^2} = \dfrac{t \cdot i}{V}$$

or, $\qquad \dfrac{L}{R^2} = \dfrac{t \cdot i}{V} \tag{1}$

Dimension of C:

$$C = \dfrac{i \cdot t}{V} \quad \left[\because i = C \dfrac{dV}{dt} \right] \tag{2}$$

So,   equation 1 = equation 2

So, dimension of L/R^2 is equal to the dimension of capacitance. **Ans.**

Problem 1.16.   The following set of problems refers to the elements and the waveforms shown in Fig. P. 1.16 (a-F). For each part of this problem, sketch the required quantity, carefully making the time scale, significant amplitudes, slopes, and so on. Give enough detail to permit the waveform to be constructed from the data alone.

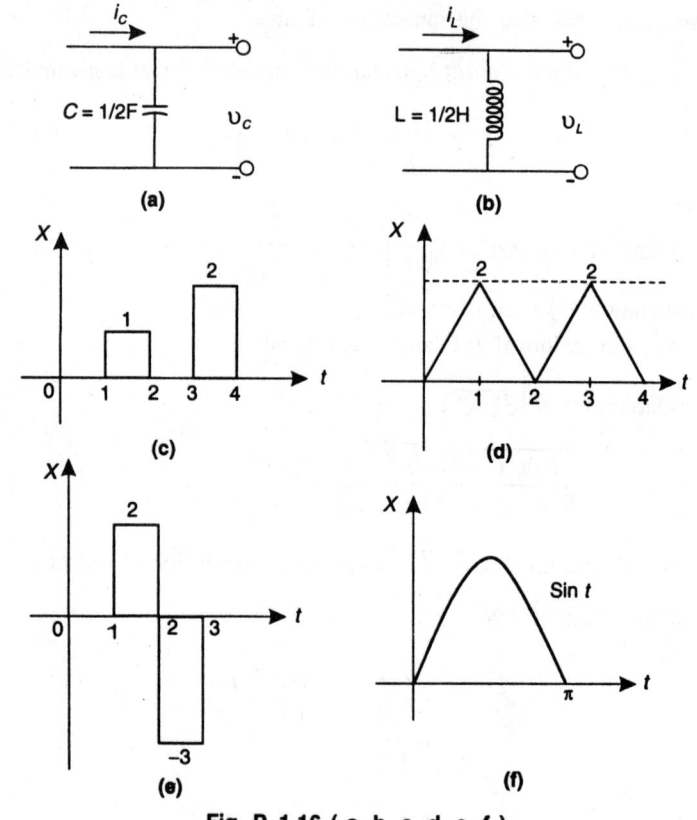

Fig. P. 1.16 (a, b, c, d, e, f)

	Network of	Given x is	Shown in	Sketch	Initial condition
A	a	v_c	d	i_c	none
B	a	v_c	f	i_c	none
C	a	i_c	c	v_c	$v_c(0) = 0$

Network	Given of	Shown x is	in	Initial Sketch	condition
D	a	i_c	d	v_c	$v_c(0) = 0$
E	a	i_c	f	v_c	$v_c(0) = 0$

Solution:

From Fig. 1.16 (a) $\quad i_c = c\dfrac{dV_c}{dt} = 0.5\dfrac{dV_c}{dt}$

$$V_c = \begin{cases} 2t, 0 < t < 1 \text{ and } 2 < t < 3 \\ -2t, 1 < t < 2 \text{ and } 3 < t < 4 \end{cases}$$

So, $\qquad i_c = \begin{cases} 0.5\dfrac{d}{dt}(2t), 0 < t < 1 \text{ and } 2 < t < 3 \\ \\ 0.5\dfrac{d}{dt}(-2t), 1 < t < 2 \text{ and } 3 < t < 4 \end{cases}$

$$i_c = \begin{cases} 1, 0 < t < 1 \text{ and } 2 < t < 3 \\ -1, 1 < t < 2 \text{ and } 3 < t < 4 \end{cases}$$

So, waveform of i_c:

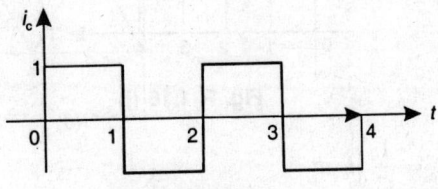

Fig. P.1 16 (g)

From Fig. 1.16 (a) $\qquad i_c = C\dfrac{dV_c}{dt} = 0.5\dfrac{dV_c}{dt}$

From Fig. 1.16 (f)

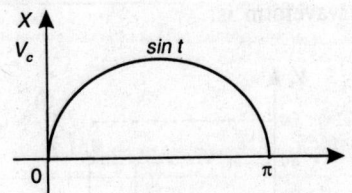

Fig. P. 1.16 (h)

$V_c = \sin t$

or, $\qquad \dfrac{dV_c}{dt} = \cos t \therefore i_c\ 0.5\cos t.$

So, waveform of i_c is:

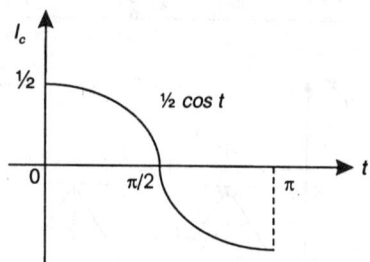

Fig. P. 1.16 (i)

(B) From Fig. 1.16 (a) $i_c = C \dfrac{dV_c}{dt}$ or, $V_c = \dfrac{1}{C} \displaystyle\int_0^t i_c \, dt$

From Fig. 1.16 (c)

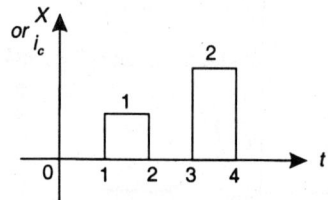

Fig. P. 1.16 (j)

So, $V_c = \dfrac{1}{C} \displaystyle\int_0^t i_c \, dt$

$$V_c = \frac{1}{C} \left[\int_1^2 dt + \int_3^4 2\, dt \right] = 2\left[[2-1] + 2[4-3] \right] = 2[1+2] = 6\,\text{V}$$

So, output voltage waveform is:

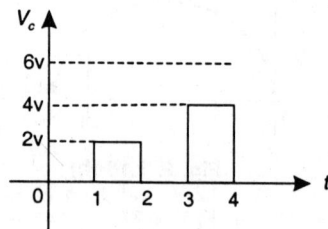

Fig. P. 1.16 (k)

(C) From Fig. (a) $\qquad V_c = \dfrac{1}{C} \displaystyle\int_0^t i_c \, dt$

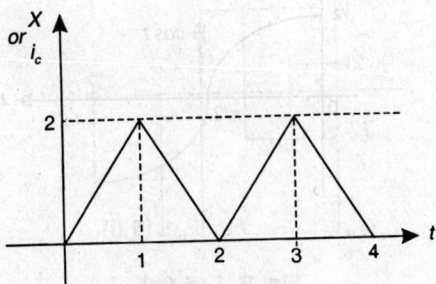

Fig. P. 1.16 (I)

$$i_c = \begin{cases} 2t, \, 0 < t < 1 \text{ and } 2 < t < 3 \\ -2t, \, 01 < t < 2 \text{ and } 3 < t < 4 \end{cases}$$

$\therefore \qquad V_c = \begin{cases} \dfrac{1}{C} \displaystyle\int_0^1 2t \, dt \\[4pt] \text{and } \dfrac{1}{C} \displaystyle\int_2^3 2t \, dt \\[4pt] -\dfrac{1}{C} \displaystyle\int_1^2 2t \cdot dt \\[4pt] \text{and } -\dfrac{1}{C} \displaystyle\int_3^4 2t \, dt \end{cases}$

$= \begin{cases} 2 \times 2 \left[\dfrac{t^2}{2} \right]_0^1 \text{ and } 2 \times 2 \left[\dfrac{t^2}{2} \right]_2^3 \\[6pt] -2 \times 2 \left[\dfrac{t^2}{2} \right]_1^2 \text{ and } -2 \times 2 \left[\dfrac{t^2}{2} \right]_3^4 \end{cases}$

$= \begin{cases} 2 \text{ V and } 10 \text{ V} \\ -6 \text{ V and } -14 \text{ V} \end{cases}$

So, waveform of V_c is:

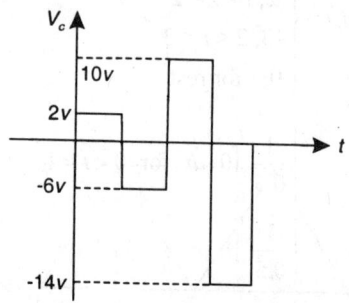

Fig. P. 1.16 (m)

(D) From Fig. 1.16 (a) $V_c = \dfrac{1}{C} \displaystyle\int_0^t i_c \, dt$

From Fig. 1.16 (f) $i_c = \sin t$

$\therefore$ $V_c = \dfrac{1}{C} \displaystyle\int_0^t \sin t \cdot dt.$

$$= 2 \int_0^t \sin t \cdot dt.$$

$$= -2 \left[\cos t\right]_0^t = -2\left[\cos t - 1\right]$$

or, $V_c = 1 \, (1 - \cos t)$

Its waveform is given as:

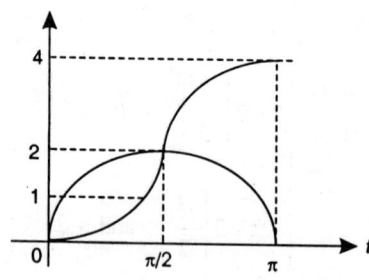

Fig. P. 1.16 (n)

(E) From Fig. P. 1.16 (a) $V_c = \dfrac{1}{C} \displaystyle\int_0^t i_c \, dt$

From Fig. P. 1.16 (e)$i_c = \begin{cases} 0, & 0 < t < 1 \\ 2, & 1 < t < 2 \\ -3, & 2 < t < 3 \\ 0, & \text{for rest} \end{cases}$

So, $V_c = \begin{cases} \dfrac{1}{0.5} \displaystyle\int_0^t 0 \cdot dt & \text{for } 0 < t < 1 \\[2mm] \dfrac{1}{0.5} \displaystyle\int_0^t 1 \cdot dt & \text{for } 1 < t < 2 \\[2mm] \dfrac{1}{0.5} \displaystyle\int_0^t -3 \cdot dt & \text{for } 2 < t < 3 \\[2mm] \dfrac{1}{0.5} \displaystyle\int_0^t 0 \cdot dt & \text{for rest} \end{cases}$

$V_c = \begin{cases} 0 & \text{for } 0 < t < 1 \\ 2 \cdot [2-1] & \text{for } 1 < t < 2 \\ -2[3-2] & \text{for } 2 < t < 3 \\ 0 & \text{for rest} \end{cases}$

$V_c = \begin{cases} 0 & \text{for } 0 < t < 1 \\ 2 & \text{for } 1 < t < 2 \\ -2 & \text{for } 2 < t < 3 \\ 0 & \text{for rest} \end{cases}$

So waveform of V_c is:

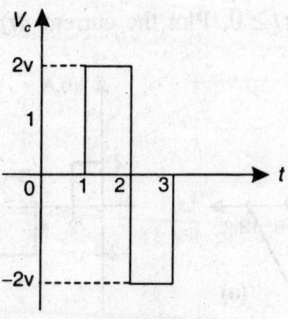

Fig. P. 1.16 (o)

Problem 1.17.

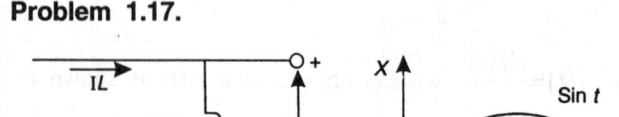

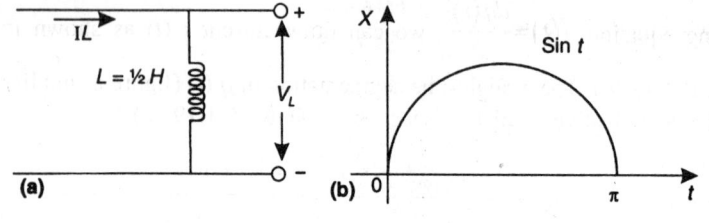

Fig. P. 1.17 (a) & (b)

Given that x is i_L, initial condition is zero. Sketch V_L.

Solution:

$$\because \qquad V_L = L\,\frac{di_L}{dt} = \frac{1}{2}\,\frac{d}{dt}\sin t$$

or,

$$V_L = \frac{1}{2}\cos t$$

So, its waveform is given as:

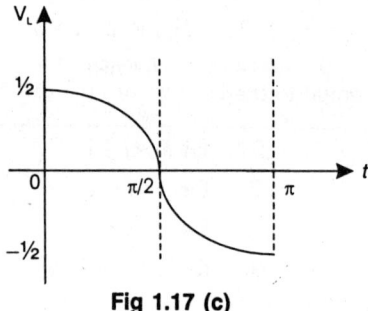

Fig 1.17 (c)

Problem 1.18.   The charge crossing a boundary in a wire is given in Fig. P. 1.18 (a) for $t \geq 0$. Plot the current $i(t)$ through the wire.

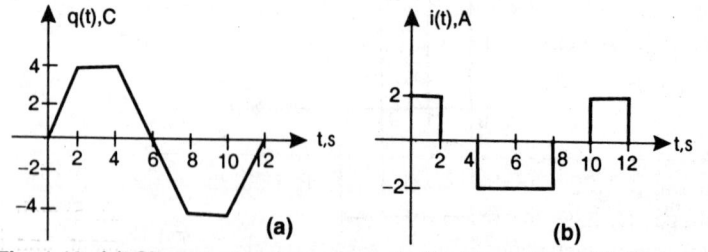

Fig. 1.18. (a) Charge crossing a hypothetical boundary. (b) Current flow associated with the charge plot of (a).

Solution:

Using equation $i(t) = \dfrac{dq(t)}{dt}$, we can draw current i (t) as shown in

Fig. P. 1.18 (b). The straight-line segmentation of q (t) (figure a) implies that the derivative is piecewise constant (Fig. P. 1.19 (a).

Problem 1.19. Find q (t), the charge transported through a cross section of a conductor over $[0, t]$, and also the total charge Q transported, if the current through the conductor is given by the waveform of figure 1.19 a

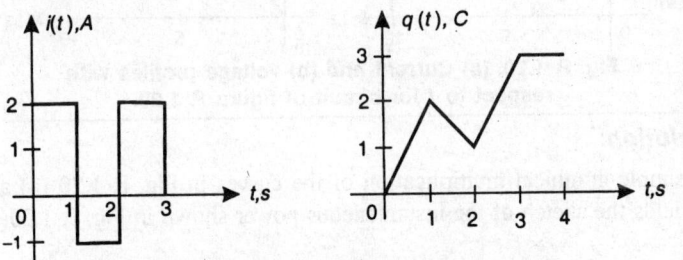

Fig P. 1.19. (a) Square-wave current signal.
(b) q (t) equal to the integral of i (t) given in (a).

Solution:

As,
$$q \ (t) = \int_0^t i(\tau) d_\tau \ \text{for } t \geq 0,$$

we have $q(t)$ drawn in Fig (b).

Thus $q(t)$ is the running area under the $i(t)$ versus t curve. Since i (t) is piecewise constant, the integral is piecewise linear because the area either increases or decreases linearly with time, as shown in figure b. Since q (t) is constant for $t \geq 3$, the total charge transported is

$Q = q$ $(3) = 3$ C.

Problem 1.20. In the circuit of Fig. P. 1.20, the current i (t) and voltage $v(t)$ have the waveforms graphed in Fig. P. 1.20 (a). Sketch p (t), the instantaneous power absorbed by the circuit element, and then sketch $W(0, t)$, the energy absorbed over the interval $[0, t]$.

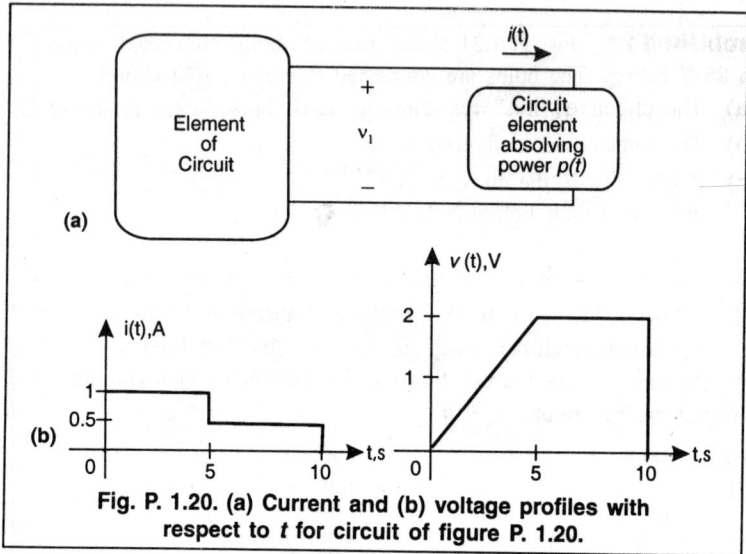

Fig. P. 1.20. (a) Current and (b) voltage profiles with respect to *t* for circuit of figure P. 1.20.

Solution:

A simple graphical multiplication of the curves in Fig. P. 1.20 (a) and b yields the sketch of the instantaneous power shown in Fig. P. 1.20(a).

Using equation $W\left(t_0, t\right) = \int_{t_0}^{t} p\left(\tau\right) d\tau$, we have for $0 \le t \le 5$,

$$W\left(0, t\right) = \int_{0}^{t} p\left(\tau\right) d\tau = \int_{0}^{t} \frac{2\tau}{5} d\tau = \frac{t^2}{5} \quad [\because t_0 = 0]$$

and for $t \ge 5$, $\quad W\left(0, t\right) = \int_{0}^{t} p\left(\tau\right) d\tau = \int_{0}^{5} p\left(\tau\right) d\tau + \int_{5}^{t} p\left(\tau\right) d\tau$

$$= 5 + \int_{5}^{t} d\tau = 5 + (t - 5) = t$$

Fig. P. 1.20(d) illustrates the resulting graph.

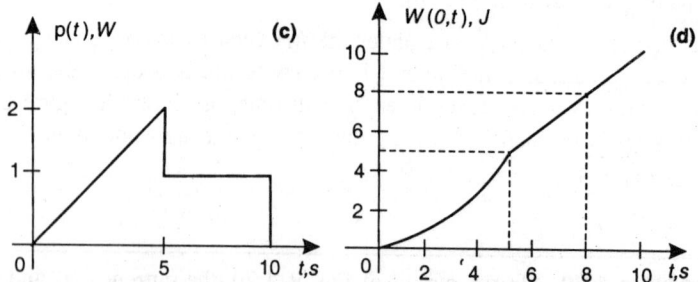

Fig. P. 1.20 (c) Profile of the instantaneous power $p(t) = v(t)\, i(t)$ for the current and voltage waveforms of figure 1.20.
(d) Associated profile of energy versus time.

Problem 1.21. Fig. P. 1.21 shows four lightbulbs connected **across** an 85-V battery. The bulbs are connected in parallel. Then, find
(a) The effective "hot" resistance of each bulb
(b) The total power delivered by the source
(c) If at $t = 10$ h, the current supplied by the source is 0.6471 A, discover which lightbulb has burned out.

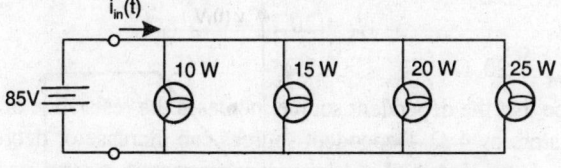

Fig. P. 1.21 Parallel connection of lightbulbs.

Solution:

(a) $P = V^2/R$, The resistances of each bulb are,

$$R_{10W} = \frac{85^2}{10} = 722.5 \, \Omega \qquad R_{15W} = \frac{85^2}{15} = 481.667 \, \Omega$$

$$R_{20W} = \frac{85^2}{20} = 361.25 \, \Omega \qquad R_{25W} = \frac{85^2}{25} = 289 \, \Omega$$

(b) The power delivered by the source equals the sum of the power consumed by each bulb. Then, the total power delivered by the source is 70 W.

(c) If at $t = 10$ h the current supplied by the source is 0.6471 A, then the power delivered by the source at $t = 10$ h is $85 \times 0.6471 = 55$ W. The bulb that burned out was using $70 - 55 = 15$ W of power. Hence the 10-W bulb has burned out.

Problem 1.22. Find the equivalent resistance seen by the source and the voltages v_1 and v_2 for the circuit of Fig. P. 1.22. What is the power dissipated in the 14-Ω resistor if $v_{in}(t) = 2$ V?

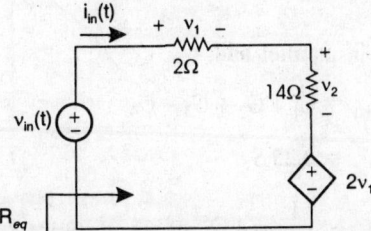

Fig. P. 1.22 Series circuit containing a dependent voltage source.

Solution:

From KVL, we have $v_{in} = v_1 + v_2 + 2v_1 = 3v_1 + v_2$ ... (1)

To express v_1 and v_2 in terms of i_{in}, observe that i_{in} is the current through each resistor (KCL or definition of two-terminal circuit element) and use Ohm's law: $v_1 = 2i_{in}$ and $v_2 = 14\ i_{in}$. Substituting into equation (1), we have,

$$v_{in} = 20\ i_{in} = R_{eq}\ i_{in}$$

Hence, $R_{eq} = 20\ \Omega$.

Notice that the dependent source increases the resistance of the two series resistors by 4 Ω. Dependent sources can increase or decrease the resistance of the circuit. With dependent sources it is even possible to make the equivalent resistance negative.

Now to find power absorbed by the 14-Ω resistor when $v_{in} = 2$ V, first compute i_{in} via Ohm's law: $i_{in} = v_{in}/R_{eq} = 2/20 = 0.1$ A. It follows that $P = I^2R = 0.01 \times 14 = 0.14$ W.

Problem 1.23. For the circuit of Fig. P. 1.23, find the input voltage $v_{in}(t)$, the current $i_2(t)$ through R_2, and the instantaneous power absorbed by R_2 when

$$i_{in}\ (t) = \begin{cases} 5e^{-t}\ A & t \geq 0 \\ 0 & t > 0 \end{cases}$$

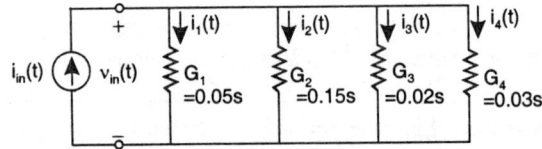

Fig. P. 1.23   Parallel connection of four resistors.

Solution:

Since conductances in parallel add,

$$G_{eq} = G_1 + G_2 + G_3 + G_4$$

$$= 0.25\ S$$

and then, $$R_{eq} = \frac{1}{G_{eq}} = 4\ \Omega$$

From Ohm's law, the voltage across the current source is

$$v_{in}(t) = R_{eq}\, i_{in}(t) = \begin{cases} 20e^{-t} \text{ V} & t \ge 0 \\ 0 & t < 0 \end{cases}$$

Using the current division formula,

We have, $\quad i_2(t) = \dfrac{G_2}{G_{eq}} i_{in}(t) = \dfrac{0.15}{0.25} i_{in}(t) = \begin{cases} 3e^{-t} \text{ A} & t \ge 0 \\ 0 & t < 0 \end{cases}$

To compute the power absorbed by R_2 for $t \ge 0$,

$$P_2(t) = v_{in}(t) \times i_2(t) = [i_2(t)]^2 R_2 = 60e^{-2t} \text{ W}$$

Problem 1.24. Find the equivalent resistance and the voltage across the source for the circuit of Fig. P. 1.24.

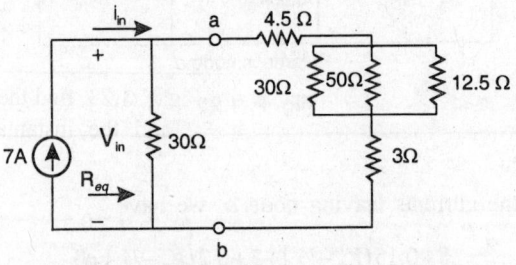

Fig. P. 1.24. **Series-parallel resistive circuit**

Solution:

Developing the equivalent resistance of parallel branches by R_{eq1},

We have $\quad R_{eq1} = \dfrac{1}{G_{eq1}} = \dfrac{1}{\dfrac{1}{30} + \dfrac{1}{50} + \dfrac{1}{12.5}} = 7.5\,\Omega$

Then, equivalent resistance to the right of terminal ab,

We have, $\quad R_{eq2} = 4.5 + 7.5 + 3 = 15\,\Omega$

Finally, the equivalent resistance

$$R_{eq} = \dfrac{1}{\dfrac{1}{30} + \dfrac{1}{15}} = 10\,\Omega$$

and, the voltage across the supply,

$$V_{in} = R_{eq} I_{in} = 10 \times 7 = 70 \text{ Volts.}$$

Problem 1.25. Find the node voltages V_a, V_b and V_c in the circuit of Fig. P. 1.25. Take the bottom node as reference.

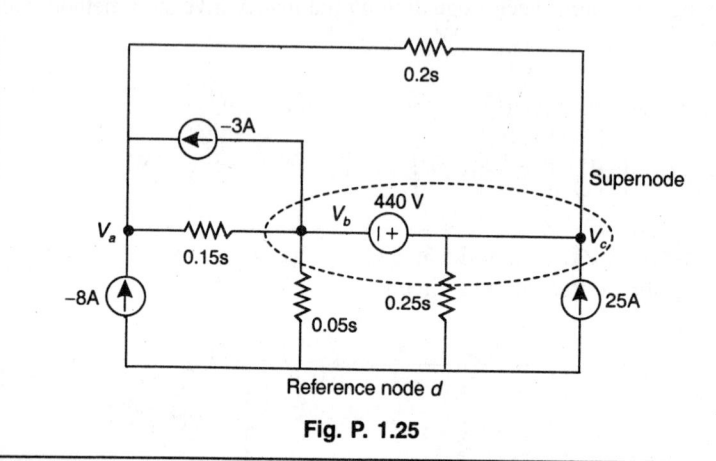

Fig. P. 1.25

Solution:

Summing the currents leaving node a, we have

$$8 + 0.15(V_a - V_b) + 3 + 0.2(V_a - V_c) = 0 \qquad (1)$$

Rewriting eq (1) as $\quad (0.15 + 0.2)V_a - 0.15 V_b - 0.2 V_c + 8 + 3 = 0$

or, $\qquad\qquad 0.35 V_a - 0.15 V_b - 0.2 V_c = -11 \qquad (2)$

Since the curve enclosing the supernode is a Gaussian curve, the sum of the currents leaving the curve is zero. Hence,

$$-3 + 0.15(V_b - V_a) + 0.05 V_b + 0.25 V_c - 25 + 0.2(V_c - V_a) = 0$$

or, $\qquad -(0.15 + 0.2)V_a - (0.15 + 0.05)V_b + (0.25 + 0.2)V_c = 28$

Simplifying this expression, we have

$$-0.35 V_a + 0.2 V_b + 0.45 V_c = 28 \qquad (3)$$

The voltages V_b and V_c inside the supernode are constrained by the voltage source. Mathematically, this constraint is $V_c - V_b = 440$, i.e.,

$$-V_b + V_c = 440 \qquad (4)$$

In matrix form equations 2, 3 and 4 can be written as,

$$\begin{bmatrix} 0.35 & -0.15 & -0.2 \\ -0.35 & 0.2 & 0.45 \\ 0 & -1 & 1 \end{bmatrix}\begin{bmatrix} V_a \\ V_b \\ V_c \end{bmatrix} = \begin{bmatrix} -11 \\ 28 \\ 440 \end{bmatrix}$$

Solving this matrix-vector equation by the matrix inversion method, we have

$$\begin{bmatrix} V_a \\ V_b \\ V_c \end{bmatrix} = \begin{bmatrix} 0.35 & -0.15 & -0.2 \\ -0.35 & 0.2 & 0.45 \\ 0 & -1 & 1 \end{bmatrix}^{-1}\begin{bmatrix} -11 \\ 28 \\ 440 \end{bmatrix}$$

$$= \begin{bmatrix} 6.1905 & 3.3333 & -0.2619 \\ 3.3333 & 3.3333 & -0.8333 \\ 3.3333 & 3.3333 & 0.1667 \end{bmatrix}\begin{bmatrix} -11 \\ 28 \\ 440 \end{bmatrix} = \begin{bmatrix} -90 \\ -310 \\ 130 \end{bmatrix}$$

Hence, $V_a = -90$ volts

$V_b = -130$ volts

$V_c = -130$ volts

Problem 1.26. The circuit of Fig. P. 1.26 contains a floating independent and a floating dependent voltage source. Find the node voltages V_a, V_b and V_c. Then find the power delivered by the 30-V source.

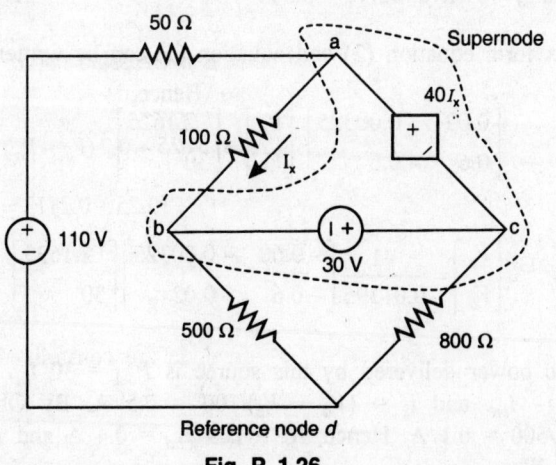

Fig. P. 1.26

Solution:

At the indicated supernode,

$$\frac{1}{50}(V_a - 110) + i_x - i_x + \frac{1}{500}V_b + \frac{1}{800}V_c = 0 \tag{1}$$

Using $V_c = V_b + 30$ in equation (1), we get

$$\frac{1}{50}(V_a - 110) + \left(\frac{1}{500} + \frac{1}{800}\right)V_b + \frac{30}{800} = 0$$

Rearranging terms and simplifying, we have

$$0.02 V_a + 0.00325 V_b = 2.1625 \tag{2}$$

Also,

$$i_x = \frac{1}{100}(V_a - V_b) \tag{3}$$

and, $0.02 V_a + 0.00325 V_b = 2.1625$ $V_a - V_b - 30 = 40 i_x$ $\tag{4}$

Then,

$$V_a - V_b - 30 = \frac{40}{100}(V_a - V_b)$$

[Using equation (3) in eqn (4)]

Simplifying, we have $0.6 V_a - 0.6 V_b = 30$ $\tag{5}$

In matrix form equation (2) and equation (5) can be written as,

$$\begin{bmatrix} 0.02 & 0.00325 \\ 0.6 & -0.6 \end{bmatrix} \begin{bmatrix} V_a \\ V_b \end{bmatrix} = \begin{bmatrix} 2.1625 \\ 30 \end{bmatrix}$$

Then,

$$\begin{bmatrix} V_a \\ V_b \end{bmatrix} = \frac{-1}{0.01395} \begin{bmatrix} -0.60 & -0.00325 \\ -0.6 & -0.02 \end{bmatrix} \begin{bmatrix} 2.1625 \\ 30 \end{bmatrix} = \begin{bmatrix} 100 \\ 50 \end{bmatrix}$$

Now, the power delivered by this source is $P_{del} = 30\ i_{bc}$. By KCL $i_{bc} = i_x - i_{bd}$. and $i_x = (V_a - V_b)/100 = 0.5$ A. By Ohm's law, $i_{bd} = V_b/500 = 0.1$ A. Hence, $i_{bc} = i_x - i_{bd} = 0.4$ A and $P_{del} = 30\ i_{bc} = 12$ W.

Problem 1.27. In the circuit of Fig. P. 1.27 the currents for each loop are again denoted by I_1, I_2 and I_3. Find all the loop currents, the voltage V_s, and the power delivered by the 8-A source.

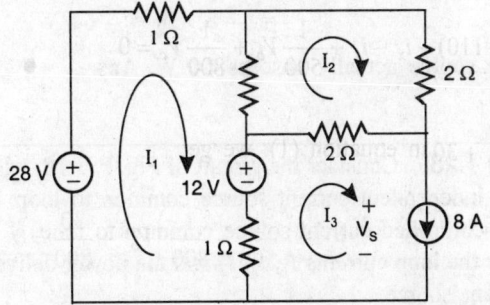

Fig. P. 1.27 Resistive circuit containing an independent current source on the perimeter of loop 3, forcing $I_3 = 8$ A.

Solution:

Because I_3 is the only loop current circulating through the branch containing the independent 8-A current source, $I_3 = 8$ A.

Now, summing the voltages around loop 1 using Ohm's law and the designated loop currents produces

$$28 = I_1 + 4(I_1 - I_2) + 12 + (I_1 - 8)$$
$$= 6I_1 - 4I_2 + 4$$

Hence, we have, $6I_1 - 4I_2 = 24$ **(1)**

Applying KVL and Ohm's law to loop 2 produces

$$0 = 4(I_2 - I_1) + 2I_2 + 2(I_2 - 8) = -4I_1 + 4I_2 - 16 \qquad \textbf{(2)}$$

Rewriting equation 2 as, $-4I_1 + 8I_2 = 16$ **(3)**

The matrix form of equations 1 and 3 as

$$\begin{bmatrix} 6 & -4 \\ -4 & 8 \end{bmatrix} \begin{bmatrix} I_1 \\ I_2 \end{bmatrix} = \begin{bmatrix} 24 \\ 16 \end{bmatrix}$$

Using the inverse matrix techinque to compute the solution, we have

$$\begin{bmatrix} I_1 \\ I_2 \end{bmatrix} = \begin{bmatrix} 6 & -4 \\ -4 & 8 \end{bmatrix}^{-1} \begin{bmatrix} 24 \\ 16 \end{bmatrix}$$

$$= \frac{1}{32} \begin{bmatrix} 8 & 4 \\ 4 & 6 \end{bmatrix} \begin{bmatrix} 24 \\ 16 \end{bmatrix} = \begin{bmatrix} 8 \\ 6 \end{bmatrix} \text{A}$$

By KVL, we have $V_s = 2(I_2 - 8) + 12 + (I_1 - 8) = 8$ V

Observe that the 8-A current source is labeled according to the passive sign convention, in which case

$$P_{del} = - P_{abs}$$

$$= - 8 \times 8 = - 64 \text{ W}$$

Hence, the source actually absorbs 64 W. **Ans.**

Problem 1.28. Consider the circuit of Fig. P. 3.18, which contains a 0.06-A independent current source common to loop 1 and 3 and a voltage-controlled current source common to loops 1 and 2. Find values for the loop currents I_1, I_2, I_3 and the power delivered by each independent source.

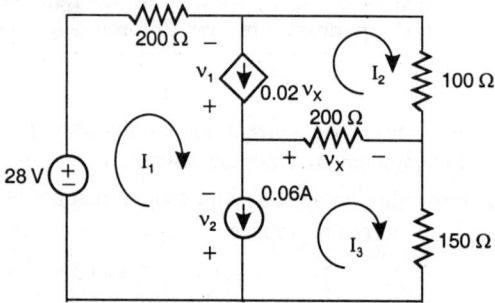

Fig. P. 1.28 Circuit containing a current source between loops.

Solution:

Using KVL, we have $28 = 200 \ I_1 - v_1 - v_2$ (1)

Applying KVL and Ohm's law to loop 2,

We obtain $100 \ I_2 + 200 \ (I_2 - I_3) + v_1 = 0.$

Rearranging, $300 \ I_2 - 200 \ I_3 + v_1 = 0$ (2)

Applying KVL to loop 3 yields $150 \ I_3 + v_2 + 200 \ (I_3 - I_2) = 0.$

Equivalently, it can be written as, $- 200 \ I_2 + 350 I_3 + v_2 = 0$ (3)

Here loops 1 and 3 are incident on the independent current source so that $0.06 = I_1 - I_3$ (4)

In a straightforward manner, we have

$$I_1 - I_2 = 0.02 \ v_x$$

$$= 0.02 \ [200 \ (I_3 - I_2)]$$

$$= 4I_3 - 4I_2$$

and, After simplification, $I_1 + 3I_2 - 3I_3 = 0$ (5)

The matrix form of the equations 1, 2, 3, 4, and 5 can be written as,

$$
\begin{bmatrix}
200 & 0 & 0 & -1 & -1 \\
0 & 300 & -200 & 1 & 0 \\
0 & -200 & 350 & 0 & 1 \\
1 & 0 & -1 & 0 & 0 \\
1 & 3 & -4 & 0 & 0
\end{bmatrix}
\begin{bmatrix}
I_1 \\ I_2 \\ I_3 \\ v_1 \\ v_2
\end{bmatrix}
=
\begin{bmatrix}
28 \\ 0 \\ 0 \\ 0.06 \\ 0
\end{bmatrix}
\tag{6}
$$

Solving equation 6 by the matrix inverse method or by an available software package yields the solution,

$$
\begin{bmatrix}
I_1 \\ I_2 \\ I_3 \\ v_1 \\ v_2
\end{bmatrix}
=
\begin{bmatrix}
200 & 0 & 0 & -1 & -1 \\
0 & 300 & -200 & 1 & 0 \\
0 & -200 & 350 & 0 & 1 \\
1 & 0 & -1 & 0 & 0 \\
1 & 3 & -4 & 0 & 0
\end{bmatrix}^{-1}
\begin{bmatrix}
28 \\ 0 \\ 0 \\ 0.06 \\ 0
\end{bmatrix}
=
\begin{bmatrix}
0.1 \\ 0.02 \\ 0.04 \\ 2 \\ -10
\end{bmatrix}
\tag{7}
$$

First, the power delivered by the independent voltage source is

$$P_{V\ source} = 28\ I_1 = 2.8\ \text{W}$$

and, the power delivered by the independent current source is

$$P_{I\ source} = 28\ v_2 = -0.6\ \text{W}$$

This last value indicates that the independent current source actually absorbs power from the circuit.

2

Conventions of Decribing Networks

Problem 2.1. For the controlled source shown in Fig. P. 2.1, prepare a plot.

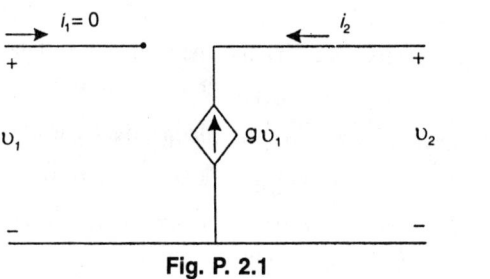

Fig. P. 2.1

Solution:

Since the given figure shows the voltage controlled current source. So $i_2 = g v_1$ and i_2 is independent of v_2.

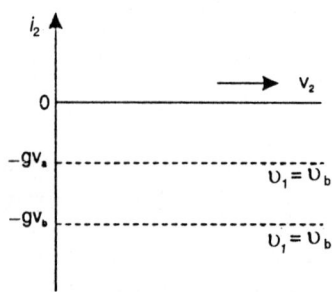

Fig. P. 2.1 (a)

Problem 2.2. Repeat Problem 2.1 for the controlled source giving in the accompanying figure.

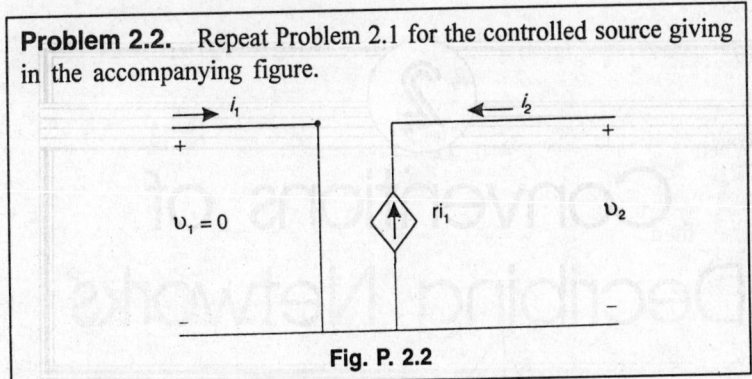

Fig. P. 2.2

Solution:

Since the given Fig. P. 2.2 shows a current controlled voltage source.
so $v_2 = r i_1$ and v_2 is independent of i_2.

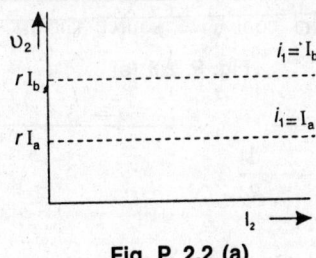

Fig. P. 2.2 (a)

Problem 2.3. The network of the accompanying figure is a model for a battery of open circuit terminal voltage v and internal resistance R_b. For this network plot i as a function v. Identify features of the plot such as slopes, intercepts and so on.

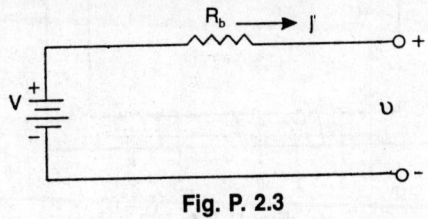

Fig. P. 2.3

Solution:

Applying KVL in the network, $v = V - iR_b$

or, $i = \dfrac{V - v}{R_b}$

When $v = 0, \ i = \dfrac{V}{R_b}$

1 ∝ then $v = V, \ i = 0$

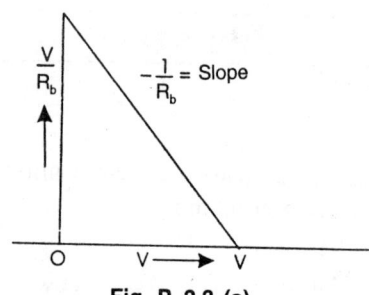

Fig. P. 2.3 (a)

Slop $\dfrac{di}{dv} = -\dfrac{1}{R_b}$

Problem 2.4. The magnetic system shown in the figure has three windings marked 1-1', 2-2' and 3-3'. Using three different forms of dots, establish polarity markings for these windings.

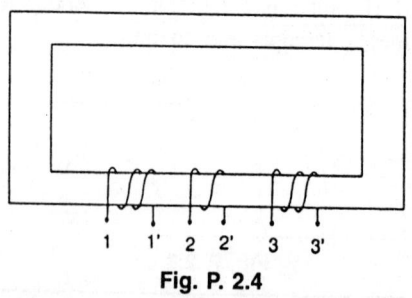

Fig. P. 2.4

Solution:

Applying dot convention: 1 effect 2, 2 effect 3 and 3 effect 1.

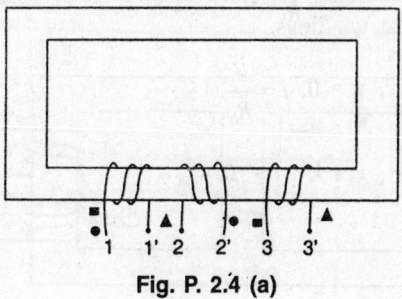

Fig. P. 2.4 (a)

Problem 2.5. Place three windings on the core shown for problem 2.4 with winding senses such that the following termials have the same mark (a) 1 and 2, 2 and 3, 3 and 1 (b) 1′ and 2′, 2′ and 3′, 3′ and 1′

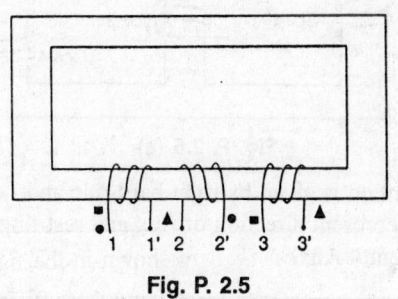

Fig. P. 2.5

Solution:

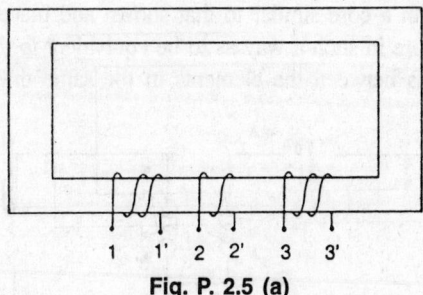

Fig. P. 2.5 (a)

Since the windings on the core shown for problem 2.4 has the same dot convention as shown in the problem 2.5, so winding placed is same as problem 2.4.

Problem 2.6. The figure shows four windings on a magnetic flux conducting core. Using different shaped dots, establish polarity marking for these windings.

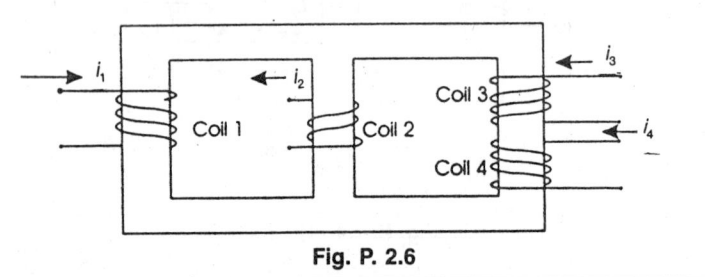

Fig. P. 2.6

Solution:
By applying dot convention or Lenz law:

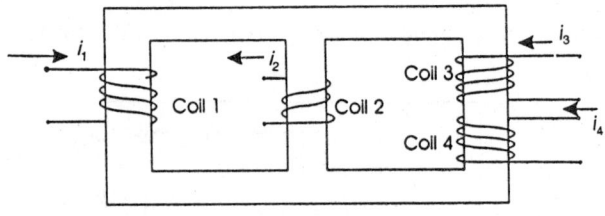

Fig. P. 2.6 (a)

[*Note:* Dot convention is given by right hard rule and Lenz law. In right hard rule thumb represent direction of flux and rest finger curled shows direction of current.] **Ans.**

Problem 2.7. The accompanying schematic shows the equivalent circuit of a shown with polarity marks on the three coupled coil. Draw a transformer with a core similar to that shown and place winding on the legs of the core in such a way as to be equivalent to the schematic show connections between the elements in the same drawing.

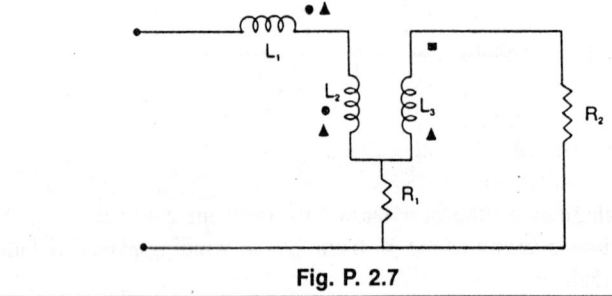

Fig. P. 2.7

Solution:

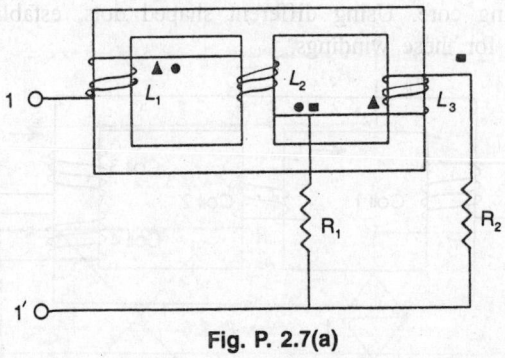

Fig. P. 2.7(a)

Problem 2.8. The accompanying schematics each show two inductors with coupling but with different dot markings. For each of the two systems determine the equivalent inductance of the system at terminals $1-1'$ by combining inductances.

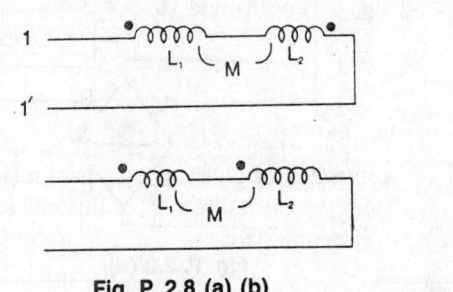

Fig. P. 2.8 (a) (b)

Solution:

From Fig. P. 2.8 (a) both the inductance are series additive so,

$$L_{eq} = L_1 + L_2 + 2\,M$$

From Fig. P. 2.8 (b), dot convention shows that both the inductances are series opposing. So $L_{eq} = L_1 + L_2 + - 2\,M$

Problem 2.9. In 2.9 (a) of the figure is shown a resistive network. In (b) and (c) are shows graph with two of the four nodes identified. For these two graphs, assign resistors to the branches and identify the two remaining nodes such that the resulting networks are topologically indentical to that shown (a).

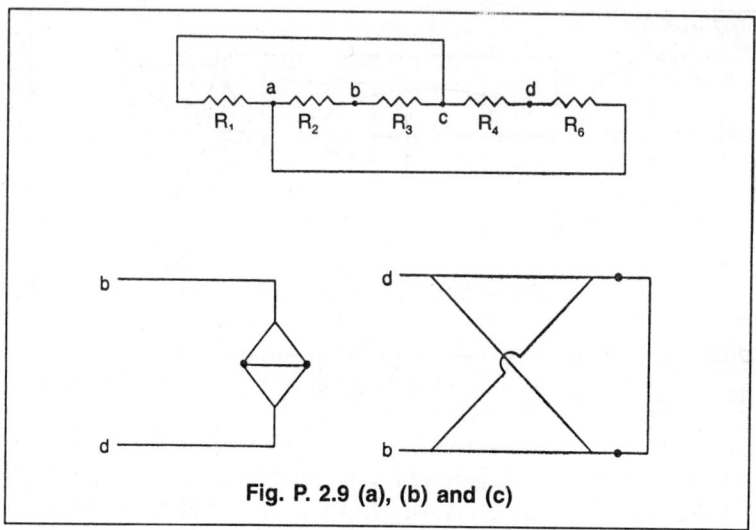

Fig. P. 2.9 (a), (b) and (c)

Solution:

By comparing Fig. P. 2.9 (a) and (b)

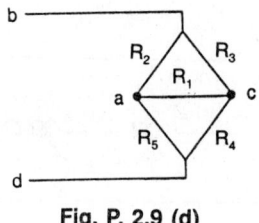

Fig. P. 2.9 (d)

By comparing Fig. P. 2.9 (a) and (c).

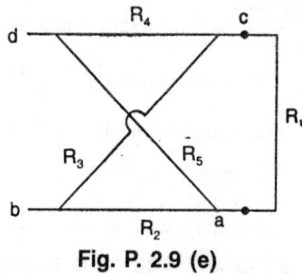

Fig. P. 2.9 (e)

Problem 2.10. Three graphs are shown in the Fig. P. 2.10. Classify each of the graph as planer or non-planer.

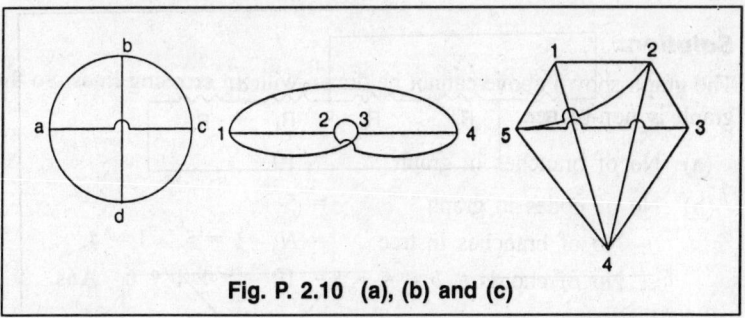

Fig. P. 2.10 (a), (b) and (c)

Solution:

After stretching Fig. P. 2.10 (a), (b) and (c) can be drawn as:

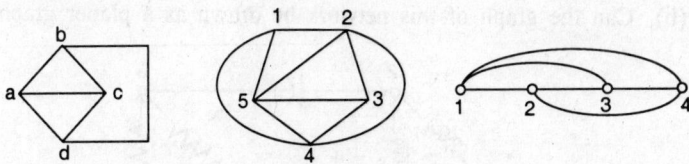

Fig. P. 2.10 (d), (e), (f)

From this stretched Fig. P. 2.10 (d), fig. (e) and fig (f) no one branch is crossing other branch. So all these three figs are planer.

Problem 2.11. For the graph of figure, classify as planer or non planer and determire the quantities specified in the equation.

Number of branches in tree = Number of nodes −1 and, Number of chords = $b - n + 1$

Where 'b' is the branch and n is nodes.

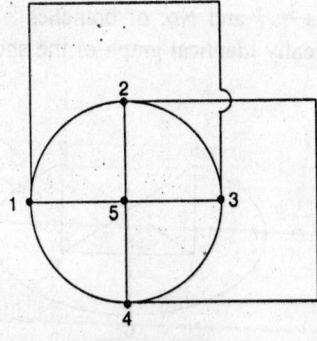

Fig. P. 2.11

Solution:

The graph shown above cannot be drawn without crossing lines. So the graph is non-planer.

(a) No of branches in graph   = 10

(b) No of nodes in graph   = 5

  ∴ No of branches in tree   = $N - 1 = 5 - 1 = 4$.

   No of chords = $b - n + 1 = 10 - 5 + 1 = 6$ **Ans.**

Problem 2.12.   The figure shows a network with elements arranged along the edge of a cube

(a) Determine the number of nodes and branches in the network

(b) Can the graph of this network be drawn as a planer graph.

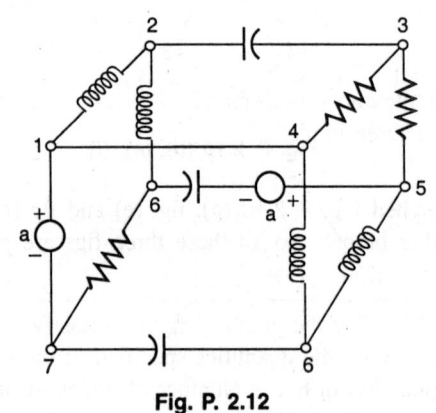

Fig. P. 2.12

Solution:

(a) No. of nodes = 7 and No. of branches = 11

(b) The topologically identical graph of the shown figure is drawn as:

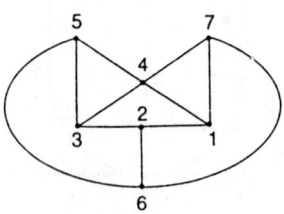

Fig. P. 2.12 (a)

In this graph no one branch crosses each other so the graph is planer graph.

Problem 2.13. The figure shows a graph of six nodes and connecting branches. You have to add non-parallel branches to this basic structure in order to accomplish the following different objectives

(a) What is the minimum number of branches that may be added to make the resulting structure non-planer?

(b) What is the maximum number of branches you may add before the resulting structure becomes non-planer.

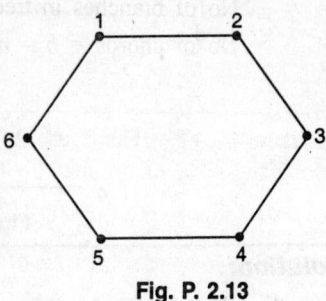

Fig. P. 2.13

Solution:

(a) To make the above planer graph to non-planer, we have to add minimum three branches.

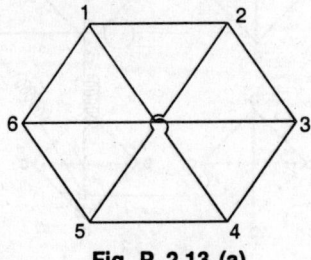

Fig. P. 2.13 (a)

(b) The maximum number of branches needed to make the planer graph non-planer is six.

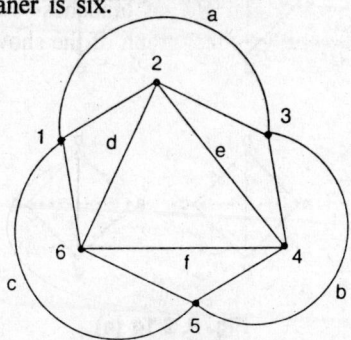

Fig. P. 2.13 (b)

Problem 2.14. Determine all trees for the graph shown in figure below. Uses solid lines for tree branches and dotted lines for chords.

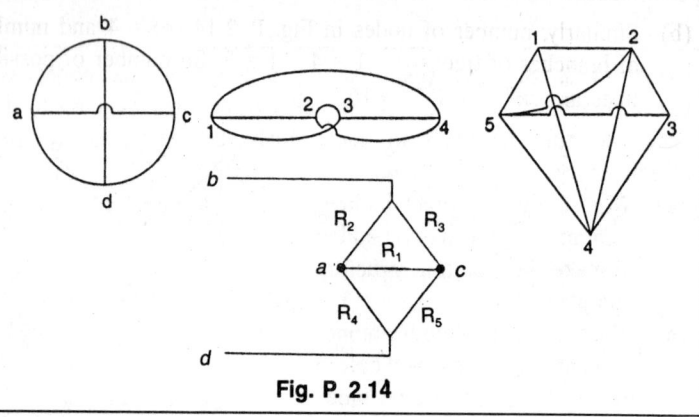

Fig. P. 2.14

Solution:

From Fig. P. 2.14 (a), No of nodes = 4, No of branches of tree = $n - 1 = 3$.

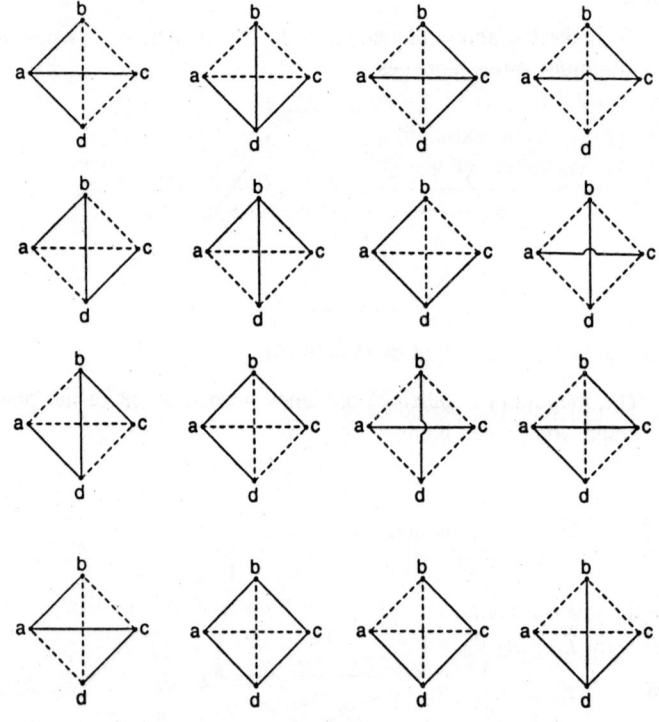

Fig. P. 2.14 (a)

Number of trees $= n^{n-2} = 4^{4-2} = 4^2 = 16$

(b) Similarly, number of nodes in Fig. P. 2.14 (b) = 4 and number of branches of tree $= n - 1 = 4 - 1 = 3$. So number of possible trees are: $n^{n-2} = 4^{4-2} = 16$

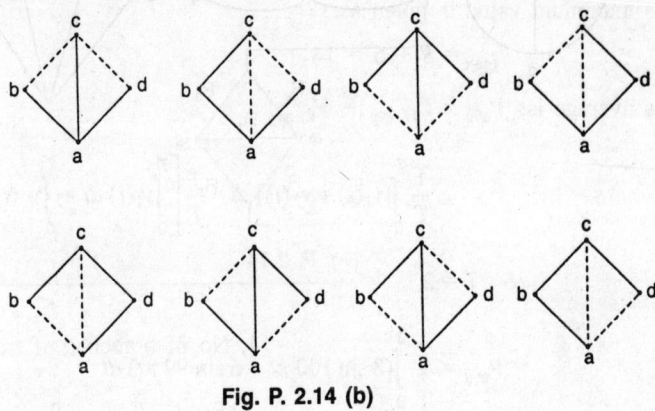

Fig. P. 2.14 (b)

Problem 2.15. Let $v_1 = 8 \sin 100 \pi t$ and $v_2 = \sin 99 \pi t$. Show that $v = v_1 + v_2$ is periodic. Find the period and the maximum, average and effective values of v.

Solution:

$$v = v_1 + v_2$$
$$v = 8 \sin 100 \pi t + 6 \sin 99 \pi t$$

For v to be periodic the ratio of period of $8 \sin 100 \pi t$ and $6 \sin 99 \pi t$ should be rational. The period of $\sin 100 \pi t$ is $T_1 = \dfrac{1}{50}$ and the period of $\sin 99 \pi t$ is 2/99.

So, $\dfrac{T_1}{T_2} = \dfrac{1 \times 99}{50 \times 2} = \dfrac{99}{100}$ which is a rational number.

Hence the given function v is a periodic function and its period is given by $T = n_1 T_1 = n_2 T_2$

Where $\qquad n_1 T_1 = n_2 T_2$

or, $\qquad \dfrac{T_1}{T_2} = \dfrac{n_2}{n_1} = \dfrac{99}{100}$

$\therefore$ $\qquad\qquad\qquad n_1 = 100, \ n_2 = 99.$

$\therefore$ $\qquad\qquad\qquad T = n_1\,T_1 = 100 \times \dfrac{1}{50} = 2 \text{ sec.}$

$\therefore$ $\qquad\qquad\qquad T = 2$ sec **Ans.**

Its maximum value is given as:

$$V_{\max} = 8 + 6 = 14$$

Its average is: $V_{\text{avg}} = V_{1,\ \text{avg}} + V_{2,\ \text{avg}}$

$$= \frac{1}{T}\int_0^T [v_1(t) + v_2(t)]\,dt = \frac{1}{T}\left[\int_0^T v_1(t)\,dt + \int_0^T v_2(t)\,dt\right]$$

$\therefore$ $\qquad\qquad T = 2$

$\therefore$ $\qquad\qquad V_{\text{avg}} = \dfrac{1}{2}\int_0^2 (8\sin 100\,\pi t + 6\sin 99\,\pi t)\,dt$

$$= \frac{1}{2}\left[-8\cos 100\,\pi t - 6\cos 99\,\pi t\right]_0^2$$

$$= -\frac{1}{2}[(8+6) - (8+6)] = 0.$$

$\therefore$ $\qquad\qquad V_{\text{avg}} = 0$

Its effective value is

$$V_{\text{eff}} = \left[\frac{1}{2}\int_0^2 (8\sin 100\,\pi t + 6\sin 99\,\pi t)\,dt\right]^{1/2}$$

$$= \left[\frac{1}{2}\int_0^2 (8^2\sin^2 100\,\pi t + 6^2\sin^2 99\,\pi t\right.$$

$$\left. + 2\times 8\times 6\cdot\sin 100\,\pi t\cdot\sin 99\,\pi t)\,dt\right]^{1/2}$$

$$= \left[\frac{1}{2}\left[8^2\left|\frac{t}{2} - \frac{\sin 100\,\pi t}{200}\right|_0^2 + 6^2\left|\frac{t}{2} - \frac{\sin 99\,\pi t}{198}\right|_0^2 + 0\right]\right]^{1/2}$$

From fourier series.

$$\begin{cases} \therefore \displaystyle\int\limits_{\alpha}^{\alpha+2\pi} \sin^2 nx\, dx = \left|\dfrac{x}{2} - \dfrac{\sin^2 nx}{2\times 2n}\right|_{\alpha}^{\alpha+2\pi} = \pi \\[3ex] \text{and } \displaystyle\int\limits_{\alpha}^{\alpha+2\pi} \sin mx \cdot \cos x\, dx = -\dfrac{1}{2}\left[\dfrac{\cos(m-n)x}{m-x} + \dfrac{\cos(m+n)x}{m+x}\right] = 0 \end{cases}$$

$$\therefore \qquad v_{eff} = \left[\dfrac{1}{2}(64+36)\right]^{1/2} = \sqrt{50} = 5\sqrt{2}$$

$$\therefore \qquad v_{eff} = 5\sqrt{2} \textbf{ Ans.}$$

Problem 2.16. Find period, frequency, phase angle in degrees and maximum, minimum, average and effective values of
$$v(t) = 2 + 6\cos(10\pi t + \pi/6)$$

Solution:

Its period is given as: $\qquad \dfrac{2\pi}{T} = 10\pi \quad \left(as\ \omega = \dfrac{2\pi}{T}\right)$

$$\therefore \qquad\qquad T = \dfrac{1}{5} = 0.2 \text{ sec}$$

Its frequency is given as:

$$f = \dfrac{1}{T} = \dfrac{1}{0.2} = 5\,\text{Hz}$$

Its phase angle is: $\qquad \phi = \dfrac{\pi}{6} = \dfrac{180°}{6} = 30°$

Its maximum and minimum value is given as:

$$V_{max} = 2 + 6 = 8$$

and $\qquad\qquad V_{min} = 2 - 6 = -4$

Its average value is given as:

$$V_{avg} = \dfrac{1}{T}\int\limits_0^T v(t)\, dt$$

$$= \frac{1}{0.2}\left[\int_0^{0.2} 2+6\cos(10\pi t+\pi/6)\,dt\right]$$

$$= \frac{1}{0.2}\left[2t+6\sin(10\pi t+\pi/6)\right]_0^{0.2}$$

$$= \frac{1}{0.2}\left[2\times0.2+6\sin(2\pi+\pi/6)-6\sin\pi/6\right]$$

$$= \frac{1}{0.2}\times2\times0.2=2$$

$$\therefore \qquad V_{avg}=2$$

Its effective value is given as:

$$V_{eff}=\sqrt{V_{d.c}^2+V_{a.c}^2}=\sqrt{2^2+\left(\frac{6}{\sqrt{2}}\right)^2}=\sqrt{4+\frac{36}{2}}=\sqrt{22}$$

$$\therefore \qquad V_{eff}=\sqrt{22}\ \ \textbf{Ans.}$$

Problem 2.17. Reduce $v(t) = 2\ \cos(\omega t + 30°) + 3\ \cos(\omega t$ to. $v(t) = A\ \sin(\omega t + \theta)$.

Solution:

$$v(t)= 2\cos(\omega t+30°)+3\cos\omega t$$

$$= 2(\cos\omega t\cdot\cos30°-\sin\omega t\cdot\sin30°)+3\cos\omega t$$

$$= 2\left(\frac{\sqrt{3}}{2}\cos\omega t-\frac{1}{2}\sin\omega t\right)+3\cos\omega t$$

$$= \sqrt{3}\cos\omega t-\sin\omega t+3\cos\omega t$$

$$v(t)= \left(3+\sqrt{3}\right)\cos\omega t-\sin\omega t \qquad (1)$$

$$\therefore \qquad v(t)= A\sin(\omega t+\theta)$$

$$v(t)= A(\sin\omega t\cos\theta+\cos\omega t\cdot\sin\theta) \qquad (2)$$

Comparing equations (1) and (2)

$$A\cos\theta=-1$$

and $\qquad A\sin\theta = (3+\sqrt{3})$

$\therefore \qquad (A\cos\theta)^2 + (A\sin\theta)^2 = A^2$

$\therefore \qquad A^2 = (-1)^2 + (3+\sqrt{3})^2 = 23.39$

$\therefore \qquad A = 4.83$

$\because \qquad A\cos\theta = -1$

or $\qquad \cos\theta = -\dfrac{1}{A} = -\dfrac{1}{4.83}$

or $\qquad \theta = \cos^{-1}\left(-\dfrac{1}{4.83}\right)$

$\qquad \theta = 101.9 \approx 102°$

$\therefore \qquad v(t) = 4.83\sin(\omega t + 120°)$ **Ans.**

Problem 2.18. Find $V_{2.\text{avg}}$ and $V_{2.\text{eff}}$ in the gragh of Fig. P. 2.18 for $V_1 = V_2 = 3$ and $T = 4T^{1/3}$

Fig. P. 2.18

Solution:

Its average value is given as:

$$V_{2,\text{avg}} = \frac{1}{T}\left[\int_0^{T_1} v_1\,dt + \int_{T_1}^{T} v_2\,dt\right] = \frac{1}{T}\left[v_1\int_0^{T_1} dt + v_2\int_{T_1}^{T} dt\right]$$

$$= \frac{1}{T}\left[3[t]_0^{T_1} + (-3)[t]_{T_1}^{T}\right] = \frac{1}{T}\left[3[T_1] - 3[T - T_1]\right]$$

$$= \frac{1}{T}[6T_1 - 3T] = \frac{1}{T}\left[6\times\frac{3}{4}T - 3T\right] \quad \left[\because T = \frac{4}{3}T_1\right]$$

$$= \frac{1}{T}[4.5T - 3T] = \frac{1.5T}{T} = 1.5$$

$$\therefore \qquad V_{2,\,avg} = 1.5$$

Its effective value is given as:

$$V_{2,\,eff} = \left[\frac{1}{T} \left[\int_0^{T_1} v_1^2 \, dt + \int_{T_1}^{T} v_2^2 \, dt \right] \right]^{1/2} = \left[\frac{1}{T} \left[v_1^2 \, [t]_0^{T_1} + v_2^2 \, [t]_{T_1}^{T} \right] \right]^{1/2}$$

$$= \left[\frac{1}{T} \left[3^2 \, [T_1] + (-3)^2 \, [T - T_1] \right] \right]^{1/2}$$

$$= \left[\frac{1}{T} \left[9T_1 + 9T - 9T_1 \right] \right]^{1/2} = (9)^{1/2} = 3$$

$$\therefore \quad V_{2,\,eff} = 3 \text{ **Ans.**}$$

Problem 2.19. Repeat problem 2.18 for $V_1 = 0$, $V_2 = -4$ and $T = 2T_1$

Solution:

$$V_{2,\,avg} = \frac{1}{T} \left[v_1 \int_0^{T_1} dt + v_2 \int_{T_1}^{T} dt \right] = \frac{1}{T} \left[0 \cdot [t]_0^{T_1} + v_2 \, [t]_{T_1}^{T} \right]$$

$$= \frac{1}{T} [0 + v_2 [T - T_1]] = \frac{1}{T} \left[-4 \left[T - \frac{T}{2} \right] \right] = -\frac{4}{T} \left[\frac{T}{2} \right] = -2$$

$$\therefore \qquad V_{2,\,avg} = -2$$

and $\qquad V_{2,\,eff} = \left[\frac{1}{T} \left[\int_0^{T_1} v_1^2 \, dt + \int_{T_1}^{T} v_2^2 \, dt \right] \right]^{1/2} = \left[\frac{1}{T} \left[0 + v_2^2 \, [t]_{T_1}^{T} \right] \right]^{1/2}$

$$V_{2,\,eff} = \left[\frac{1}{T} \left[v_2^2 [T - T_1] \right] \right]^{1/2} = \left[\frac{1}{T} \left[(-4)^2 [T - T/2] \right] \right]^{1/2}$$

$$= \left[\frac{1}{T} \left[16 \times \frac{T}{2} \right] \right]^{1/2} = \sqrt{8} = 2\sqrt{2} \quad \therefore V_{2,\,eff} = 2\sqrt{2} \text{ **Ans.**}$$

Problem 2.20. The waveform in Fig. P. 2.20 in sinusoidal. Express it in the form $v = A + B\sin(wt + \theta)$ and find its mean and r.m.s. values.

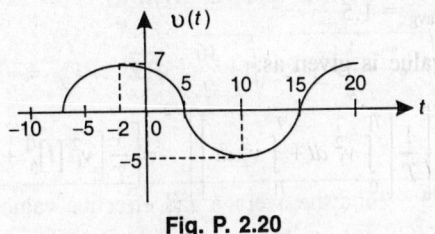

Fig. P. 2.20

Solution:

The time between two positive peaks, $T = 24s$, is one period corresponding to a frequency $f = 0.041$ Hz. The signal is a sinusoidal function with amplitude B added to a constant value A.

$$B = \frac{1}{2}(V_{max} - V_{min}) = \frac{1}{2}(7+5) = 6$$

$$A = V_{max} - B = V_{min} + B = 1$$

The sinusoidal is shifted by 2s to the left which corresponds to a phase lead of $\left(\dfrac{-2}{24}\right) \times 360° = -30°$

Therefore, the signal is expressed by

$$v(t) = A + B\sin(\omega t + \theta) = 1 + 6\sin\left(\frac{2\pi}{T}t + (-30°)\right) = 1 + 6\sin\left(\frac{\pi}{12}t - 30°\right)$$

$$v(t) = 1 + 6\sin\left(\frac{\pi}{12}t + 120°\right) \textbf{ Ans.}$$

Its average value is given as:

$$V_{ave} = \frac{1}{T}\int_0^T v(t)\,dt = \frac{1}{T}\left[\int_0^T \left(1 + 6\sin\left(\frac{\pi}{12}t + 120°\right)\right)dt\right]$$

$$= \frac{1}{T}\left[\left[t + (-6)\sin\left(\frac{\pi T}{12} + 120°\right)\right]_0^T\right]$$

$$= \frac{1}{T}\left[T - 6\sin\left(\frac{\pi T}{12} + 120°\right) + 6\sin\left(\frac{\pi T}{12} + 120°\right)\right] = \frac{T}{T} = 1$$

$\therefore \quad V_{avg} = 1$

Its r.m.s. value or effective value is given as:

$$V_{eff} = \sqrt{V^2{}_{d \cdot c} + \frac{1}{2}V^2{}_{d \cdot c}} = \sqrt{1 + \frac{36}{2}} = \sqrt{19} \quad \therefore \ V_{eff} = \sqrt{19} \ \textbf{Ans.}$$

Problem 2.21. Find the average and effective values of $v_1(t)$ in Fig. P. 2.21 (a) and v_2 in Fig. P. 2.21 (b).

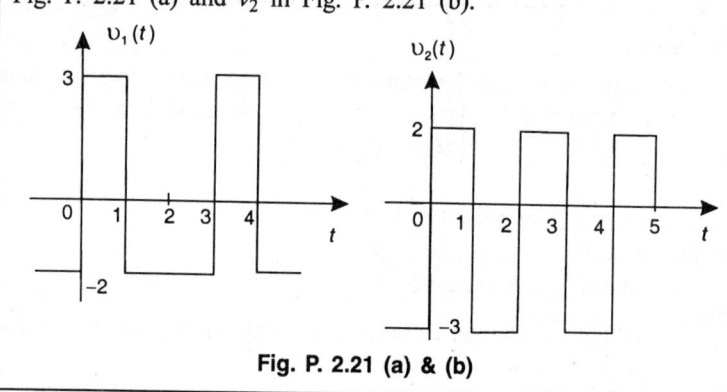

Fig. P. 2.21 (a) & (b)

Solution:

Average value in Fig. P. 2.21 (a) of period $T = 3$

$$V_{1avg} = \frac{1}{3}\left[\int_0^1 3\,dt + \int_1^3 -2\,dt\right] = \frac{1}{3}\Big[3[1] - 2[3-1]\Big]$$

$$= \frac{1}{3}[3-4] = \frac{-1}{3} \quad \therefore \ V_{1avg} = -\frac{1}{3} \ \textbf{Ans.}$$

Effective value of Fig. P. 2.21 (a) of period $T = 3$

$$V_{1eff} = \left[\frac{1}{3}\left[\int_0^1 3^2\,dt + \int_1^3 (-2)^2\,dt\right]\right]^{1/2} = \left[\frac{1}{3}9[1-0] + 4[3-1]\right]^{1/2}$$

$$= \frac{1}{3}[9+8]^{1/2} = \left[\frac{17}{3}\right]^{1/2}$$

$$V_{1eff} = \sqrt{\frac{17}{3}} \ \textbf{Ans.}$$

Average value of Fig. P. 2.21 (b) whose period T = 2.

$$V_{2,avg} = \frac{1}{2}\left[\int_0^1 2dt + \int_1^2 (-3)\,dt\right] = \frac{1}{2}\left[2\left[1-0\right]-3\left[2-1\right]\right]$$

$$= \frac{1}{2}[2-3] = -\frac{1}{2} \quad \therefore \quad V_{2,avg} = -\frac{1}{2} \textbf{ Ans.}$$

Effective value of Fig. P. 2.21 (b) whose period is 2.

$$V_{2,eff} = \left[\frac{1}{2}\left[\int_0^1 (2)^2\,dt + \int_1^2 (-3)^2\,dt\right]\right]^{1/2} = \left[\frac{1}{2}\left[4\left[1-0\right]+\left[2-1\right]\right]\right]^{1/2}$$

$$= \left[\frac{1}{2}[4+9]\right]^{1/2} = \left[\frac{13}{2}\right]^{1/2} \quad \therefore \quad V_{2,eff} = \sqrt{\frac{13}{2}} \textbf{ Ans.}$$

Problem 2.22. The current through a series *RL* circuit with $R = 5\ \Omega$ and $L = 10\ H$ is given in Fig. P. 2.22 Where $T = 1$s. Find the voltage across *RL*.

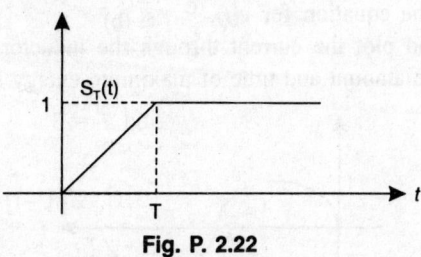

Fig. P. 2.22

Solution:

Voltage across *RL*:

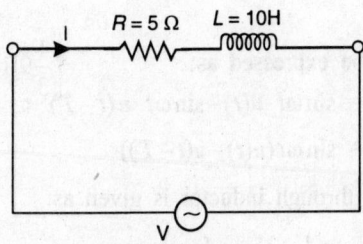

Fig. P. 2.22 (a)

$$\therefore \qquad V = IR + L\frac{di}{dt}$$

At time $t < 0$, current $I = 0$

$$\therefore \qquad V = 0 + 0 = 0\,\text{V}$$

At time $0 < t < 1$, current $I = t$ (a ramp function)

$$\therefore \qquad V = Rt + L\frac{dt}{dt} = 5t + 10.1 \quad [r = 5\,\Omega, L = 10\,\text{H}]$$

$$V = 10 + 5t\,\text{V}$$

At time $t > 1$, current $I = 1$ (constant)

$$\therefore \qquad V = 5 \times 1 + 10 \times 0\,\text{V} = 5\,\text{V}$$

$$\therefore \qquad V = \begin{cases} 0, & \text{for } t < 0 \\ 5t + 10, & \text{for } 0 < t < 1 \\ 5, & \text{for } t > 1 \end{cases}$$

Problem 2.23. The voltage v across a 1 H inductor consists of one cycle of a sinusoidal waveform as shown in Fig. P. 2.23.
(a) Write the equation for $v(t)$.
(b) Find and plot the current through the inductor.
(c) Find the amount and time of maximum energy in the inductor.

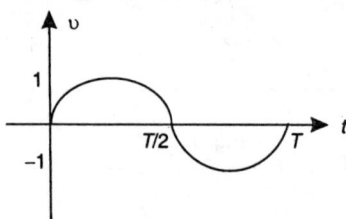

Fig. P. 2.23

Solution:

(a) $v(t)$ can be expressed as:

$$v(t) = \sin \omega t \cdot u(t) - \sin \omega t \cdot u(t - T)$$

$$v(t) = \sin \omega t \, (u(t) - u(t - T))$$

(b) Current i_L through inductor is given as:

$$i = \frac{1}{L}\int_0^t v\,dt = \frac{1}{L}\int_0^t \sin \omega t \,(u(t) - u(t - T))\,dt$$

$$= \frac{1}{\omega L} \left[-\cos\omega t \left(u(t) - u(t-T) \right) \right]_0^t$$

$$= \frac{1}{\omega L} \left[-\cos\omega t \right]_0^t \left(u(t) - u(t-T) \right)$$

$$i = \frac{1}{\omega L} \left(1 - \cos\omega t \right) \left(u(t) - u(t-T) \right)$$

So $i(t)$ can be plotted as:

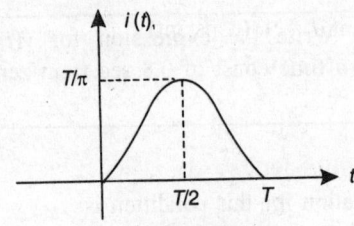

Fig. P. 2.23 (a)

Maximum value of $i(t)$ occurs at $t = T/2$.

(c) Amount of energy is given by:

$$w_{max} = \frac{1}{2} L i_{max}^2$$

$$i_{max} = \frac{1}{\omega L} \left(1 - \cos\frac{2\pi \times T}{T \times 2} \right), \text{ at } t = T/2 = \frac{1}{\omega L}(1 - (-1)) = \frac{2}{\omega L}$$

$$\therefore \quad w_{max} = \frac{1}{2} L \left(\frac{2}{\omega L} \right)^2 = \frac{L}{2} \times \left(\frac{2}{\frac{2\pi}{T} L} \right)^2 = \frac{L}{2} \left(\frac{T}{\pi L} \right)^2$$

$$w_{max} = \frac{1}{2} \left(\frac{T}{\pi} \right)^2 \quad [\because L = 1H]$$

or $\quad w_{max} = \dfrac{T^2}{2\pi^2}$ **Ans.**

Problem 2.24. Write the expression for $v(t)$ which decays exponentially from 1 at $t = 0$ to 3 at $t = \infty$ with a time constant of 200 ms.

Solution:

The generalized equation is:

$$v(t) = \text{(intial value } - \text{ final value)} \, e^{-t/\tau} + \text{final value}$$

where τ = time const

$$\therefore \quad v(t) = (7-3) \, e^{-\left(\frac{t}{200ms}\right)} + 3 = 4e^{-5t} + 3$$

$$v(t) = 3 + 4e^{-5t} \quad \text{for } t > 0 \text{ **Ans.**}$$

Problem 2.25. Write the expression for $v(t)$ which grows exponentially with a time const of 0.8 sec from zero at $t = -\infty$ to 9 at $t = 0$

Solution:

The generalized equation for this condition is:

$$v(t) = \text{(value at } t = 0) \, e^{-t/\tau}$$

$$v(t) = 9e^{-\frac{t}{0.8}} = 9e^{-\frac{5}{4}t}$$

$$\therefore \quad v(t) = 9e^{-\frac{5}{4}t} \quad \text{for } t < 0 \text{ **Ans.**}$$

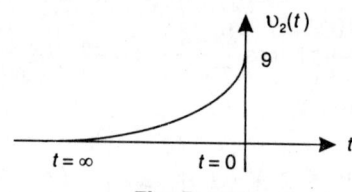

Fig. P. 2.25

Problem 2.26. Express the current of Fig. P. 2.26 in terms of step-function.

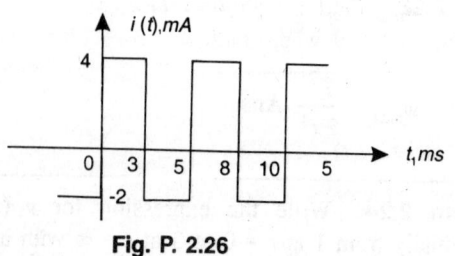

Fig. P. 2.26

Solution:

The above function can be expressed in terms of step-function as:

$$i(t) = 4u(t) - 6u(t-3) + 6u(t-5) - 6u(t-8) + 6u(t-10)\cdots$$

$$i(t) = 4u(t) + 6\sum_{k=1}^{\infty} \left[u(t-5k) - u(t-5k+2) \right] \textbf{ Ans.}$$

Problem 2.27. In Fig. 2.27 Let $T = 1$s and call the waveform S_1 (t). Express S_1 (t) and its first two derivatives dS_1/dt and d^2S_1/dt^2, using step and impulse functions.

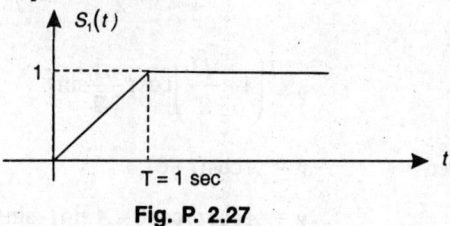

Fig. P. 2.27

Solution:

S_1 (t) can be expressed as:

$$S_1(t) = t(u(t) - u(t-1)) + u(t-1)$$

$$\frac{dS_1(t)}{dt} = u(t) - u(t-1) + \delta(t-1)$$

$$\frac{d^2S_1(t)}{dt^2} = \delta(t) - \delta(t-1)$$

Problem 2.28. Find an impulse voltage which creates a 1-A current jump at $t = 0$ when applied across a 10 mH inductor.

Solution:

$$i(t) = u(t)$$

Voltage across inductor

$$V_L = L\frac{di}{dt} = 10\,\text{mH} \cdot \frac{du(t)}{dt} \quad V_L = 10^{-2}\,\delta(t) \textbf{ Ans.}$$

Problem 2.29. (a) Given $v_1 = \cos t$, $v_2 = \cos (t + 30°$ and $v = v_1 + v_2$, write v in the form of a single cosine function $v = A \cos (t + \theta)$. (b) Find effective value of v_1, v_2 and v. Discuss why $v_{\text{eff}}^2 > (v_{1,\text{eff}}^2 + v_{2,\text{eff}}^2)$

Solution:

(a)
$$v = v_1 + v_2 = \cos t + \cos(t + 30°)$$
$$= \cos t + \cos t \cdot \cos 30° - \sin t \cdot \sin 30°$$
$$= \cos t + \cos t \cdot \frac{\sqrt{3}}{2} - \sin t \cdot \frac{1}{2}$$
$$v = \left(1 + \frac{\sqrt{3}}{2}\right)\cos t - \frac{1}{2}\sin t \qquad (1)$$

Given
$$v = A\cos(t + \theta)$$
$$v = A\cos t \cdot \cos\theta - A\sin t \cdot \sin\theta \qquad (2)$$

Comparing equation (1) with equation (2).

$$A\cos\theta = \left(1 + \frac{\sqrt{3}}{2}\right), \quad A\sin\theta = \frac{1}{2}$$

or,
$$A^2 \cos^2\theta + A^2 \sin^2\theta = A^2(1) = A^2$$

or,
$$\left(1 + \frac{\sqrt{3}}{2}\right)^2 + \left(\frac{1}{2}\right)^2 = A^2$$

∴
$$A = 1.93$$

∴
$$A\sin\theta = \frac{1}{2}$$

∴
$$\theta = \sin^{-1}\frac{1}{2A} = 15°$$

∴
$$\theta = 15°$$

∴
$$v = a \cos (t + \theta) = 1.93 \cos (t + 15°)$$
$$v = 1.93 \cos (t + 15°)$$

(b) $v_{1,\text{eff}} = \left[\dfrac{1}{2\pi}\displaystyle\int_0^{2\pi}\cos^2 t\, dt\right]^{1/2}$ [where 2π = period of $\cos t$]

$= \left[\dfrac{1}{2\pi}\displaystyle\int_0^{2\pi}\dfrac{1+\cos 2t}{2}\, dt\right]^{1/2} = \left[\dfrac{1}{4\pi}+[t+\dfrac{\sin 2t}{2}]_0^{2\pi}\right]^{1/2}$

$= \left[\dfrac{1}{4\pi}[2\pi+0+0]\right]^{1/2} = \dfrac{1}{\sqrt{2}}$

$\therefore\ v_{2,\text{eff}} = \left[\dfrac{1}{2\pi},\displaystyle\int_0^{2\pi}\cos^2(t+30°)\, dt\right]^{1/2} = \left[\dfrac{1}{4\pi}\displaystyle\int_0^{2\pi}1+\cos 2(t+30°)\, dt\right]^{1/2}$

$= \left[\dfrac{1}{4\pi}[2\pi+0.866-0.866]\right]^{1/2} = \left(\dfrac{1}{2}\right)^{1/2} \therefore v_{2,\text{eff}} = \left(\dfrac{1}{\sqrt{2}}\right)$

or, $v_{\text{eff}} = \left[\dfrac{1}{2\pi}\displaystyle\int_0^{2\pi}(1.93\cos(t+15°))^2\, dt\right]^{1/2}$

$v_{\text{eff}} = \left[\dfrac{3.724}{2\pi\times 2}[2\pi+0.5-0.5]\right]^{1/2} = [1.862]^{1/2} = 1.366$

$\therefore\ v_{\text{eff}} = 1.366$

This shows that $v_{\text{eff}}^2 > \left(v_{1,\text{eff}}^2 + v_{2,\text{eff}}^2\right)$

$\therefore\quad v_{\text{eff}}^2 = <v^2> = <(v_1+v_2)^2> = <v_1^2+v_2^2+2v_1v_2>$

$= <v_1^2> + <v_2^2> + 2<v_1 v_2>$

Since v_1 and v_2 have the same frequency and are 30° out of phase,

We get $<v_1 v_2> = \dfrac{1}{2}\cos 30°$

$\therefore\qquad <v_1 v_2> = \sqrt{3}/4$ which is $+ve$

Hence $\qquad v_{\text{eff}}^2 > (v_{1,\text{eff}}^2 + v_{2,\text{eff}}^2)$ **Ans.**

Probleme. 2.30. (a) Show that $v_1 = \cos t + \cos \sqrt{2}\, t$ is not periodic.
(b) Replace $\sqrt{2}$ by 1.4 and show that $v_1 = \cos t + \cos 1.4t$ is periodic and find its period T_2. (c) Replace $\sqrt{2}$ by 1.41 and find the period T_3 of $v_3 = \cos t + \cos 1.41\, t$. (d) replace $\sqrt{2}$ by 1.4142 and find the period T_4 of $v_4 = \cos t + \cos 1.4142\, t.$

Solution:

(a) $v_1 = \cos t + \cos \sqrt{2}\, t$ period of $\cos t$ $(T_1 = 2\pi)$ and that of $\cos \sqrt{2}\, t$ is $(T_2 = \pi)$

Since $\dfrac{T_1}{T_2} = \dfrac{2\pi}{\sqrt{2}\,\pi} = \sqrt{2}$ is not a rational number, hence v_1 is not periodic.

(b) $v_1 = \cos t + \cos 1.4t$. period of $\cos t$ $(T_1 = 2\pi)$ and that of cos

$1.4t$ is $\left(T_2 = \dfrac{2\pi}{1.4} \right)$, $\therefore$ $\dfrac{T_1}{T_2} = \dfrac{2\pi}{2\pi /1.4} = 1.4 = \dfrac{14}{10}$ is a rational number.

Hence it is a periodic function. Its period is given as

$$T = n_1 T_1 = n_2 T_2$$

$\therefore$ $n_1 T_1 = n_2 T_2$

or, $\dfrac{n_1}{n_2} = \dfrac{T_2}{T_1} = \dfrac{10}{14} = \dfrac{5}{7} \therefore n_1 = 5$

$\therefore$ $T = n_1 T_1 = 5 \times 2\pi = 10\pi$

$\therefore$ $T = 10\pi$ sec

(c) for, $v_1 = \cos t + \cos 1.41\, t$. period of $\cos t = 2\pi = T_1$ and

that of $\cos 1.41\, t + $is $T_2 = \dfrac{2\pi}{1.41}$

$\because$ $\dfrac{T_1}{T_2} = \dfrac{2\pi}{2\pi /1.41} = \dfrac{141}{100} = $ rational number

Hence $v_1 = $ periodic function, its period is given as:

$$\frac{T_1}{T_2} = \frac{n_2}{n_1} = \frac{141}{100}$$

$\therefore$ $\qquad n_1 = 100$ (minimum integer)

$\therefore$ $\qquad T = n_1 T_1 = 100 \times 2\pi = 200\pi$

$\therefore$ $\qquad T = 200\pi$ sec

(d) For $\qquad v_1 = \cos t + \cos 1.4142 \, t$

period of $\cos t = 2\pi = T_1$ and that of $\cos 1.4142 \, t$ is

$$T_2 = \frac{2\pi}{1.4142}$$

$\therefore$ $\qquad \dfrac{T_1}{T_2} = \dfrac{2\pi}{2\pi / 1.4142} = 1.4142 = \dfrac{14142}{10000}$

which is a rational number. Hence v_1 is a periodic function. Its period is given as

$$\frac{n_1}{n_2} = \frac{T_2}{T_1} = \frac{10000}{14142} = \frac{5000}{7071}$$

$\therefore$ $\qquad n_1 = 5000.$

$\therefore$ $\qquad T = n_1 T_1 = 5000 \times 2\pi$

$\therefore$ $\qquad T = 10000\pi$ sec **Ans.**

Promlem 2.31. Determine the branch current in terms of the **loop** current, for a tree of the network given below.

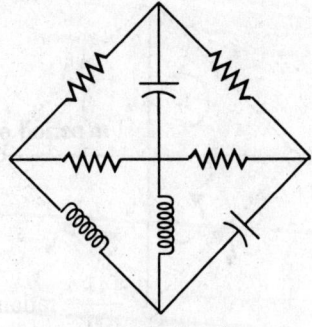

Fig. P. 2. 31

Solution:

The oriented graph or directed graph of network is

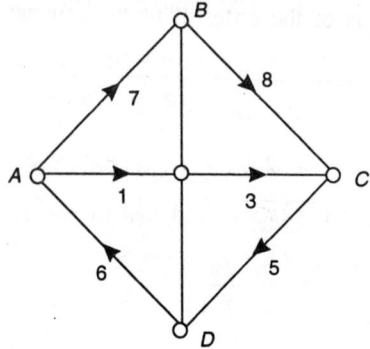

Fig. P. 2. 31 (a)

and the connected subgraph or tree graph is

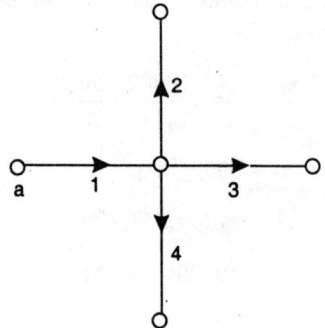

Fig. P. 2. 31 (b)

Tie set of the graph is

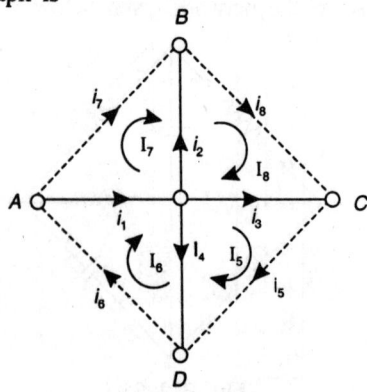

Fig. P. 2. 31 (c)

From above fig. the twigs are [1, 2, 3, 4] and links [5, 6, 7, 8]. The no of loops is 4, i.e. I_7, I_8, I_5 and I_6 are the loop current which flows in the four loops and these current should coincide with the direction of the corresponding links.

The tie set graph is of the order of (4 × 8) tie set matrix.

Loop current or Link current	Branch Current							
	1	2	3	4	5	6	7	8
5	0	0	1	−1	1	0	0	0
6	1	0	0	1	0	1	0	0
7	−1	−1	0	0	0	0	1	0
8	0	1	−1	0	0	0	0	1

Arranging in matrix form,

$$\beta = \begin{bmatrix} 0 & 0 & 1 & -1 & 1 & 0 & 0 & 0 \\ 1 & 0 & 0 & 1 & 0 & 1 & 0 & 0 \\ -1 & -1 & 0 & 0 & 0 & 0 & 1 & 0 \\ 0 & 1 & -1 & 0 & 0 & 0 & 0 & 1 \end{bmatrix}$$

I_b is branch current matrix

$$[I_b]^T = [i_1\ i_2\ i_3\ i_4\ i_5\ i_6\ i_7\ i_8]$$

and I_L is loop current matrix

$$[I_L]^T = [i_5\ i_6\ i_7\ i_8]$$

∴ Since $I_b = \beta^T I_L$

$$\begin{bmatrix} i_1 \\ i_2 \\ i_3 \\ i_4 \\ i_5 \\ i_6 \\ i_7 \\ i_8 \end{bmatrix} = \begin{bmatrix} 0 & 1 & -1 & 0 \\ 0 & 0 & -1 & 1 \\ 1 & 0 & 0 & -1 \\ -1 & 1 & 0 & 0 \\ 1 & 0 & 0 & 0 \\ 0 & 1 & 0 & 0 \\ 0 & 0 & 1 & 0 \\ 0 & 0 & 0 & 1 \end{bmatrix} \begin{bmatrix} I_5 \\ T_6 \\ I_7 \\ I_8 \end{bmatrix}$$

After solving, the branch currents in terms of loop currents are

$$i_1 = I_6 - I_7$$
$$i_2 = -I_7 + I_8$$
$$i_3 = I_5 - I_8$$
$$i_4 = -I_5 - I_6$$
$$i_5 = I_5$$
$$i_6 = I_6$$
$$i_7 = I_7$$
$$i_8 = I_8$$

Probmlem 2.32. Draw matrix considering a case of a network consisting of three resistors, three inductors and three capacitors. Each branch of the network contains only one element, which may be a resistance, inductance or capacitance.

Solution:

Suppose the branches consisting of inductors be numbered 1, 2, 3, branches consisting of resistors be 4, 5, 6 and those consisting of capacitor are 7, 8, 9. Then we can write branch impedance matrix Z_b in the form of

$$
\begin{bmatrix}
SL_{11} & SL_{12} & SL_{13} & 0 & 0 & 0 & 0 & 0 & 0 \\
SL_{21} & SL_{22} & SL_{23} & 0 & 0 & 0 & 0 & 0 & 0 \\
SL_{31} & SL_{32} & SL_{33} & 0 & 0 & 0 & 0 & 0 & 0 \\
0 & 0 & 0 & R_4 & 0 & 0 & 0 & 0 & 0 \\
0 & 0 & 0 & 0 & R_5 & 0 & 0 & 0 & 0 \\
0 & 0 & 0 & 0 & 0 & R_6 & 0 & 0 & 0 \\
0 & 0 & 0 & 0 & 0 & 0 & 1/SC_7 & 0 & 0 \\
0 & 0 & 0 & 0 & 0 & 0 & 0 & 1/SC_8 & 0 \\
0 & 0 & 0 & 0 & 0 & 0 & 0 & 0 & 1/SC_9
\end{bmatrix}
$$

Y_b is branch admittance matrix is written as

$$Y_b = [Z_b]^{-1} = \begin{bmatrix} SL_b & 0 & 0 \\ 0 & R_b & 0 \\ 0 & 0 & 1/SC_b \end{bmatrix} = \begin{bmatrix} [SL_b]^{-1} & 0 & 0 \\ 0 & [R_b]^{-1} & 0 \\ 0 & 0 & [1/SC_b]^{-1} \end{bmatrix}$$

$$= \begin{bmatrix} Y_{11} & \cdots & \cdots & Y_{19} \\ \cdots & \cdots & \cdots & \cdots \\ Y_{91} & \cdots & \cdots & Y_{99} \end{bmatrix}$$

In the above, $\quad SL_b = \begin{bmatrix} SL_{11} & SL_{12} & SL_{13} \\ SL_{21} & SL_{22} & SL_{23} \\ SL_{31} & SL_{32} & SL_{33} \end{bmatrix}$

$$R_b = \begin{bmatrix} R_4 & 0 & 0 \\ 0 & R_5 & 0 \\ 0 & 0 & R_6 \end{bmatrix}$$

$$\frac{1}{SC_b} = \begin{bmatrix} \dfrac{1}{SC_7} & 0 & 0 \\ 0 & 1/SC_8 & 0 \\ 0 & 0 & 1/SC_9 \end{bmatrix}$$

Problem 2.33. Draw the graph of the network shown in figure below and draw all possible trees of the network with the help of matrix.

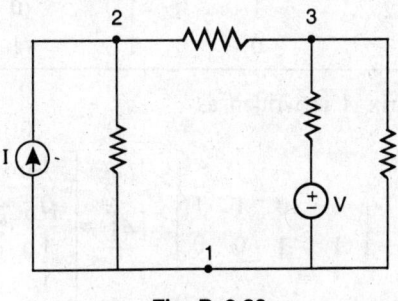

Fig. P. 2.33

Solution:

Network is transformed into directed graph. In the form of directed graph current source is replaced by an open circuit, voltage source is replaced by short circuit and all linear elements are shown as lines in the graph.

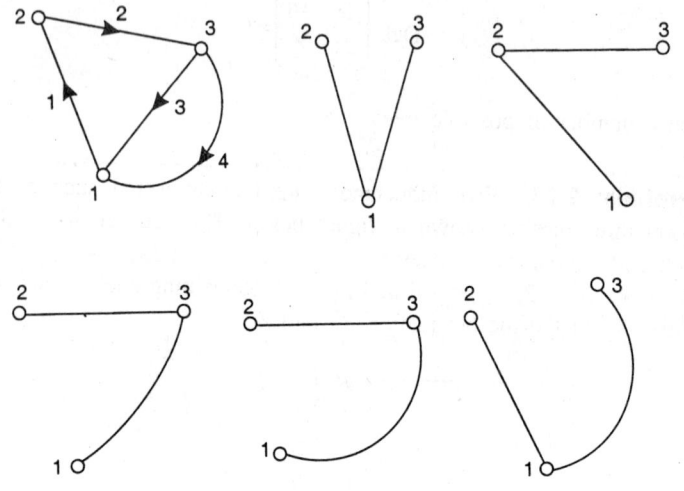

Fig. P. 2.33 (a)

In the above directed graph branches $b = 4$, nodes $n = 3$ therefore tree branches will be $(n - 1) = 2$ and links $(b - n + 1) = 2$.

The complete incidence matrix is

	Nodes	Branches			
		1	2	3	4
	1	−1	0	1	1
$A_a =$	2	1	−1	0	0
	3	0	1	−1	−1

The reduced matrix A is written as

$$A = \begin{bmatrix} -1 & 0 & 1 & 1 \\ 1 & -1 & 0 & 0 \end{bmatrix}, \quad [A]^T = \begin{bmatrix} -1 & 1 \\ 0 & -1 \\ 1 & 0 \\ 1 & 0 \end{bmatrix}$$

Number of possible trees = Det $\{[A]\,[A]^T\}$

$$\text{Det}\left\{ \begin{bmatrix} -1 & 0 & 1 & 1 \\ 1 & -1 & 0 & 0 \end{bmatrix} \begin{bmatrix} -1 & 1 \\ 0 & -1 \\ 1 & 0 \\ 1 & 0 \end{bmatrix} \right\}$$

$$= \text{Det} \begin{bmatrix} 3 & -1 \\ -1 & 2 \end{bmatrix} = 5$$

Hence number of possible trees = 5

Problem 2.34. Five inductances are interconnected such as to form three meshes shown in figure below. The coefficients of the self and mutual inductances are $L_{11} = 2$, $L_{22} = 2$, $L_{33} = 4$, $L_{44} = 2$, $L_{55} = 4$, $L_{12} = 2$, $L_{15} = -4$ and $L_{34} = 2$. Remaining coefficients are Zero. Write the mesh equation in matrix form.

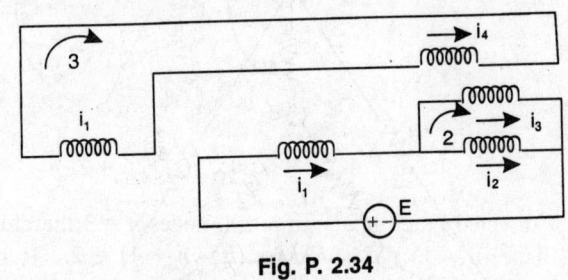

Fig. P. 2.34

Solution:

The tie matrix β for above network is

$$\beta = \begin{bmatrix} 1 & 1 & 0 & 0 & 0 \\ 0 & -1 & 1 & 0 & 0 \\ 0 & 0 & 0 & -1 & -1 \end{bmatrix}$$

We know that $Z = \beta Z_b \beta^T$

$$= \begin{bmatrix} 1 & 1 & 0 & 0 & 0 \\ 0 & -1 & 1 & 0 & 0 \\ 0 & 0 & 0 & -1 & -1 \end{bmatrix} \begin{bmatrix} 2 & 2 & 0 & 0 & -4 \\ 2 & 2 & 0 & 0 & 0 \\ 0 & 0 & 4 & 2 & 0 \\ 0 & 0 & 2 & 2 & 0 \\ -4 & 0 & 0 & 0 & 4 \end{bmatrix} \begin{bmatrix} 1 & 0 & 0 \\ 1 & -1 & 0 \\ 0 & 1 & 0 \\ 0 & 0 & -1 \\ 0 & 0 & -1 \end{bmatrix}$$

$$= \begin{bmatrix} 1 & 1 & 0 & 0 & 0 \\ 0 & -1 & 1 & 0 & 0 \\ 0 & 0 & 0 & -1 & -1 \end{bmatrix} \begin{bmatrix} 4 & -2 & 4 \\ 4 & -2 & 0 \\ 0 & 4 & -2 \\ -4 & 0 & -4 \end{bmatrix} = \begin{bmatrix} 8 & -4 & 4 \\ -4 & 6 & -2 \\ 4 & -2 & 6 \end{bmatrix}$$

We know that $Z_{hK} = Z_{kh}$ means $[Z]$ is symmetric. Also, we know that
$E = ZI_L$

$$\begin{bmatrix} E \\ 0 \\ 0 \end{bmatrix} = \begin{bmatrix} 8 & -4 & 4 \\ -4 & 6 & -2 \\ 4 & -2 & 6 \end{bmatrix} \begin{bmatrix} I_1 \\ I_2 \\ I_3 \end{bmatrix}$$

Problem 2.35. Draw the directed graph for the circuit shown below. Select loop current variables and write the network equilibrium eqn matrix form by using kVL.

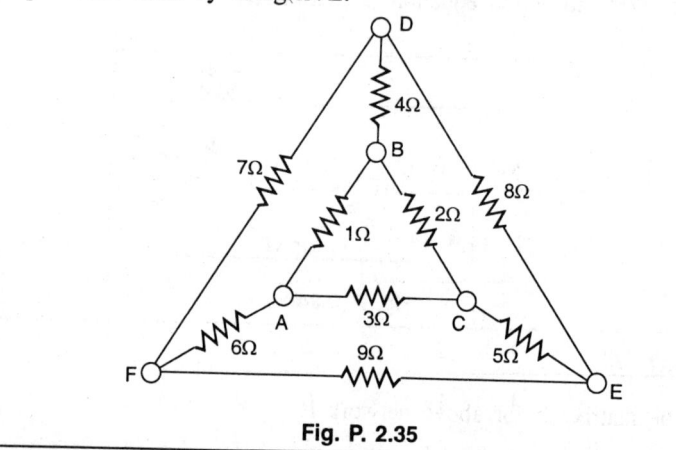

Fig. P. 2.35

Solution:

Network is transformed into directed graph.

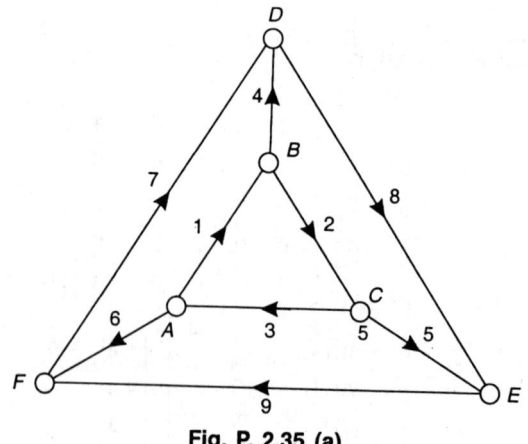

Fig. P. 2.35 (a)

Possible tree in directed graph

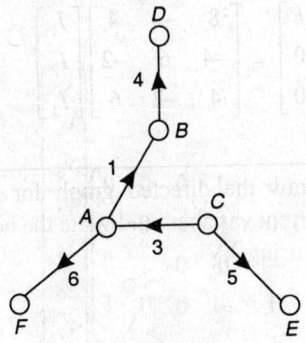

Fig. P. 2.35 (b)

In the above Figure set of twigs is (1, 3, 4, 5, 6) and set of links is (2, 7, 8, 9).

Tie set matrix β is written as in the form

Loop current					Branch				
	1	2	3	4	5	6	7	8	9
2	1	1	1	0	0	0	0	0	0
7 $\beta =$	−1	0	0	−1	0	1	1	0	0
8	1	0	1	1	−1	0	0	1	0
9	0	0	−1	0	1	−1	0	0	1

Hence β^T can be written as

$$\beta^T = \begin{bmatrix} 1 & -1 & 1 & 0 \\ 1 & 0 & 0 & 0 \\ 1 & 0 & 1 & -1 \\ 0 & -1 & 1 & 0 \\ 0 & 0 & -1 & 1 \\ 0 & 1 & 0 & -1 \\ 0 & 1 & 0 & 0 \\ 0 & 0 & 1 & 0 \\ 0 & 0 & 0 & 1 \end{bmatrix}$$

We know that $[\beta][V_b] = [0]$

or,

$$\begin{bmatrix} 1 & 1 & 1 & 0 & 0 & 0 & 0 & 0 & 0 \\ -1 & 0 & 0 & -1 & 0 & 1 & 1 & 0 & 0 \\ 1 & 0 & 1 & 1 & -1 & 0 & 0 & 1 & 0 \\ 0 & 0 & -1 & 0 & 1 & -1 & 0 & 0 & 1 \end{bmatrix} \begin{bmatrix} v_1 \\ v_2 \\ v_3 \\ v_4 \\ v_5 \\ v_6 \\ v_7 \\ v_8 \\ v_9 \end{bmatrix} = \begin{bmatrix} 0 \\ 0 \\ 0 \\ 0 \end{bmatrix}$$

We know that $[I_b] = [\beta^T][I_L]$

So,

$$\begin{bmatrix} i_1 \\ i_2 \\ i_3 \\ i_4 \\ i_5 \\ i_6 \\ i_7 \\ i_8 \\ i_9 \end{bmatrix} = \begin{bmatrix} 1 & -1 & 1 & 0 \\ 1 & 0 & 0 & 0 \\ 1 & 0 & 1 & -1 \\ 0 & -1 & 1 & 0 \\ 0 & 0 & -1 & 1 \\ 0 & 1 & 0 & -1 \\ 0 & 1 & 0 & 0 \\ 0 & 0 & 1 & 0 \\ 0 & 0 & 0 & 1 \end{bmatrix} \begin{bmatrix} I_2 \\ I_7 \\ I_8 \\ I_9 \end{bmatrix}$$

We know that $[V_b] = [Z_b][I_b] - [V_s]$

We have $[V_s] = 0$

So we can write

$$
\begin{bmatrix} V_1 \\ V_2 \\ V_3 \\ V_4 \\ V_5 \\ V_6 \\ V_7 \\ V_8 \\ V_9 \end{bmatrix} =
\begin{bmatrix}
1 & - & - & - & - & - & - & - & 0 \\
- & 2 & - & - & - & - & - & - & - \\
- & - & 3 & - & - & - & - & - & - \\
- & - & - & 4 & - & - & - & - & - \\
- & - & - & - & 5 & - & - & - & - \\
- & - & - & - & - & 6 & - & - & - \\
- & - & - & - & - & - & 7 & - & - \\
- & - & - & - & - & - & - & 8 & - \\
0 & - & - & - & - & - & - & - & 9
\end{bmatrix}
\begin{bmatrix} i_1 \\ i_2 \\ i_3 \\ i_4 \\ i_5 \\ i_6 \\ i_7 \\ i_8 \\ i_9 \end{bmatrix}
$$

We know that, $[V_b] = [Z_b][I_b] = [Z_b][\beta^T][I_L]$

and, $[\beta][V_b] = [\beta][Z_b][\beta^T][I_L] = [0]$

Hence, $[\beta][Z_b][\beta^T][I_L] = [0]$

$[Z][I_L] = [0]$ where $[Z] = [\beta][Z_b][\beta^T]$

So we can write $[Z]$ as

$$
Z =
\begin{bmatrix}
1 & 1 & 1 & 0 & 0 & 0 & 0 & 0 & 0 \\
-1 & 0 & 0 & -1 & 0 & 1 & 1 & 0 & 0 \\
1 & 0 & 1 & 1 & -1 & 0 & 0 & 1 & 0 \\
0 & 0 & -1 & 0 & 1 & -1 & 0 & 0 & 1
\end{bmatrix} \times
$$

$$
\begin{bmatrix}
1 & & & & & & & & 0 \\
& 2 & & & & & & & \\
& & 3 & & & & & & \\
& & & 4 & & & & & \\
& & & & 5 & & & & \\
& & & & & 6 & & & \\
& & & & & & 7 & & \\
& & & & & & & 8 & \\
0 & & & & & & & & 9
\end{bmatrix}
\begin{bmatrix}
1 & -1 & 0 & 0 \\
1 & 0 & 0 & 0 \\
1 & 0 & 1 & -1 \\
0 & -1 & 0 & 0 \\
0 & 0 & -1 & 1 \\
0 & 1 & 1 & -1 \\
0 & 1 & 0 & 0 \\
0 & 0 & 0 & 0 \\
0 & 0 & 1 & 1
\end{bmatrix}
$$

After solving it,

$$Z = \begin{bmatrix} 6 & -1 & 4 & -3 \\ -1 & 18 & -5 & -6 \\ 4 & -5 & 5 & -8 \\ -3 & -6 & -8 & 23 \end{bmatrix}$$

we know that $E = Z\,I_L$ so Kirochoff's voltage law eqn can be written as

$$[Z][I_L] = \begin{bmatrix} 6 & -1 & 4 & -3 \\ -1 & 18 & -5 & -6 \\ 4 & -5 & 5 & -8 \\ -3 & -6 & -8 & 23 \end{bmatrix}\begin{bmatrix} I_2 \\ I_7 \\ I_8 \\ I_9 \end{bmatrix} = [0]$$

After soving it KVL eqn can be written as

$$6I_2 - I_7 + 4I_8 - 3I_9 = 0$$

$$-I_2 + 18I_7 - 5I_8 - 6I_9 = 0$$

$$4I_2 - 5I_7 + 5I_8 - 8I_9 = 0$$

$$-3I_2 - 6I_7 - 8I_8 - 23I_9 = 0$$

Problem 2.36. For the network shown below, write down the tie set matrix and obtain the network equilibrium eqn in matrix form using KVL. Calculate the loop currents to calculate branch voltage.

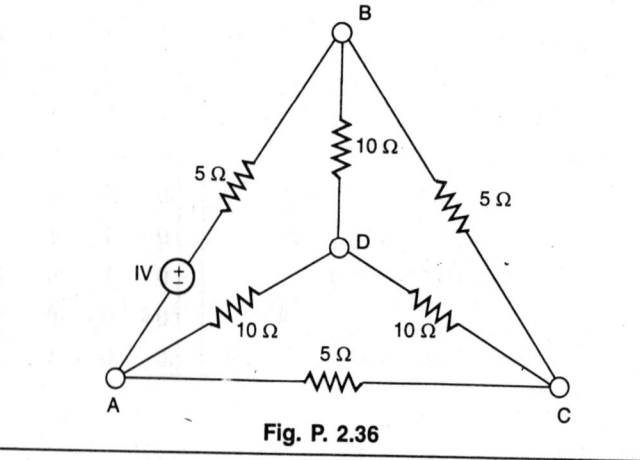

Fig. P. 2.36

Solution

The network is transformed into directed graph and then draw the possible trees. In directed graph voltage source is denoted as short circuit line and linear element is denoted by line.

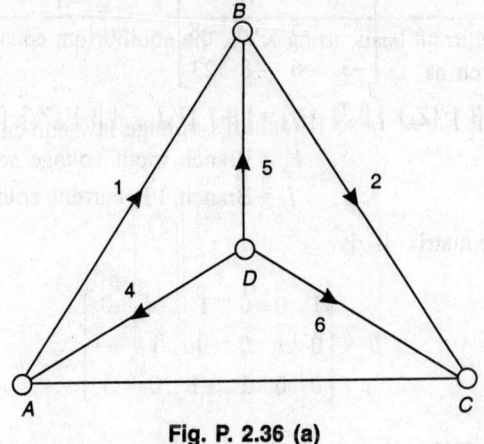

Fig. P. 2.36 (a)

In the above Figure set of twigs are (4, 5, 6) and links are (1, 2, 3). In the above Figure.

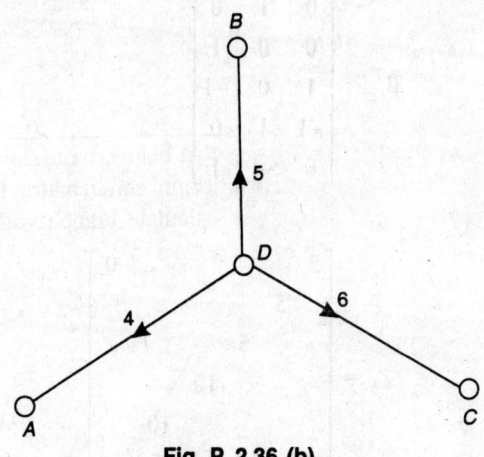

Fig. P. 2.36 (b)

No. of branches = 6

No. of nodes = 4

No. of links = 3

So, the tie matrix is

Loop cureent	Branches					
	1	2	3	4	5	6
I_1	1	0	0	1	-1	0
I_2	0	1	0	0	1	-1
I_3	0	0	1	-1	0	1

From loop current basis, using KVL, the equilibrium equation in matrix form is given as

$$[\beta] \ [Z_b] \ [\beta^T] \ [I_L] = [\beta] \ [V_s] - [\beta] \ [Z_b] \ [I_s]$$

$$V_s = \text{Branch input voltage source vector}$$
$$I_s = \text{Branch I/P current source vector}$$

So, Tie set matrix β is

$$B = \begin{bmatrix} 1 & 0 & 0 & 1 & -1 & 0 \\ 0 & 1 & 0 & 0 & 1 & -1 \\ 0 & 0 & 1 & -1 & 0 & 1 \end{bmatrix}$$

Transpose matrix of β is

$$\beta^T = \begin{bmatrix} 1 & 0 & 0 \\ 0 & 1 & 0 \\ 0 & 0 & 1 \\ 1 & 0 & -1 \\ -1 & 1 & 0 \\ 0 & -1 & 1 \end{bmatrix}$$

$$Z_b = \begin{bmatrix} 5 & & & & & 0 \\ & 5 & & & & \\ & & 5 & & & \\ & & & 10 & & \\ & & & & 10 & \\ 0 & & & & & 10 \end{bmatrix}$$

Loop current $\qquad I_L = \begin{bmatrix} I_1 \\ I_2 \\ I_3 \end{bmatrix}$

$$V_s = \begin{bmatrix} 1 \\ 0 \\ 0 \\ 0 \\ 0 \\ 0 \end{bmatrix} \text{ and } I_s \begin{bmatrix} 0 \\ 0 \\ 0 \\ 0 \\ 0 \\ 0 \end{bmatrix}$$

So, $$[\beta][Z_b][\beta^T][I_L] = [\beta][V_s]$$

or, $$\begin{bmatrix} 25 & -10 & -10 \\ -10 & 25 & -10 \\ -10 & -10 & 25 \end{bmatrix} \begin{bmatrix} I_1 \\ I_2 \\ I_3 \end{bmatrix} = \begin{bmatrix} 1 \\ 0 \\ 0 \end{bmatrix}$$

After solving it, $I_1 = \dfrac{\Delta_1}{\Delta} = 3/35\,A$

$$I_2 = \dfrac{\Delta_2}{\Delta} = 2/35\,A$$

$$I_3 = \dfrac{\Delta_3}{\Delta} = 2/35\,A$$

Branch current is written as

$$[I_b] = \left[\beta^T\right][I_L]$$

Branch voltage is $[V_b] = [Z_b][I_b] - [V_s]$

$$[V_b] = [Z_b]\left[\beta^T\right][I_L] - [V_s]$$

So, $$\begin{bmatrix} V_1 \\ V_2 \\ V_3 \\ V_4 \\ V_5 \\ V_6 \end{bmatrix} = \begin{bmatrix} 5 & & & & & \\ & 5 & & & & \\ & & 5 & & & \\ & & & 10 & & \\ & & & & 10 & \\ & & & & & 10 \end{bmatrix} \begin{bmatrix} 1 & 0 & 0 \\ 0 & 1 & 0 \\ 0 & 0 & 1 \\ 1 & 0 & -1 \\ -1 & 1 & 0 \\ 0 & -1 & 1 \end{bmatrix} \begin{bmatrix} 3/35 \\ 2/35 \\ 2/35 \end{bmatrix} - \begin{bmatrix} 1 \\ 0 \\ 0 \\ 0 \\ 0 \\ 0 \end{bmatrix}$$

After solving it

$V_1 = -4/7V, V_2 = 2/7V, V_3 = 2/7V, V_4 = 2/7V, V_5 = -2/7V, V_6 = 0\,V$

Problem 2.37. Draw the directed graph for the circuit shown below and write the network equilibrium eqn.

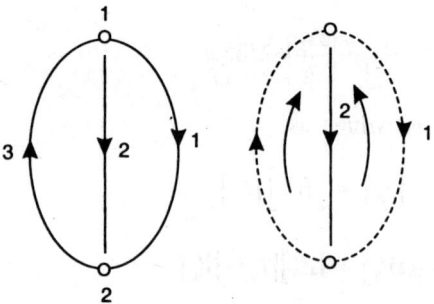

Fig. P. 2.37

Solution

The network is transformed into directed graph. In the directed graph voltage source is replaced by short circuited line and linear elements are tranformed by line. After drawing directed graph possible tree is drawn.

Fig. P. 2.37 (a)

in the above figure links are {1, 3}
So, the tie set matrix is

Loop	Branches		
	1	2	3
1	1	-1	0
3	0	1	1

So,

$$\beta = \begin{bmatrix} 1 & -1 & 0 \\ 0 & 1 & 1 \end{bmatrix}$$

Transpose matrix β is

$$[\beta^T] = \begin{bmatrix} 1 & 0 \\ -1 & 1 \\ 0 & 1 \end{bmatrix}$$

Branch impedance matrix

$$[Z_b] = \begin{bmatrix} 1 & 0 & 0 \\ 0 & S & 0.5S \\ 0 & 0.5S & 2S \end{bmatrix}$$

We know that
$$[Z] = [\beta][Z_b][\beta^T] = \begin{bmatrix} 1+S & -1.5S \\ -1.5S & 4S \end{bmatrix}$$

The network equilibruim equation on loop basis is

$$[Z][I_L] = [B]\{[V_s] - [Z_b][I_s]\} = [B][V_s]$$

here
$$[I_s] = [0]$$

So,
$$\begin{bmatrix} 1+S & -1.5S \\ -1.5S & 4S \end{bmatrix}\begin{bmatrix} I_1 \\ I_3 \end{bmatrix} = \begin{bmatrix} 1 & -1 & 0 \\ 0 & 1 & 1 \end{bmatrix}\begin{bmatrix} 0 \\ 0 \\ 1 \end{bmatrix} = \begin{bmatrix} 0 \\ 1 \end{bmatrix}$$

Hence equations are,

$$(1+S)I_1 - 1.5S\,I_3 = 0$$

$$-1.5SI_1 + 4S\,I_3 = 1$$

Peroblem 2.38. For the network given below, write f-cut set matrix and obtain equilibruim equation on node basis.

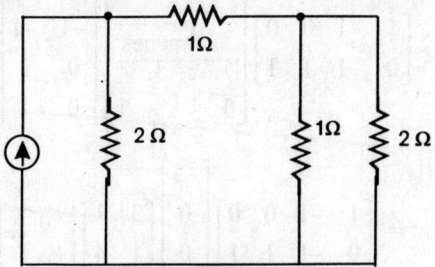

Fig. P. 2.38

Solution:

The network is transofrmed into directed graph. In the directed graph current source is denoted by open circuit line and linear circuit is simply a line, then we draw possible tree.

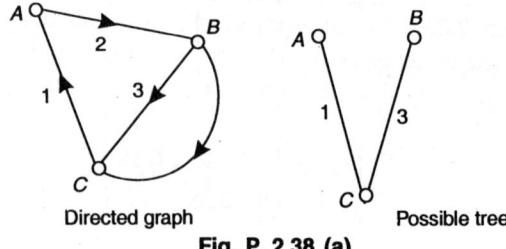

Directed graph   Possible tree

Fig. P. 2.38 (a)

Cut set matrix Q is written as:

Cut set		Branches		
	1	2	3	4
Q 1	1	−1	0	0
3	0	−1	1	1

Transpose cut set matrix

$$Q^T = \begin{bmatrix} 1 & 0 \\ -1 & -1 \\ 0 & 1 \\ 0 & 1 \end{bmatrix}, Y_b = \begin{bmatrix} 2 & & & 0 \\ & 1 & & \\ & & 1 & \\ 0 & & & 2 \end{bmatrix}, I_s = \begin{bmatrix} -5 \\ 0 \\ 0 \\ 0 \end{bmatrix}, V_s = [0]$$

We know that, $[Q][Y_b][Q^T][V_t] = [Q][I_s]$

$$\begin{bmatrix} 1 & -1 & 0 & 0 \\ 0 & -1 & 1 & 1 \end{bmatrix} \begin{bmatrix} 2 & & & 0 \\ & 1 & & \\ & & 1 & \\ 0 & & & 1 \end{bmatrix} \begin{bmatrix} 1 & 0 \\ -1 & -1 \\ 0 & 1 \\ 0 & 1 \end{bmatrix} \begin{bmatrix} V_{t1} \\ V_{t3} \end{bmatrix}$$

$$= \begin{bmatrix} 1 & -1 & 0 & 0 \\ 0 & -1 & 1 & 1 \end{bmatrix} \begin{bmatrix} -5 \\ 0 \\ 0 \\ 0 \end{bmatrix} \begin{bmatrix} 3 & 1 \\ 1 & 4 \end{bmatrix} \begin{bmatrix} V_{t1} \\ V_{t3} \end{bmatrix} = \begin{bmatrix} -5 \\ 0 \end{bmatrix}$$

After solving, $V_{t1} = -20/11V$ or $V_{AC} = \dfrac{20}{11}V$

$V_{t3} = 5/11V$ or $V_{BC} = \dfrac{5}{11}V$

Problem 2.39. For the network shown below, draw the directed graph write tie set equation and hence obtain equilibrium equation on loop basis. Calculate the values of branch currents and branch voltages.

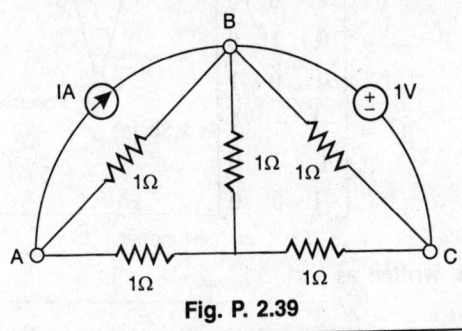

Fig. P. 2.39

Solution:

The above network is transformed into directed graph where current source is represented by open circuited and voltage source is represented by short circuited then we draw possible tree.

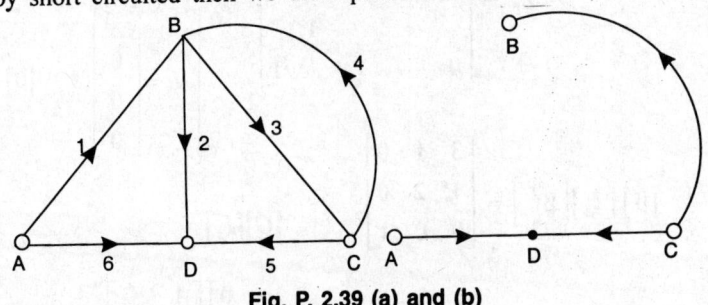

Fig. P. 2.39 (a) and (b)

In the above graph tree twigs are 4, 5, 6 and links are 1, 2, 3
Tie set matrix is written as.

Link current	Branches					
	1	2	3	4	5	6
1	1	0	0	– 1	1	– 1
2	0	1	0	– 1	1	0
3	0	0	1	1	0	0

Hence tie set matrix is

$$\beta = \begin{bmatrix} 1 & 0 & 0 & -1 & 1 & -1 \\ 0 & 1 & 0 & -1 & 1 & 0 \\ 0 & 0 & 1 & 1 & 0 & 0 \end{bmatrix}$$

β The transpose matrix is

$$\beta^T = \begin{bmatrix} 1 & 0 & 0 \\ 0 & 1 & 0 \\ 0 & 0 & 1 \\ -1 & -1 & 1 \\ 1 & 1 & 0 \\ -1 & 0 & 0 \end{bmatrix}$$

and Z_b can be written as

$$Z_b = \begin{bmatrix} 1 & & & & & 0 \\ & 1 & & & & \\ & & 1 & & & \\ & & & 0 & & \\ & & & & 1 & \\ 0 & & & & & 1 \end{bmatrix}$$

$$[\beta][Z_b][\beta^T] = \begin{bmatrix} 3 & 1 & 0 \\ 1 & 2 & 0 \\ 0 & 0 & 1 \end{bmatrix}$$

and

$$[V_s]-[Z_b][I_s] = \begin{bmatrix} 0 \\ 0 \\ 0 \\ 1 \\ 0 \\ 0 \end{bmatrix} - \begin{bmatrix} 1 & & & & & \\ & 1 & & & & \\ & & 1 & & & \\ & & & 0 & & \\ & & & & 1 & \\ & & & & & 1 \end{bmatrix} \begin{bmatrix} -1 \\ 0 \\ 0 \\ 0 \\ 0 \\ 0 \end{bmatrix} = \begin{bmatrix} 1 \\ 0 \\ 0 \\ 1 \\ 0 \\ 0 \end{bmatrix}$$

on loop basis equilibrium equation can be written as

$$[\beta][Z_b][\beta^T][I_L] = [\beta][V_s] - [Z_b][I_L]$$

$$
\begin{bmatrix} 3 & 1 & 0 \\ 1 & 2 & 0 \\ 0 & 0 & 1 \end{bmatrix}
\begin{bmatrix} I_1 \\ I_2 \\ I_3 \end{bmatrix} =
\begin{bmatrix} 1 & 0 & 0 & -1 & 1 & -1 \\ 0 & 1 & 0 & -1 & 1 & 0 \\ 0 & 0 & 1 & 1 & 0 & 0 \end{bmatrix}
\begin{bmatrix} 1 \\ 0 \\ 0 \\ 0 \\ 1 \\ 0 \\ 0 \end{bmatrix}
\begin{bmatrix} 0 \\ -1 \\ 1 \end{bmatrix}
$$

After solving,

$$I_1 = \frac{\Delta_1}{\Delta} = \frac{1}{5}A, I_2 = \frac{\Delta_2}{\Delta} = \frac{-3}{5}A, I_3 = \frac{\Delta 3}{\Delta} = 1A$$

Branch currents can be written as

$$[I_b] = [\beta^T][I_L] + [I_s]$$

$$
\begin{bmatrix} i_1 \\ i_2 \\ i_3 \\ i_4 \\ i_5 \\ i_6 \end{bmatrix} =
\begin{bmatrix} 1 & 0 & 0 \\ 0 & 1 & 0 \\ 0 & 0 & 1 \\ -1 & -1 & 1 \\ 1 & 1 & 0 \\ -1 & 0 & 0 \end{bmatrix}
\begin{bmatrix} 1/5 \\ -3/5 \\ 1 \end{bmatrix} +
\begin{bmatrix} -1 \\ 0 \\ 0 \\ 0 \\ 0 \\ 0 \end{bmatrix} =
\begin{bmatrix} -4/5 \\ -3/5 \\ 1 \\ 7/5 \\ -2/5 \\ -1/5 \end{bmatrix}
$$

We know that $[V_b] = [Z_b][I_b] - [V_s]$

$$
\begin{bmatrix} v_1 \\ v_2 \\ v_3 \\ v_4 \\ v_5 \\ v_6 \end{bmatrix} =
\begin{bmatrix} 1 & & & & & \\ & 1 & & & & \\ & & 1 & & & \\ & & & 0 & & \\ & & & & 1 & \\ 0 & & & & & 1 \end{bmatrix}
\begin{bmatrix} -4/5 \\ -3/5 \\ 1 \\ 7/5 \\ -2/5 \\ -1/5 \end{bmatrix}
\begin{bmatrix} 0 \\ 0 \\ 0 \\ 1 \\ 0 \\ 0 \end{bmatrix} =
\begin{bmatrix} -4/5 \\ -3/5 \\ 1 \\ -1 \\ -2/5 \\ -1/5 \end{bmatrix}
$$

We know that $\beta V_b = 0$, i.e. sum of voltage in any loop in zero.

$$\begin{bmatrix} 1 & 0 & 0 & -1 & 1 & -1 \\ 0 & 1 & 0 & -1 & 1 & 0 \\ 0 & 0 & 1 & 1 & 0 & 0 \end{bmatrix} \begin{bmatrix} -4/5 \\ -3/5 \\ 1 \\ -1 \\ -2/5 \\ -1/5 \end{bmatrix} = \begin{bmatrix} 0 \\ 0 \end{bmatrix}$$

Problem 2.40. Calculate the branch voltage and branch currents using the voltage variable method of the network shown below.

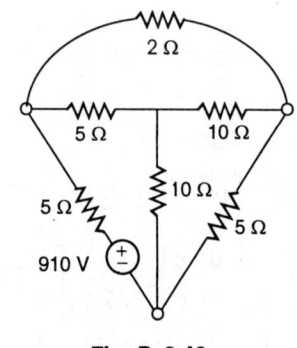

Fig. P. 2.40

Solution

Network is transformed into directed graph. In the directed graph voltage source is represented as short circuited line. Then fundamental cut set graph a drawn.

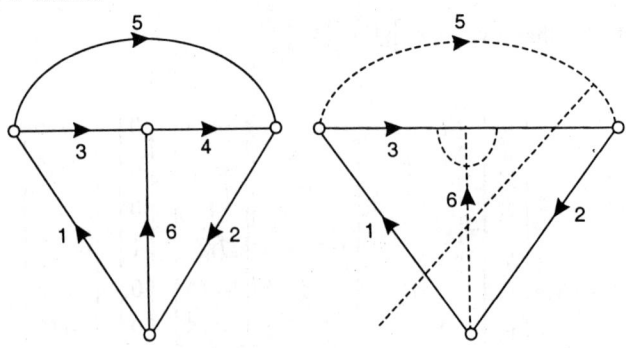

Fig. P. 2.40 (a) and (b)

f-cut set matrix can be written as.

Tree branch no.	Branches					
i.e f-cut set	1	2	3	4	5	6
1	1	0	0	-1	-1	1
	0	1	0	-1	-1	0
	0	0	1	-1	0	1

$$Q = \begin{bmatrix} 1 & 0 & 0 & -1 & -1 & 1 \\ 0 & 1 & 0 & -1 & -1 & 0 \\ 0 & 0 & 1 & -1 & 0 & 1 \end{bmatrix}$$

Transpose matrix Q is written as $Q^T = \begin{bmatrix} 1 & 0 & 0 \\ 0 & 1 & 0 \\ 0 & 0 & 1 \\ -1 & -1 & -1 \\ -1 & -1 & 0 \\ 1 & 0 & 1 \end{bmatrix}$

$$Y_b = \begin{bmatrix} 0.2 & & & & & 0 \\ & 0.2 & & & & \\ & & 0.2 & & & \\ & & & 0.1 & & \\ & & & & 0.5 & \\ 0 & & & & & 0.1 \end{bmatrix}$$

We have $V_s = \begin{bmatrix} 910 \\ 0 \\ 0 \\ 0 \\ 0 \\ 0 \end{bmatrix}$ & $I_s = [0]$

On node basis the network equilibrium equation can be written as

$$[Q][Y_b][Q^T][V_t] = Q\{I_s - Y_b V_s\} = -QY_b V_s \text{ and } I_s = 0$$

After putting the values, we get

$$\begin{bmatrix} 0.9 & 0.6 & 0.2 \\ 0.6 & 0.8 & 0.1 \\ 0.2 & 0.1 & 0.3 \end{bmatrix} \begin{bmatrix} V_{t1} \\ V_{t2} \\ V_{t3} \end{bmatrix} = \begin{bmatrix} -182 \\ 0 \\ 0 \end{bmatrix}$$

Solving,

$$V_{t1} = \frac{\Delta_1}{\Delta} = -460 \text{ V}, \ V_{t2} = \frac{\Delta_2}{\Delta} = 320 \text{ V}, \ V_{t3} = \frac{\Delta_3}{\Delta} = 200 \text{ V}$$

V_b is the branch voltage, that can be written as.

$$V_b = Q^T V_t + V_s = \begin{bmatrix} 1 & 0 & 0 \\ 0 & 1 & 0 \\ 0 & 0 & 1 \\ -1 & -1 & -1 \\ -1 & -1 & 0 \\ 1 & 0 & 1 \end{bmatrix} \begin{bmatrix} -460 \\ 320 \\ 200 \end{bmatrix} + \begin{bmatrix} 910 \\ 0 \\ 0 \end{bmatrix}, \begin{bmatrix} V_1 \\ V_2 \\ V_3 \\ V_4 \\ V_5 \\ V_6 \end{bmatrix} = \begin{bmatrix} 450 \\ 320 \\ 200 \\ -60 \\ 160 \\ -260 \end{bmatrix}$$

We know that
$$I_b = Y_b V_b - I_s$$

After putting the values, it is

$$\begin{bmatrix} i_1 \\ i_2 \\ i_3 \\ i_4 \\ i_5 \\ i_6 \end{bmatrix} = \begin{bmatrix} 0.2 & & & & & 0 \\ & 0.2 & & & & \\ & & 0.2 & & & \\ 0 & & & 0.1 & & \\ & & & & 0.5 & \\ & & & & & 0.1 \end{bmatrix} \begin{bmatrix} 450 \\ 320 \\ 200 \\ -60 \\ 140 \\ 260 \end{bmatrix} - \begin{bmatrix} 0 \\ 0 \\ 0 \\ 0 \\ 0 \\ 0 \end{bmatrix} = \begin{bmatrix} 90 \\ 64 \\ 20 \\ -6 \\ 70 \\ -26 \end{bmatrix}$$

Problem 2.41. Draw the graph of the network shown below. Formulate the f-cut set matrix and write the equilibrium equation in matrix form on node basis.

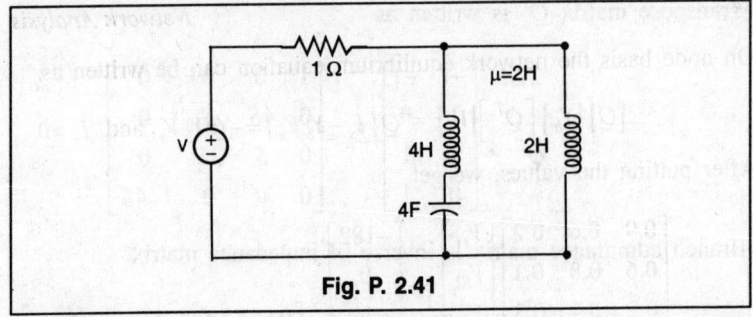

Fig. P. 2.41

Solution

The network is transformed into directed graph and the directed graph voltage source is replaced by short-circuited line and linear elements resistance inductance and capacitance is represented as line. After drawing directed graph possible tree is drawn and the fundamental cut set is drawn.

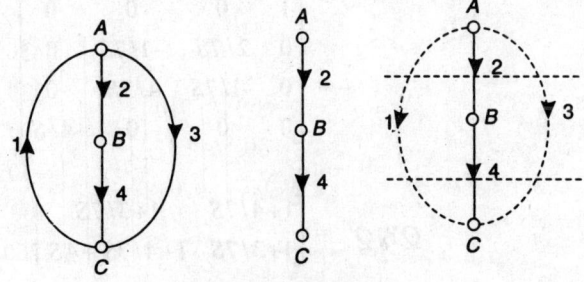

Fig. P. 2.41 (a)

From the above figure

No. of branches, $b = 4$

No. of nodes, $n = 3$

Hence no. of tree branches $= n - 1 = 2$.
and no. of links $= b - n + 1$
The f-cut set matrix is written as

Tree branch no, i.e f-cut set		Branch no.			
		1	2	3	4
Q =	2	-1	1	1	0
	4	-1	0	1	1

$$Q = \begin{bmatrix} -1 & 1 & 1 & 0 \\ -1 & 0 & 1 & 1 \end{bmatrix}$$

Transpose matrix Q^T is written as

$$Q^T = \begin{bmatrix} -1 & -1 \\ 1 & 0 \\ 1 & 1 \\ 0 & 1 \end{bmatrix}, Z_b = \begin{bmatrix} 1 & 0 & 0 & 0 \\ 0 & 4S & S & 0 \\ 0 & S & 2S & 0 \\ 0 & 0 & 0 & 1/4S \end{bmatrix}$$

Branch admittance matrix is inverse of impedance matrix.

$$Y_b = [Z_b]^{-1} = \begin{bmatrix} 1 & 0 & 0 & 0 \\ 0 & 4S & S & 0 \\ 0 & S & 2S & 0 \\ 0 & 0 & 0 & 1/4S \end{bmatrix}^{-1} = \begin{bmatrix} [1]^{-1} & 0 & 0 & 0 \\ 0 & \begin{bmatrix} 4S & S \\ S & 2S \end{bmatrix}^{-1} & 0 \\ 0 & 0 & 0 & [1/4S]^{-1} \end{bmatrix}$$

$$= \begin{bmatrix} 1 & 0 & 0 & 0 \\ 0 & 2/7S & -1/7S & 0 \\ 0 & -1/7S & 4/7S & 0 \\ 0 & 0 & 0 & 4/S \end{bmatrix}$$

$$QY_b Q^T = \begin{bmatrix} 1+4/7S & 1+3/7S \\ 1+3/7S & 1+4/7S+4S \end{bmatrix}$$

From network,

$$V_s = \begin{bmatrix} 1 \\ 0 \\ 0 \\ 0 \\ 0 \\ 0 \end{bmatrix} \text{ and } I_S = [0]$$

The network equilibrium equation on node basis can be written as

$$QY_b Q^T V_t = Q\{I_s - Y_b V_s\} = -QY_b V_s$$

From network $I_s = 0$ and $V_0 = V_t$ is the tree voltage vector after putting the values

$$QY_b Q^T V_t = -QY_b V_s$$

The network equilibrium eqn. on node basis can be written as,

$$\begin{bmatrix} 1+4/7S & 1+3/7S \\ 1+3/7S & 1+4/7S+4S \end{bmatrix} \begin{bmatrix} V_{t1} \\ V_{t2} \end{bmatrix} \begin{bmatrix} 1 \\ 1 \end{bmatrix}$$

Problem 2.42. For the resistance network shown below, calculate the branch voltages and branch currents using voltage variable method.

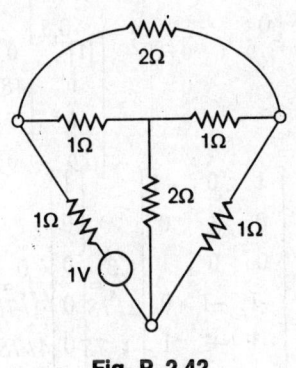

Fig. P. 2.42

Solution:

The network is replaced by directed graph. In directed graph voltage source is replaced by short circuited line and then we draw f-cut set.

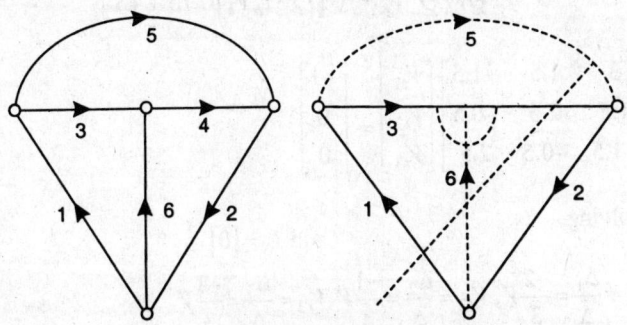

Fig. P. 2.42 (a), (b)

The f-cut set matrix Q is written as

Tree branch f-cut set				Branches			
		1	2	3	4	5	6
	6	1	0	0	−1	−1	1
$Q =$	2	0	1	0	−1	−1	0
	3	−1	0	1	0	1	0

Branch admittance matrix is written as

$$Y_b = \begin{bmatrix} 1 & & & & & 0 \\ & 1 & & & & \\ & & 1 & & & \\ & & & 1 & & \\ & & & & 0.5 & \\ 0 & & & & & 0.5 \end{bmatrix}$$

Transpose matrix Q^T

$$Q^T = \begin{bmatrix} 1 & 0 & -1 \\ 0 & 1 & 0 \\ 0 & 0 & 1 \\ -1 & -1 & 0 \\ -1 & -1 & 1 \\ 1 & 0 & 0 \end{bmatrix}, V_s = \begin{bmatrix} 1 \\ 0 \\ 0 \\ 0 \\ 0 \\ 0 \end{bmatrix}, I_s = [0]$$

The network equilibruim eqn. on node basis is written as,

$$QV_b Q^T V_t = Q[I_s - Y_b V_s]$$

$$\begin{bmatrix} 3 & 1.5 & -1.5 \\ 1.5 & 2.5 & -0.5 \\ -1.5 & -0.5 & 2.5 \end{bmatrix} \begin{bmatrix} V_{t1} \\ V_{t2} \\ V_{t3} \end{bmatrix} = \begin{bmatrix} -1 \\ 0 \\ 0 \end{bmatrix}$$

After solving,

$$V_{t1} = \frac{\Delta_1}{\Delta} = \frac{-2}{3} V, \ V_{t2} = \frac{\Delta_2}{\Delta} = \frac{-1}{3} V, V_{t3} = \frac{\Delta_3}{\Delta} = \frac{-1}{3} V$$

Network Equations

Problem 3.1. What must be the relationship between C_{eq} and C_1 and C_2 in (a) of Fig. P. 3.1 of the networks if (a) and (c) are equivalent? Repeat for the network shown in (b).

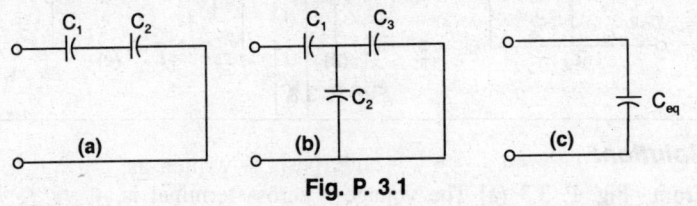

Fig. P. 3.1

Solution: From Fig. P. 3.1 (a)

Equivalent capacitance, $C_{eq} = C_1 + C_2$ and,

from Fig. 3.1 (b). Equivalent capacitance,

$$C_{eq} = C_1 /\!/ (C_2 + C_3) = \frac{C_1(C_2 + C_3)}{C_1 + C_2 + C_3}$$

Problem 3.2. What must be the relationship between L_{eq} and L_1, L_2 and M for the networks of (a) and of (b) to be equivalent to that of (c)?

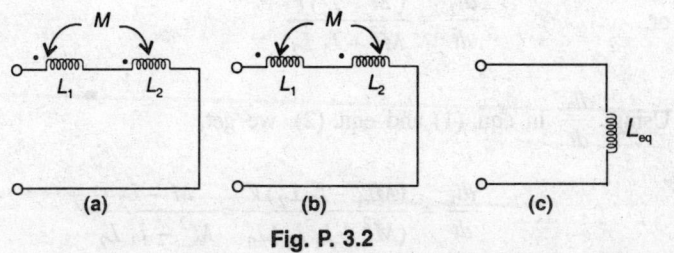

Fig. P. 3.2

Solution:

In Fig. P. 3.2 (a)

L_1 and L_2 are series adding.

Then, $L_{eq} = L_1 + L_2 + 2M$

In Fig. P. 3.2 (b), L_1 and L_2 are series opposing.

Then, $L_{eq} = L_1 + L_2 - 2M$

Problem 3.3. Repeat Problem 3.2 for the three networks shown in Fig. P. 3.3

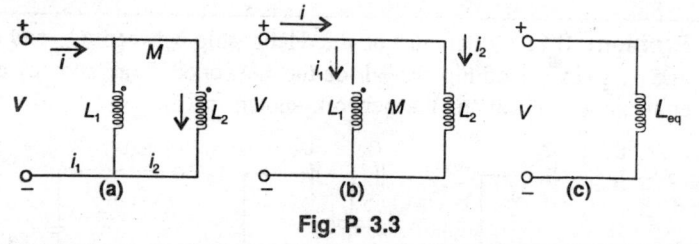

Fig. P. 3.3

Solution:

From Fig. P. 3.3 (a) The voltage v across terminal is,

$$v = L_1 \frac{di_1}{dt} + M \frac{di_2}{dt} \tag{1}$$

and, the same across the terminal of Fig. P. 3.3 (b) is,

$$v = M \frac{di_1}{dt} + L_2 \frac{di_2}{dt} \tag{2}$$

Now, (ii) × L_1 – (i) × M, gives

$$(M - L_1) = (M^2 - L_1 L_2) \frac{di_2}{dt}$$

or, $$\frac{di_2}{dt} = \frac{(M - L_1) V}{M^2 - L_1 L_2} \tag{3}$$

Using $\frac{di_2}{dt}$ in eqn. (1) and eqn. (2), we get,

$$\frac{di_1}{dt} = \frac{(ML_1 - L_1 L_2) V}{(M^2 - L_1 L_2) L_1} = \frac{M - L_2}{M^2 - L_1 L_2} V \tag{4}$$

Now, $\dfrac{di}{dt} = \dfrac{di_1}{dt} + \dfrac{di_2}{dt}$

From eqn. (3) and eqn. (4), be obtain,

$$\frac{di}{dt} = \left(\frac{L_1 + L_2 - 2M}{L_1 L_2 - M^2}\right) V$$

or, $$V = \frac{di/dt\,(L_1 L_2 - M^2)}{(L_1 + L_2 - 2M)} \qquad (5)$$

Since, from Fig. P. 3.3 (C), $V = L_{eq} \cdot \dfrac{di}{dt}$ $\qquad (6)$

Combining equations (5) and (6)

$$L_{eq} = \frac{L_1 L_2 - M^2}{L_1 + L_2 - 2M}$$

(b) In figure (b) inductances are connected in series opposing therefore substituting $-M$ in place of M in the result derived in part (a), we get

$$L_{eg} = \frac{L_1 L_2 - M^2}{L_1 + L_2 + 2M}$$

Problem 3.4. The network of inductors shown in the Fig. P. 3.4 is composed of a 1-H inductor on each edge of a cube with the inductors connected to the vertices of the cube as shown. Show that, with respect to vertices a and b, the network is equivalent to that in (b) of the figure when $L_{eq} = \dfrac{5}{6}$ H. Make use of symmetry in working this problem, rather than writing Kirchhoff laws.

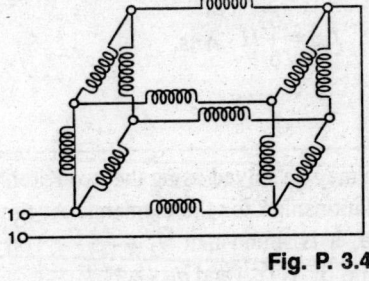

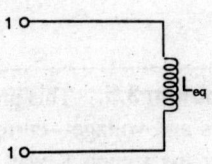

Fig. P. 3.4

Solution:

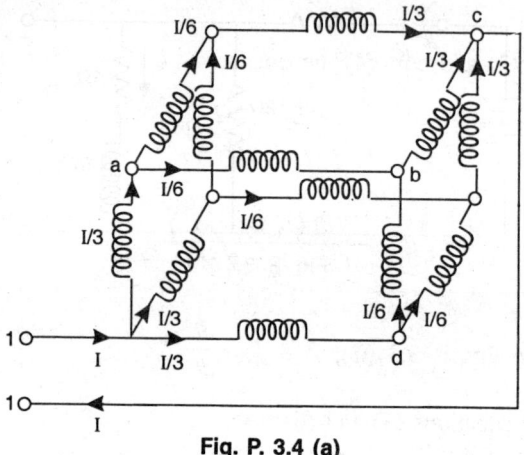

Fig. P. 3.4 (a)

Let V be the voltage applied at terminate $1 - 1'$.

So in the given network the current I enter at terminal 1 and distributed into three equal parts I/3. When I/3 reached at point a it again breaks into two equal part I/6 and at point b, this I/6 (ab) and I/6 (bd) is sum up and flows I/3 current in (bc).

So, by taking a path $(1 - a - b - c - 1')$:

$$V - \frac{I}{3}(j\omega) - \frac{I}{6}(j\omega) - \left(\frac{I}{3}\right)(j\omega) = 0$$

or, $$V = \left(\frac{1}{3} + \frac{1}{6} + \frac{1}{3}\right)I(j\omega) = \left(\frac{5}{6}j\omega\right)I$$

So, $$X_{L_{eq}} = \frac{5}{6}j\omega \therefore L_{eq} = \frac{5}{6}H \quad \textbf{Ans.}$$

Problem 3.5. This problem may be solved using the two Kirchhoff laws and voltage—current relationships for the elements. At time t_0 after the switch K was closed, it is found that $v_2 = +5V$. You are required to determine the value of $i_2(t_0)$ and $di_2(t_0)/dt$.

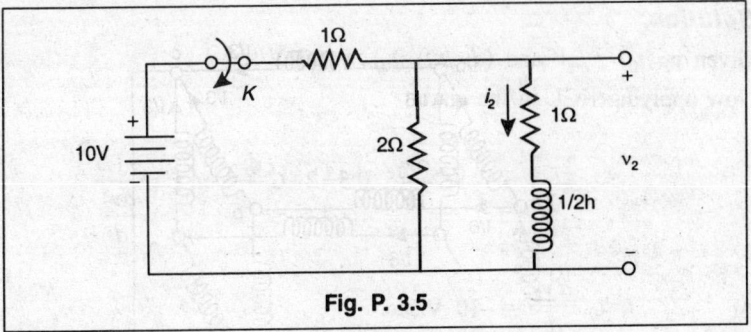

Fig. P. 3.5

Solution:

At time $t = t_0, V_2 = 5V.$ So, $i_2 \cdot 1 + \dfrac{1}{2}\dfrac{di_2(t)}{dt} = 5V$

$\therefore \qquad i_2(t) = C\, e^{-(R/L)t}$

At time $t = t_0, i_2(t_0) = \left(\dfrac{10-5}{1} - \dfrac{5}{2}\right) amp = (5-2.5)\ amp$

$\qquad\qquad\qquad = 2.5\ amp.$

$\qquad\qquad i_2(t_o) = C = 2.5\ amp \qquad\qquad\qquad (1)$

So, $\qquad i_2(t) = 2.5\, e^{-(R/L)t} = 2.5\, e^{-2t}$

$\qquad \dfrac{di_2(t)}{dt} = 2.5(-2)e^{-2t}, \dfrac{di_2}{dt}\bigg|_{t=t_0} = -5\ A/sec \qquad (2)$

So, $\qquad i_2(t_o) = 2.5\ amp$ and $\dfrac{di_2(t)}{dt}\bigg|_{t=t_0} = -5 A/sec$ **Ans.**

Problem 3.6. This problem is similar to Prob. 3.5. In the network given in the figure, it is given that $v_2(t_0) = 2$ V, and (dv_2/dt) $(t_0) = -10$ V/ Sec, where t_0 is the time after the switch K was closed. Determine the value of C.

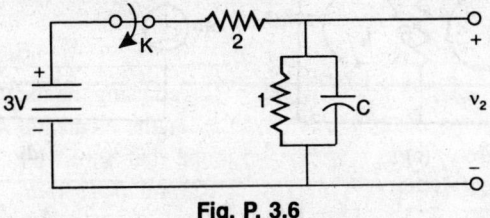

Fig. P. 3.6

Solution:

Given $v_2(t_0) = 2V$ and $(dv_2/dt)(t_0) = -10$ V/sec.

Now applying KCL in the circuit

$$\frac{3-2}{2} = \frac{2}{1} + C\frac{dv_2}{dt}, \quad 0.5 = 2 + C\frac{dv_2}{dt} \text{ or, } C = \frac{-1.5}{\dfrac{dv_2}{dt}}$$

At $\quad\quad t = t_0 \quad \dfrac{dv_2}{dt} = -10$ V/sec.

$\therefore \quad\quad\quad\quad C = \dfrac{-1.5}{-10} = 0.15$ F $\therefore c = 0.15$ F **Ans.**

Problem 3.7. For each of the four networks shown in the Fig. P. 3.7, determine the number of independent loop currents, and the number of independent node-to-node voltage that may be used in writing equilibrium equations using the Kirchhoff laws.

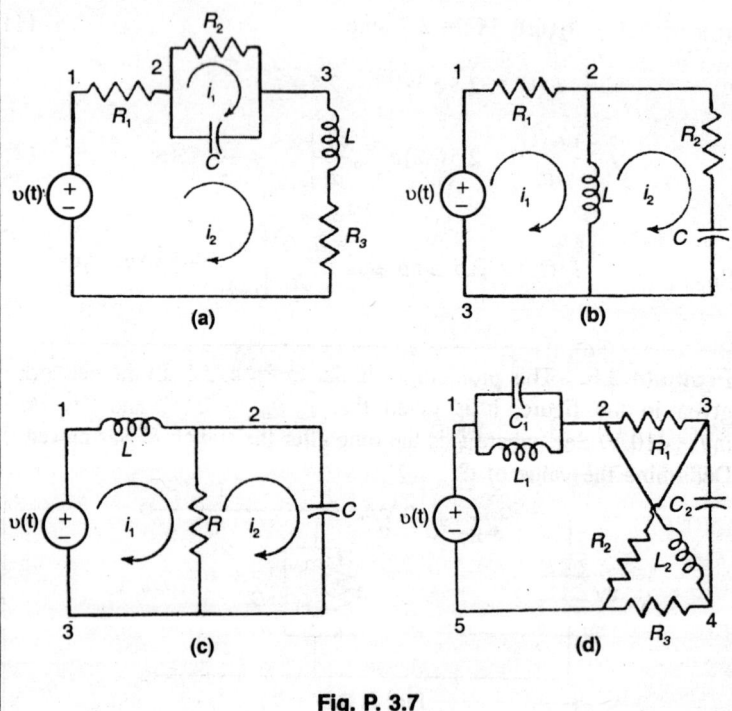

Fig. P. 3.7

Solution:

From Fig. P. 3.7

(a) No. of branches $b = 6$. No. of nodes = 5. Hence No. of independent loop currents = $b - n + 1 = 6 - 5 + 1 = 2$ and No. of independent node-to-node voltages is equal to No. of branches in tree, i.e. $n - 1 = 5 - 1 = 4$

(b) From Fig. (b), No. of branches = 5, No. of nodes $n = 4$. Hence the numbers of independent loop current = $b - n + 1 = 5 - 4 + 1 = 2$ and No. of independent node to node voltage = $n - 1 = 4 - 1 = 3$.

(c) From Fig. (c), No. of branches $b = 4$, No. of nodes $n = 3$. Hence the number of independent loop current = $b - n + 1 = 4 - 3 + 1 = 2$ and No. of independent node to node voltage = $n - 1 = 3 - 1 = 2$.

(d) From Fig. (d), No. of branches $b = 8$, No. of nodes $n = 5$. Hence the number of independent loop current = $b - n + 1 = 8 - 5 + 1 = 4$ and No. of independent node to node votages = $n - 1 = 5 - 1 = 4$.

Problem 3.8. Repeat Prob. 3.7 for each of the four networks shown in the Fig. P. 3.8 (a), (d).

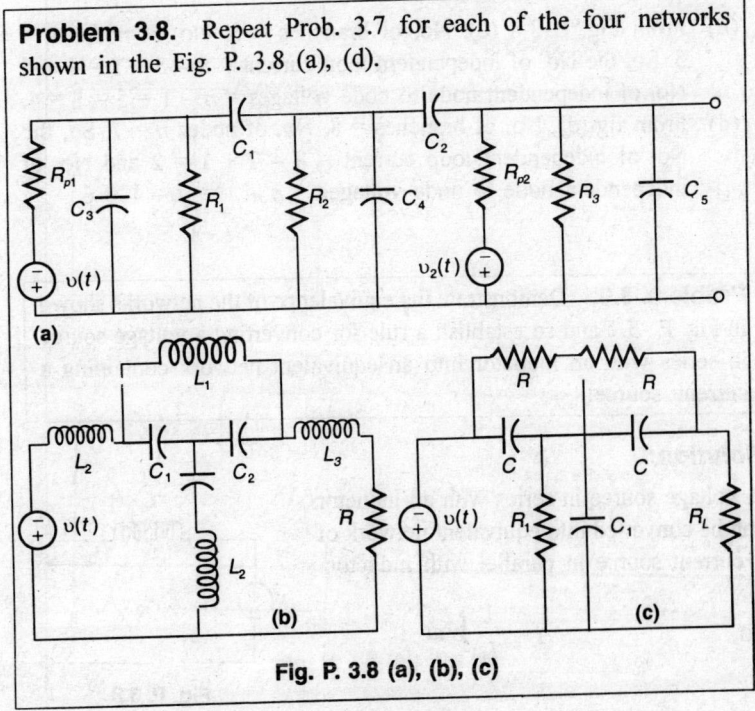

Fig. P. 3.8 (a), (b), (c)

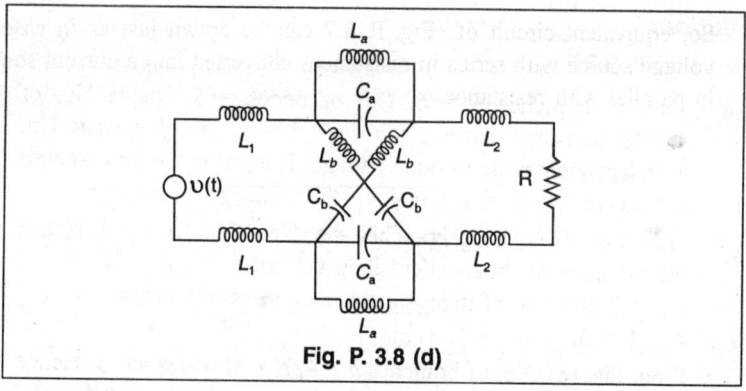

Fig. P. 3.8 (d)

Solution:

(a) From Fig. P. 3.8 (a), No. of branches = 10, No. of nodes n = 4. So, the No. of independent loop currents = $10 - 4 + 1 = 7$ and No. of independent node to node voltage = $n - 1 = 4 - 1 = 3$.

(b) From the Fig. P. 3.8 (b), No. of branches = 7, No. of nodes n = 5. So, the No. of independents loop currents = $7 - 5 + 1 = 3$ and No. of independent node to node voltages = $n - 1 = 5 - 1 = 4$

(c) From Fig. P. 3.8 (c), No. of branches = 8, No. of nodes n = 5. So, the No. of independent loop current = $8 - 5 + 1 = 4$ and No. of independent node to node voltages = $n - 1 = 5 - 1 = 4$.

(d) From fig (d), No. of branches = 8, No. of nodes n = 7. So, the No. of independent loop current = $8 - 7 + 1 = 2$ and No. of independent node to node voltages = $n - 1 = 7 - 1 = 6$.

Problem. 3.9. Demonstrate the equivalence of the networks shown in Fig. P. 3.7 and so establish a rule for converting a voltage source in series with an inductor into an equivalent network containing a current source.

Solution:

A voltage source in series with an inductor can be converted into equivalent network of a current source in parallel with inductor.

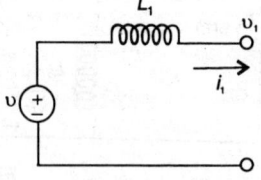

So, $i = \dfrac{1}{L_1} \int v\, dt$

Fig. P. 3.9

So, equivalent circuit of Fig. P. 3.7 can be drawn just as in case of voltage source with series in resistances converted into a current source in parallel with resistance.

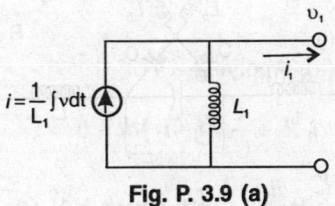

Fig. P. 3.9 (a)

Problem. 3.10. Demonstrate that the two networks shown in are equivalent.

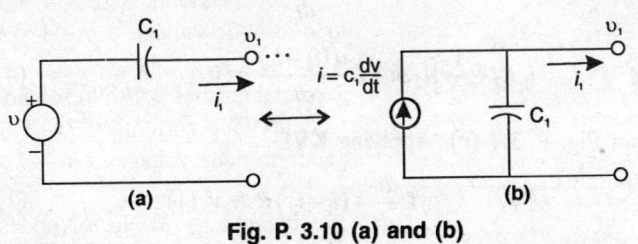

Fig. P. 3.10 (a) and (b)

Solution:
Applying KVL in Fig. P. 3.10 (a) and KCL in Fig. P. 3.10 (b).

$$v = \frac{1}{C_1} \int i_1 \, dt + v_1 \tag{1}$$

and

$$C_1 \frac{dv}{dt} = C_1 \frac{dv_1}{dt} + i_1 \tag{2}$$

Differentiating *eq* (1) we get.

$$\frac{dv}{dt} = \frac{1}{C_1} i_1 + C_1 \frac{dv_1}{dt}$$

or,

$$C_1 \frac{dv}{dt} = i_1 + C_1 \frac{dv_1}{dt} \tag{3}$$

eq (2) and *eq* (3) are similar hence both the network are equivalent **Ans.**

Problem. 3.11.   Write a set of equations using the Krichhoff voltage law in terms of appropriate loop-current variables for the four network of Problem 3.7.

Solution:

(a)   From Fig. P. 3.7 (a), Applying KVL:

$$i_1 R_2 + \frac{1}{C} \int (i_1 - i_2) \, dt = 0 \tag{1}$$

and $$(R_1 + R_3) i_2 + L \frac{di_2}{dt} + \frac{1}{C} \int (i_2 - i_1) \, dt = V \, (t) \tag{2}$$

(b)   From Fig. P. 3.7 (b), Applying KVL:

$$i_1 R_1 + L \frac{d(i_1 - i_2)}{dt} = v(t) \tag{3}$$

and $$i_1 R_2 + \frac{1}{C} \int i_2 \, dt + L \frac{d(i_2 - i_1)}{dt} = 0 \tag{4}$$

(c)   From Fig. P. 3.7 (c), Applying KVL

$$L \frac{di_1}{dt} + (i_1 - i_2) R = v \, (1) \tag{1}$$

and $$\frac{1}{C} \int i_2 \, dt \, (i_2 - i_1) R = 0 \tag{2}$$

(d)   From Fig. P. 3.17 (d), Applying KVL

Since there is four loop current

$$i_1 = \text{Through } L_1, \, L_2, \, R_3 \text{ and } v(t)$$
$$i_2 = \text{Through } C_1, \, L_1$$
$$i_3 = \text{Through } R_1, \, C_2, \text{ and } L_2$$
$$i_4 = \text{Through } C_2, \, R_3, \, R_2, \text{ so from}$$

Loop 1. $$L_1 \frac{d}{dt} (i_1 - i_2) + L_2 \frac{d}{dt} (i_1 - i_3 - i_4) + R_3 (i_1 - i_4) = v(t) \tag{1}$$

Loop 2. $$L_1 \frac{d(i_2 - i_1)}{dt} + \frac{1}{C_1} \int i_2 \, dt = 0 \tag{2}$$

Loop 3. $$R_1 i_3 \frac{1}{C_2} \int (i_3 - i_4) \, dt + L_2 \frac{d(i_3 - i_1 + i_4)}{dt} = 0 \tag{3}$$

Loop 4. $$R_2 (i_4) + R_3 (i_4 + i_1) + \frac{1}{C_2} \int (i_4 + i_3) \, dt = 0 \tag{4}$$

Problem 3.12. Write a set of equilibrium equations on the loop basis to describe the network in the following figure. Note that the network contains one controlled source.

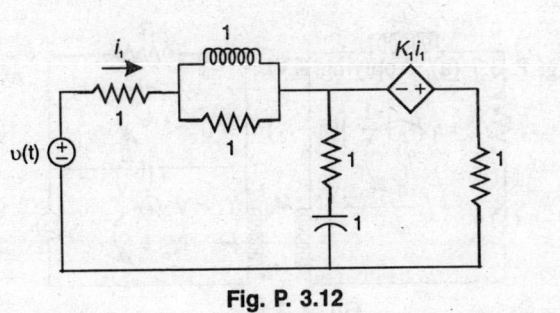

Fig. P. 3.12

Solution:

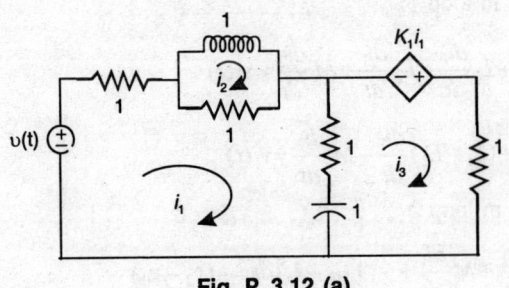

Fig. P. 3.12 (a)

For loop 1. Applying KVL

$$i_1(1+1+1) - i_2 - i_3 + \int(i_1 - i_3)\,dt = v_1(t)$$

or,

$$3i_1 - i_2 - i_3 + \int(i_1 - i_3)\,dt = v_1(t) \qquad (1)$$

For loop 2. Applying KVL

$$1 \cdot \frac{di_2}{dt} + 1\int(i_2 - i_1)\,dt = 0 \qquad (2)$$

For loop 3. Applying KVL

$$k_i\, i_1 = 1 \cdot i_3 + 1\int(i_3 - i_1)\,dt + 1 \cdot (i_3 - i_1)$$

or $\quad 2i_3 - i_1 + \int(i_3 - i_1)\,dt = k_1\, i_1 \quad$ **Ans.** $\qquad (3)$

Problem 3.13. For the coupled network of the figure, write loop equation using the Kirchhoff voltage law. In your formulation, use the three loop currents which are identified.

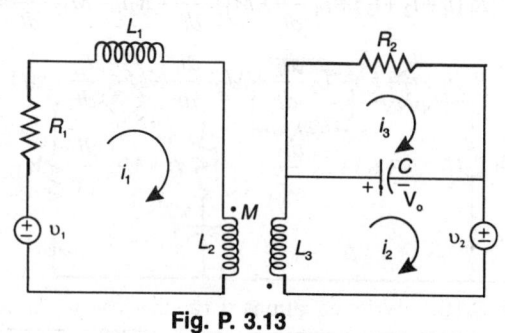

Fig. P. 3.13

Solution:

Applying KVL in loop 1:

$$i_1 R_1 + L_1 \frac{di_1}{dt} + L_2 \frac{di_1}{dt} + M \frac{di_2}{dt} = v_1(t)$$

or $\qquad i_1 R_1 + (L_1 + L_2)\dfrac{di_1}{dt} + M \dfrac{di_2}{dt} = v_1(t)$ \hfill (1)

Applying KVL in loop 2

$$L_3 \frac{di_2}{dt} + M \frac{di_1}{dt} + \frac{1}{C} \int (i_2 - i_3)\,dt = -(v_2 + v_0) \qquad (2)$$

Applying KVL in loop 3 $\quad R_2 i_3 + \dfrac{1}{C} \int (i_3 - i_2)\,dt = v_0$ \hfill (3) **Ans.**

Problem 3.14. The network of the figure is that of but with different loop-current-variables chosen. Using the specified current, write the Kirchhoff voltage law equations for this network.

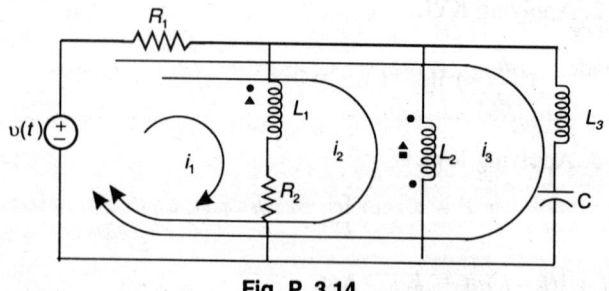

Fig. P. 3.14

Solution:

Applying KVL: In loop 1

$$R_1(i_1+i_2+i_3)+L_1\frac{di_1}{dt}+M_{12}\frac{di_2}{dt}+R_2i_1-M_{13}\frac{di_3}{dt}=v(t) \qquad (1)$$

In loop 2 $\quad R_1(i_1+i_2+i_3)+L_2\frac{di_2}{dt}+M_{21}\frac{di_1}{dt}-M_{23}\frac{di_3}{dt}=v(t) \qquad (2)$

In loop 3 $\quad R_1(i_1+i_2+i_3)+L_3\frac{di_3}{dt}+\frac{1}{C}\int i_3\,dt+M_{31}\frac{di_1}{dt}-M_{32}\frac{di_2}{dt}=v(t) \ (3)$

Ans.

Problem 3.15. Write equations using the Kirchhoff current law in terms of node-to-datum voltage variables for the four networks of Problem 3.7.

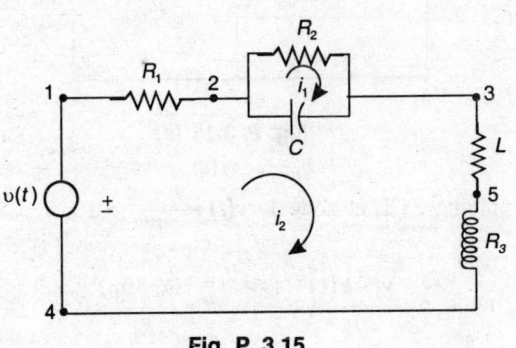

Fig. P. 3.15

Solution.

(a) Considering 4 as datum node. So $v_4 = 0$.

Let the voltage at node node 1 = $v(t)$

Node $\quad 2 = v_2(t)$

Node $\quad 3 = v_3(t)$

Node $\quad 5 = v_5(t)$

Applying KCL at node

Node 1 $\quad i_2+\dfrac{v_1-v_2}{R_1}=0$ or, $i_2+\dfrac{v-v_2}{R_1}=0$

At node 2 $\quad \dfrac{v_2-v_1}{R_1}+\dfrac{v_2-v_3}{R_2}+C\dfrac{d(v_2-v_3)}{dt}=0$

or, $\dfrac{v_2-v}{R_1}+\dfrac{v_2-v_3}{R_2}+C\dfrac{d}{dt}(v_2-v_3)=0$

At node 3 $\dfrac{v_3-v_2}{R_2}+C\dfrac{d(v_3-v_2)}{dt}+\dfrac{1}{L}\int(v_3-v_5)dt=0$

At node 5 $\dfrac{1}{L}\int(v_3-v_5)dt+\dfrac{v_5-v_4}{R_3}=0$

(b) Taking 4 as datum node, here $v_1 = v(t)$

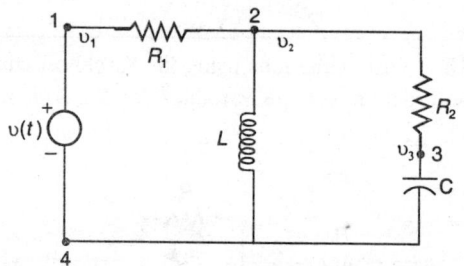

Fig. P. 3.15 (a)

Applying KCL at node 1. $v(t)+\dfrac{v_t-v_2}{R_1}=0$ (1)

At node 2. $\dfrac{v_2-v(t)}{R_1}+\dfrac{1}{L}\int v_2\,dt+\dfrac{v_2-v_3}{R_2}=0$ (2)

At node 3. $\dfrac{v_3-v_2}{R_2}+C\dfrac{d(v_3-v_4)}{dt}=0$ (3)

(c) From Fig. P. 3.15 (b)
Taking 3 as datum node. Applying KCL at node 1

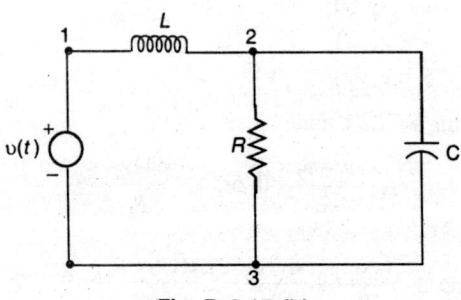

Fig. P. 3.15 (b)

$$v(t)+\frac{1}{2}\int(v(t)-v_2)\,dt=0 \tag{1}$$

At node 2. $\dfrac{1}{L}\int(v_2-v(t))\,dt+\dfrac{v_2}{R}+C\dfrac{dv_2}{dt}=0$ (2)

Problem 3.16. For the given network, write the node-basis equations using the node-to-datum voltage as variables.

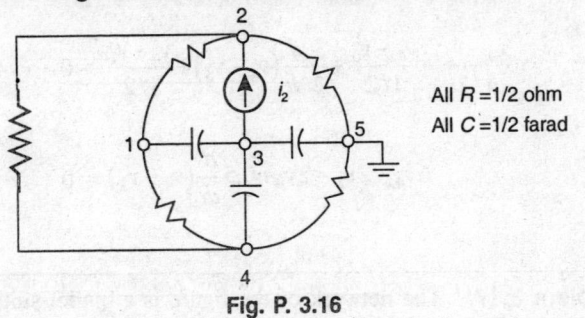

All $R=1/2$ ohm
All $C=1/2$ farad

Fig. P. 3.16

Solution:

Take 5 as a datum node so $v_5=0$

At node 1. $\dfrac{v_1-v_4}{R}+\dfrac{v_1-v_2}{R}+C\dfrac{d}{dt}(v_1-v_3)=0$

$$\frac{v_1-v_4}{1/2}+\frac{v_1-v_2}{1/2}+\frac{1}{2}\frac{d}{dt}(v_1-v_3)=0$$

or, $4v_1-2v_2-2v_4+0.5\dfrac{d}{dt}(v_1-v_3)=0$ (1)

At node 2. $\dfrac{v_2-v_4}{R}+\dfrac{v_2-v_1}{R}+i_2+\dfrac{v_2-v_5}{R}=0$

or, $\dfrac{v_2-v_4}{1/2}+\dfrac{v_2-v_1}{1/2}+i_2+\dfrac{v_2-v_5}{1/2}=0$

or, $6v_2-2v_4-2v_1-2v_5+i_2=0$ (2) $[v_5=0]$

At node 3

$$C\frac{d}{dt}(v_3-v_1)+C\frac{d}{dt}(v_3-v_5)+C\frac{d}{dt}(v_3-v_4)+i_2=0$$

or, $\dfrac{1}{2}\dfrac{d}{dt}(v_3-v_1)+\dfrac{1}{2}\dfrac{d}{dt}(v_3-v_5)+\dfrac{1}{2}\dfrac{d}{dt}(v_3-v_4)+i_2=0$

or, $1.5\dfrac{d}{dt}v_3-0.5\dfrac{d}{dt}v_1-0.5\dfrac{d}{dt}v_5-0.5\dfrac{d}{dt}v_4+i_2=0$ (3) $\left[v_5=0\right]$

At node 4 $\dfrac{v_4-v_2}{R}+\dfrac{v_4-v_5}{R}+C\dfrac{d}{dt}(v_4-v_3)+\dfrac{v_4-v_1}{R}=0$

or, $\dfrac{v_4-v_2}{1/2}+\dfrac{v_4-v_0}{1/2}+\dfrac{1}{2}\dfrac{d}{dt}(v_4-v_3)+\dfrac{v_4-v_1}{1/2}=0$

or, $6v_4-2v_2-2v_1+0.5\dfrac{d}{dt}(v_4-v_3)=0$ (4) **Ans.**

Problem 3.17. The network of the figure is a model suitable for "midband" operation of the "cascade-connected" MOS transistor amplifier. Analyze the network on

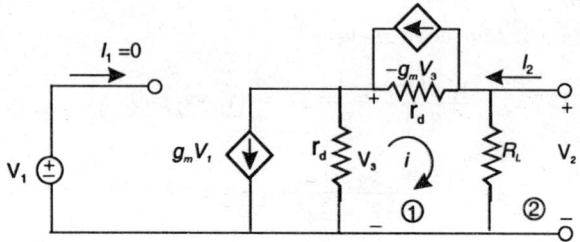

Fig. P. 3.17

(a) the loop basis and

(b) the node basis and write the resulting equations in matrix form, but do not solve them.

Solution:

(a) Let i be the loop current in the above circuit. So voltage v_3.

$$v_3=-(g_m v_1+i)r_d \text{ volt.}$$

Now converting the current source $(-g_m v_3$ and $r_d)$ into equivalent voltage source as. $v_{eq}=g_m r_d v_3$ in series with r_d.

Now applying KVL in loop 1 we get:

$$v_3 - v_{eq} - i(r_d) - R_L(i + I_2) = 0$$

or, $\quad -(g_m v_1 + i)r_d - g_m r_d v_3 - r_d i - R_L(i + I_2) = 0$

or, $\quad g_m v_1 r_d + 2i r_d + g_m r_d v_3 + R_L i + R_L I_2 = 0$

or, $\quad (2r_d + R_L)i + g_m r_d v_1 + g_m r_d v_3 + R_L I_2 = 0$

or, $\quad (2r_d + R_L)i + g_m r_d v_1 + g_m r_d \left(-(g_m v_1 + i)r_d\right) + R_L I_2 = 0$

or, $\quad \left(2r_d + R_L + g_m r_d^2\right)i + R_2 I_2 = g_m r_d (g_m r_d - 1)v_1 \qquad (1)$

From loop 2 $\quad R_L(I_2 + i) = v_2$

or, $\qquad\qquad R_L I_2 + R_L i = v_2 \qquad\qquad\qquad (2)$

So in the matrix form

$$\begin{bmatrix} 2r_d + R_L + g_m r_d^2 & R_L \\ R_L & R_L \end{bmatrix} \begin{bmatrix} i \\ I_2 \end{bmatrix} = \begin{bmatrix} g_m r_d (g_m r_d - 1)v_1 \\ v_2 \end{bmatrix}$$

(b) At node 1 applying KCL:

$$\frac{v_3}{r_d} + g_m v_1 + \frac{v_3 - v_2}{r_d} + g_m v_3 = 0$$

or, $\quad \left[\dfrac{2}{r_d}\right] v_3 + g_m v_3 + g_m v_1 - \dfrac{v_2}{r_d} = 0$

or, $\quad -\left[\dfrac{2}{r_d} + g_m\right] v_3 + \dfrac{v_2}{r_d} = g_m v_1 \qquad\qquad (1)$

At node 2, applying KCL:

$$\frac{-v_3}{r_d} + \left(\frac{1}{r_d} + \frac{1}{R_L}\right)v_2 - g_m v_3 = I_2$$

or,

$$\left(\frac{1}{r_d} + \frac{1}{R_L}\right)v_2 - \left(g_m + \frac{1}{r_d}\right)v_3 = I_2 \qquad (2)$$

So in the matrix form:

$$\begin{bmatrix} \dfrac{1}{r_d} & -\left(\dfrac{2}{r_d} + g_m\right) \\ \left(\dfrac{1}{r_d} + \dfrac{1}{R_L}\right) & -\left(\dfrac{1}{r_d} + g_m\right) \end{bmatrix} \begin{bmatrix} v_2 \\ v_3 \end{bmatrix} = \begin{bmatrix} g_m v_1 \\ I_2 \end{bmatrix} \text{ Ans.}$$

Problem 3.18. In the network of the following figure, each branch contains a 1-Ω resistor, and four branches contain a 1-V voltage source. Analyze the network on the loop basis

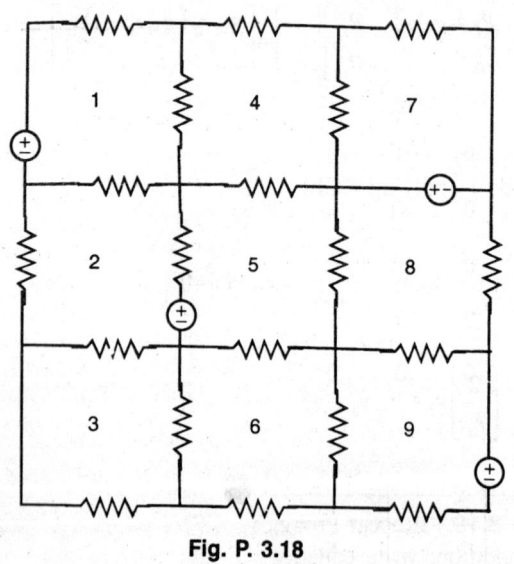

Fig. P. 3.18

Solution:

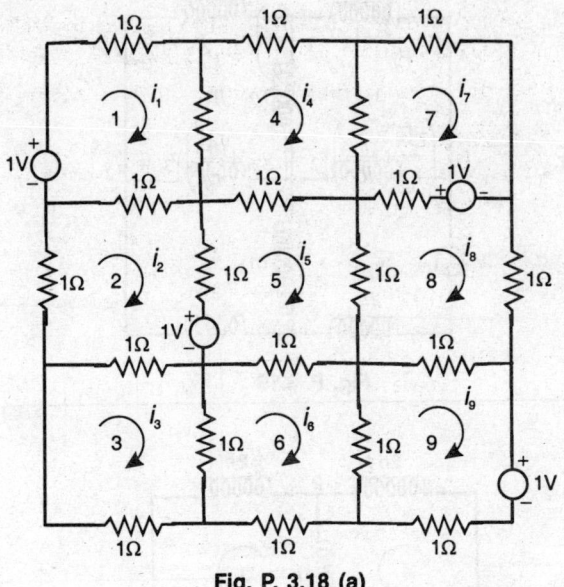

Fig. P. 3.18 (a)

$$\begin{bmatrix} 3 & -1 & 0 & 1 & 0 & 0 & 0 & 0 & 0 \\ 1 & 4 & 1 & 0 & 1 & 0 & 0 & 0 & 0 \\ 0 & 1 & 1 & 0 & 0 & 1 & 0 & 0 & 0 \\ 1 & 0 & 3 & 4 & 1 & 0 & 1 & 1 & 0 \\ 0 & 1 & 0 & 1 & 4 & 1 & 0 & 1 & 0 \\ 0 & 1 & 0 & -1 & 4 & 0 & 0 & 1 & 0 \\ 0 & 0 & 0 & 1 & 0 & 0 & 3 & 1 & 0 \\ 0 & 0 & 0 & 0 & 1 & 0 & 1 & 4 & 1 \\ 0 & 0 & 0 & 0 & 0 & 1 & 0 & 1 & 3 \end{bmatrix} \begin{bmatrix} i_1 \\ i_2 \\ i_3 \\ i_4 \\ i_5 \\ i_6 \\ i_7 \\ i_8 \\ i_9 \end{bmatrix} = \begin{bmatrix} 1 \\ -1 \\ 0 \\ 0 \\ 1 \\ 1 \\ -1 \\ -1 \\ -1 \end{bmatrix} \quad \text{Ans.}$$

Problem 3.19. Repeat Problem 3.18 for the network of Problem 3.19. In addition, write equations on the node basis.

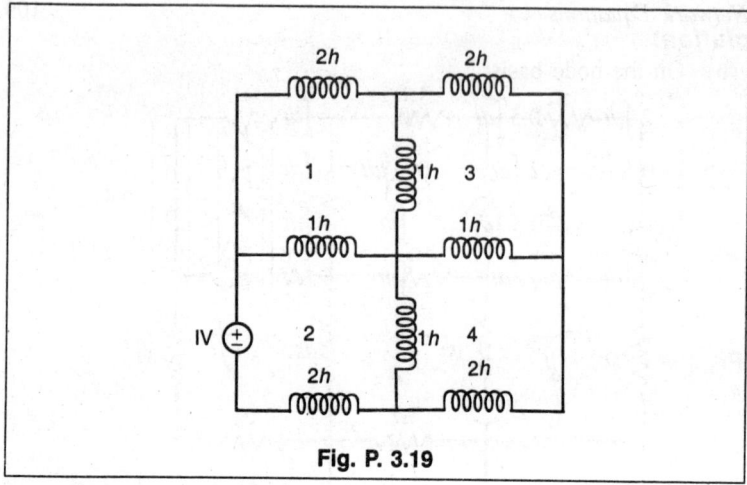

Fig. P. 3.19

Solution:

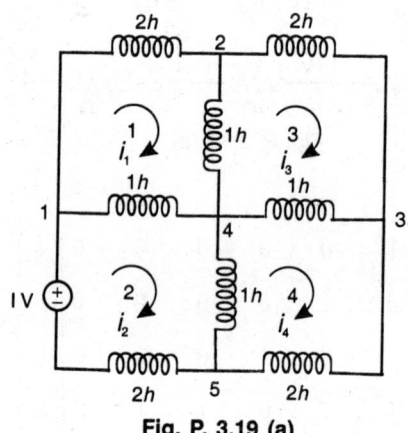

Fig. P. 3.19 (a)

Let the datum node be 5, so that $v_5 = 0$

(a) On the loop basis:

$$
\begin{bmatrix}
4\dfrac{d}{dt} & -\dfrac{d}{dt} & -\dfrac{d}{dt} & 0 \\[2mm]
-\dfrac{d}{dt} & 4\dfrac{d}{dt} & 0 & -\dfrac{d}{dt} \\[2mm]
-\dfrac{d}{dt} & 0 & 4\dfrac{d}{dt} & -\dfrac{d}{dt} \\[2mm]
0 & -\dfrac{d}{dt} & -\dfrac{d}{dt} & 2\dfrac{d}{dt}
\end{bmatrix}
\begin{bmatrix}
i_1 \\ i_2 \\ i_3 \\ i_4
\end{bmatrix}
=
\begin{bmatrix}
0 \\ 1 \\ 0 \\ 0
\end{bmatrix}
$$

(b) On the node basis:

$$\begin{bmatrix} 2\int dt & -0.5\int dt & 0 & -\int dt \\ -0.5\int dt & 2\int dt & -0.5\int dt & -\int dt \\ 0 & -0.5\int dt & -\int dt & -\int dt \\ -\int dt & -\int dt & -\int dt & 4\int dt \end{bmatrix} \begin{bmatrix} V_1 \\ V_2 \\ V_3 \\ V_4 \end{bmatrix} = \begin{bmatrix} V \\ 0 \\ 0 \\ 0 \end{bmatrix} \quad \textbf{Ans.}$$

Problem 3.20. In the network of the following figure, $R = 2\ \Omega$ and $R_1 = 1\ \Omega$. Write equations on (a) the loop basis, and (b) the node basis.

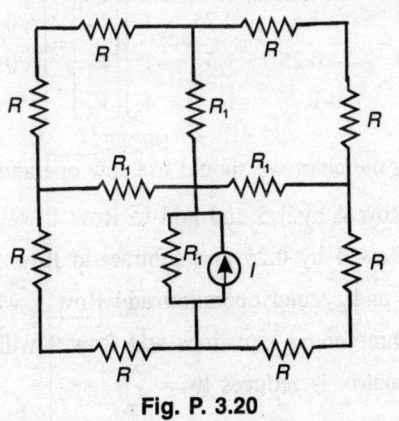

Fig. P. 3.20

Solution:

On converting the current sources into equivalent voltage sources. The above circuit can be drawn as shown in fig. Let the datum node be 5 that $V_5 = 0$. Here $V = I.R_1 = I\ [\because\ R_1 = 1]$

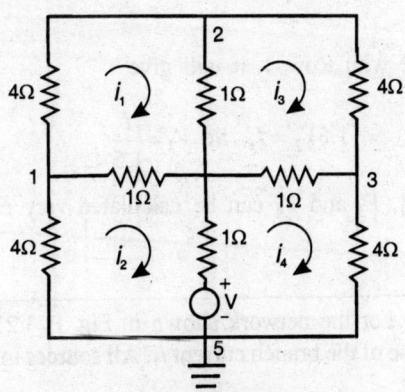

Fig. P. 3.20 (a)

(a) On the loop basis:

$$\begin{bmatrix} 6 & -1 & 0 & -1 \\ -1 & 6 & -1 & 0 \\ 0 & -1 & 6 & -1 \\ -1 & 0 & -1 & 6 \end{bmatrix} \begin{bmatrix} i_1 \\ i_2 \\ i_3 \\ i_4 \end{bmatrix} = \begin{bmatrix} 0 \\ 0 \\ +1 \\ -1 \end{bmatrix}$$

(b) On the node basis:

$$\begin{bmatrix} 1.5 & -0.25 & 0 & -1 \\ -0.25 & 1.5 & -0.25 & -1 \\ 0 & -0.25 & 1.5 & -1 \\ -1 & -1 & -1 & 4 \end{bmatrix} \begin{bmatrix} V_1 \\ V_2 \\ V_3 \\ V_4 \end{bmatrix} = \begin{bmatrix} 0 \\ 0 \\ 0 \\ V=I \end{bmatrix}$$

For simplyinfying the chart we should use row operation in the matrix.

(1) Multiply Row 4 by 1.5 and add to Row 1

(2) Multiply Row 4 by 0.25 and subtract to Row 2

(3) After first and second operation add Row 1 with Row 2

(4) After all three above equations add Row 3 with Row 1

Now the matrix is reduces to:

$$\begin{bmatrix} 0 & 0 & -1.5 & 3 \\ 0 & 1.75 & 0 & -2 \\ 0 & -0.25 & 0 & 2 \\ -1 & -1 & -1 & 4 \end{bmatrix} \begin{bmatrix} V_1 \\ V_2 \\ V_3 \\ V_4 \end{bmatrix} = \begin{bmatrix} 1.25\,I \\ -0.25\,I \\ 1.25\,I \\ I \end{bmatrix}$$

Add Row 2 with Row 3, it will give

$$1.5 V_2 = I, \quad \text{So } V_2 = \frac{I}{1.5}$$

Similarly V_1, V_3 and V_4 can be calculated very easily. **Ans.**

Problem 3.21. For the network shown in Fig. P. 3.21, determine the numerical value of the branch current i_1. All sources in the network are time invariant.

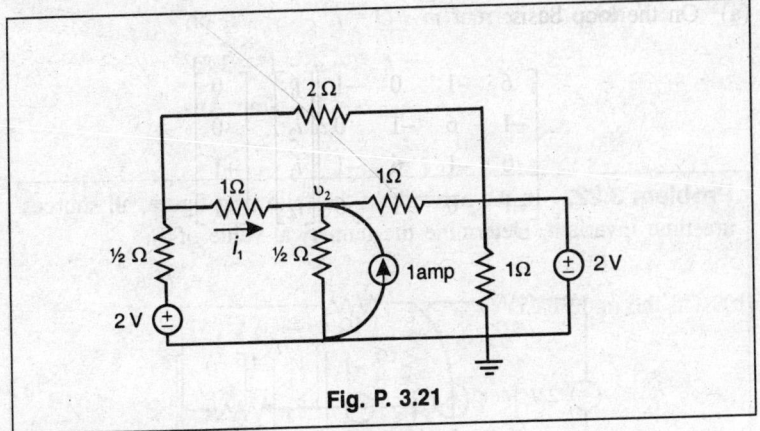

Fig. P. 3.21

Solution:

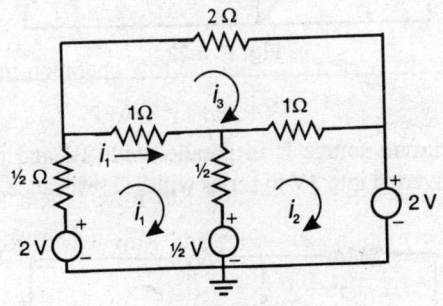

Fig. P. 3.21. (a)

The above figure can be drawn as:

Applying KVL in mesh 1, 2, and 3.

From mesh 1
$$\frac{1}{2}i_1 + (i_1 - i_3) + \frac{1}{2}(i_1 - i_2) + 1/2 = 2$$

or,
$$4i_1 - i_2 - 2i_3 = 3 \tag{1}$$

From mesh 2
$$\frac{1}{2}(i_2 - i_1) + (i_2 - i_3) + 2 = 1/2$$

or,
$$\frac{-i_1}{2} + \frac{3}{2}i_2 - i_3 = -3/2 \tag{2}$$

From mesh 3
$$4i_3 - i_2 - i_1 = 0 \tag{3}$$

Solving these equation we get, $i_1 = -4$ amp, $i_2 = \dfrac{-18}{5}$ amp and

$i_3 = -3.8$ amp, So current in $PQ = i' = i_1 - i_3$ or,

$$i_1' = -4 - (-3.8)$$
$$= -0.2 \text{ amp } \textbf{Ans.}$$

Problem 3.22. In the network of the following figure, all sources are time invariant. Determine the numerical value of i_2.

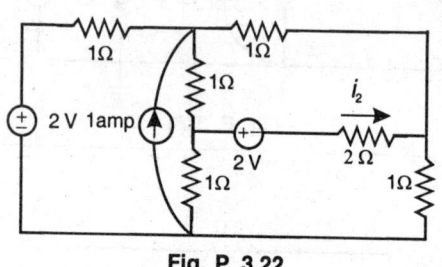

Fig. P. 3.22

Solution:

Since 1 amp current source is in parallel with 2V and resistance 1Ω. So it can be converted into 3V in series with 1Ω resistance. So equivalent circuit is:

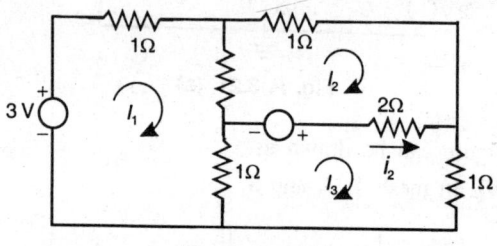

Fig. P. 3.22 (a)

Now applying mesh current I_1, I_2 and I_3. Here $i_2 = I_3 - I_2$

So, Applying KVL in mesh (1), we get

$$I_1 + I_1 - I_2 + I_1 - I_3 = 3 \text{ V}$$

or $3I_1 - I_2 - I_3 = 3 \text{ V}$ (1)

In mesh (2) $I_2 + (I_2 - I_3) \times 2 + 2(I_2 - I_1) = 0$

or, $4I_2 - 2I_3 - I_1 = -2$

or, $$-I_1 + 4I_2 - 2I_3 = -2 \tag{2}$$

In mesh (3) $$(I_3 - I_1) + 2(I_3 - I_2) + I_3 = 0$$

or, $$-I_1 - 2I_2 + 4I_3 = 2 \tag{3}$$

In matrix form:
$$\begin{bmatrix} 3 & -1 & -1 \\ -1 & 4 & -2 \\ -1 & -2 & 4 \end{bmatrix} \begin{bmatrix} I_1 \\ I_2 \\ I_3 \end{bmatrix} = \begin{bmatrix} 3 \\ -2 \\ 2 \end{bmatrix} V$$

∴
$$I_2 = \frac{\begin{vmatrix} 3 & 3 & -1 \\ -1 & -2 & -2 \\ -1 & 2 & 4 \end{vmatrix}}{\begin{vmatrix} 3 & -1 & -1 \\ -1 & 4 & -2 \\ -1 & -2 & 4 \end{vmatrix}} = \frac{11}{24}, \quad I_3 = \frac{\begin{vmatrix} 3 & -1 & -3 \\ -1 & -4 & -2 \\ -1 & -2 & 2 \end{vmatrix}}{24} = \frac{26}{24}$$

So, $$i_1 = I_3 - I_2 = \frac{26}{24} - \frac{11}{24} = \frac{15}{24} = \frac{5}{8}$$

or, $$i_2 = \frac{5}{8} \text{ amp } \textbf{Ans.}$$

Problem 3.23. In the network of the figure, all voltage sources and current source are time invariant, and all resistors have the value $R = \frac{1}{2}\Omega$. Solve for the four node-to-datum voltages.

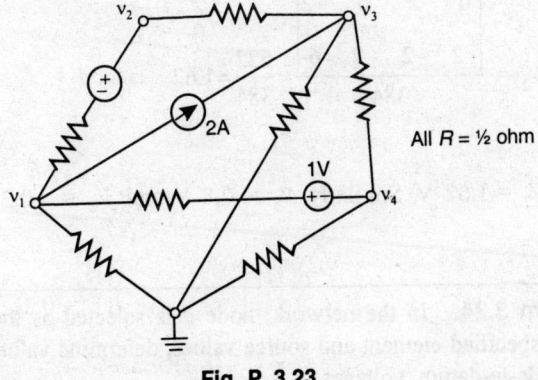

Fig. P. 3.23

Solution:

$$
\begin{bmatrix} 6 & -2 & 0 & -2 \\ -2 & 4 & -2 & 0 \\ 0 & -2 & 6 & -2 \\ -2 & 0 & -2 & 6 \end{bmatrix} \begin{bmatrix} V_1 \\ V_2 \\ V_3 \\ V_4 \end{bmatrix} = \begin{bmatrix} -6 \\ 6 \\ 2 \\ -2 \end{bmatrix}
$$

So,
$$
V_1 = \frac{\begin{vmatrix} -6 & -2 & 0 & 2 \\ 6 & 4 & -2 & 0 \\ 2 & -2 & 6 & -2 \\ -2 & 0 & -2 & 6 \end{vmatrix}}{\begin{vmatrix} 6 & -2 & 0 & -2 \\ -2 & 4 & -2 & 0 \\ 0 & -2 & 6 & -2 \\ -2 & 0 & -2 & 6 \end{vmatrix}} = \frac{-208}{384} = -0.541
$$

So, $V_1 = -0.541$ V

$$
V_2 = \frac{\begin{vmatrix} 6 & -6 & 0 & -2 \\ -2 & 6 & -2 & 0 \\ 0 & 2 & 6 & -2 \\ -2 & -2 & -2 & 6 \end{vmatrix}}{384} = \frac{622}{384} = 1.62
$$

or $V_2 = 1.62$ V, Similarly $V_3 = 0.8$ V and $V_4 = -0.25$ V **Ans.**

Problem 3.24.   In the network, node *d* is selected as the datum. For the specified element and source values, determine values for the four node-to-datum voltages.

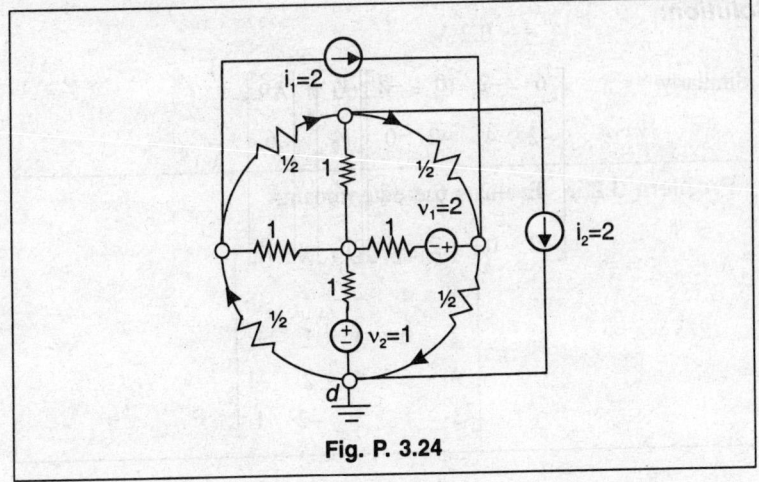

Fig. P. 3.24

Solution:

Applying KCL at each node.

$$\begin{bmatrix} 5 & -2 & 0 & -1 \\ -2 & 5 & -2 & -1 \\ 0 & -2 & 5 & -1 \\ -1 & -1 & -1 & 4 \end{bmatrix} \begin{bmatrix} V_a \\ V_b \\ V_c \\ V_d \end{bmatrix} = \begin{bmatrix} -2 \\ -1 \\ 4 \\ -1 \end{bmatrix}$$

$$V_a = \frac{\begin{vmatrix} -2 & -2 & 0 & -1 \\ -1 & 5 & -2 & -1 \\ 4 & -2 & 5 & -1 \\ -1 & -1 & -1 & 4 \end{vmatrix}}{\begin{vmatrix} 5 & -2 & 0 & -1 \\ -2 & 5 & -2 & -1 \\ 0 & -2 & 5 & -1 \\ -1 & -1 & -1 & 4 \end{vmatrix}} = \frac{-120}{225} = -0.533$$

$$V_a = -0.533 \text{ V}$$

$$V_b = \frac{\begin{vmatrix} 5 & -2 & 0 & -1 \\ -2 & -1 & -2 & -1 \\ 0 & 4 & 5 & -1 \\ -1 & -1 & -1 & 4 \end{vmatrix}}{225} = \frac{-45}{225} = -0.2$$

$\therefore$ $\qquad$ $V_b = -0.2$ V

Similarly $\qquad$ $V_c = 4$ V, $V_d = -0.266$ V **Ans.**

Problem 3.25. Evaluate the determinant:

$$\begin{bmatrix} 1 & -2 & 0 & 3 & 4 \\ -1 & 4 & -1 & 1 & 0 \\ 2 & 0 & 1 & 1 & 3 \\ 4 & -2 & 4 & 2 & -1 \\ 3 & 1 & 3 & -2 & 1 \end{bmatrix}$$

Solution:

(1) Multiply column 1 by 2 and add to column 2.

(2) Multiply column 1 by 3 and substract to column 4

(3) Multiply column 1 by 4 and substract to column 5

We get:

$$\begin{bmatrix} 1 & 0 & 0 & 0 & 0 \\ -1 & 2 & -1 & 4 & 4 \\ 2 & 4 & 1 & -5 & -5 \\ 4 & 6 & 4 & -10 & -17 \\ 3 & 7 & 3 & -11 & -11 \end{bmatrix} = 1 \cdot \begin{bmatrix} 2 & -1 & 4 & 4 \\ 4 & 1 & -5 & -5 \\ 6 & 4 & -10 & -17 \\ 7 & 3 & -11 & -11 \end{bmatrix}$$

Now, substract column 4 from column 3 of new determinant we get:

$$\begin{vmatrix} 2 & -1 & 0 & 4 \\ 4 & 1 & 0 & -5 \\ 6 & 4 & 7 & -17 \\ 7 & 3 & 0 & -11 \end{vmatrix}$$

Now, again, $2\,C_2 + C_1$ in column 1, and $4\,C_2 + C_4$ in column 4, we get

$$\begin{vmatrix} 0 & -1 & 0 & 0 \\ 6 & 1 & 0 & -1 \\ 14 & 4 & 7 & -1 \\ 13 & 3 & 0 & 1 \end{vmatrix} = -(-1)\begin{vmatrix} 6 & 0 & -1 \\ 14 & 7 & -1 \\ 13 & 0 & 1 \end{vmatrix}$$

$$= 1 \cdot \left[6(7) + 0 - 1(-13 \times 7)\right] = [42 + 91] = 133 \text{ Ans.}$$

Problem 3.26. Solve the following system of equations for i_1, i_2, and i_3, using Cramer's rule.

$$3i_1 - 2i_2 + 0i_3 = 5$$

$$-2i_1 + 9i_2 - 4i_3 = 0$$

$$0i_1 - 4i_2 + 9i_3 = 10$$

Solution:

In matrix form:

$$\begin{bmatrix} 3 & -2 & 0 \\ -2 & 9 & -4 \\ 0 & -4 & 9 \end{bmatrix} \begin{bmatrix} i_1 \\ i_2 \\ i_3 \end{bmatrix} = \begin{bmatrix} 5 \\ 0 \\ 10 \end{bmatrix}$$

So,

$$\Delta = \begin{bmatrix} 3 & -2 & 0 \\ -2 & 9 & -4 \\ 0 & -4 & 9 \end{bmatrix} = 3(81 - 16) - (-2)(-18)$$
$$= 195 - 36 = 159.$$

∴

$$i_1 = \frac{\begin{vmatrix} 5 & -2 & 0 \\ 0 & 9 & -4 \\ 10 & -4 & 9 \end{vmatrix}}{159} = \frac{5(81 - 16) - (-2)(40)}{159}$$

$$= \frac{405}{159} = 2.547 \text{ amp}$$

So, $\quad i_1 = 2.547$ amp

$$i_2 = \frac{\begin{vmatrix} 3 & 5 & 0 \\ -2 & 0 & -4 \\ 0 & 10 & 9 \end{vmatrix}}{159} = \frac{3(40) - 5(-18)}{159} = \frac{210}{159} = 1.32 \text{ amp}$$

So, $i_2 = 1.32$ amp

Similarly, $i_3 = \dfrac{\begin{vmatrix} 3 & -2 & 5 \\ -2 & 9 & 0 \\ 0 & -4 & 10 \end{vmatrix}}{159} = \dfrac{270}{159} = 1.70$ amp

So, $i_3 = 1.70$ amp **Ans.**

Problem 3.27. Determine i_1, i_2, i_3, and i_4 from the following system of equations.

$$6i_1 - 8i_2 - 10i_3 + 12i_4 = 8$$

$$2i_1 - 4i_2 + 5i_3 + 6i_4 = 33$$

$$-8i_1 + 20i_2 + 14i_3 - 16i_4 = 10$$

$$5i_1 + 7i_2 + 2i_3 - 10i_4 = -15$$

Solution:

$$\begin{bmatrix} 6 & -8 & -10 & 12 \\ 2 & -4 & 5 & 6 \\ -8 & 20 & 14 & -16 \\ 5 & 7 & 2 & -10 \end{bmatrix} \begin{bmatrix} i_1 \\ i_2 \\ i_3 \\ i_4 \end{bmatrix} = \begin{bmatrix} 8 \\ 33 \\ 10 \\ -15 \end{bmatrix}$$

$$\Delta = \begin{vmatrix} 6 & -8 & -10 & 12 \\ 2 & -4 & 5 & 6 \\ -8 & 20 & 14 & -16 \\ 5 & 7 & 2 & -10 \end{vmatrix} = 2622$$

$$i_1 = \dfrac{\begin{vmatrix} 8 & -8 & -10 & 12 \\ 33 & -4 & 5 & 6 \\ 10 & 20 & 14 & -16 \\ -15 & 7 & 2 & -10 \end{vmatrix}}{\Delta} = \dfrac{2622}{2622} = 1$$

$\therefore \qquad i_1 = 1 \text{ amp}$

$$i_2 = \frac{\begin{vmatrix} 6 & 8 & -10 & 12 \\ 2 & 33 & 5 & 6 \\ -8 & 10 & 14 & -16 \\ 5 & -15 & 2 & -10 \end{vmatrix}}{\Delta} = \frac{5244}{2622} = 2$$

So, $\qquad i_2 = 2 \text{ amp}$

Similarly $\qquad i_3 = 3 \text{ amp and } i_4 = 4 \text{ amp } \textbf{Ans.}$

Problem 3.28. Consider the equations
$$3x - y - 3z = 1$$
$$x - 3y + z = 1$$
$$4x + 0y - 5z = 1$$
(a) Is (4, 2, 3) a solution? Is (−1, −1, −1) a solution? (b) Can these equations be solved by determinants? Why? (c) What can you conclude regarding the three represented by these equations?

Solution:

(a) (1) Putting the value of (4, 2, 3), i.e (x, y, z) in above equation:
$$3 \times 4 - 2 - 3 \times 3 = 1$$
$$4 - 3 \times 2 + 3 = 1$$
$$4 \times 4 + 0 - 5 \times 3 = 1$$

So, (4, 2, 3) is a solution of the above equation.

(2) Putting $(x, y, z) = (-1, -1, -1)$ in the above equation.
$$-3 + 1 + 3 = 1$$
$$-1 + 3 - 1 = 1$$
$$-4 + 0 + 5 = 1$$

hence (−1, −1, −1) is also a solution of the above equation.

(b) Determinant of these equation is:

$$\Delta = \begin{vmatrix} 3 & -1 & -3 \\ 1 & -3 & 1 \\ 4 & 0 & -5 \end{vmatrix} = 0$$

So, the above equation can not be solved by determinants rule, because there is no unique solution when $\Delta = 0$ in cramer's rule.

(c) From the above three equations (which is the equation of planes) shows that they have a common line of intersection and any point on this line will satisfy all the three equations. **Ans.**

Problem 3.29. The element represented in the network is a gyrator which is described by the equations

$$v_1 = R_0 i_2$$

$$v_2 = -R_0 i_1$$

Find the two-element equivalent network shown in (b) of the figure.

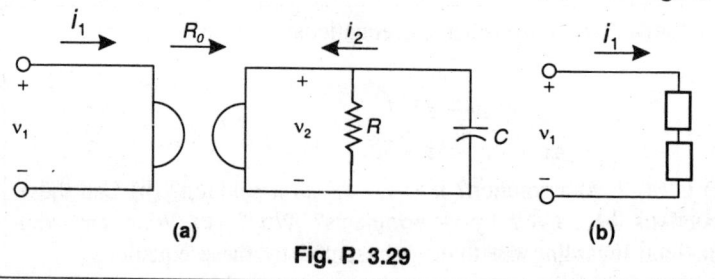

(a)   (b)

Fig. P. 3.29

Solution:

Given $v_1 = R_0 i_2$ (1)

$$v_2 = -R_0 i_1$$ (2)

Now, differentiating eq (2) we get

$$\frac{d v_2}{dt} = -R_0 \frac{di_1}{dt}$$ (3)

From Fig. (a), writing node equation for i_2

$$i_2 = \frac{v_2}{R} - c \frac{dv_2}{dt}$$ (4)

Putting the value of $\dfrac{d v_2}{dt}$ in eq (4) we get

$$i_2 = \frac{v_2}{R} = -C\left(-R_0 \frac{di_1}{dt}\right)$$

$$i_2 = \frac{R_0 i_1}{R} + R_0 C \frac{di_1}{dt}$$ (5)

Putting the value of eq (5) in eq (1), we get

$$v_1 = \frac{R_0^2\, i_1}{R} + R_0^2\, C \frac{d\,i_1}{dt} \qquad (6)$$

So, from eq (6) and fig. P. 3.29 (b). the equivalent diagram is: **Ans.**

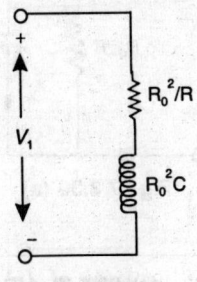

Fig. P. 3.29 (a)

Problem 3.30. For the gyrator-*RL* network of the following figure, write the differential equation relating v_1 to i_1. Find a two-element equivalent network.

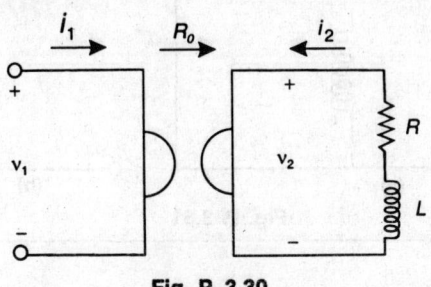

Fig. P. 3.30

Solution:

For the gyrator $\qquad\qquad v_1 = R_0\, i_2 \qquad\qquad\qquad (1)$

$$v_2 = R_0\, i_1 \qquad\qquad\qquad (2)$$

From the loop equation in the output

$$i_2\, R + L \frac{di_2}{dt} = v_2 \qquad (3)$$

Putting the value of i_2 from eq (1) in eq (3), we get:

$$\frac{v_1}{R_0} R + L \frac{d}{dt}\left(v_1 / R_0\right) = v_2$$

or
$$\frac{v_1}{R_0}R+\frac{L}{R_0}\frac{d}{dt}v_1 = R_0 i_1 \quad \text{So, } i_1 = \frac{Rv_1}{R_0^2}+\frac{L}{R_0^2}\frac{dv_1}{dt} \qquad (4)$$

So, its equivalent circuit is: **Ans.**

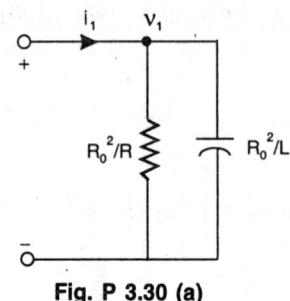

Fig. P 3.30 (a)

Problem 3.31. In the network of (a) of the figure, all self inductance values are 1 H, and mutual inductance values are $\frac{1}{2}$ H. Find L_{eq}, the equivalent inductance, shown in (b) of the figure.

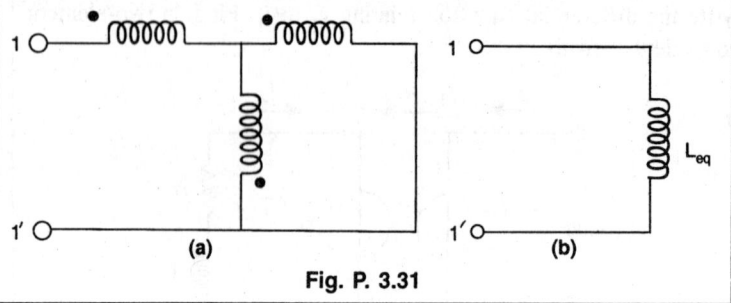

Fig. P. 3.31

Solution:

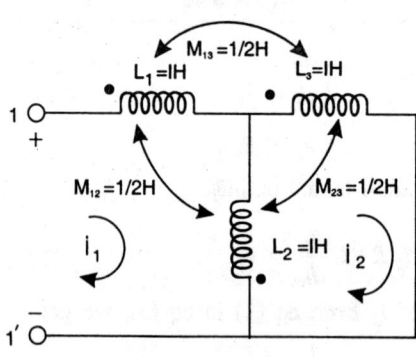

Fig. P. 3.31 (a)

Taking loop equation in mesh (1) we get

$$L_1 \frac{di_1}{dt} + L_2 \frac{d(i_1 - i_2)}{dt} + M_{12} \frac{d(i_2 - i_1)}{dt} - M_{21} \frac{di_i}{dt} + M_{13} \frac{di_2}{dt} - M_{23} \frac{di_2}{dt} = v_1$$

or,
$$\frac{di_1}{dt} + \frac{d(i_1 - i_2)}{dt} + \frac{1}{2} \frac{d(i_2 - i_1)}{dt} - \frac{1}{2} \frac{di_1}{dt} + \frac{1}{2} \frac{di_2}{dt} - \frac{1}{2} \frac{di_2}{dt} = v_1$$

or,
$$\frac{di_1}{dt} - \frac{1}{2} \frac{di_2}{dt} = v_1 \qquad (1)$$

Taking loop equation from mesh (2), we get

$$L_2 \frac{d(i_2 - i_1)}{dt} + L_3 \frac{di_2}{dt} + M_{32} \frac{d(i_2 - i_1)}{dt} + M_{31} \frac{di_1}{dt} + M_{23} \frac{di_2}{dt} - M_{21} \frac{di_1}{dt} = 0$$

or
$$\frac{d(i_2 - i_1)}{dt} + \frac{di_2}{dt} + \frac{1}{2} \frac{d(i_2 - i_1)}{dt} + \frac{1}{2} \frac{di_1}{dt} + \frac{1}{2} \frac{di_2}{dt} - \frac{1}{2} \frac{di_1}{dt} = 0$$

or
$$\frac{-3}{2} \frac{di_1}{dt} + 3 \frac{di_2}{dt} = 0 \qquad (2)$$

or
$$\frac{1}{2} \frac{di_1}{dt} = \frac{di_2}{dt}$$

Putting the value of $\dfrac{di_2}{dt}$ in eq (1) we get

$$\frac{di_1}{dt} - \frac{1}{2} \left(\frac{1}{2} \frac{di_1}{dt} \right) = v_1$$

or
$$\frac{3}{4} \frac{di_1}{dt} = v_1$$

or
$$v_1 = \frac{3}{4} \frac{di_1}{dt}$$

$\therefore$
$$L_{eq} = \frac{3}{4} H$$

or
$$L_{eq} = 0.75 \ H \ \textbf{Ans.}$$

Problem 3.32.   It is intended that the two networks of the following figure be equivalent with respect to the pair of terminals which are identified. What must be the values for C_1, L_2, and L_3?

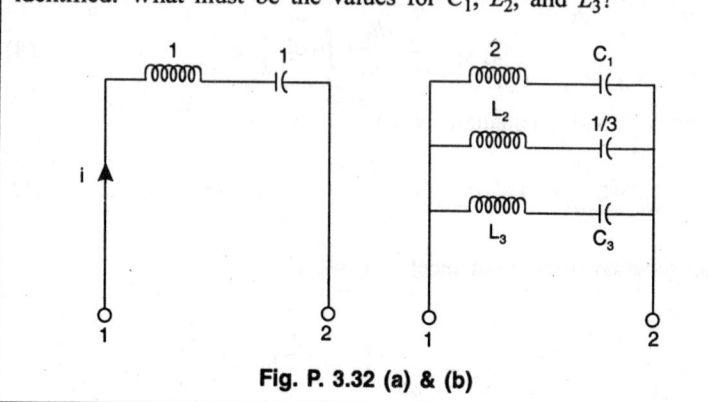

Fig. P. 3.32 (a) & (b)

Solution:

From Fig. P. 3.32   (a) $V_{1-1'} = \dfrac{di}{dt} + \int i\, dt$ $\qquad$ (1)

From Fig. P. 3.32 (b) $V_{1-1'} = L_3 \dfrac{di_3}{dt} + \dfrac{1}{C_3} \int i_3 dt$

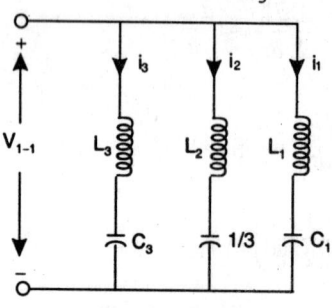

Fig. P. 3.32 (c)

or $\qquad\qquad\qquad\qquad C_3\, v_{1-1'} = L_3\, C_2 \dfrac{di_3}{dt} + \int i_3\, dt$ $\qquad$ (2)

and similarly. $\qquad\qquad C_1\, v_{1-1'} = L_2\, C_2 \dfrac{di_2}{dt} + \int i_2\, dt$ $\qquad$ (3)

Let $\qquad\qquad\qquad\qquad L_1\, C_1 = L_2\, C_2 = L_3\, C_3 = 1$ $\qquad$ (5)

So, $\qquad\qquad\qquad\qquad C_1\, V_{1-1'} = \dfrac{di_1}{dt} + \int i_1\, dt$ $\qquad$ (6)

$$C_2 v_{1-1'} = \frac{di_2}{dt} + \int i_2\, dt \tag{7}$$

$$C_3 v_{1-1'} = \frac{di_3}{dt} + \int i_3\, dt \tag{8}$$

Adding 6, 7 and 8 equation we get

$$(C_1 + C_2 + C_3) v_{1-1'} = \frac{d}{dt}(i_1 + i_2 + i_3) + \int (i_1 + i_2 + i_3)\, dt \tag{9}$$

Since equation (9) is equivalent to eq (1).

So, $$(C_1 + C_2 + C_3) = 1 \tag{10}$$

From (5) and (10) and given value $L_1 = 2, C_2 = 1/3$

$$L_1 C_1 = 1$$

or $$2C_1 = 1$$

or $$C_1 = 1/2$$

$$\therefore \quad C_3 = 1 - (C_1 + C_2) = 1 - (1/2 + 1/3) = 1 - 5/6 = \frac{1}{6}$$

$$L_2 C_2 = 1$$

$$\therefore \quad L_2 = \frac{1}{C_2} = \frac{1}{1/3} = 3.$$

$$\therefore \quad L_2 = 3, \text{ and } L_3 C_3 = 1 \therefore L_3 = 1/C_3 = \frac{1}{1/6}$$

or $$L_3 = 6$$

$$\therefore \quad C_1 = \frac{1}{2}, C_3 = \frac{1}{6}, L_2 = 3, L_3 = 6 \text{ Ans.}$$

Problem 3.33. It is intended that the two networks of the figure be equivalent with respect to two pairs of terminals pair 1–1' and terminal pair 2–2'. For the equivalence to exist, what must be the values for $C_1, C_2,$ and C_3?

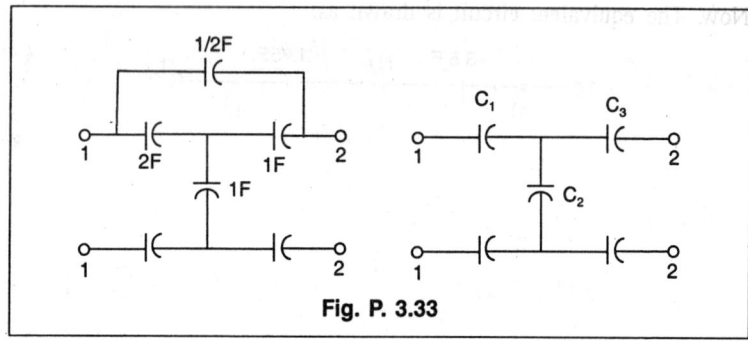

Fig. P. 3.33

Solution:

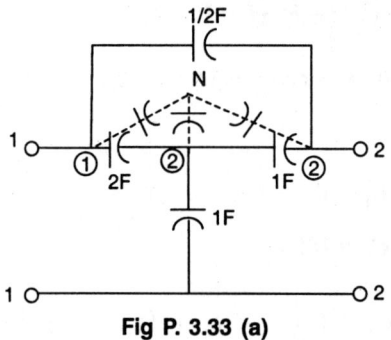

Fig P. 3.33 (a)

Converting delta connected capacitor into equivalent star connected capacitor.

$$C_{1N} = \frac{1}{2}+2+\frac{2\times 1/2}{1}$$

$$C_{1N} = \frac{1}{2}+2+1=3.5$$

$$C_{2N} = \frac{1}{2}+1+\frac{1/2}{2}=0.5+1+0.25$$

$$C_{2N} = 1.75$$

$$C_{3N} = 2+1+\frac{2\times 1}{1/2}$$

$$C_{3N} = 2+1+4$$

or $\qquad C_{3N} = 7$

Now, The equivalent circuit is drawn as:

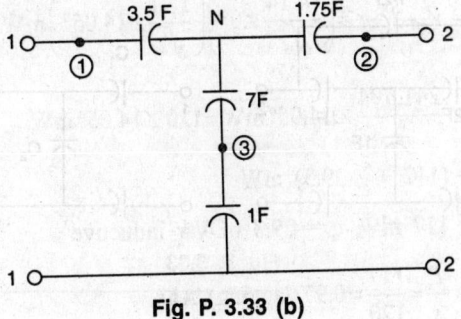

Fig. P. 3.33 (b)

Since capacitor C_{3N} and 1F capacitor is in series so equivalent capacitance is given as

$$c_{eq} = \frac{7 \times 1}{1+7} = \frac{7}{8}$$

Now, the equivalent circuit is given as:

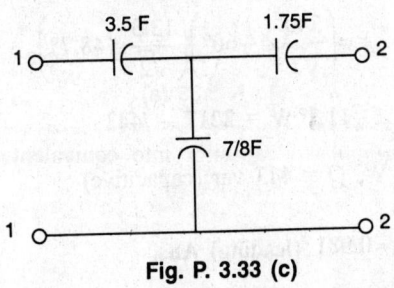

Fig. P. 3.33 (c)

Now, comparing this equivalent circuit with second figure given in problem. So

$$C_1 = 3.5 \ \text{F}$$

$$C_3 = 1.75 \ \text{F}$$

$$C_2 = \frac{7}{8} \ \text{F Ans.}$$

Problem 3.34. Given a circuit with an applied voltage $v = 14.14 \cos \omega t$ (V) and a resulting current $i = 17.1 \cos (wt - 14.05°)$ (mA), determine the complete power triangle.

Solution:

$$S = V_{eff}\, I^*_{eff} = \left(\frac{14.14}{\sqrt{2}} \angle 0°\right)\left(\frac{17.1}{\sqrt{2}} \angle 14.05°\right) mW$$

$$= \frac{241.794}{2} \angle 14.05° \,mW = 120 \angle 14.05° \,mW$$

$$= (117 + j\ 29.3)\ mW$$

$\therefore \qquad P = 117$ mW, q = 29.3 m Var inductive

$$p.f = \frac{P}{s} = \frac{117}{120} = 0.97 \,(\text{lagging}) \textbf{ Ans.}$$

Problem 3.35. Given a circuit with an applied voltage $v = 340 \sin$ $(\omega t - 60°)$ (V) and a resulting current $i = 13.3 \sin (\omega t - 48.7°)$ (A), determine the complete power triangle.

Solution:

$$S = V_{eff}\, I^*_{eff} = \left(\frac{340}{\sqrt{2}} \angle -60°\right)\left(\frac{13.3}{\sqrt{2}} \angle 48.7°\right)$$

$$= 2261 \angle -11.3°\,W = 2217 - j443$$

$\therefore \qquad P = 2217$ W, $Q = 443$ var (capacitive)

$$p.f. = \frac{2217}{2261} = 0.981 \ (\text{leading}) \textbf{ Ans.}$$

Problem 3.36. A two element series circuit with $R = 5.0$ W and $X_L = 15.0$ W, has an effective voltage 31.6 V across the resistance find the complex power and the power factor.

Solution:

$$V_R = 31.6 \text{ K.} \quad R = 5\ \Omega$$

$$\therefore \qquad I_{eff} = \frac{31.6}{5} = 6.32 \text{ amp.}$$

$$\therefore \qquad P = I^2_{eff}R = (6.32)^2 \times 5 = 200 \text{ W}$$

$$\theta = I_{eff}^2 \, X_L = (6.32)^2 \times 15 = 600 \, \text{Var}.$$

$$\therefore \qquad S = (200 + j \, 600) \, \text{VA}.$$

$$\text{p.f} = \frac{200}{632.45} = 0.316 \quad \text{lagging}.$$

Problem 3.37. A circuit with impedance $Z = 8.0 - j6.0 \ \Omega$ has an applied phasor voltage $70.7 \angle -90.0° \ \text{V}$. Obtain the complete power triangle.

Solution:

$$Z = 8 - j6 \ \Omega = 10 \angle -36.86°$$

$$I_{eff} = \frac{V_{eff}}{Z} = \frac{70.7 \angle -90°}{10 \angle -36.86°} = 7.07 \angle -53.13°$$

$$\therefore \qquad S = V_{eff} \cdot I^*_{eff}$$

$$= \left(\frac{70.7}{\sqrt{2}} \angle -90° \right) \left(\frac{7.07}{\sqrt{2}} \angle 53.13° \right) = 250 \angle -36.87°$$

$$= 200 - j \, 150$$

$$\therefore \qquad P = 200 \ \text{W}, \ Q = 150 \ \text{Var (capacitive)}$$

$$\text{p.f.} = \frac{200}{250} = 0.80 \quad \text{leading}$$

Problem 3.38. Determine the circuit impedance which has a complex power $S = 5031 \angle -26.57° \ \text{VA}$ for an applied phasor voltage $212.1 \angle 0° \ \text{V}$.

Solution:

$$S = 5031 \angle -26.57° = 4499.66 - j \, 2250.32$$

$$\therefore \qquad P = 4499.66 \ \text{W}$$

$$Q = 2250.32 \ \text{Var}.$$

$$S = V_{eff} \, I^*_{eff}$$

$$= 5031\angle-26.57°\left(212.1/\sqrt{2}\angle0°\right)\left(I_{eff}\right)^*$$

or, $I^*_{eff} = \sqrt{2}\times\dfrac{5031\angle-26.57°}{212.1\angle0°}=33.54\angle-26.54°$

∴ $I_{eff} = 33.54$

∴ $P = I_{eff}^2 R$

or $4499.66 = \left(33.54\right)^2 \times R$

or, $R = 3.998$

or, $R = 4Ω. \quad Q = I_{eff}^2\left(\omega c\right)$

or, $(\omega c) = \dfrac{2250.32}{\left(33.54\right)^2}=2Ω$

∴ $Z = 4 - j\ 2$ **Ans.**

Problem 3.39.   Determine   the   impedance corresponding to apparent power 3500 VA, power factor 0.76 lagging and effective current 18.0 A.

Solution:

$S = 3500$ VA, $\cos\theta = 0.76$.

∴ $P = S\cos\theta = 3500 \times 0.76 = 2660$ W

or, $p = I_{eff}^2 R$

or, $R = \dfrac{2660}{\left(18\right)^2}=8.2Ω$

$\theta = S\sin\theta = 3500 \times 0.64 = 2274$ Var

$\theta = I_{eff}^2 X_L$

or, $X_L = \dfrac{2274}{\left(18\right)^2}=7$

∴ $Z = 8.2 + j\ 7$

∴ $Z = 10.78\ \angle40.54°\ Ω$ **Ans.**

Problem 3.40. A two branch parallel circuit, with $Z_1 = 10 \angle 0°$ Ω and $Z_2 = 8.0 \angle -30.0° \ \Omega$, has a total current $i = 7.07 \cos (\omega t - 90°)$ (A). Obtain the complete power triangle.

Solution:

$$Z_1 = 10 \angle 0°, \quad Z_2 = 8.0 \angle -30.0°$$

$$Z_{eq} = \frac{Z_1 Z_2}{Z_1 + Z_2} = \frac{10 \angle 0 \times 8.0 \angle -30.0°}{10 \angle 0 + 8.0 \angle -30°} = \frac{80 \angle -30.0°}{17 \angle -13.3°}$$

$$Z_{eq} = 4.6 \angle -16.7°$$

$$Z_{eq} = 4.4 - j1.32$$

$$I_{eff} = \frac{7.07}{\sqrt{2}}$$

$$P = \left(I_{eff}\right)^2 \times 4.4 = \left(\frac{7.07}{\sqrt{2}}\right)^2 \times 4.4 = 109.96 \text{ W}$$

$$Q = \left(I_{eff}\right)^2 \times 1.32 = \left(\frac{7.07}{\sqrt{2}}\right)^2 \times 1.32 = 32.9 \text{ Var (inductive)}$$

$$S = 109.96 - j32.9 = 114.77 \angle -16.65$$

p.f. = cos 16.65

p.f. = 0.958 leading

Problem 3.41. A two branch parallel circuit has branch impedances $Z_1 = 2 - j5$ W and $Z_2 = 1 + j1 \ \Omega$. Obtain the complete power triangle for the circuit if the 2.02 Ω resistor consumes 20 W.

Solution:

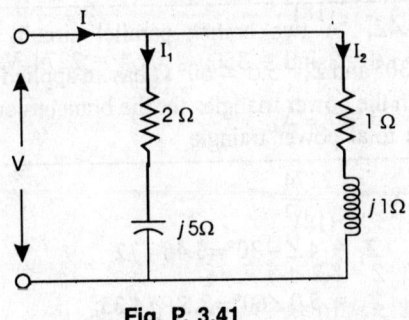

Fig. P. 3.41

$$Z_1 = 5.38\angle -68°, \quad Z_2 = 1.41\angle 45°$$

$$Z_{eq} = \frac{Z_1 Z_2}{Z_1 + Z_2} = \frac{5.38\angle -68° \times 1.41\angle 45°}{2 - j5 + 1 + j1} = \frac{7.58\angle -23°}{5\angle -53.13°}$$

$$Z_{eq} = 1.516\angle 30.13° = 1.31 + j\, 0.76$$

Given $\qquad I_{1eff}^2\, 2 = 20$

or, $\qquad I_{1eff} = 3.16$ amp

$\therefore \qquad I_{1eff}^2 = \dfrac{I_{eff}\, Z_2}{Z_1 + Z_2}$

$$3.16 = \frac{I_{eff} \times 1.41\angle 45°}{5\angle -53.13°}$$

$\therefore \qquad I_{eff} = \dfrac{3.16 \times 5\angle -53.13°}{1.41\angle 45°}$

$$I_{eff} = 11.21 \angle -98.13°$$

$\therefore \qquad P = I_{eff}^2\, R_{eq} = \left(11.121\right)^2 \times 1.31 = 165$ W

$$Q = I_{eff}^2\, X_{eq} = \left(11.121\right)^2 \times 0.76 = 95 \text{ Var (inductive)}$$

$$\text{p.f} = \frac{165(P)}{190(s)} = 0.866 \text{ (lagging)}$$

Problem 3.42. A two branch parallel circuit, with impedance $Z_1 = 4.0 \angle -30°$ and $Z_2 = 5.0 \angle 60° \,\Omega$ has an applied effective voltage of 20 V obtain the power triangles for the branches and combine them to obtain the total power traingle.

Solution:

$$Z_1 = 4\angle -30° = 3.46 - j2$$

$$Z_2 = 5.0\angle 60° = 2.5 + j\, 4.33$$

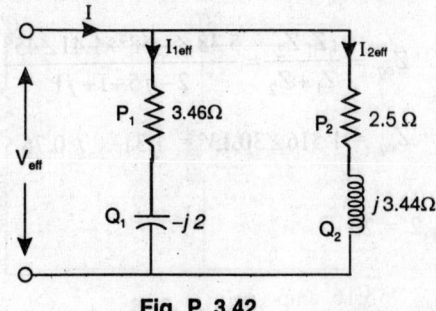

Fig. P. 3.42

$$I_{1\,eff} = \frac{V_{eff}}{4} = \frac{20}{4} = 5\,\text{amp}$$

$$I_{2\,eff} = \frac{20}{5} = 4\,\text{amp}$$

$$P_1 = (5)^2 \times 3.46 = 86.5$$

$$Q_1 = -(5)^2 \times 2 = -50$$

$$P_2 = (4)^2 \times 2.5 = 40$$

$$Q_2 = (4)^2 \times 4.33 = 69.28$$

∴ $$P_T = P_1 + P_2 = (86.5 + 40)\,\text{W}$$

$$Q_T = -\theta_1 + \theta_2 = (-50 + 69.28)\,\text{Var}$$

∴ $$P_T = 126.5\,\text{W}$$

$$Q_T = 19.28\,\text{Var}.$$

∴ $$S_T = \sqrt{P^2 + Q^2}\,\text{VA} = \sqrt{(126.5)^2 + (19.28)^2}\,\text{VA}$$

$$S_T = 128\,\text{VA}$$

$$\text{p.f} = \frac{P_T}{S_T} = \frac{126.5}{128} = 0.989\,(\text{lagging})$$

Problem 3.43. Obtain the complex power for the complete circuit of Fig 3.43 if branch 1 takes 8.0 kVar.

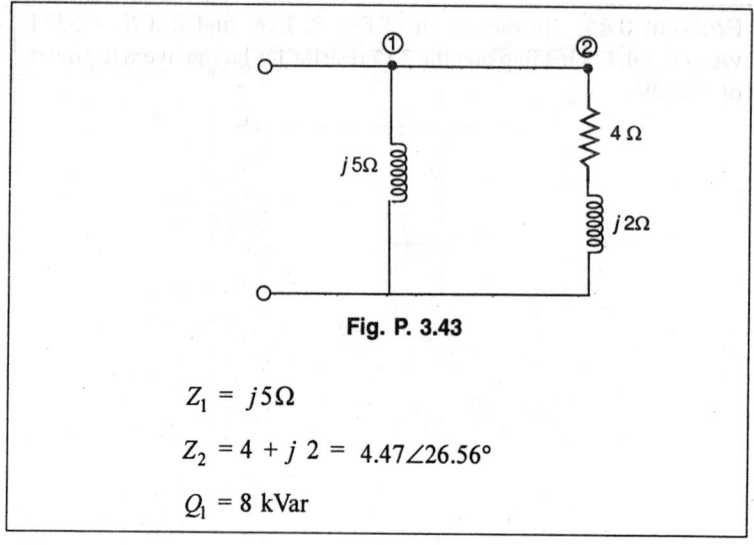

Fig. P. 3.43

$$Z_1 = j5\Omega$$

$$Z_2 = 4 + j\,2 = 4.47\angle26.56°$$

$$Q_1 = 8 \text{ kVar}$$

Solution:

Branch 1 takes 8.0 kVar

$$\therefore \qquad \frac{v_{eff}^2}{5} = 8 \text{ kVar}$$

or, $\qquad v_{eff}^2 = 40 \times 1000 \text{ V}$

or, $\qquad v_{eff} = 200 \text{ V}$

Effective current in branch 2 is:

$$I_{2,eff} = \frac{v_{eff}}{4.47} = \frac{200}{4.47} = 44.74$$

$$\therefore \qquad P_2 = \left(I_{2,eff}\right)^2 \times 4$$

$$P_2 = 8 \text{ kW}$$

$$Q_2 = \left(I_{2,eff}\right)^2 \times 2 = (44.74)^2 \times 2 = 4\text{kVar}.$$

$$\therefore \qquad P_T = P_2 = 8\text{kW}, Q_T = \theta_1 + \theta_2 = 12\text{kVar}$$

$$\therefore \qquad S_T = (8 + j12)\text{k VA, p.f.} = \frac{8}{14.41} = 0.555\left(\text{lagging}\right) \textbf{ Ans.}$$

Problem 3.44. In the circuit of Fig. P. 3.44, find Z if $S_T = 3373$ vA, p.f. = 0.938 leading and the 3 Ω resistor for has an average power of 666 W.

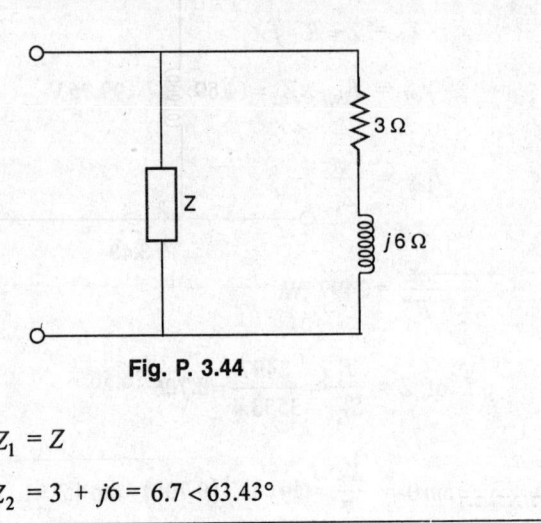

Fig. P. 3.44

$$Z_1 = Z$$

$$Z_2 = 3 + j6 = 6.7 < 63.43°$$

Solution:

$$S_T = 3373\,\text{VA}, \text{p.f.} = 0.938\,\text{leading}$$

$$S_T = P_T - jQ_T$$

$$P_T = 3375 \times 0.938 = 3163 \text{ W}$$

$$Q_T = 3375 \times 0.3466 = 1170 \text{ Var.}$$

$$P_T = P_1 + P_2 = P_1 + 666 = 3163$$

∴ $$P_1 = 3163 - 666$$

or, $$P_1 = 2497 \text{ W}$$

$$3I_{eff,2}^2 = 666$$

or $$I_{eff,2} = (222)^{1/2} = 14.89 \text{ amp.}$$

∴ $$I_{eff,2}^2 \times 6 = 1330.27 \text{ Var.} = Q_2$$

∴ $$Q_T = Q_1 + Q_2, -1170 = Q_1 + 1330$$

∴ $$Q_1 = -2500 \text{ Var}$$

Then, $\qquad S_{T,1} = \sqrt{P_1^2 + Q_1^2} = 3533.4$ W

$\therefore \qquad\qquad Z_1 = Z = R - jX_c$

$\qquad\qquad\quad v_{eff} = I_{2\,eff} \times Z_2 = 14.89 \times 6.7 = 99.76 \text{ V}$

$\therefore \qquad\qquad I_{1\,eff} = \dfrac{v_{eff}}{Z_1}$

$\qquad\qquad \dfrac{v_{eff}^2}{R} = 2497$ W.

$\qquad$ p.f. of $Z = \dfrac{P_1}{S_{T,1}} = \dfrac{2497}{3533.4} = 0.706$

$\therefore \qquad\qquad \tan\theta = \dfrac{X_c}{R} = \tan\cos^{-1}(0.706) = \tan 45° = 1$

$\therefore \qquad\qquad X_c = R$

$\qquad\qquad I_{eff}^2 R = 2497$ W

$\qquad\qquad \left(\dfrac{v_{eff}}{Z}\right)^2 R = 2497$ W

$\qquad\qquad \dfrac{v_{eff}^2 R}{\left(R^2 + X_c^2\right)} = 2497, \dfrac{(99.76)^2 R}{2R^2} = 2497$

or, $\qquad\qquad R = \dfrac{(99.76)^2}{2 \times 2497} = 1.99$

or, $\qquad\qquad R = 2\,\Omega$

$\therefore \qquad\qquad X_c = 2\,\Omega$

$\therefore \qquad\qquad Z = R - jX_c$

$\qquad\qquad Z = 2 - j2$ **Ans.**

Problem 3.45. The parallel circuit in Fig. P. 3.45 has a total average power of 1500 W. Obtain the total power traingle in formation.

$$Z_1 = 2 + j3 \ \text{W} = 3.6\angle 56°$$

$$Z_2 = 3 + j6 \ \text{W} = 6.7\angle 63.43°$$

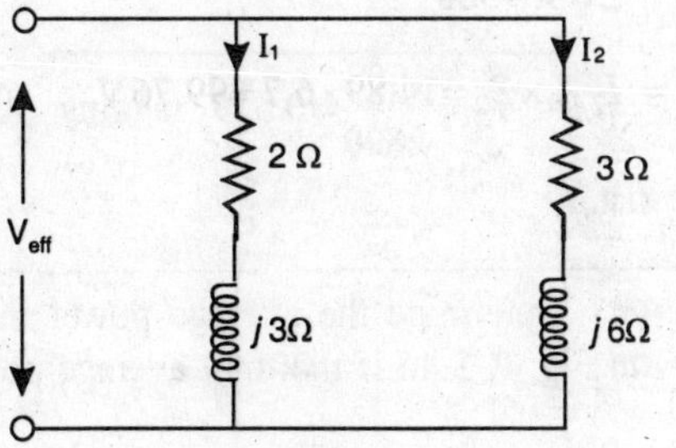

Fig. P. 3.45

Solution:

$$I_{1,eff} = \frac{V_{eff}}{3.6}$$

$$I_{2,eff} = \frac{V_{eff}}{6.7}$$

∴ $$P_1 + P_2 = 1500 \ \text{W (given)}$$

or, $$\left(I_{1,eff}\right)^2 \times 2 + \left(I_{2,eff}\right)^2 \times 3 = 1500$$

or, $$\left(\frac{v_{eff}}{3.6}\right)^2 \times 2 + \left(\frac{v_{eff}}{6.7}\right)^2 \times 3 = 1500$$

or, $$\left(v_{eff}\right)^2 = 6783 \ \text{V}^2$$

or, $$v_{eff} = 82.3 \ \text{V}$$

∴ $$I_{1,eff} = \frac{v_{eff}}{3.6} = \frac{82.36}{3.60} = 22.87 \, \text{amp}$$

$$I_{2,eff} = \frac{v_{eff}}{6.7} = \frac{82.36}{6.7} = 12.29 \, \text{amp}$$

∴ $$Q_1 + Q_2 = I_{1,eff}^2 \times 3 + I_{2,eff}^2 \times 6 = (22.87)^2 \times 3 + (12.29)^2 \times 6$$

$$= 1569 + 906$$

$$Q_T = 2475 \text{ VA}$$

∴ $$S = P_T + jQ_T$$

$$S = 1500 + j\ 2475$$

$$\text{p.f} = \frac{P_T}{S} = \frac{1500}{2890} = 0.519 \text{ lagging } \textbf{Ans.}$$

Problem 3.46. Determine the average power in the 15 Ω and 8 Ω resistances in Fig. P. 3.46 if the total average power in the circuit is 2000 W.

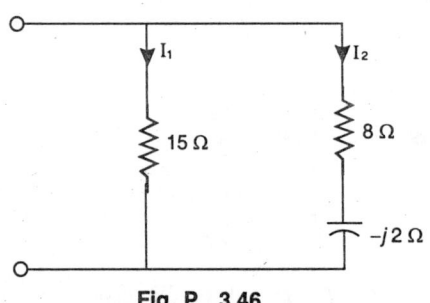

Fig. P. 3.46

$$Z_1 = 15 \text{ W}$$

$$Z_2 = 8 - j\ 2 \text{ W} = 8.24 \ \angle -14.03°$$

Solution:

$$P_1 + P_2 = 2000 \text{ W}$$

$$P_1 = \frac{V_{eff}^2}{R} = \frac{V_{eff}^2}{15}$$

$$P_2 = \left(I_{2,eff}\right)^2 \times 8 = \left(\frac{V_{eff}}{8.24}\right)^2 \times 8$$

or, $$\frac{\left(V_{eff}\right)^2}{15} + \frac{\left(V_{eff}\right)^2}{\left(8.24\right)^2} \times 8 = 2000$$

$$\left(V_{eff}\right)^2 (0.185) = 2000$$

or, $\qquad V_{eff} = 104.11$ V

$\therefore \qquad P_1 = \dfrac{V_{eff}^2}{15} = \dfrac{(104.11)^2}{15} = 722.70$ W

$$P_2 = \dfrac{(V_{eff})^2}{(8.24)^2} \times 8 = \dfrac{(104.11)^2}{(8.24)^2} \times 8 = 1277.00 \text{ W}$$

$\therefore \qquad P_1 = 722.7$ W

$\qquad\qquad P_2 = 1277$ W **Ans.**

Problem 3.47. A three branch parallel circuit with $Z_1 = 25 \angle 15°$ Ω, $Z_2 = 15 \angle -60°$ Ω, and $Z_3 = 15 \angle 90°$ Ω, has an applied voltage V = 339.4 $\angle -30°$ V. Obtain the total apparent power and the overall power factor.

Solution:

$$v_{eff} = \dfrac{339.4}{\sqrt{2}} = 340 \text{ V}$$

$$Z_1 = 25 \angle 15° = 24.14 + j6.4 \ \Omega$$

$$Z_2 = 15 \angle -60° = 7.5 - j13 \ \Omega$$

$$Z_3 = 15 \angle 90° = 0 + j15 \ \Omega$$

$$I_{1,eff} = \dfrac{340}{25} = 9.6.$$

$$I_{2,eff} = \dfrac{340}{15} = 16.$$

$$I_{3,eff} = \dfrac{340}{15} = 16.$$

$$P_1 = (I_{1,eff})^2 \, 24.14 = (9.6)^2 \times 24.14$$

or, $\qquad P_1 = 2224.74$ W.

or, $\qquad P_2 = (I_{2,eff})^2 \, 7.5 = (16)^2 \times 7.5$

or, P_2 = 1920 W

                       P_3 = 0.

∴ P_T = $P_1 + P_2 + P_3 = 2224.74 + 1920 + 0$

or, P_T = 4144 W

$$\theta_1 = \left(I_{1,eff}\right)^2 6.4 = (9.6)^2 \times 6.4$$

$$\theta_1 = 589.84 \text{ Var}$$

$$Q_2 = -\left(I_{2,eff}\right)^2 \times 13 = -(16)^2 \times 13 = -3328 \text{ Var}$$

$$Q_3 = \left(I_{3,eff}\right)^2 \times 15 = -(16)^2 \times 15 = -3840 \text{ Var}$$

∴ Q_T = $Q_1 + Q_2 + Q_3$ = 589.84 − 3328 + 3840 = 589.84 + 512

          Q_T = 1101.84

∴ S_T = $P_T + jQ_T$ = 4144 + j 1101.84

          S_T = 4289 $\angle$ 14.8°

∴ $\cos\theta$ = cos 14.8° = 0.966 lagging. **Ans.**

Problem 3.48.   Obtain the complete power trangle for the following parallel connected loads: load # 1, 200 VA, p.f. = 0.70 lagging: load # 2, 350 VA, p.f = 0.50 lagging: load # 3, 275 VA, p.f = 1.00.

Solution:

$$S_1 = 200 \text{ VA} \angle 46.7° \text{ lagging}$$

$$S_2 = 350 \text{ VA} \angle 60° \text{ lagging}$$

$$S_3 = 275 \text{ VA} \angle 0° \text{ lagging}$$

$$S_1 = P_1 + j\theta_1 = 140 + j142.82$$

$$S_2 = P_2 + j\theta_2 = 175 + j303.10$$

$$S_3 = P_3 + j\theta_3 = 275 + j0$$

∴ P_T = $P_1 + P_2 + P_3$ = 140 + 175 + 275 = 590 W

$$\theta_T = \theta_1 + \theta_2 + \theta_3 = 142.82 + 303.10 + 0 = 445.92$$

$$\therefore \qquad S_T = P_T + j\theta_T = 590 + j\,445.92$$

$$S_T = 740 \;\angle 37°$$

$$\therefore \qquad \text{p.f.} = \frac{590}{740} = 0.799 \text{ lagging } \textbf{Ans.}$$

Problem 3.49. The addition of a 20 k Var capacitor bank improved the power factor of a certain load to 0.90 lagging. Determine the complex power before the addition of the capacitors, if the final apparent power is 185 kVA.

Solution:

$$S_{Final} = 185\,\text{kVA} \angle \cos^{-1} 0.9$$

$$= 185 \; \angle 25.84° \text{ kVA}$$

$$S_F = 166.5 + j\,80.63. \text{ kVA}$$

Value of capacitor bank $= j\,20$ kVar

$$\therefore \qquad S_{Initial} = S_{Final} + j\,20\,\text{k Var}$$

$$= (166.5 + j\,80.63 + j\,20) \text{ kVA}$$

$$S_{Initial} = (166.5 + j\,100.63) \text{ kVA } \textbf{Ans.}$$

Problem 3.50. A 25 kVA load with power factor 0.80 lagging has a group of resistive heating units added at unity power factor. How many kW do these units takes, if the new overall power factor is 0.85 lagging?

Solution:

$$S = 25 \text{ kVA } \angle 36.86°$$

$$S = 20 + j\,15 \text{ kVA}$$

Let the kW taken by the unit be p'

New power factor $= 0.85$ lagging.

$$\therefore \qquad \theta = \cos^{-1}(0.85) = 31.87°$$

or, $$\tan\theta = \frac{15}{20 + p'} = 0.619$$

[as new $s = (20+p') + j15)$]

or, $\qquad 0.619 = \dfrac{15}{20+p'}$

$\qquad 20+p' = \dfrac{15}{0.619}$

$\qquad 20+p' = 24.20$

or, $\qquad p' = 24.20 - 20$

or, $\qquad p' = 4.2$ kW **Ans.**

Problem 3.51. A 500 kVA transformer is at full load and 0.60 lagging power factor. A capacitor bank is added, improving the power factor to 0.90 lagging. After improvement, what percent of rated kVA is the transformer carrying?

Solution:

$$S = 500 \angle \cos^{-1} 0.6 \, \text{kVA}$$
$$S = 300 + j\,400 \text{ kVA}$$

When capacitor bank is added then kvar rating is decreases. Let Q_c be the kVar of capacitor bank. Then new kVA rating is

$$S' = 300 + j\,400 - jQ.$$
$$= 300 + j\,(400 - Q).$$

Since New power factor = 0.90 (lagging)

$\therefore \qquad \tan\theta = \dfrac{400-Q}{300} = \tan\left(\cos^{-1} 0.9\right)$

or, $\qquad \tan 25.84 = \dfrac{400-Q}{300}$

or, $\qquad Q = 400 - 145.29$

$\qquad Q = 254.70 \text{ kVar}$

$\therefore \qquad S' = 300 + j\,(400 - 254.70)$

$\qquad S' = 300 + j\,145.29$

$\qquad S' = 333.33 \angle 25.84 \text{ kVA}$

Hence after improvement, percentage of rated kVA is:

$$\frac{333.33}{500} \times 100 = 66.66\% \textbf{ Ans.}$$

Problem 3.52. A 100 kVA transformer is at 80% of rated load at power factor 0.85 lagging. How many kVA in additional load at 0.60 lagging power factor will bring the transformer to full rated load?

Solution:

Let the additional load at 0.60

Lagging be $\qquad S_1 = S_1 \angle 53.13° = 0.6 S_1 + j 0.8 S_1$

At 80% of rated load and at 0.85 lagging power factor, kVA rating of transformer is $\qquad S = 80 \times 0.85 + j\ 80 \times 0.52$

$$S = 68 + j\ 42$$

The total kVa is $\quad S_T = S + S_1 = 68 + j 42 + 0.6 S_1 + j 0.8 S_1$

$$S_T = (68 + 0.6 S_1) + j(42 + 0.8 S_1)$$

Since the transformer comes at full rated load

$\therefore \qquad\qquad S_T = 100 \text{ kVA}$

or, $\qquad 100 \text{ kVA} = \sqrt{(68 + 0.6 S_1)^2 + (42 + 0.8 S_1)^2}$

or, $\qquad (100)^2 = (68 + 0.6 S_1)^2 + (42 + 0.8 S_1)^2$

gives $\qquad S_1 = 21.2 \text{ kVA }\textbf{Ans.}$

Problem 3.53. A 65 kVA load with a lagging power factor is combined with a 25 kVA synchronous motor load which operates at p.f. = 0.60 leading. Find the power factor of the 65 kVA load, if the overall power factor is 0.85 lagging.

Solution:

Rating of synchronous motor is

$$S_m = 25 \text{kVA} \angle \cos^{-1} 0.6 \text{ leading} = 25 \angle -53.13°$$

$$S_m = 15 - j\ 20$$

Let the load p.f. be $\cos\theta$ (lagging)

$\therefore$ $\qquad\qquad S_L = 65\angle 0° = 65\cos\theta + j65\sin\theta.$

New power factor = 0.85 lagging

Total power $\qquad S_T = (65\cos\theta+15)+j(65\sin\theta-20)$

New $\qquad\qquad \cos\theta' = 0.85$

$\therefore$ $\qquad\qquad\qquad \theta' = \cos^{-1}(0.85)$

$\therefore$ $\qquad\qquad\qquad \theta' = 31.87°$

Now, $\qquad\qquad \tan\theta' = \dfrac{65\sin\theta-20}{65\cos\theta+15}=0.619$

$\qquad\qquad\qquad\quad = \dfrac{65\sin\theta-20}{65\cos\theta+15}=0.619.$

or, $\qquad\qquad\qquad \theta = 54.19°$

$\therefore$ $\qquad\qquad\qquad \cos\theta = 0.585$ **Ans.**

Prolem 3.54. For the circuit of Fig. P. 3.54, find the outputs I_A and V_B in terms of the sources I_{S1} an V_{S2}.

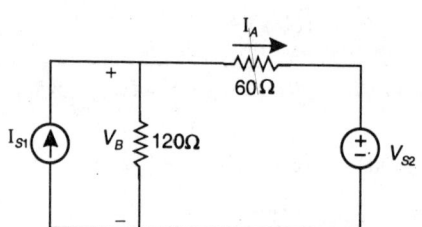

Fig. P. 3.54 Resistive circuit driven by current and voltage sources.

Solution:

A node equation at the top of the current sources is given by,

$$\frac{V_B}{120}+\frac{1}{60}(V_B-V_{S2})=I_{S1} = I_{S1}$$

Solving for V_B, we have $V_B = 40I_{S1}+\dfrac{2}{3}V_{S2}$

From the above we know V_B. The current I_A satisfies

$$I_A = \frac{1}{60}(V_B-V_{S2})=\frac{1}{60}\left(40I_{S1}+\frac{2}{3}V_{S2}-V_{S2}\right)=\frac{2}{3}I_{s1}-\frac{1}{180}V_{S2}$$

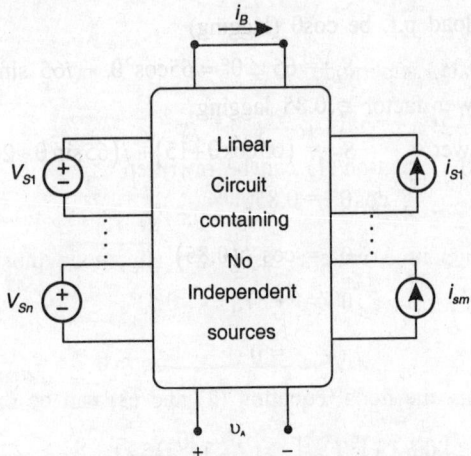

Fig. P. 3.54. Linear circuit driven by *n* independent
voltage sources and *m* indepenent current
sources with outputs of v_A and i_B.

containing all four types of controlled sources. One obtains a matrix
equation of the form

$$[\;\;] = [M^{-1}][\;\;] \leftarrow \text{Vector of independent source or zero values.}$$
$\quad\uparrow$

Vector of calculated voltages and currents containing *y*.

M is almost always invertible, meaning that the solution has the form

$$[\;\;][\;\;] = [\;\;] \leftarrow \text{Vector of Independent source values or zero values.}$$
$\qquad\uparrow$

Vector of unknown voltages and currents containing *y*.

This matrix equation means that each calculated voltage or current is
a linear combination of the independent source values.

Problem 3.55. Express V_{out} in terms of the input excitations I_{S1}
and V_{S2}. for the circuit given in Fig. P. (3.55)

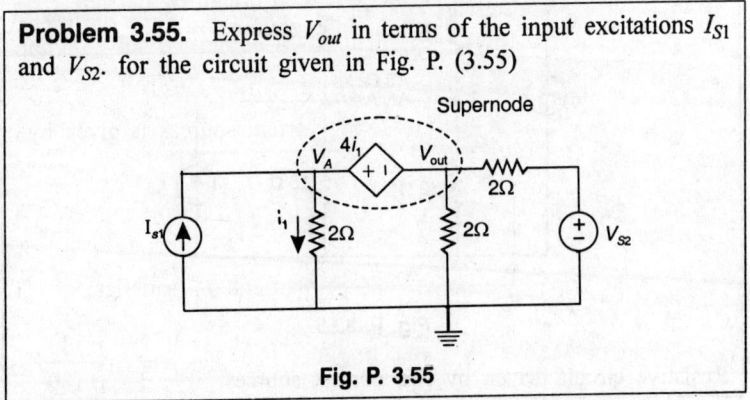

Fig. P. 3.55

Solution:

For the indicated supernode,

$$0.5V_A + 0.5V_{out} + 0.5(V_{out} - V_{S2}) = I_{S1} \tag{1}$$

or, equivalently, equation (1) can be rewritten as,

$$0.5V_A + V_{out} = I_{S1} + 0.5V_{S2} \tag{2}$$

The depenent source voltage inside the supernode must satisfy the equation, $V_A + V_{out} = 4i_1 = 4 \times 0.5V_A$

Rewriting as, $V_A + V_{out} = 0$ (3)

In matrix form, the nodal equation (2) and (3) can be expresed as,

$$\begin{bmatrix} 0.5 & 1 \\ 1 & 1 \end{bmatrix}\begin{bmatrix} V_A \\ V_{out} \end{bmatrix} = \begin{bmatrix} I_{S1} + 0.5V_{S2} \\ 0 \end{bmatrix} \tag{4}$$

Solving equation 4 for V_A ans V_{out} we obtain,

$$\begin{bmatrix} V_A \\ V_{out} \end{bmatrix} = \begin{bmatrix} 0.51 & 1 \\ 1 & 1 \end{bmatrix}^{-1}\begin{bmatrix} I_{S1} + 0.5V_{S2} \\ 0 \end{bmatrix} = \begin{bmatrix} -2 & 2 \\ 2 & -1 \end{bmatrix}\begin{bmatrix} I_{S1} + 0.5V_{S2} \\ 0 \end{bmatrix}$$

Then, it follows that, $V_{out} = 2I_{S1} + V_{S2}$

which shows that V_{out} is linearly related to I_{s1} and V_{s2}

Problem 3.56.   For the circuit of Fig. P. 3.56, find the node voltage V_2 in terms of I_{s1} and I_{s2}. If $I_{s2} = 0$, compute the equivalent resistance R_{eq} seen by the current source I_{s1}.

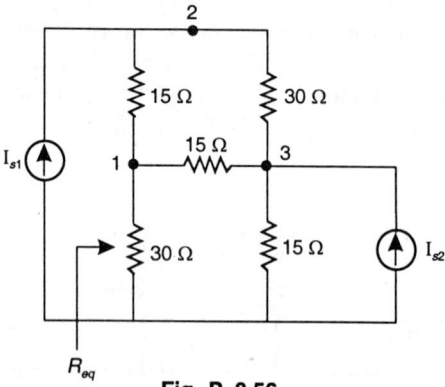

Fig. P. 3.56

Resistive circuit driven by two current sources.

Solution:

Since only resistance and independent current sources are present, the value of the entries in the coefficient matrix of the nodal equations can be computed by inspection: the 1-1 entry is the sum of the conductances at node 1: the 2-2 entry is the sum of the conductances at node 2, and so on. Further, the $i\text{-}j$ entry is the negative of the sum of the conductances between nodes i and j and the $j\text{-}i$ entry has the same value. On the right-hand side of the entry is the sum of the independent current sources injected into node i. By inspection, we have

$$\frac{1}{30}\begin{bmatrix} 5 & -2 & -2 \\ -2 & 3 & -1 \\ -2 & -1 & 5 \end{bmatrix}\begin{bmatrix} V_1 \\ V_2 \\ V_3 \end{bmatrix} = \begin{bmatrix} 0 \\ I_{S1} \\ I_{S2} \end{bmatrix} \tag{1}$$

Solving equation (1) we get

$$\begin{bmatrix} V_1 \\ V_2 \\ V_3 \end{bmatrix} = \begin{bmatrix} 14 & 12 & 8 \\ 12 & 21 & 9 \\ 8 & 9 & 11 \end{bmatrix}\begin{bmatrix} 0 \\ I_{S1} \\ I_{S2} \end{bmatrix}$$

Then, $$V_2 = 21I_{S1} + 9I_{S2} \tag{2}$$

Equation (2) shows that V_2 is linear in the sources current

From equation (2) and the definition of equivalent resistance,

$$R_{eq} = \frac{V_2}{I_{s1}} = 21\Omega$$

Problem 3.57. A linear resistive circuit has two inputs V_{s1} and V_{s2} with output V_{out} as shown in Fig. P. 3.57. Find V_{out} by the principle of superposition. Compute the power absorbed by the 24-Ω resistor.

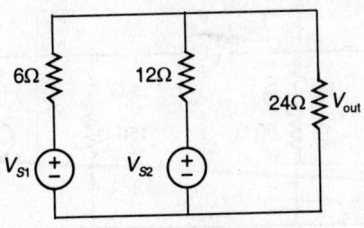

Fig. P. 3.57

Linear resistive circuit driven by two voltage sources.

Solution:

With $V_{s2} = 0$, the equivalent circuit is shown in Fig. P. 3.57. (a) Here the 12-Ω and 24-Ω resistors are in parallel, yielding an equivalent resistance of $8 = 12 \times 24/ (12 + 24)\ \Omega$. By voltage division, we have,

$$V_{out}^1 = \frac{8}{8+6} V_{S1} = \frac{4}{7} V_{S1}$$

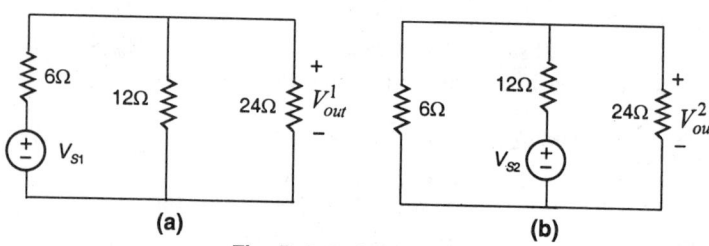

(a)   **(b)**

Fig. P. 3.57 (a) and (b)

[Circuits equivalent to Fig. P. 3.57 when (a) $V_{s2} = 0$ and (b) $V_{s1} = 0$.]

With $V_{s1} = 0$, the equivalent circuit is shown in Fig. P. 3.57 b. Here the 6-Ω and 24-Ω resistors are in parallel, yielding an equivalent resistance of $4.8 = 6 \times 24/ (6 + 24)\ \Omega$. By voltage division,

$$V_{out}^2 = \frac{4.8}{4.8+12} V_{S2} = \frac{2}{7} V_{S2}$$

Using superposition, Theorem, we get

$$V_{out} = V_{out}^1 + V_{out}^2 = \frac{4}{7} V_{S1} + \frac{2}{7} V_{S2}$$

Hence, the power absorbed by the 24-Ω resistor is

$$P_{24\,\Omega} = \frac{(V_{out})^2}{24} = \frac{1}{24}\left(\frac{16}{24}(V_{S1})^2 + \frac{16}{49} V_{S1} V_{S2} + \frac{4}{49}(V_{S2})^2 \right)$$

Problem 3.58. Consider the three-source circuit of Fig. P. 3.58. Compute I_{out}.

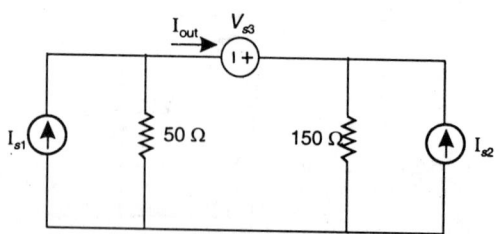

Fig. P. 3.58

Three-source circuit for illustrating superposition.

Solution:

With $I_{S2} = V_{S3} = 0$, the equivalent circuit is shown in Fig. P. 3.58 (a) Using current division, we get,

$$I_{out}^1 = \frac{1/150}{1/150+1/50} I_{S1} = \frac{50}{50+150} I_{S1} = 0.25 I_{S1}$$

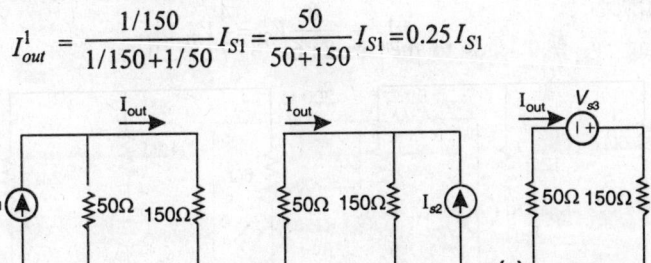

Fig. P. 3.58 (a) (b) (c) [Circuits equivalent to Fig. P. 3.58 when (a) I_{S1} is working alone, (b) I_{S2} is working alone, (c) V_{S3} is working alone.]

With $I_{S1} = V_{S3} = 0$, the equivalent circuit is shown in Fig. P. 3.58 b. Again, using current division, we get

$$I_{out}^2 = -\frac{50}{50+150} I_{S2} = -0.25 I_{S2}$$

With $I_{S1} = I_{S2} = 0$, the equivalent circuit is shown in Fig. P. 3.58c.

By Ohm's law, $I_{out}^3 = \dfrac{V_{S3}}{50+150} = 0.005 V_{S3}$

By superposition there, we get

$$I_{out} = I_{out}^1 + I_{out}^2 + I_{out}^3 = 0.25 I_{S1} - 0.25 I_{S2} + 0.005 V_{S3}$$

Prolem 3.59. For the circuit of Fig. P. 3.59, use superposition to find v_{out} in terms of i_{S1}, v_{S2} and m.

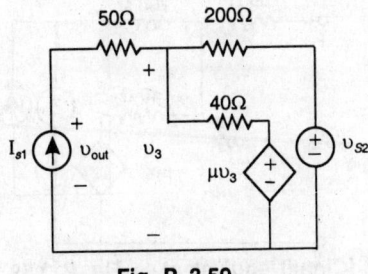

Fig. P. 3.59

Circuit containing a dependent source for illustrating the principal of superposition.

Solution:

Setting $v_{S2} = 0$ leads to the circuit of Fig. P. 3.59 (a)

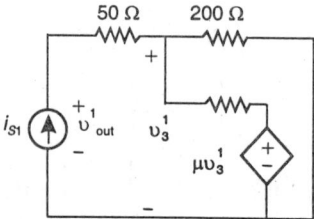

Fig. 3.59 (a) [Circuit equivalent to Fig. P. 3.59 when $v_{S2} = 0$.]

Applying KCL to the top center node associated with v_3^1 we get

$$\frac{v_3^1}{200} + \frac{v_3^1 - \mu v_3^1}{40} = i_{s1}$$

from which $v_3^1 = 200\, i_{S1} / (6 - 5\mu)$. From Ohm's law and KVL applied to Fig. P. 3.59 (a), we have,

$$v_{out}^1 = V_3^1 = \frac{200 i_{S1}}{(6 - 5\mu)}$$

$$v_{out}^1 = 50 i_{S1} + v_3^1 = \frac{500 - 250\mu}{6 - 5\mu} i_{s1}$$

Setting $i_{S1} = 0$ in Fig. P. 3.59 leads to the circuit of Fig. P. 3.59 (b) where v_{out}^2 and v_3^2 denote voltages due only to v_{S2}; these superscripts are not squares.

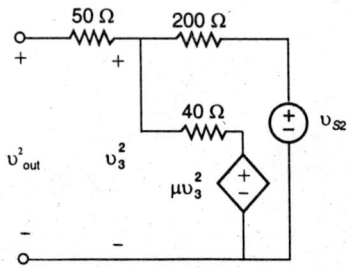

Fig. P. 3.59 (b) [Circuit equivalent to Fig. P. 3.59 when $i_{S1} = 0$.]

Now we apply KCL to the top center node to obtain

$$\frac{v_3^2 - v_{S2}}{200} + \frac{v_3^2 - \mu v_3^2}{40} = 0$$

Solving this equation and noting that no current flows through the 50-Ω resistor in Fig. P. 3.59 (b), we get

$$v_{out}^2 = v_3^2 = \frac{1}{6 - 5\mu} v_{S2}$$

Now, Applying the principle of superposition, we obtain

$$v_{out} = v_{out}^1 + v_{out}^2 = \frac{500 - 250\mu}{6 - 5\mu} i_{S1} + \frac{1}{6 - 5\mu} v_{S2}$$

Problem 3.60. This example illustrates the principle of superposition for the difference amplifier (Fig. P. 3.60 (a)) Find V_{out} in terms of V_{S1} and V_{S2}.

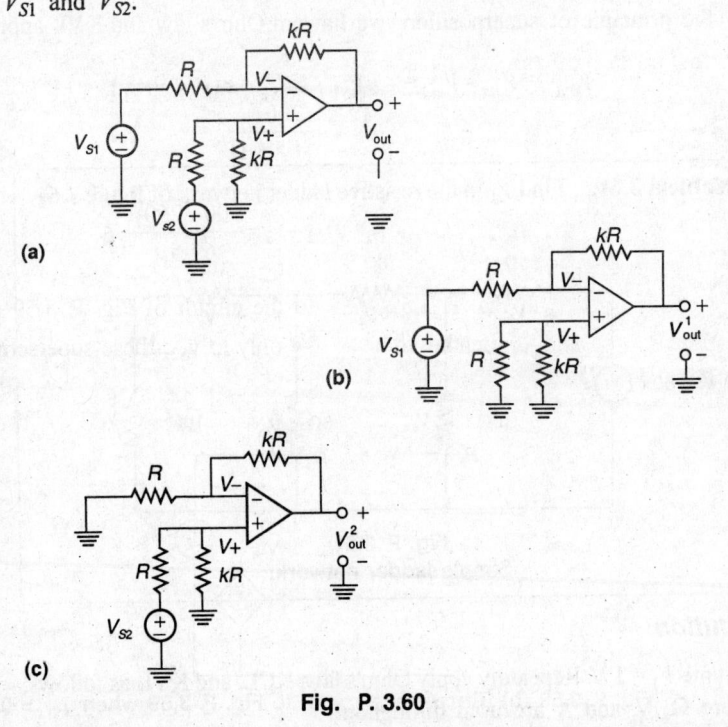

(a)

(b)

(c)

Fig. P. 3.60

Solution:

With $V_{S2} = 0$, the equivalent circuit is shown in Fig. P. 3.60 (a). Because no current flows into or out of the noninverting terminal of the op amp, no voltage develops across the parallel combination of R and kR. Therefore $V_+ = 0$. Hence the noninverting terminal is at virtual ground. As such, the circuit of Fig. P. 3.60 (a) acts like an inverting amplifier whose output is

$$V_{out}^1 = -k V_{S1}$$

With $V_{S1} = 0$, the equivalent circuit is shown in Fig. P. 3.60 (b) From voltage division,

$$V_+ = \frac{kR}{R+kR} V_{S2} = \frac{k}{1+k} V_{S2}$$

Fig. P. 3.60 (b) is a noninverting amplifier configuration, and

$$V_{out}^2 = \left(1+\frac{kR}{R}\right) V_+ = (1+k)\left(\frac{k}{1+k}\right) V_{S2} = k V_{S2}$$

By the principle of superposition, we have,

$$V_{out} = V_{out}^1 + V_{out}^2 = -k V_{S1} + k V_{S2} = k\left(V_{S2} - V_{S1}\right)$$

Problem 3.61. Find V_1 in the resistive ladder network of figure 3.61.

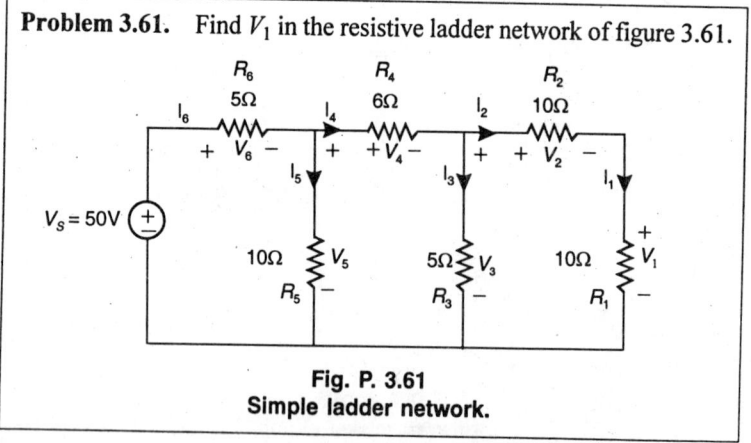

Fig. P. 3.61
Simple ladder network.

Solution:

Assume $V_1 = 1$ V. Repeatdly apply Ohm's law, KCL, and KVL as follows, where Ω, V, and A are used throughout:

$I_1 = \dfrac{V_1}{R_1} = \dfrac{1}{10} = 0.1$ (Ohm's law)
$\begin{cases} V_4 = R_4\,I_4 = 6 \times 0.5 = 3 \quad \text{(Ohm's law)} \\ V_5 = V_3 + V_4 = 2 + 3 = 5 \ \text{(KVL)} \\ I_5 = \dfrac{V_5}{R_5} = \dfrac{5}{10} = 0.5 \ \text{(Ohm's law)} \\ I_6 = I_4 + I_5 = 0.5 + 0.5 = 1 \ \text{(KCL)} \end{cases}$

$I_2 = I_1 = 0.1$ (KCL)

$\begin{cases} V_2 = R_2\,I_2 = 10 \times 0.1 = 1 \quad \text{(Ohm's law)} \\ V_3 = V_1 + V_2 = 1 + 1 = 2 \ \text{(KVL)} \\ I_3\,\dfrac{V_3}{R_3} = \dfrac{2}{5} = 0.4 \ \text{(Ohm's law)} \\ I_4 = I_2 + I_3 = 0.1 + 0.4 = 0.5 \ \text{(KCL)} \end{cases}$

$V_6 = R_6\,I_6 = 5 \times 1$ (Ohm's law)
$V_S = V_5 + V_6 = 5 + 5 = 10$ (KVL)

We conclude that if $V_1 = 1$ V, the source voltage must be $V_S = 10$ V. But the source voltage is actually 50 V. Define $K = 50/10 = 5$. By the proportionally property, if $V_S = 50$ V, then $V_1 = K = 5$ V.

Problem 3.62. For the circuits of Fig. P. 3.62. find V_S and R so that N_1 and N_2 are equivalent.

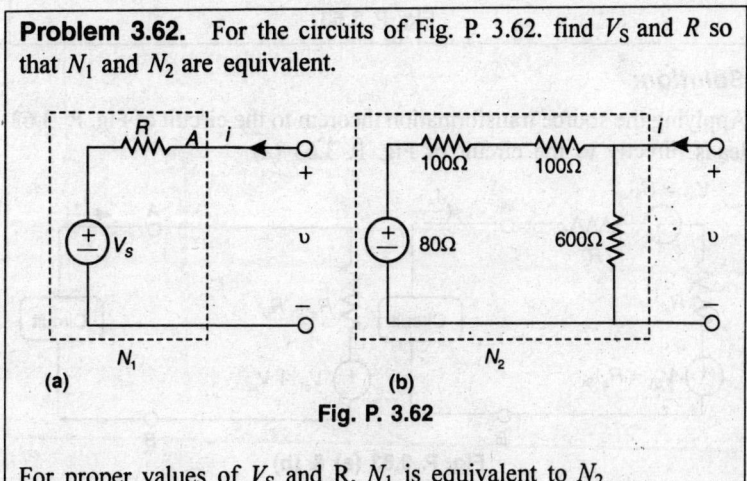

Fig. P. 3.62

For proper values of V_S and R, N_1 is equivalent to N_2

Solution:

For N_2, a node equation at A is given by,

$$i = \frac{v}{600} + \frac{v-80}{200} = \frac{1}{150}v - 0.4 \qquad (1)$$

and, for N_1, at node A, node equation is,

$$i = \frac{v - V_S}{R} = \frac{1}{R}v - \frac{V_S}{R} \qquad (2)$$

Equation coefficients in equations (1) and (2) implies that R = 150 Ω and V_S = 60 V.

Here, It can be seen here that N_1 is the so-called Therenin equivalent circuit of N_2.

Problem 3.63. Simplify the circuit to the left of A–B in Fig. P. 3.63 to a single voltage source in series with a single resistor.

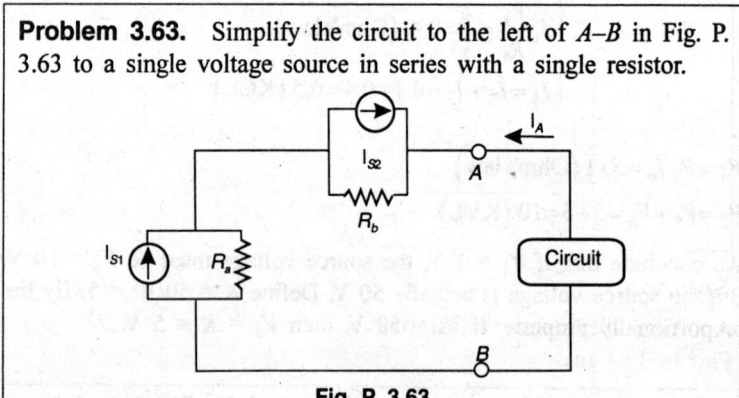

Fig. P. 3.63

Solution:

Applying the source transformation theorem to the circuit of Fig. P. 3.63 leads directly to the circuit of Fig. P. 3.63 (a)

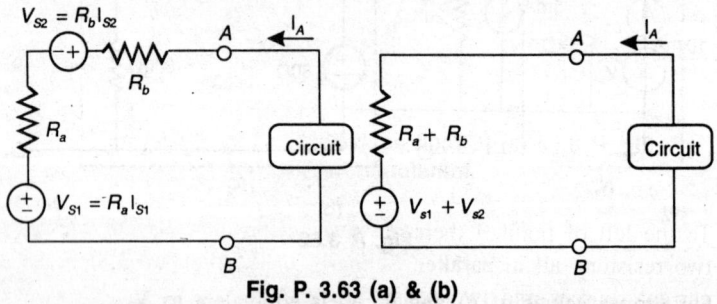

Fig. P. 3.63 (a) & (b)

Intuitively one may combine the two voltage sources into a single source and sum the resistances to form a single resistor of value $R_a + R_b$, as shown in Fig. P. 3.63 (b).

For Fig. P. 3.63 (a) application of KVL to the left of terminals *A–B* gives,

$$V_{AB} = R_a I_A + V_{S1} + R_b I_A + V_{S2} = (R_a + R_b)I_A + (V_{S1} + V_{S2})$$

and for Fig. P. 3.63 (b), application of KVL leads directly to

$$V_{AB} = (R_a + R_b)I_A + (V_{S1} + V_{S2})$$

Problem 3.64. Find I_{AB} in Fig. P. 3.64 by repeated applications of the source transformation theorem.

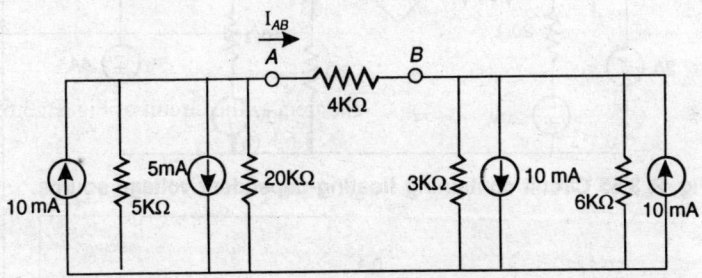

Fig. P. 3.64

Solution:

Applying the source transformation theorem four times results in Fig. P. 3.64 (a)

Fig. P. 3.64 (a) [Circuit equivalent to Fig. P. 3.64 by source transformation theorem.]

To the left of point A there are two independent current sources and two resistors, all in parallel. Similarly, to the right of B there are two current sources and two resistors in parallel.

Combining current sources and resistors to the left of A results in a single current source of 5 mA directed upward and an equivalent resistance of 4 kΩ. To the right of, the current sources cancel each other out and

the equivalent resistance is 2Ω. The resulting circuit is illustrated in
Fig. P. 3.64 (b).

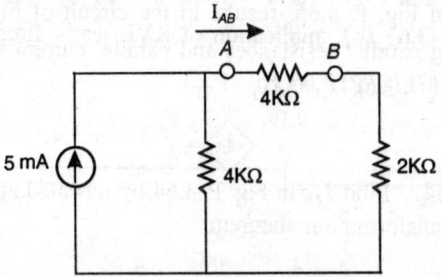

Fig. P. 3.64 (b) Simplification of the circuit of Fig. P. 3.64 (a)

Using the current division formula we get

$$I_{AB} = \frac{4\,k\Omega}{4\,k\Omega + 6\,k\Omega} 5\,\text{mA}$$

$$= 2\text{mA}$$

Problem 3.65. For the circuit of Fig. P. 3.65, apply source
transformations and then find v_1 and v_2 by nodal analysis.

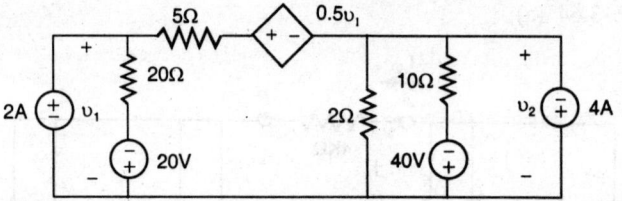

Fig. P. 3.65 Circuit containing floating-dependent voltage source.

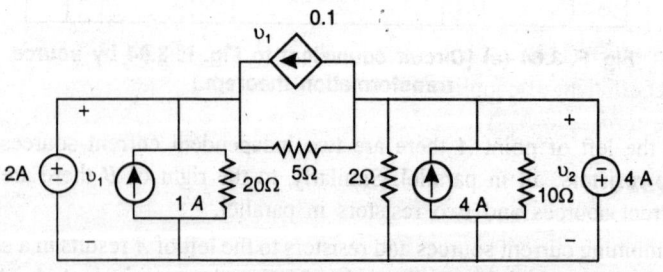

Fig. P. 3.65 (a) Circuit resulting after source transformations.

Solution:

Employing the independent and dependent voltage source transformation to the circuit of Fig. P. 3.65. results in the circuit of Fig. P. 3.65 (a).

Now combining parallel resistances and parallel current sources results in the circuit of Fig. P. 3.65 (b)

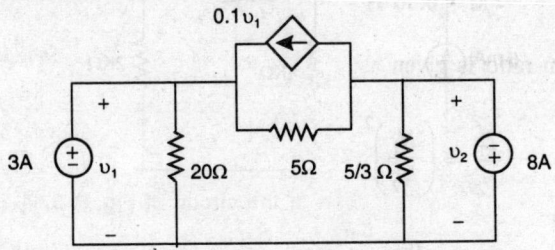

Fig. P. 3.65 (b) [Simplification of the circuit of Fig. P. 3.65 (a)]

Writing a node equation at the left node of the circuit of Fig. P. 3.65 (b) we have

$$3 = 0.05v_1 - 0.1v_1 + 0.2(v_1 - v_2) = 0.15v_1 - 0.2v_2 \qquad (1)$$

and similarly, at the right node,

$$8 = 0.1v_1 + 0.2(v_2 - v_1) + 0.6v_2 = -0.1v_1 - 0.8v_2 \qquad (2)$$

In matrix form, the nodal equations (1) and (2) can be written as,

$$\begin{bmatrix} 0.15 & -0.2 \\ -0.1 & 0.8 \end{bmatrix} \begin{bmatrix} v_1 \\ v_2 \end{bmatrix} = \begin{bmatrix} 3 \\ 8 \end{bmatrix} \qquad (3)$$

Solving the matrix eqn. (3), we get $v_1 = 40$ V and $v_2 = 15$ V.

Problem 3.66. Two coupled coils $L_1 = 0.8$ H and $L_2 = 0.2$ H, have a coefficient of coupling $k = 0.90$, find the mutual inductance M and the turn ratio N_1/N_2.

Solution:

$$L_1 = 0.8\,\text{H}$$

$$L_2 = 0.2\,\text{H}$$

$$k = 0.90$$

Mutual inductance

$$M = k\sqrt{L_1 L_2}$$

$$\therefore \qquad M = 0.9\sqrt{0.8 \times 0.2}$$

$$M = 0.36 \text{ H}$$

and for turn ratio is given as:

$$\frac{L_1}{L_2} = \left(\frac{N_1}{N_2}\right)^2$$

or,

$$\frac{N_1}{N_2} = \sqrt{\frac{L_1}{Ll2}} = \sqrt{\frac{0.8}{0.2}} = \sqrt{4} = 2$$

$$\therefore \qquad \frac{N_1}{N_2} = 2 \quad \textbf{Ans.}$$

Problem 3.67. Two coupled coils $N_1 = 100$ and $N_2 = 800$, have a coupling coefficient $k = 0.85$ with coil 1 open and a current of 5.0 A in coil 2, the flux is $\phi = 0.35$ mWb. Find L_1, L_2 and M.

Solution:

$$L_2 = \frac{N_2 \phi_2}{I_2} = \frac{800 \times 0.35 \times 10^{-3}}{5} = 56 \text{ mH}$$

$$\therefore \qquad \left(\frac{N_1}{N_2}\right)^2 = \left(\frac{L_1}{L_2}\right)$$

or,

$$L_1 = L_2\left(\frac{N_1}{N_2}\right)^2 = 56\left(\frac{100}{800}\right)^2 \text{ mH} = \frac{56}{64} = 0.875 \text{ mH}$$

$$M = k\sqrt{L_1 L_2} = 0.85\sqrt{0.875 \times 56} \text{ mH}$$

$$M = 5.95 \text{ mH} \quad \textbf{Ans.}$$

Problem 3.68. Two identical coupled coils have an equivalent inductance of 800 mH when connected in series aiding, and 35 mH in series opposing. Find L_1, L_2, M and K.

Solution:

$$L_{eq} = L_1 + L_2 + 2M \quad \text{in series adding}$$

$$L'_{eq} = L_1 + L_2 + 2M \quad \text{in series opposing}$$

$\therefore$ $$80 \text{ mH} = L_1 + L_2 + 2M$$

$$35 \text{ mH} = L_1 + L_2 - 2M$$

$\therefore$ $$M = \frac{1}{4}\left(L_{80} - L_{35}\right)\text{mH} = \frac{1}{4}(80 - 35)\text{mH}$$

$$M = 11.25 \text{ mH}$$

Since $$L_1 = L_2$$

$\therefore$ $$80 \text{ mH} = 2L_1 + 22.5$$

or, $$2L_1 = 80 - 22.5$$

or, $$L_1 = \frac{57.5}{2}\text{mH} = 28.75 \text{ mH}$$

or, $$L_2 = 28.75 \text{ mH}$$

$$K = \frac{M}{\sqrt{L_1 L_2}} = \frac{M}{L_1}$$

or, $$K = \frac{11.25}{28.75} = 0.392$$

$\therefore$ $$K = 0.392 \text{ Ans.}$$

Problem 3.69. Two couples with $L_1 = 20$ mH, $L_2 = 10$ mH and $k = 0.50$, are connected four different ways series aiding, series opposing and parallel with both arragments of winding sense. Obtain the equivalent inductances of the four connections.

Solution:

$$L_1 = 200\,\text{mH}, L_2 = 10\,\text{mH}, k = 0.50.$$

$$M = k\sqrt{L_1 L_2}$$

or, $M = 0.5\sqrt{20 \times 10}$ mH

              $M = 7.07$ mH

(a) In series aiding:

           $L_{sa} = L_1 + L_2 + 2M = (20 + 10 + 2 \times 7.07)$ mH

           $L_{sa} = 44.142$ mH

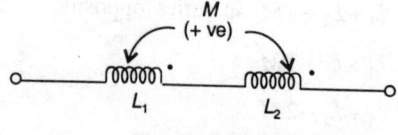

Fig. P. 3.69 (a)

(b) In series opposing:

           $L_{so} = L_1 + L_2 - 2M = 20 + 10 - 2 \times 7.07$ mH

           $L_{so} = 15.86$ mH

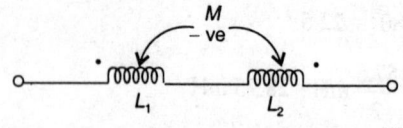

Fig P. 3.69 (b)

(c) In parallel aiding:

$$L_{pa} = \frac{L_1 L_2 - M^2}{L_1 + L_2 - 2M} = \frac{20 \times 10 - (7.07)^2}{15.86} = 9.45 \, \text{mH}$$

∴ $L_{pa} = 9.45$ mH

(d) In parallel opposing:

$$L_{po} = \frac{L_1 L_2 - M^2}{L_1 + L_2 + 2M} = \frac{20 \times 10 - (7.07)^2}{44.142} = 3.398 \, \text{mH}$$

∴ $L_{po} = 3.398$ mH **Ans.**

Problem 3.70. Write the mesh current equations for the coupled circuit shown in Fig. P. 3.70. Obtain the dotted equivalent circuit and write the same equations.

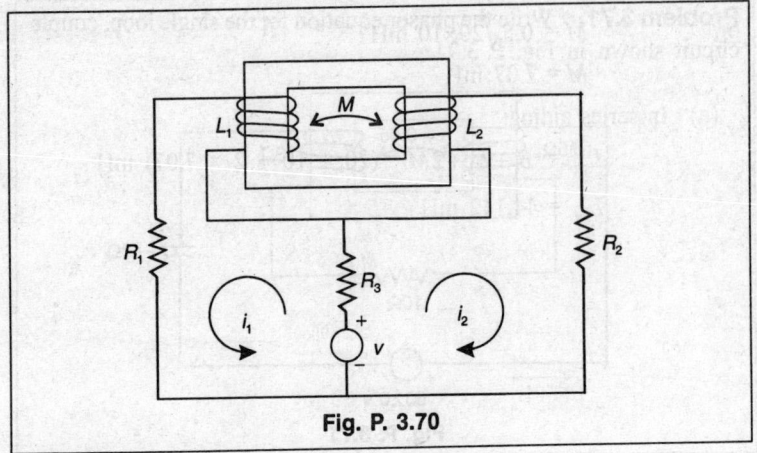

Fig. P. 3.70

Solution:

In loop 1: Applying KVL

$$i_1 R_1 + L_1 \frac{di_1}{dt} + M \frac{di_2}{dt} + (i_1 + i_2) R_3 = v$$

In loop 2: Applying KVL

$$i_2 R_2 + L_2 \frac{di_2}{dt} + M \frac{di_2}{dt} + (i_1 + i_2) R_3 = v$$

or, $$(R_1 + R_3) i_1 + R_3 i_2 + L_1 \frac{di_1}{dt} + M \frac{di_2}{dt} = v \qquad (1)$$

and $$(R_3) i_1 + (R_2 + R_3) i_2 + L \frac{di_2}{dt} + M \frac{di_1}{dt} = v \qquad (2)$$

Its dotted equivalent circuit is:

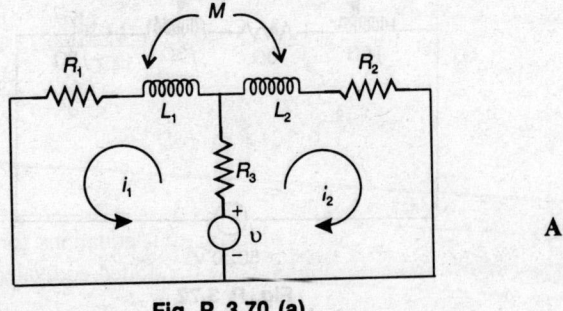

Ans.

Fig. P. 3.70 (a)

Problem 3.71. Write the phasor equation for the single loop, couple circuit shown in Fig. P 3.71

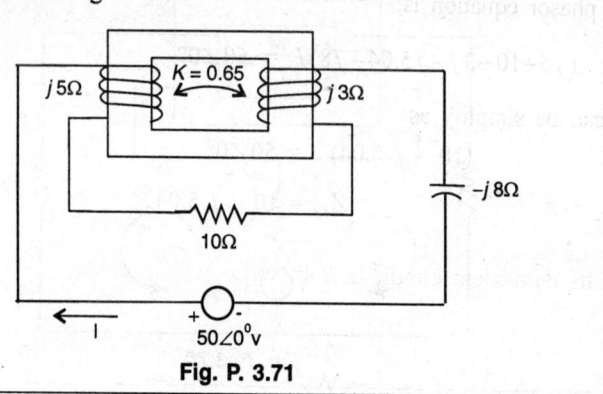

Fig. P. 3.71

Solution:

In the above circuit mutual inductance is given as:

$$M = k\sqrt{L_1 L_2}$$

or, $\omega M = k\sqrt{\omega L_1 \cdot \omega L_2}$

$$X_m = k\sqrt{X_{L_1} \cdot X_{L_2}} = -j\,0.65\sqrt{5 \times 3} = -j\,0.65 \times 3.87 = -j\,2.517\,\Omega$$

Applying KVL in the single loop:

$$j5I + 10I + (-j2\cdot517)I + j3I - j2.517I - j8I = 50\angle0°$$

or, $(10 - j5.03)\,I = 50\angle0°$ **Ans.**

Problem 3.72. Obtain the dotted equivalent circuit for the coupled circuit of Fig. P. 3.72.

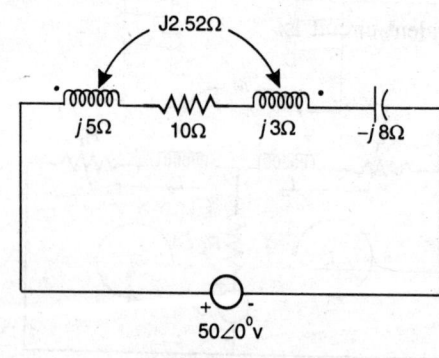

Fig. P. 3.72

Solution:

Its phasor equation is:

$$(j5+10+3j-j5.04-j8)I = 50\angle 0°$$

It can be simplify as:

$$(10 - j\,5.04)\,I = 50\angle 0°$$

So,
$$Z_{eq} = 10 - j\,5.04.$$

∴ Its equivalent circuit is R in series with capacitor.

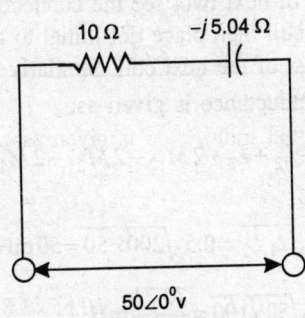

$50\angle 0°$v

Fig. P. 3.72 (a)

Problem 3.73. The three coupled coils shown in Fig. P. 3.73 have coupling coefficients of 0.50. Obtain the equivalent inductance between AB terminal.

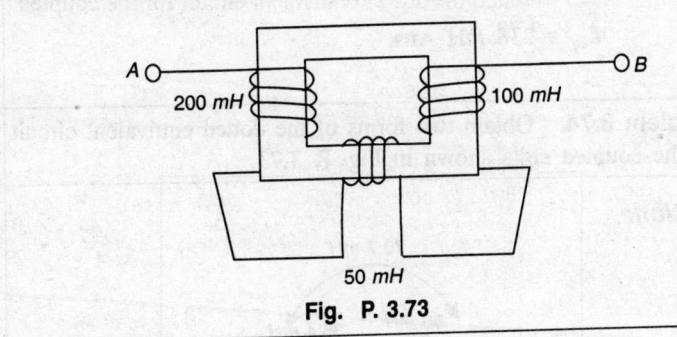

A ○——— 200 *mH*   100 *mH* ———○ *B*

50 *mH*

Fig. P. 3.73

Solution:

These three coupling coils can be drawn in simple equivalent inductance circuit:

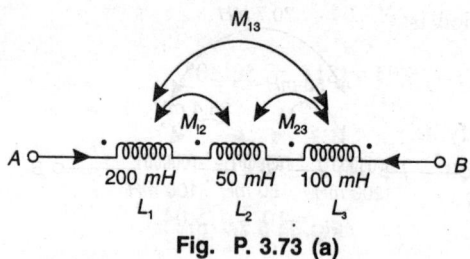

Fig. P. 3.73 (a)

Drive a current into the first coil and place a dot where this current enters. For the dot convention of next two, see the connection. If the coil turn is continue in the next coil then place dot initial to the next coil, if not then place dot at the last of the next coil. Similarly with the third coil. Now, total equivalent inductance is given as:

$$L_{eq} = L_1 + L_2 + L_3 + 2M_{12} - 2M_{23} - 2M_{13}$$

∵ $$k = 0.5$$

∴ $$M_{12} = 0.5\sqrt{L_1 L_2} = 0.5\sqrt{200 \times 50} = 50\,\text{mH}$$

$$M_{23} = 0.5\sqrt{50 \times 100} = \frac{70.71}{2}\,\text{mH}$$

$$M_{13} = 0.5\sqrt{200 \times 100} = \frac{1441.42}{2}\,\text{mH}$$

∴ $$L_{eq} = 200 + 50 + 100 + 2 \times 50 - 2 \times \frac{70.71}{2} - \frac{2 \times 141.42}{2}$$

$$= 350 + 100 - 70.71 - 141.42$$

$$L_{eq} = 238\,\text{mH}\ \textbf{Ans.}$$

Problem 3.74. Obtain two forms of the dotted equivalent circuit for the coupled coils shown in Fig. P. 3.73.

Solution:

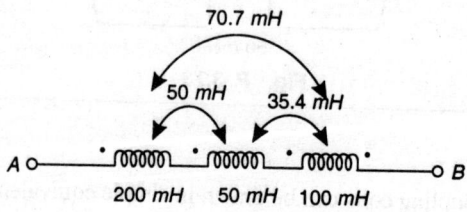

Fig. P. 3.74

Solution:

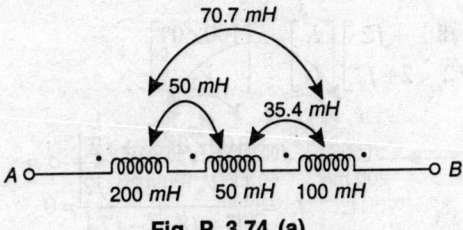

Fig. P. 3.74 (a)

(1) First one is shown in the previous question when current entering terminal is considered as dot. While the second is the dot convention of current leaving terminal.

(2) When current leaving in coil one is taken as dot convention, then coil first and second is in additive manner while first and last (3rd) and second and last are in opposite sign convention. **Ans.**

Problem 3.75. In the couple circuit shown in Fig. P. 3.75, find v_2 for which $I_1 = 0$. What voltage appears at the 8 Ω inductive reactance under this condition?

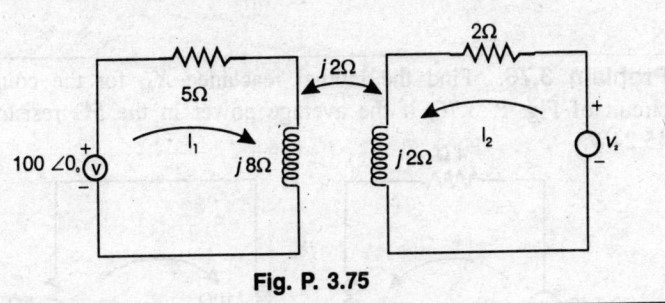

Fig. P. 3.75

Solution:

Applying KVL in both the mesh:

$$5I_1 + j8I_1 + j2I_2 = 100\angle0°$$

$$(5 + j8)I_1 + j2I_2 = 100\angle0° \tag{1}$$

In mesh 2

$$2I_2 + j2I_2 + j2I_1 = V_2$$

or, $$(2 + j2)I_2 + j2I_1 = V_2 \tag{2}$$

In matrix form:

$$\begin{bmatrix} 5+j8 & +j2 \\ +j2 & 2+j2 \end{bmatrix} \begin{bmatrix} I_1 \\ I_2 \end{bmatrix} = \begin{bmatrix} 100\angle0° \\ V_2 \end{bmatrix}$$

Now:

$$I_1 = \frac{\begin{vmatrix} 100\angle0° & +j2 \\ V_2 & 2+j2 \end{vmatrix}}{\begin{vmatrix} 5+j8 & -j2 \\ -j2 & 2+j2 \end{vmatrix}} = 0$$

or, $100\angle0°(2.828\angle45°) - j2V_2 = 0$

or, $\qquad\qquad 2.828\angle45° = j2V_2$

or, $\qquad\qquad 2.828\angle45° = 2\angle90°V_2$

or, $\qquad\qquad V_2 = \dfrac{282.8\angle45°}{2\angle90°}$

or, $\qquad\qquad V_2 = 141.4\angle-45°$

When $I_1 = 0$, there will be no drop in 5Ω resistance. Hence $100\angle0°$ V appears across $j8Ω$ inductive impedance. **Ans.**

Problem 3.76. Find the mutual reactance X_M for the coupled circuit of Fig. P. 3.76, if the average power in the 5Ω resistor is 45.24W.

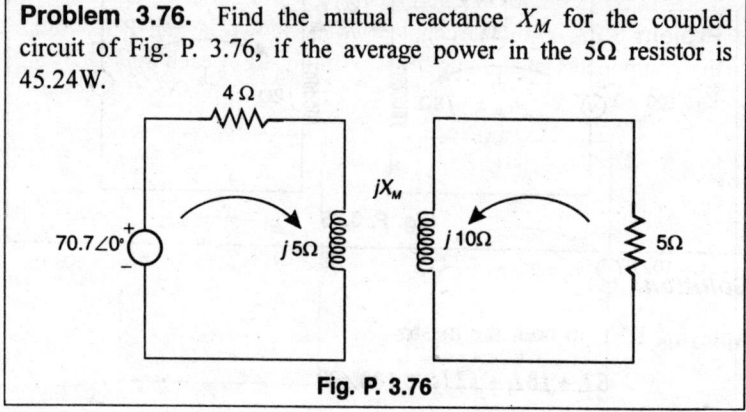

Fig. P. 3.76

Solution:

Since the secondary side is passive circuit. Hence there is no need for sign convention. Now the corresponding equation in matrix form is:

$$\begin{bmatrix} 4+j5 & -jX_M \\ -jX_M & 5+j10 \end{bmatrix} \begin{bmatrix} I_1 \\ I_2 \end{bmatrix} = \begin{bmatrix} 70.7\angle0° \\ 0 \end{bmatrix}$$

Now, $I_{2,eff}^2 \times 5 = 45.24$ W

$\therefore$ $I_{2,eff} = 3$ amp

Now, $I_2 = \dfrac{\begin{vmatrix} 4+j5 & 70.7\angle 0^\circ \\ -jX_M & 0 \end{vmatrix}}{\begin{vmatrix} 4+j5 & -jX_M \\ -jX_M & 5+j10 \end{vmatrix}}$

or, $3\sqrt{2} = \dfrac{-70.7\,X_M\,\angle 90^\circ}{-30+j65+X_M^2}$

$4.25\left(-30+X_M^2+j65\right) = -j70.7\,X_M$

or, $\left(-127.6+4.25\,X_M^2+j276.25\right) = -j70.7\,X_M$

Comparing both sides

     $276.25 = 70.7\,X_M$

or, $X_M = 4\,\Omega$ **Ans.**

Prolem 3.77. For the coupled circuit shown in Fig. P. 3.77, find the components of the current I_2 resulting from each source V_1 and V_2.

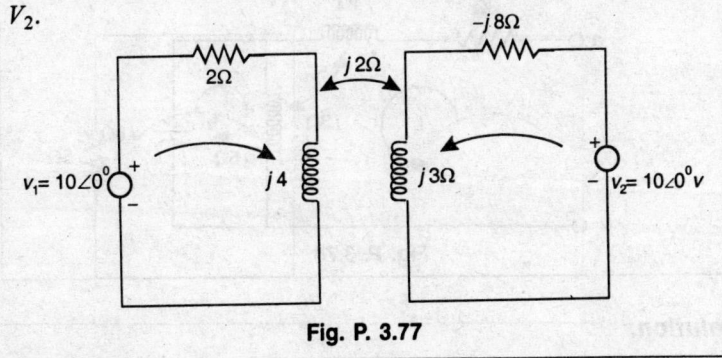

Fig. P. 3.77

Solution:

In matrix form:

$$\begin{bmatrix} 2+j4 & -j2 \\ -j2 & 0-j5 \end{bmatrix} \begin{bmatrix} I_1 \\ I_2 \end{bmatrix} = \begin{bmatrix} V_1 \\ V_2 \end{bmatrix}$$

$$I_2 = \frac{\begin{vmatrix} 2+j4 & v_1 \\ -j2 & v_2 \end{vmatrix}}{\begin{vmatrix} 2+j4 & -j2 \\ -j2 & -j5 \end{vmatrix}} = \frac{\begin{vmatrix} 2+j4 & v_1 \\ -j2 & v_2 \end{vmatrix}}{26\angle-22.61°}$$

I_2' when $V_1 = 0$

∴ $$I_2' = \frac{(4.47\angle 63.43)(10\angle 0)}{26\angle -22.61°}$$

∴ $I_2' = 1.72 \ \angle 86.04° \ A$

I_2'' when $V_1 = 0$

$$I_2'' = \frac{20\angle 90°}{26\angle -22.61°}$$

$$= 0.769\angle 112.61°$$

or, $I_2'' = 0.769 <112.61°$ **Ans.**

Problem 3.78. For the coupled circuit shown in Fig. P. 3.78, find the input impedance at terminal ab.

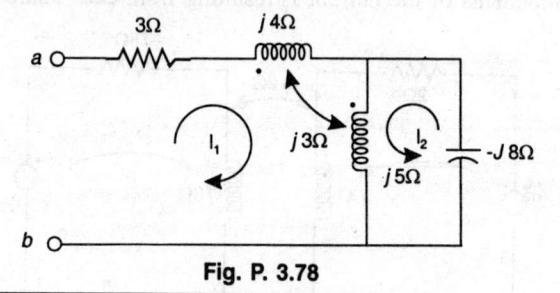

Fig. P. 3.78

Solution:

Let I_1, I_b br the mesh current and voltage across a–b is $v_{ab} \angle 0°$. Now applying mesh current law:

$$3I_1 + j4I_1 + j3I_1 + j5(I_1+I_2) + j3I_1 + j3I_2 = V_{ab}\angle 0° \qquad (1)$$

or, $(3+j15)I_1 + j8I_2 = V_{ab}\angle 0°$.

In mesh 2:

$$j3I_1 + j5(I_1 + I_2) - j8I_2 = 0$$

or, $$\qquad j8I_1 - j3I_2 = 0 \qquad (2)$$

or, In matrix form:

$$\begin{bmatrix} 3+j15 & j8 \\ j8 & -j3 \end{bmatrix} = \begin{bmatrix} V_{ab} \angle 0° \\ 0 \end{bmatrix}$$

$$I_1 = \frac{\begin{vmatrix} V_{ab} \angle 0 & 8\angle 90° \\ 0 & 3\angle -90° \end{vmatrix}}{\begin{vmatrix} 3+j15 & j8 \\ j8 & -j3 \end{vmatrix}}$$

$$I_1 = \frac{3V_{ab} \angle -90°}{109.37 \angle -4.72°}$$

or, $$Z_{in} = \frac{V_{ab}}{I_1}$$

∴ $$\frac{V_{ab}}{I_1} = Z_{in} = \frac{109.37 \angle -4.72°}{3\angle -90°} = 36.45 \angle 85.28°$$

$$Z_{in} = 3 + j36.3\text{W} \text{ Ans.}$$

Problem 3.79. Find the input impedance at terminals a-b of the coupled circuit shown in Fig. P. 3.79

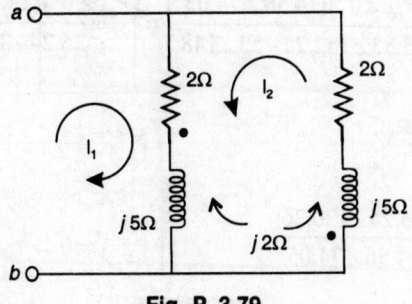

Fig. P. 3.79

Let voltage across a–b termial is $V_{ab} \angle 0°$.

Solution:
For solving such type of circuit, the direction of current should be considered in the same direction in different loops. Let I_1 and I_2 be the current in mesh 1 and mesh 2 respectively. And their direction as shown in fig for convenient.

Now applying KVL in mesh 1:

$$2(I_1+I_2)+j5(I_1+I_2)+j2I_2 = V_{ab} \angle 0°$$

or, $(2+j5)I_1 +(2+j7)I_2 = V_{ab} \angle 0°$ (1)

In mesh 2:

$$j5I_2 +2I_2 +2(I_2 +I_1)+j5(I_2 +I_1)+j2(I_1 +I_2)+j2I_2 = 0$$

or, $(2+j7)I_1 +(4+j14)I_2 = 0$ (2)

The above two equations can be written in the matrix form:

$$\begin{bmatrix} 2+j5 & 2+j7 \\ 2+j7 & 4+j14 \end{bmatrix} \begin{bmatrix} I_1 \\ I_2 \end{bmatrix} = \begin{bmatrix} V_{ab} \angle 0° \\ 0 \end{bmatrix}$$

Now, $I_1 = \dfrac{\begin{vmatrix} V_{ab} \angle 0° & 2+j7 \\ 0 & 4+j14 \end{vmatrix}}{\begin{vmatrix} 2+j5 & 2+j7 \\ 2+j7 & 4+j14 \end{vmatrix}}$

or, $I_1 = \dfrac{(V_{ab} \angle 0°)(14.56 \angle 74.05°)}{78.33 \angle 142.24 -53 \angle 148.1} = \dfrac{(V_{ab} \angle 0°)(14.56 \angle 74.05°)}{26.24 \angle 130.36°}$

or, $Z_{in} = \dfrac{V_{ab}}{I_1}$

$\therefore$ $Z_{in} = \dfrac{26.24 \angle 130.36°}{14.56 \angle 74.05°}$

$Z_{in} = 1.8 \angle 56.31°$

or, $Z_{in} = 1+j1.5 \ \Omega$ **Ans.**

Problem 3.80. Obtain the thevenin and Norton equivalent circuit at terminals ab of the coupled circuit shown in Fig. P. 3.80.

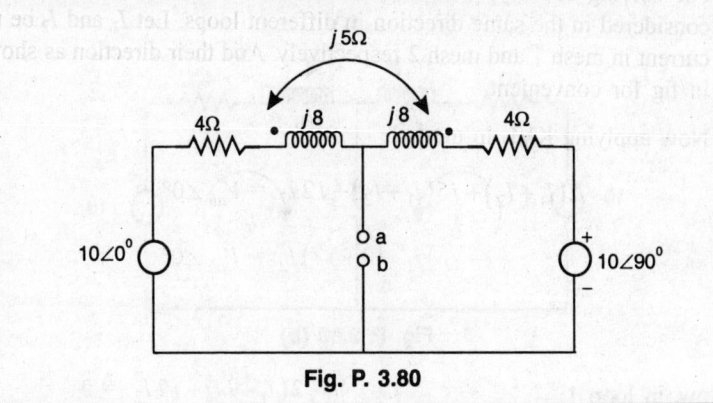

Fig. P. 3.80

Solution:

For finding open circuit voltage across ab, let I be the current in the circuit:

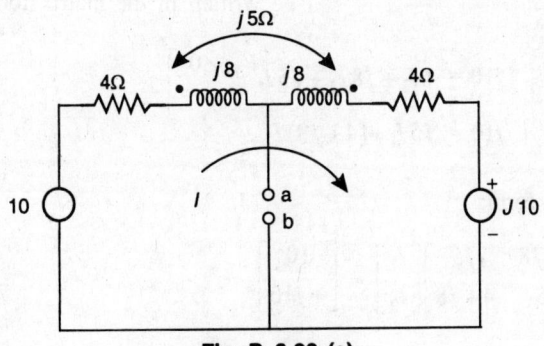

Fig. P. 3.80 (a)

Now, applying KVL in the above circuit.

$$4I + j8I + j8I + 4I - j5I - j5I = 10 - j10$$

or,
$$I(8 + j6) = 10 (1 - j)$$

or, $I = \dfrac{10 - j10}{8 + j6} = \dfrac{14.14 \angle -45.0°}{10 \angle 36.86} = 1.414 \angle -81.8$

Now,
$$v_{ab} = v' = 10 - (4 + j8 - j5)I$$

or,
$$v' = 10 - (4 + j3)(1.414 \angle -81.87°)$$

$$v' = 7.07 \angle 45°$$

For finding short circuit current $I_{sc} = I'$, the circuit is:
Let I_1 and I_2 be the loop current.

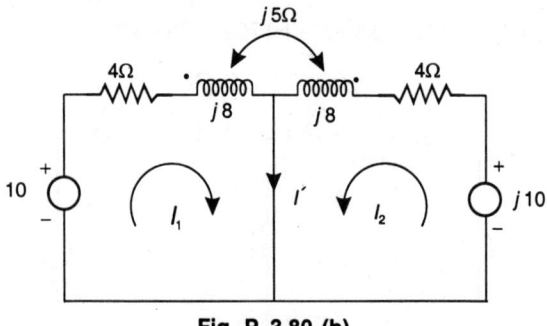

Fig. P. 3.80 (b)

Now, in loop 1:

$$10 = 4I_1 + j8I_1 + j5I_2$$

or,
$$10 = (4 + j8)I_1 + j5I_2 \qquad (1)$$

In loop 2:

$$j10 = 4I_2 + j8I_2 + j5I_1$$

$$j10 = j5I_1 + (4 + j8)I_2 \qquad (2)$$

In matrix form:

$$\begin{bmatrix} 4+j8 & j5 \\ j5 & 4+j8 \end{bmatrix}\begin{bmatrix} I_1 \\ I_2 \end{bmatrix} = \begin{bmatrix} 10 \\ +j10 \end{bmatrix}$$

$\therefore$
$$I_1 = \frac{\begin{vmatrix} 10 & j5 \\ j10 & 4+j8 \end{vmatrix}}{\begin{vmatrix} 4+j8 & J5 \\ j5 & 4+j8 \end{vmatrix}} = \frac{120\angle 41.63°}{68\angle 109.75°}$$

$$I_1 = 1.77\angle -68.12°$$

$$I_2 = \frac{\begin{vmatrix} 4+j8 & 10 \\ j5 & j10 \end{vmatrix}}{\begin{vmatrix} 4+j8 & j5 \\ j5 & 4+j8 \end{vmatrix}} = \frac{80\angle -172.87°}{68\angle 109.75°}$$

or, $$I_2 = 1.176\angle -282.62°$$

∴ $$I' = I_1 + I_2 = 1.77\angle -68.12° + 1.176\angle -282.64°$$

$$I' = 1.04\angle -28.0°$$

∴ $$Z' \text{ (The venin resistance)} = \frac{V'}{I'}$$

or, $$Z' = \frac{7.07\angle 45°}{1.04\angle -28°}$$

$$Z' = 6.80\angle 73° \textbf{ Ans.}$$

Problem 3.81. A Y-connected load, with $Z_A = 10\angle 0°$ Ω, $Z_B = 10\angle 60°$ Ω and $Z_C = 10\angle -60°$ Ω, is connected to a three-phase, three-wire, ABC system having effective line voltage 141.4 V. Find the load voltage V_{A0}, V_{Co} and the displacement neutral voltage V_{0N}. Construct a phasor diagram.

Solution:

Maximum line voltage = $141.4 \times \sqrt{2} = 200$ V.

$$V_{AB} = 200\angle 120°$$

$$V_{BC} = 200\angle 0°$$

$$V_{CA} = 200\angle 240°$$

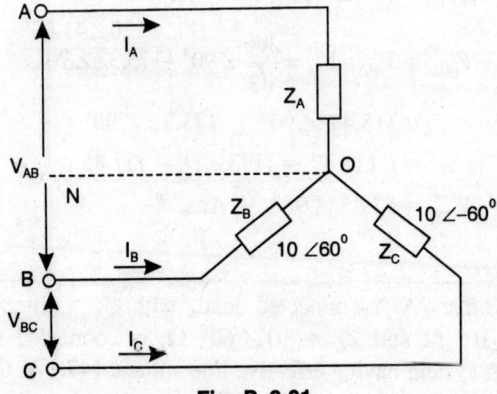

Fig. P. 3.81

In ABC sequences $\qquad V_{AN} = \dfrac{200}{\sqrt{3}} \angle 90°$

Appling node-voltage equation:

$$\frac{V_{OA} - V_{AB}}{Z_B} + \frac{V_{OA}}{Z_A} + \frac{V_{OA} + V_{AC}}{Z_C} = 0$$

or, $\qquad V_{OA}\left(\dfrac{1}{Z_A} + \dfrac{1}{Z_B} + \dfrac{1}{Z_C}\right) = \dfrac{V_{AB}}{Z_B} - \dfrac{V_{AC}}{Z_C}$

or, $\quad V_{OA}\left(\dfrac{1}{10\angle 0°} + \dfrac{1}{10\angle 60°} + \dfrac{1}{10\angle -60°}\right) = \dfrac{200\angle 120°}{10\angle 60°} - \dfrac{200\angle 240°}{10\angle -60°}$

or, $\qquad V_{OA}\left(\dfrac{1}{5\angle 0°}\right) = 20 \angle 60° - 20 \angle 300°$

$$= + 10 + j17.32 - 10 + 17.32 = j34.64$$

$$\frac{V_{OA}}{5\angle 0°} = 34.64 \angle 90°$$

or, $\qquad V_{OA} = 173.32 \angle 90°$ V

$$V_{OB} = V_{OA} - V_{AB} = 173 \angle 90° - 200 \angle 120°$$

$$= j173.32 + 100 - j173.32 = 100$$

$\therefore \qquad V_{OB} = 100 \angle 0°$ V

$$V_{OC} = V_{OA} + V_{AC} = 173 \angle 90° + 200 \angle 240°$$

$$= j173.32 - 100 - j173.32 = -100$$

$\therefore \qquad V_{OC} = 100 \angle 180°$ V

$$V_{ON} = V_{AN} - V_{OA} = \frac{200}{\sqrt{3}} \angle 90° - 173.32 \angle 90°$$

$$= 115.47 \angle 90° - 173.32 \angle 90°$$

$$= j\, 115.47 - j173.32 = - j57.85$$

$\therefore \qquad V_{ON} = 57.85 \angle 90°$ V **Ans.**

Problem 3.82. A Y-connected load, with $Z_A = 10 \angle -60°\ \Omega$, $Z_B = 10 \angle 10°\ \Omega$ and $Z_C = 10 \angle 60°\ \Omega$, is connected to a 3-φ, 3-wire, CBA system having effective line voltage 147.1 V. Obtain the line current I_A, I_B and I_C.

Solution:

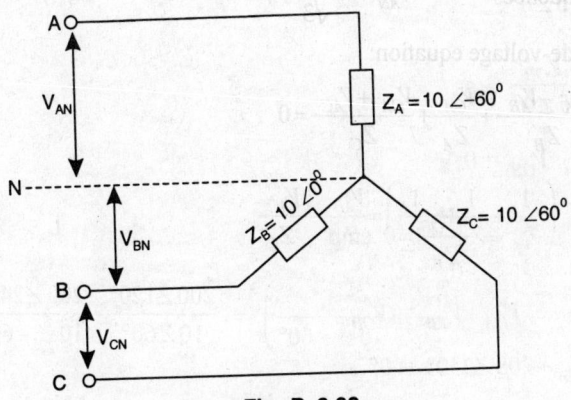

Fig. P. 3.82

In CBA system:

$$V_{AN} = \frac{V_L}{\sqrt{3}} \angle -90° = \frac{147.1 \times \sqrt{2}}{\sqrt{3}} \angle -90°$$

$$V_{BN} = \frac{V_L}{\sqrt{3}} \angle 30° = \frac{147.1 \times \sqrt{2}}{\sqrt{3}} \angle 30°$$

$$V_{CN} = \frac{V_L}{\sqrt{3}} \angle 150° = \frac{147.1 \times \sqrt{2}}{\sqrt{3}} \angle 150°$$

∴ $$V_{AN} = 120 \angle -90°, V_{BN} = 120 \angle 30°, V_{CN} = 120 \angle 150°$$

$$I_A = \frac{-V_{OA}}{Z_A}, I_B = \frac{-V_{OB}}{Z_B}, I_C = \frac{-V_{OC}}{Z_C}$$

Applying node-voltage method:

$$\frac{V_{OB} - V_{AB}}{Z_A} + \frac{V_{OB}}{Z_B} + \frac{V_{OB} + V_{BC}}{Z_C} = 0$$

or, $$V_{OB}\left(\frac{1}{Z_A} + \frac{1}{Z_B} + \frac{1}{Z_C}\right) = \frac{V_{AB}}{Z_A} - \frac{V_{BC}}{Z_C}$$

or, $$V_{OB}\left(\frac{1}{10\angle -60°} + \frac{1}{10\angle 0°} + \frac{1}{10\angle 60°}\right) = \frac{208\angle 240°}{10\angle -60°} - \frac{208\angle 0°}{10\angle 60°}$$

$$V_{OB}\left(\frac{1}{5\angle 0°}\right) = 20.8 \angle 300° - 20.8 \angle -60°$$

$$\frac{V_{OB}}{5\angle 0°} = 0$$

$\therefore$ $\qquad V_{OB} = 0$ V

$\therefore$ $\qquad I_B = \dfrac{-V_{OB}}{Z_B} = 0$ amp

From $\qquad V_{OA} + V_{AB} = V_{OB}$

or, $\qquad V_{OA} + 208 \angle 240° = 0°$

$$V_{OA} = 104 + j180$$

$$V_{OA} = 208 \angle 60°$$

$$I_A = \frac{-V_{OA}}{Z_A} = \frac{-208\angle 60°}{10\angle -60°} = \frac{208\angle 60°}{10\angle 120°}$$

$\therefore$ $\qquad I_A = 20.8 \angle -60°$

$$V_{OC} = V_{OB} - V_{CB} = 0 - 208 \angle 0° = -208$$

$$= 208 \angle 180°$$

$$I_C = +\frac{V_{OC}}{Z_C} = \frac{+208\angle 180°}{10\angle 60°} = \frac{208\angle 180°}{10\angle +60°}$$

$$I_C = 20.8 \angle 120°$$ A

or, I_C can be calculated as

$$I_A + I_B + I_C = 0$$

or, $\qquad I_C = -(I_A + I_B) = -(I_A) = -20.8 \angle -60°$

$$I_C = 20.8 \angle 120°$$ **Ans.**

Problem 3.83. A 3-ϕ, 3 wire, ABC system with a balanced load has effective line voltage 200 V and (maximum) line current $I_A = 13.61$ $\angle 60°$ A. Obtain the total power.

Solution:

$$V_{L,eff} = 200\,\text{V}, I_{L,eff} = \frac{13.61\angle 60°}{\sqrt{2}}$$

Total power $\quad P = \sqrt{3}\,V_{L,eff}\,I_{L,eff}\cos\theta = \sqrt{3}\times 200\times\frac{13.61}{\sqrt{2}}\cos 60°$

$$= 3333.75\ \cos 60° = 1666.87\ \text{W Ans.}$$

Reactive power $\quad \theta = \sqrt{3}\,V_{L,eff}\,I_{L,eff}\sin\theta = \sqrt{3}\times 200\times\frac{13.61}{\sqrt{2}}\sin 60°$

$$= 3333.75\ \sin 60° = 2887\ \text{Var Ans.}$$

Problem 3.84. Obtain the reading of two watt meters in a 3-ϕ, 3-wire system having effective line voltage 240 V and balanced, Δ-connected load impedances $20 \angle 80°$ Ω.

Solution:

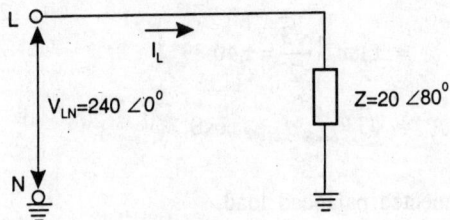

Fig. P. 3.84

$V_{L,eff} = 240$ V.

$$I_L = \frac{240\angle 0°}{20\angle 80°} = 12\angle -80°$$

$$I_L = 12\times\sqrt{3} = 20.8\ \text{amp}$$

$\therefore\qquad w_1 = V_{L,eff}\,I_{L,eff}\cos(\theta+30°)$

$$= 240 \times 20.8\ \cos(80+30°)$$

$$w_1 = -1706\ \text{W}$$

$$w_2 = V_{L,eff}\,I_{L,eff}\cos(\theta-30°)$$

$$= 240 \times 20.8\ \cos(80-30°)$$

$$w_2 = 3206\ \text{W Ans.}$$

Problem 3.85. Two watt meters in a 3-φ, 3-wire system with effective line voltage 120 V read 1500 W and 500 W. What is the impedance of the blanced Δ-connected load?

Solution:

In a two wattmeter:

$$\pm \tan\theta = \sqrt{3}\,\frac{w_1 - w_2}{w_1 + w_2}$$

where, $w_1 = 1500$ W

                $w_2 = 500$ W

$\therefore$ $\theta = \pm\tan^{-1}\left(\sqrt{3}\,\frac{w_1 - w_2}{w_1 + w_2}\right)$

                $= \pm\tan^{-1}\left(\sqrt{3}\,\dfrac{1000}{2000}\right)$

                $= \pm\tan^{-1}\dfrac{\sqrt{3}}{2} = \pm 40.89°$

$\because$ $P_T = \sqrt{3}\,V_{L,eff}\,I_{L,eff}\cos\theta$

For, delta connected balanced load.

$$Z_\Delta = \frac{\sqrt{3}\,V_{L,eff}}{I_{L,eff}}$$

$\therefore$ $P_T = \sqrt{3}\,V_{L,eff}\,\dfrac{\sqrt{3}\,V_{L,eff}}{Z_\Delta}\cos\theta$

or, $2000 = \dfrac{3\cdot(120)^2}{Z_\Delta}\cos 40.89°$

or, $Z_\Delta = \dfrac{3\times(120)^2\cos 40.89°}{2000}$

                $Z_\Delta = 16.3\ \Omega$

$\therefore$ $Z_\Delta = 16.3\angle\pm 40.89°$ **Ans.**

Problem 3.86. A three-phase, three wire, ABC system has effective line voltage 173.2 V. wattmeters in line *A* and *B* read –301 W and 1327 W, respectively. Find the impedance of the balanced Y-Connected load. (Since the sequence is specified, the sign of the impedance angle can be determined.)

Solution:

Since the phase squence is given ABC. Hence.

$$\theta = \pm \tan^{-1} \sqrt{3} \frac{w_1 - w_2}{w_1 + w_2}$$

$$\theta = \pm \tan^{-1} \sqrt{3} \frac{(-301 - 1327)}{(-301 + 1327)}$$

$$= \pm \tan^{-1} \sqrt{3} \left(\frac{-1628}{1026} \right) = \pm \tan^{-1} (-2.74)$$

$$\theta = -70°$$

For a Y connected load:

$$P_T = \frac{(V_{L,eff})^2}{Z_Y} \cos \theta$$

or, $$(-301 + 1327) = \frac{(173.2)^2}{Z_Y} \cos 70°$$

or, $$Z_Y = \frac{(173.2)^2}{(1026)} \cos 70°$$

$$Z_Y = 10 \Omega$$

∴ $$Z_Y = 10 \angle -70° \ \Omega \quad \textbf{Ans.}$$

Problem 3.87. A 3-φ, 3 wire system with a line voltage $V_{BC} = 339.4 \angle 0°$ V, has a balanced Y-connected load of $Z_Y = 15 \angle 60° \ \Omega$. The lines between the system and the load have impedances $2.24 \angle 26.57° \ \Omega$. Find the line voltage magnitude at the load.

Solution:

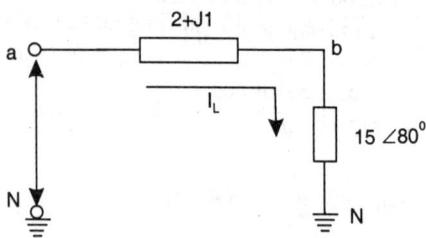

Fig. P. 3.87

Single-line equivalent circuit.

By voltage division, the voltage across the Y connected load is:

$$V_{BN} = \left(\frac{15\angle 60°}{2+j1+15\angle 60°}\right)\left(\frac{339.4\angle 0°}{\sqrt{3}}\right)$$

$$= (0.886\angle 4.16°)\left(\frac{339.4\angle 0°}{\sqrt{3}}\right)$$

$$V_{BN} = \left(\frac{301.0\angle 4.16°}{\sqrt{3}}\right)$$

$$\therefore \quad V_L = \frac{301.0}{\sqrt{3}}\times\sqrt{3} = 301.0\,\text{V}.$$

Problem 3.88. Repeat Problem 3.87 with the load impedance $Z_Y = 15\angle -60°\ \Omega$. By drawing the voltage phasor diagrams for the two cases, illustrate the effect of load impedance angle on the voltage drop for a given line impedance.

Solution:

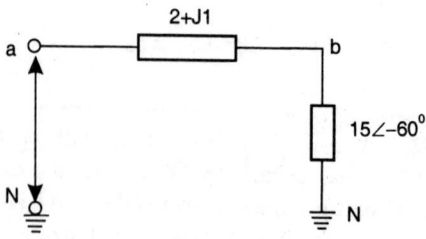

Fig. P. 3.88

Single line equivalent circuit: By voltage division, the voltage across the Y connected load is:

$$V_{BN} = \left(\frac{15\angle 60°}{2+j1+15\angle 60°}\right)\left(\frac{339.4\angle 0°}{\sqrt{3}}\right)$$

$$= \left(\frac{7.5-j13}{9.5-j12}\right)\left(\frac{339.4}{\sqrt{3}}\angle 0°\right)$$

$$= \left(\frac{15\angle -60°}{15.30\angle -51°}\right)\left(\frac{339.4}{\sqrt{3}}\angle 0°\right)$$

$$= (0.98\angle -2°)\left(\frac{339.4}{\sqrt{3}}\angle 0°\right)$$

$$V_{BN} = \frac{332.74}{\sqrt{3}}\angle -2°$$

$$\therefore \quad V_L = \frac{332.74}{\sqrt{3}}\times\sqrt{3}$$

$$\therefore \quad V_L = 332.74 \text{ V } \textbf{Ans.}$$

Problem 3.89. Apply the mesh current method to the network of Fig. P. 3.89 and write the matrix equations by inspection. Obtain current I_1 by expanding the numerator determinant about the column containing the voltage sources to show that each source supplies a current of 2.13 A.

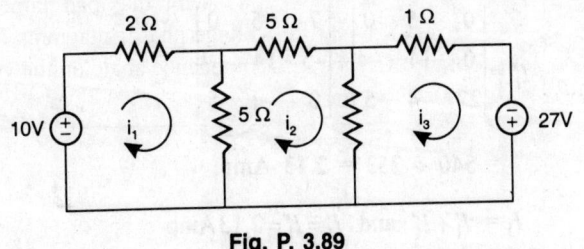

Fig. P. 3.89

Solution:

Applying KVL to each mesh:

$$2I_1+5(I_1-I_2) = 10 \text{ V} \tag{1}$$

$$5I_2+4(I_2-I_3)+5(I_2-I_1) = 0 \tag{2}$$

$$I_3 + 4(I_3 - I_2) = 27 \text{ V} \tag{3}$$

In the matrix form:

$$\begin{bmatrix} 7 & -5 & 0 \\ -5 & 14 & -4 \\ 0 & -4 & 5 \end{bmatrix} \begin{bmatrix} I_1 \\ I_2 \\ I_3 \end{bmatrix} = \begin{bmatrix} 10 \\ 0 \\ 27 \end{bmatrix}$$

Using Cramer's rule to find I_1

$$I_1 = \begin{vmatrix} 10 & -5 & 0 \\ 0 & 14 & -4 \\ 27 & -4 & 5 \end{vmatrix} \div \begin{vmatrix} 7 & -5 & 0 \\ -5 & 14 & -4 \\ 0 & -4 & 5 \end{vmatrix} = 1080 \div 253$$

$I_1 = 4.26$ Amp

When only 10 V source is acting, then

$$I_1' = \begin{vmatrix} 10 & -5 & 0 \\ 0 & 14 & -4 \\ 0 & -4 & 5 \end{vmatrix} \div \begin{vmatrix} 7 & -5 & 0 \\ -5 & 14 & -4 \\ 0 & -4 & 5 \end{vmatrix}$$

$$= 540 \div 253 = 2.13 \text{ Amp.}$$

When only 27 V source is acting, then

$$I_1'' = \begin{vmatrix} 0 & -5 & 0 \\ 0 & 14 & -4 \\ 27 & -4 & 5 \end{vmatrix} \div \begin{vmatrix} 7 & -5 & 0 \\ -5 & 14 & -4 \\ 0 & -4 & 5 \end{vmatrix}$$

$$= 540 \div 253 = 2.13 \text{ Amp.}$$

$\therefore \qquad I_1 = I_1' + I_1''$ and $I_1' = I_1'' = 2.13$ Amp

Hence each source supplies a current of 2.13 A. **Ans.**

Problem 3.90. Loop currents are shown in the network of Fig. P. 3.90. Write the matrix equation and solve for the three currents.

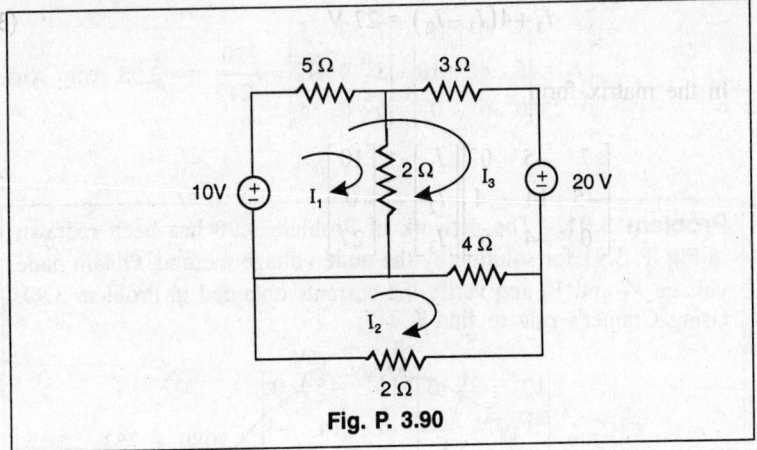

Fig. P. 3.90

Solution:

Applying KVL in each mesh:

$$7I_1 + 5I_3 = 10 \text{ V} \tag{1}$$

$$5I_1 + 4I_2 + 12I_3 + 20 = 10 \text{ V} \tag{2}$$

$$6I_2 + 4I_3 = 0 \text{ V} \tag{3}$$

In the matrix form:

$$\begin{bmatrix} 7 & 0 & 5 \\ 5 & 4 & 12 \\ 0 & 6 & 4 \end{bmatrix} \begin{bmatrix} I_1 \\ I_2 \\ I_3 \end{bmatrix} = \begin{bmatrix} 10 \\ -10 \\ 0 \end{bmatrix}$$

$$\therefore \quad I_1 = \begin{vmatrix} 10 & 0 & 0 \\ -10 & 4 & 12 \\ 0 & 6 & 4 \end{vmatrix} \div \begin{vmatrix} 7 & 0 & 5 \\ 5 & 4 & 12 \\ 0 & 6 & 4 \end{vmatrix} = (-860) \div (-242)$$

$$= \frac{-860}{-242} = 3.55 \text{A}$$

$$\therefore \quad I_1 = 3.55 \text{ A}$$

$$I_2 = \begin{vmatrix} 7 & 10 & 5 \\ 5 & -10 & 12 \\ 0 & 0 & 4 \end{vmatrix} \div \begin{vmatrix} 7 & 0 & 5 \\ 5 & 4 & 12 \\ 0 & 6 & 4 \end{vmatrix} = -480 \div (-242)$$

$$I_2 = 1.98 \text{ A}$$

$$I_3 = \begin{vmatrix} 7 & 0 & 10 \\ 5 & 4 & -10 \\ 0 & 6 & 0 \end{vmatrix} \div \begin{vmatrix} 7 & 0 & 5 \\ 5 & 4 & 12 \\ 0 & 6 & 4 \end{vmatrix} = \frac{720}{-242} = -2.98 \text{ Amp } \textbf{Ans.}$$

Problem 3.91. The network of Problem 3.90 has been redrawn in Fig. P. 3.91 for solution by the node voltage method. Obtain node voltage V_1 and V_2 and verify the currents obtained in Problem 3.90.

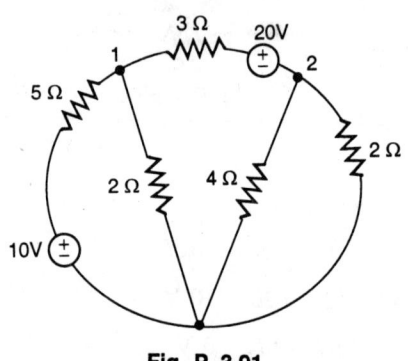

Fig. P. 3.91

Solution:

Applying KCL at node 1.

$$\frac{V_1 - 10}{5} + \frac{V_1}{2} + \frac{V_1 - 20 - V_2}{3} = 0 \qquad (1)$$

at node 2: $$\frac{V_2 + 20 - V_1}{3} + \frac{V_2}{4} + \frac{V_2}{2} = 0 \qquad (2)$$

Rearranging eqs (1) and (2), we get:

$$\frac{30}{30} V_1 - \frac{1}{3} V_2 = \frac{26}{3} \qquad (1)$$

and $$\frac{-1}{3} V_1 + \frac{13}{12} V_2 = \frac{-20}{3} \qquad (2)$$

In the matrix form:

$$\begin{bmatrix} \dfrac{31}{30} & -\dfrac{1}{3} \\ -1/3 & 13/12 \end{bmatrix} \begin{bmatrix} V_1 \\ V_2 \end{bmatrix} = \begin{bmatrix} \dfrac{26}{3} \\ \dfrac{-20}{3} \end{bmatrix}$$

Using Cramer's rule to find V_1 and V_2:

$$V_1 = \begin{bmatrix} \dfrac{26}{3} & -\dfrac{1}{3} \\ \dfrac{-20}{3} & \dfrac{13}{12} \end{bmatrix} \div \begin{bmatrix} \dfrac{31}{30} & -\dfrac{1}{3} \\ -1/3 & 13/12 \end{bmatrix}$$

$$= \left(\dfrac{169}{18} - \dfrac{20}{9} \right) \div \left(\dfrac{403}{360} - \dfrac{1}{9} \right)$$

$$V_1 = 7.10 \text{ V}$$

$$V_2 = \begin{vmatrix} \dfrac{31}{30} & \dfrac{26}{3} \\ -1/3 & \dfrac{-20}{3} \end{vmatrix} \div \begin{vmatrix} \dfrac{31}{30} & -\dfrac{1}{3} \\ -1/3 & 13/12 \end{vmatrix}$$

$$V_2 = -3.96 \text{ V}$$

For varifying current:

current in 5 Ω resistance

$$= \frac{10 - 7.10}{5} = 0.57 \text{ amp}$$

from nodal voltage method.

and from mesh current method, current in 5 Ω resistance is $= I_1 + I_3$ $= 3.55 + (-2.98)$ A $= 0.57$ amp. Hence in both the case current obtained is 5 amp. **Ans.**

Problem 3.92. In the network shown in Fig. P. 3.92, current $I_0 = 7.5$ mA. Use mesh currents to find the required source voltage V_S.

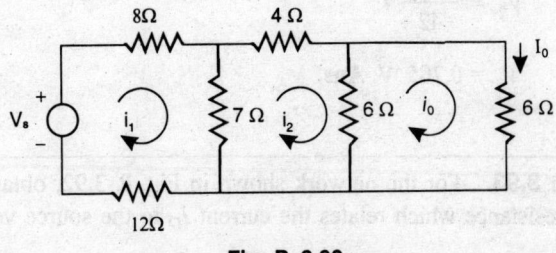

Fig. P. 3.92

Solution:

Applying KVL in the mesh 1, mesh 2 and mesh 3.

$$8i_1 + 7(i_1 - i_2) + 12i_1 = V_S \tag{1}$$

$$4i_2 + 6(i_2 - i_0) + 7(i_2 - i_1) = 0 \tag{2}$$

$$6i_0 + 6(i_0 - i_2) = 0 \tag{3}$$

In the matrix form:

$$\begin{bmatrix} 27 & -7 & 0 \\ -7 & 17 & -6 \\ 0 & -6 & 12 \end{bmatrix} \begin{bmatrix} I_1 \\ I_2 \\ I_0 \end{bmatrix} = \begin{bmatrix} V_S \\ 0 \\ 0 \end{bmatrix}$$

Using Cramer's rule:

$$I_0 = \begin{vmatrix} 27 & -7 & V_S \\ -7 & 17 & 0 \\ 0 & -6 & 0 \end{vmatrix} \div \begin{vmatrix} 27 & -7 & 0 \\ -7 & 17 & -6 \\ 0 & -6 & 12 \end{vmatrix}$$

$$= \frac{42V_S}{4536 - 588} \text{ amp} = \frac{42V_S}{3948} \text{ amp}$$

$$I_0 = 7.5\,\text{mA} = \frac{42V_S}{3948}$$

or, $$7.5 \times 10^{-3} = \frac{42V_S}{3948}$$

$$V_S = \frac{3948 \times 7.5 \times 10^{-3}}{42}\,\text{V}$$

or, $$V_S = \frac{29.61}{42}\,\text{V}$$

or, $$V_3 = 0.705 \text{ V } \textbf{Ans.}$$

Problem 3.93. For the network shown in Fig. P. 3.92, obtain the transfer resistance which relates the current I_O to the source voltage V_S.

Solution:

Transfer resistance $\qquad (r_t) = \dfrac{V_s}{I_O}$

or, $\qquad\qquad\qquad r_t = \dfrac{0.705\,V}{7.5\,mA}$

or, $\qquad\qquad\qquad r_t = 94\ \Omega\ $ **Ans.**

Problem 3.94. For the network shown in fig 3.94. Obtain the mesh current.

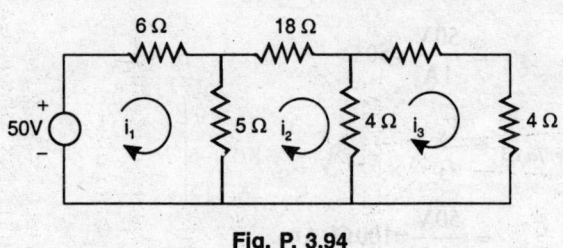

Fig. P. 3.94

Solution:

$$6i_1 + 5(i_1 - i_2) = 50 \qquad\qquad (1)$$
$$18i_2 + 4(i_2 - i_3) + 5(i_2 - i_1) = 0 \qquad\qquad (2)$$
$$4i_3 + 4(i_3 - i_2) = 0 \qquad\qquad (3)$$

In the matrix form:

$$\begin{bmatrix} 11 & -5 & 0 \\ -5 & 27 & -4 \\ 0 & -4 & 8 \end{bmatrix} \begin{bmatrix} I_1 \\ I_2 \\ I_3 \end{bmatrix} = \begin{bmatrix} 50 \\ 0 \\ 0 \end{bmatrix}$$

$$\therefore \qquad I_1 = \begin{vmatrix} 50 & -5 & 0 \\ 0 & 27 & -4 \\ 0 & -4 & 8 \end{vmatrix} \div \begin{vmatrix} 11 & -5 & 0 \\ -5 & 27 & -4 \\ 0 & -4 & 8 \end{vmatrix}$$

$$= 10{,}000 \div 2{,}000 = 5\ A$$

or, $\qquad\qquad I_1 = 5\ A$

Simiarly $\qquad I_2 = 1A$ and $I_3 = 0.5\ A$ **Ans.**

Problem 3.95. Using the matrices from Problem 3.94. Calculate $R_{input,1}$, $R_{transfer\,12}$, and $R_{transfer\,13}$.

Solution:

$$R_{input,1} = \frac{V_S}{I_1}$$

$$= \frac{50\,\text{V}}{5\,\text{A}} = 10\,\Omega$$

$$R_{transfer\,12} = \frac{V_S}{I_2}$$

$$= \frac{50\,\text{V}}{1\,\text{A}} = 50\,\Omega$$

$$R_{transfer\,1.3} = \frac{V_S}{I_3}$$

$$= \frac{50\,\text{V}}{0.5} = 100\,\Omega \quad \textbf{Ans.}$$

Problem 3.96. In the network shown in Fig. P. 3.96. Obtain the mesh currents.

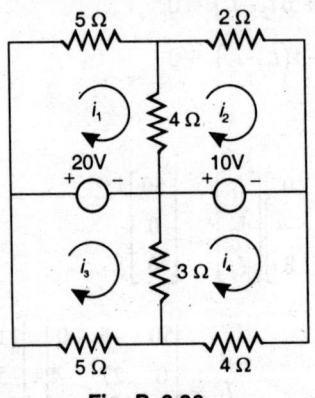

Fig. P. 3.96

Solution:

Applying KVL in all the mesh:

$$5i_1 + 4\left(i_1 - i_2\right) = 20\,\text{V} \tag{1}$$

$$2i_2 + 4(i_2 - i_1) = -10 \text{ V} \tag{2}$$

$$4(i_3 - i_4) + 5i_3 = -20 \text{ V} \tag{3}$$

$$4i_4 + 3(i_4 - i_2) = 10 \text{ V} \tag{4}$$

In the matrix form:

$$\begin{bmatrix} 9 & -4 & 0 & 0 \\ -4 & 6 & 0 & 0 \\ 0 & 0 & 8 & -3 \\ 0 & -3 & 0 & 7 \end{bmatrix} \begin{bmatrix} I_1 \\ I_2 \\ I_3 \\ I_4 \end{bmatrix} = \begin{bmatrix} 20 \\ -10 \\ -20 \\ 10 \end{bmatrix}$$

$$\Delta = \begin{bmatrix} 9 & -4 & 0 & 0 \\ -4 & 6 & 0 & 0 \\ 0 & 0 & 8 & -3 \\ 0 & -3 & 0 & 7 \end{bmatrix} = 3024 - 896 = 2128$$

$$\Delta_1 = \begin{bmatrix} 20 & -4 & 0 & 0 \\ -10 & 6 & 0 & 0 \\ -20 & 0 & 8 & -3 \\ 10 & -3 & 0 & 7 \end{bmatrix} = 6720 - 2240 = 4480$$

$$\Delta_2 = \begin{bmatrix} 9 & 20 & 0 & 0 \\ -4 & -10 & 0 & 0 \\ 0 & -20 & 8 & -3 \\ 0 & 10 & 0 & 7 \end{bmatrix} = -560$$

$$\Delta_3 = \begin{bmatrix} 9 & -4 & 20 & 0 \\ -4 & 6 & -10 & 0 \\ 0 & 0 & -20 & -3 \\ 0 & -3 & 10 & 7 \end{bmatrix} = -4980$$

$$\Delta_4 = \begin{bmatrix} 9 & -4 & 0 & 20 \\ -4 & 6 & 0 & -10 \\ 0 & 0 & 8 & -20 \\ 0 & -3 & 0 & 10 \end{bmatrix} = 906$$

$$I_1 = \frac{\Delta_1}{\Delta} = \frac{4480}{2128} = 2.10\,\text{A}$$

$$I_2 = \frac{\Delta_2}{\Delta} = \frac{-560}{2128} = -0.263\,\text{A}$$

$$I_3 = \frac{\Delta_3}{\Delta} = \frac{-4980}{2128} = -2.34\,\text{A}$$

$$I_4 = \frac{\Delta_4}{\Delta} = \frac{906}{2128} = 0.425\,\text{A} \quad \textbf{Ans.}$$

Problem 3.97. Use the node voltage method to obtain $V_{o.c}$ and $I_{s.c}$ at the terminals ab of the network shown in Fig. P. 3.97, consider 'a' positive with respect to 'b'

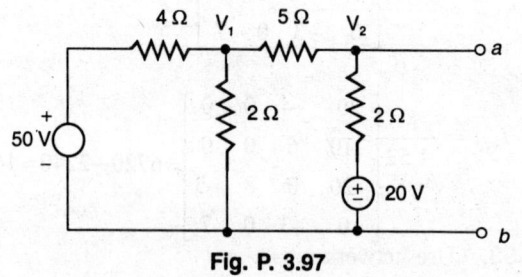

Fig. P. 3.97

Solution:

Applying KCL at node 1.

$$\frac{V_1 - 50}{4} + \frac{V_1}{2} + \frac{V_1 - V_2}{5} = 0 \tag{1}$$

and at node 2

$$\frac{V_2 - V_1}{5} + \frac{V_2 + 20}{2} = 0 \tag{2}$$

In the matrix form:

$$\begin{bmatrix} \dfrac{19}{20} & -\dfrac{1}{5} \\ -\dfrac{1}{5} & \dfrac{7}{10} \end{bmatrix} \begin{bmatrix} V_1 \\ V_2 \end{bmatrix} = \begin{bmatrix} \dfrac{50}{4} \\ -10 \end{bmatrix}$$

$$\therefore \qquad V_{o \cdot c} = V_2 = \begin{vmatrix} \dfrac{19}{20} & \dfrac{50}{4} \\ -\dfrac{1}{5} & -10 \end{vmatrix} \div \begin{vmatrix} \dfrac{19}{20} & -\dfrac{1}{5} \\ -\dfrac{1}{5} & \dfrac{7}{10} \end{vmatrix}$$

$$= (-7) \div 0.625 = -11.2 \text{ V}$$

$$I_{s \cdot c} = \dfrac{V_{o \cdot c}}{R_{Th}}$$

where R_{Th} = Thevenin resistance across a–b.

$$R_{Th} = \left((4\,\Omega \| 2\,\Omega) + 5\,\Omega \right) \| 2\,\Omega$$

$$= \dfrac{\left(\dfrac{4 \times 2}{6} + 5 \right) \times 2}{\dfrac{4 \times 2}{6} + 5 + 2} = 1.52\,\Omega$$

$$\therefore \qquad I_{s \cdot c} = \dfrac{-11.2}{1.52} = -7.368\,\text{Amp.}$$

Prolem 3.98. Use network reduction to obtain the current in each of the resistors in the circuits shown in Fig. P. 3.98.

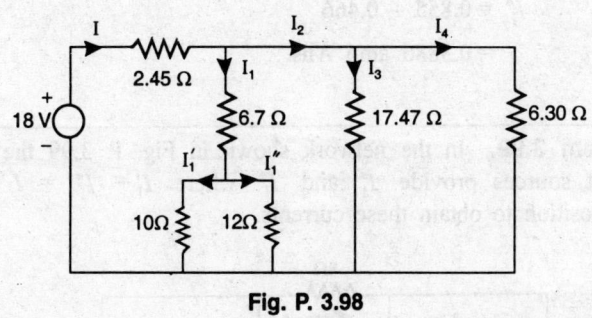

Fig. P. 3.98

Solution:

For 10 Ω and 12 Ω resistance in parallel $R'_{eq} = \dfrac{10 \times 12}{10 + 12} = 5.45\,\Omega$. This is in series with 6.7 Ω. So, $R_1 = 6.7 + 5.45 = 12.15\,\Omega$.

The 12.15 Ω, 17.44Ω and 6.30 Ω are in parallel hence $R''_{eq} = 3.344\,\Omega$. This R'_{eq} is in series with 2.45 Ω . Hence total equivalent resistance is equal to = 3.344 + 2.45 = 5.79 Ω.

$$\therefore \qquad I = \frac{18V}{5.79} = 3.10\,\text{amp}$$

at node 1, $V_1 = 18\ \text{V} - 2.45 \times 3.10\ \text{V}$

$$= (18 - 7.61)\ \text{V}$$

$$= 10.389\ \text{V}$$

$$\therefore \qquad I_3 = \frac{V_1}{17.47} = \frac{10.389}{17.47}$$

$$= 0.59\ \text{amp}$$

$$I_4 = \frac{V_1}{6.30} = \frac{10.389}{6.3}$$

$$= 1.649\ \text{amp}$$

$$I_1 = \frac{V_1}{R_1} = \frac{10.389}{12.15}$$

$$= 0.855\ \text{amp}$$

$$I'_1 = \frac{0.855}{10+12} \times 12$$

$$= 0.466\ \text{amp}$$

$$I''_1 = 0.855 - 0.466$$

$$= 0.3886\ \text{amp } \textbf{Ans.}$$

Problem 3.99. In the network shown in Fig. P. 3.99 the two current sources provide I'_1 and I''_1 where $I'_1 + I''_1 = I$. Use superposition to obtain these currents.

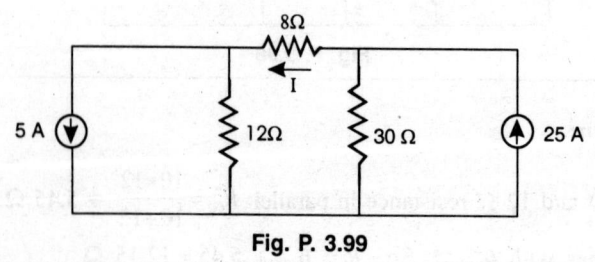

Fig. P. 3.99

Solution:

Using superposition, current in 8 Ω (I') due to 25 A is given by:

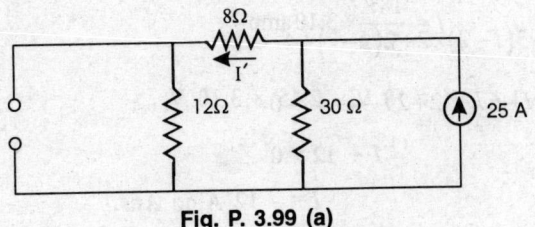

Fig. P. 3.99 (a)

$$I_1' = \frac{25}{(8+12)+30} \times 30 = 15\,\text{A}$$

and current in 8Ω (I_1'') due to 5A is given by:

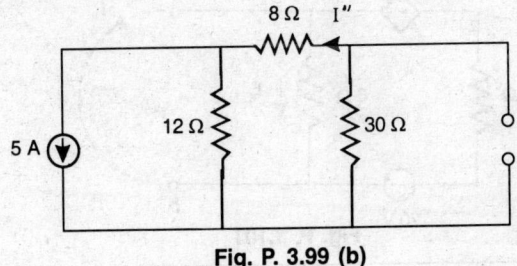

Fig. P. 3.99 (b)

$$I_1'' = \frac{5}{50} \times 12 = 1.2\,\text{A}$$

∴ Total current $\quad I = I_1' + I_1'' = 15 + 1.2 = 16.2\,\text{A}$ **Ans.**

Problem 3.100. Obtain the current I in the network shown in Fig. P. 3.100.

Fig. P. 3.100

Solution:

Applying KVL in loop 2

$$5I - 3V_R + 2(I-2) + 4 = 0$$

or, $5I - 3(I-2) \times 2 + 2(I-2) + 4 = 0$

or, $5I - 6I + 12 + 2I - 4 + 4 = 0$

$$I + 12 = 0$$

or, $I = -12$ Amp **Ans.**

Problem 3.101.   Obtain the Thevenin and Norton equivalents for the network shown in Fig. P. 3.101.

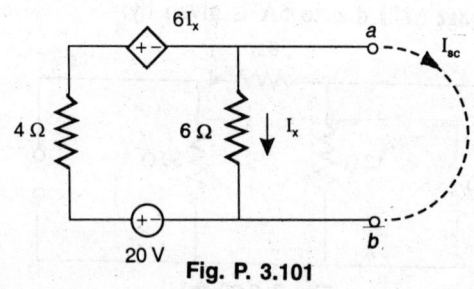

Fig. P. 3.101

Solution:

For finding Thevenin voltage:

$$V_{o \cdot c} = 6I_x \tag{1}$$

$$-20 + 4I_x - 6I_x + V_{o \cdot c} = 0 \tag{2}$$

or, $\cdot$ $-2I_x + V_{o \cdot c} = 20$ V

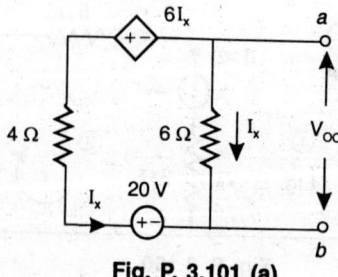

Fig. P. 3.101 (a)

from eq (1) $\qquad I_x = \dfrac{V_{o \cdot c}}{6}$

$\therefore$ Now eq (2) becomes.

$$-2\frac{V_{o \cdot c}}{6} + V_{o \cdot c} = 20 \text{ V}$$

or, $\qquad \dfrac{2V_{o \cdot c}}{3} = 20 \text{ V}$

$$= 3 \times 10 = 30 \text{ V}$$

$\therefore$ $V' = 30$ V Thevenin voltage for finding short circuit current:

When short circuit current $I_{s \cdot c}$ flows from a to b. then $I_x = 0$, i.e. $6 I_x = 0$ so voltage source of $6 I_x$ becomes 0, i.e. shorted.

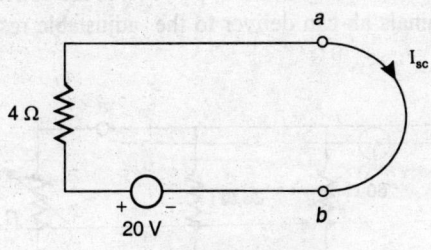

Fig. P. 3.101 (b)

$\therefore$ $\quad I_{s \cdot c} = \dfrac{20}{4}, I_{s \cdot c} = 5 \text{A}$

$\therefore$ $\quad I' = 5$ A Norton current.

So, Thevenin eq is:

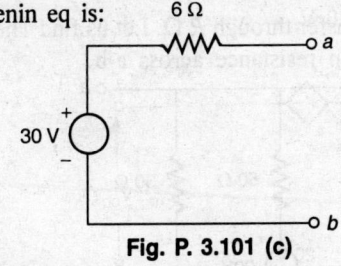

Fig. P. 3.101 (c)

For finding $R_{Th} = R'$

$$R' = \frac{V_{o \cdot c}}{I_{s \cdot c}}$$

$$= \frac{30}{5} = 6\,\Omega$$

and Norton eq is:

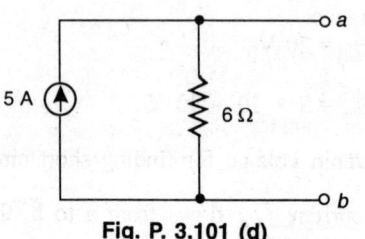

5 A   6 Ω   **Ans.**

Fig. P. 3.101 (d)

Problem 3.102.   Find the maximum power that the active network to the left of terminals ab can deliver to the adjustable resistor R in Fig. P. 3.102.

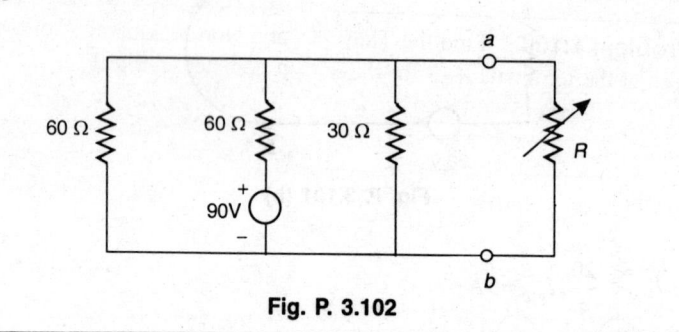

60 Ω   60 Ω   30 Ω   R

90V

Fig. P. 3.102

Solution:

For maximum power transfer through R Ω. Let us find Thevenin voltage across a–b and Thevenin resistance across a-b

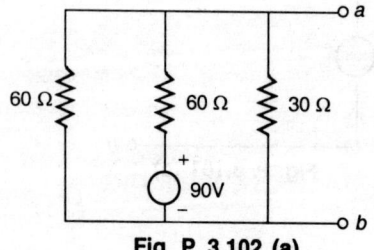

60 Ω   60 Ω   30 Ω

90V

Fig. P. 3.102 (a)

$$I_1 = \frac{90}{(60\,\|\,30)+60} = \frac{90}{80} = \frac{9}{8}\,\text{amp}$$

$$I'' = \frac{9/8}{90} \times 60 = \frac{3}{4}\,\text{amp.}$$

$$\therefore \qquad V' = \frac{3}{4} \times 30 = 22.5\,\text{V} \quad \text{Thevenin voltage.}$$

For, Thevenin resistance:

$$R' = 60\,\|\,60\,\|\,30$$

$$R' = 15\;\Omega$$

For maximum power transfer

$$R = R' = 15\;\Omega.$$

So, $$P_{max} = \frac{V'^2}{4R} = \frac{V'^2}{4R'}$$

or, $$P_{max} = \frac{(22.5)^2}{4 \times 15} = 8.43\,\text{W} \quad \textbf{Ans.}$$

Problem 3.103. Find the Thevenin and Norton equivalent circuits seen at the terminals A–B for the circuit depicted in Fig. P. 3.103 (a).

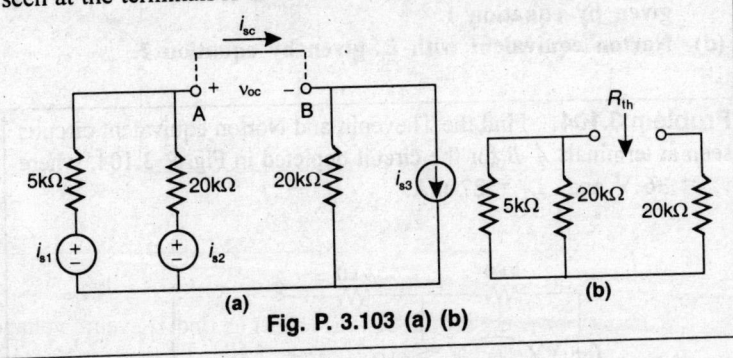

Fig. P. 3.103 (a) (b)

Solution:

To compute R_{th}, we set all source values zero. Each voltage source becomes a short and each current source becomes an open. The resultant circuit is shown in Fig. P. 3.103 (b) Since 5 kΩ in parallel with 20 kΩ is 4 kΩ, $R_{th} = 24$ kΩ.

Now. To compute $v_{oc}(t)$, we use nodal analysis. At node B,

$$v_B = -20 \times 10^3\, i_{s3}$$

and at node A, $\frac{1}{5\times10^3}(v_A-v_{s1})+\frac{1}{20\times10^3}(v_A+v_{s2})=0$

Multiplying through by 20×10^3 and manipulating leads to

$$v_A = 0.8v_{s1}-0.2v_{s2}$$

Hence, $v_{oc} = v_A-v_B=0.8v_{s1}-0.2v_{s2}+20\times10^3 i_{s3}$ (1)

Equation (1) and $R_{th} = 24$ kΩ implying the Thevenin equivalent shown in Fig. P. 3.103 (b). From the source transformation theorem, we have

$$i_{sc}(t) = \frac{v_{oc}(t)}{R_{th}}=\frac{1}{3}\times10^{-4}v_{s1}-\frac{1}{12}\times10^{-4}v_{s2}+\frac{5}{6}i_{s3}$$ (2)

Equation 3.103 (c) leads to the Norton equivalent circuit illustrated in Fig. P. 3.103 (d).

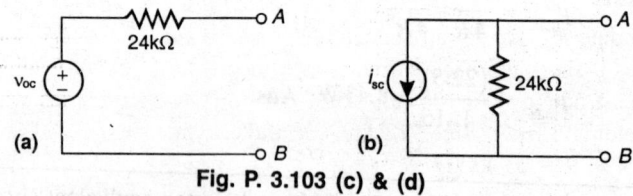

Fig. P. 3.103 (c) & (d)

(c) **Thevenin equivalent of circuit of Fig. P. 3.103 (a) with v_{oc} given by equation 1.**

(d) **Norton equivalent with i_{sc} given by equation 2.**

Problem 3.104.   Find the Thevenin and Norton equivalent circuits seen at terminals A–B for the circuit depicted in Fig. P. 3.104, where $v_{s1} = 36$ V and $i_{s2} = 27$ mA.

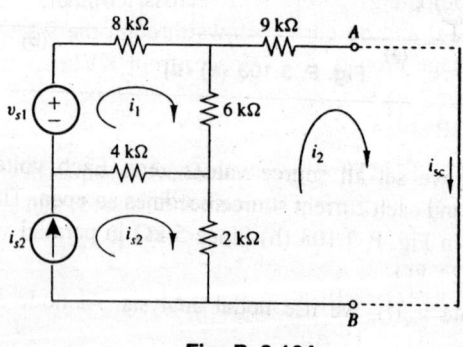

Fig. P. 3.104

Two-source circuit with loop currents shown: with 100
Currents $v_{s1} = 36$ V and $i_{s2} = 27$ mA.

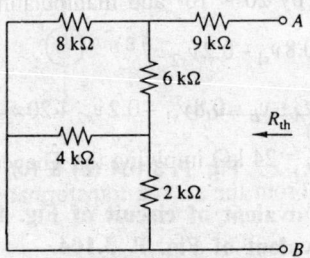

Fig. P. 3.104 (a)

Circuit of Fig. P. 3.104 with all independent sources deactivated.

Solution:

To compute R_{th}, we set all source values to zero. Each voltage source becomes a short and each current source becomes an open, leading to the circuit of Fig. P. 3.104 (a). Here we have 4 kΩ in series with 8 kΩ, yielding 12 kΩ. Since 12 kΩ in parallel with 6 kΩ is 4 kΩ, R_{th} = 2 + 4 + 9 = 15 kΩ.

To compute $v_{oc}(t)$ for this example, we use loop analysis with loops as drawn in Fig. P. 3.104. Around loop 1,

$$v_{s1} = 18 \times 10^3 i_1 - 4 \times 10^3 i_2$$

Since $v_{s1} = 36$ V and $i_{s2} = 27$ mA, we have,

$$i_1 = \frac{1}{18 \times 10^3} 36 + \frac{4 \times 10^3}{18 \times 10^3} 27 \times 10^{-3} = 8 \times 10^{-3} \text{ A}$$

Because we are computing v_{oc}, the short across terminals A–B is not present. Hence $i_3 = 0$ and no current flows through the 9-kΩ resistor. This means its voltage drop is zero. Hence, from KVL,

$$v_{oc} = 6 \times 10^3 i_1 + 2 \times 10^3 i_{s2} = 102 \text{ V} \qquad (1)$$

Equation (1) and R_{th} = 15 kΩ imply that the Thevenin equivalent is given by Fig. P. 3.104 (b). Furthermore, from the source transformation theorem, we have,

$$i_{sc}(t) = \frac{v_{oc}(t)}{R_{th}} = \frac{102}{15 \times 10^3} = 6.8 \times 10^{-3}$$

62

Equation 2 leads to Norton equivalent circuit shown in Fig. P. 3.104 (c).

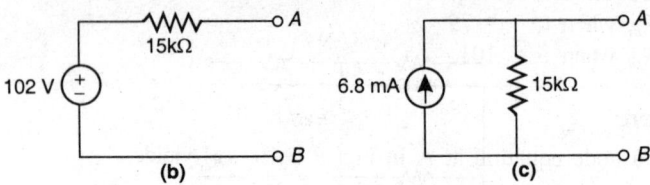

Fig. P. 3.104 (b) & (c)

(a) **Thevenin equivalent of circuit of Fig. P. 3.104.**
(b) **Norton equivalent of Fig. P. 3.104.**

Referring again to Fig. P. 3.104. and assuming that the short across A–B is present, then $i_2 = i_{sc}$. Hence, around loop 1,

$$v_{s1} = 18 \times 10^3 \, i_1 - 4 \times 10^3 \, i_{s2} - 6 \times 10^3 \, i_{sc}$$

Hence, $144 = 18 \times 10^3 \, i_1 - 6 \times 10^3 \, i_{sc}$ (36)

Summing the voltage around loop 2, we get

$$54 = -6 \times 10^3 \, i_1 + 17 \times 10^3 \, i_{sc}$$ (64)

Putting in matrix form and solving, we get

$$\begin{bmatrix} i_1 \\ i_{sc} \end{bmatrix} = \left[10^3 \begin{bmatrix} 18 & -6 \\ -6 & 17 \end{bmatrix} \right]^{-1} \begin{bmatrix} 144 \\ 54 \end{bmatrix} = \begin{bmatrix} 10.27 \\ 6.8 \end{bmatrix} \times 10^{-3}$$

This shows that $i_{sc} = 6.8$ mA

Prolem 3.105.   Find the Thevenin equivalent circuit for the two-terminal network (marked by dashed box) in Fig. P. 3.105 (a) for two cases. (a) $\mu = -79$, and (b) $\mu = 101$. (The dependent source acts as a voltage amplifier.)

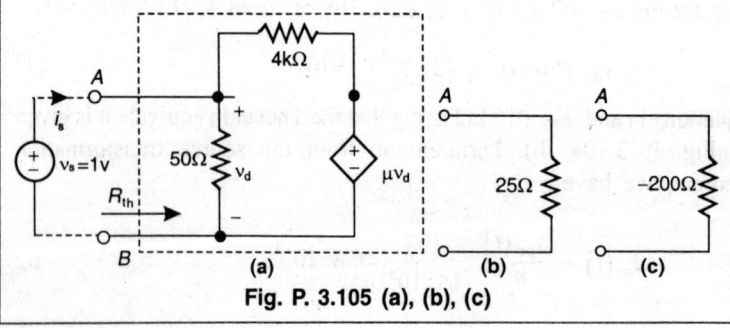

Fig. P. 3.105 (a), (b), (c)

(a) Circuit with an applied 1-V external source.
(b) R_{th} when $\mu = -79$.
(c) R_{th} when $\mu = 101$.

Solution:

Writing a node equation at A in Fig. P. 3.105 (a) yields

$$i_s = \frac{v_s}{50} + \frac{v_s - \mu v_d}{4000} = \frac{v_s}{50} + \frac{(1-\mu)v_s}{4000} = \frac{20 + 0.25(1-\mu)}{10^3}$$

Therefore, $\quad R_{th} = \frac{v_s}{i_s} = \frac{10^3}{20 + 0.25(1-\mu)}$ \hfill (1)

Using eq^n (1), we have

(a) For $\mu = -79$, $R_{th} = 25\ \Omega$, as in Fig. P. 3.105 b
(b) For $\mu = 101$, $R_{th} = -200\ \Omega$, as in Fig. P. 3.105 c

Observe that with dependent sources, R_{th} can be negative or positive. A negative resistance delivers energy to a load whereas a positive resistance absorbs energy.

Problem 3.106. Find the Thevenin equivalent circuit of the two-terminal network seen at terminals C–D of Fig. P. 3.106. In addition to finding the Thevenin equivalent, find the voltage gain v_L/v_S when a load R_L is connected acrosss C–D.

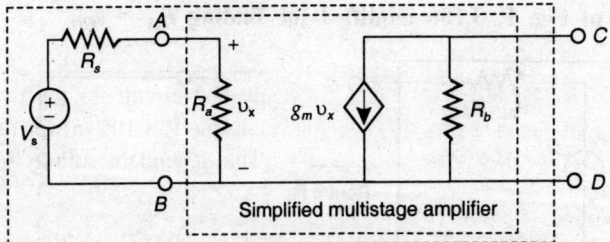

Fig. P. 3.106

Two-terminal network cosisting of a highly simplified multistage amplifier driven by a practical source that might represent a microphone.

Solution:

From Fig. P. 3.106 and voltage division,

$$v_x = \frac{R_a}{R_s + R_a} v_s$$

From Ohm's law and the definition of v_{oc}, we have,

$$v_{oc} = v_{CD} = -g_m R_b v_x = -g_m R_b \left(\frac{R_a}{R_s + R_a} \right) v_s = K v_s \quad (1)$$

Now deactivate the independent voltage source and excite terminals C–D with a fictitious 1-A current source then Fig. P. 3.106 (a) illustrates the resulting configuration.

By inspection, $v_x = 0$, implying $g_m v_x = 0$. Thus $v_{CD} = R_b \times 1 = R_b$. R_{th} is the equivalent resistance at C–D when internal independent sources are deactivated. From Ohm's law,

$$R_{th} = \frac{v_{CD}}{1} = v_{CD} = R_b$$

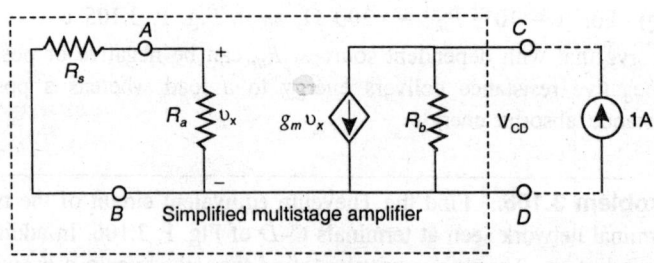

Fig. P. 3.106 (a)

Circuit of Fig. P. 3.106 modified for finding $R_{th} = v_{CD}$.

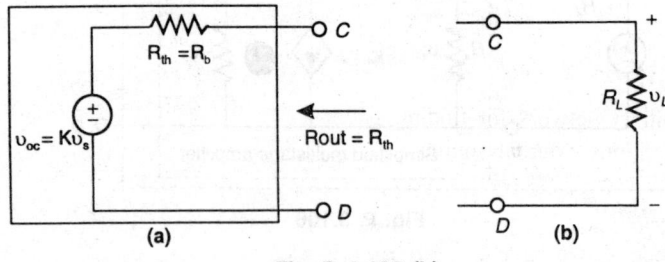

Fig. P. 3.106 (b)

(a) **Thèvenin equivalent of the circuit of Fig. P. 3.106, in which v_{oc} equals a constant K times v_S, with k given in equation 1. The resistance seen at the output terminals is $R_{out} = R_{th}$.**

(b) **Load R_L connected to the output terminals C–D of the multistage amplifier.**

From the preceding calculation, we have the Thévenin equivalent in the circuit of Fig. P. 106 (b) & (c).

Now, when a load is connected, the **voltage gain** is obtained by voltage division i.e.

$$\text{voltage gain} = \frac{v_L}{v_s} = K\frac{R_L}{R_L + R_{out}} = -g_m R_b\left(\frac{R_a}{R_s + R_a}\right)\left(\frac{R_L}{R_L + R_{out}}\right)$$

When the purpose of a circuit is to amplify voltage, then it is desirable to have small $R_{out} = R_{th}$ relative to R_L so that most of V_{oc} will appear across the load.

Problem 3.107. Find the Thévenin and Norton equivalent circuits to the left of the fictitious current source i_s in Fig. P. 3.107. There are three nodes labeled as A, B, and C, with the bottom node being the reference.

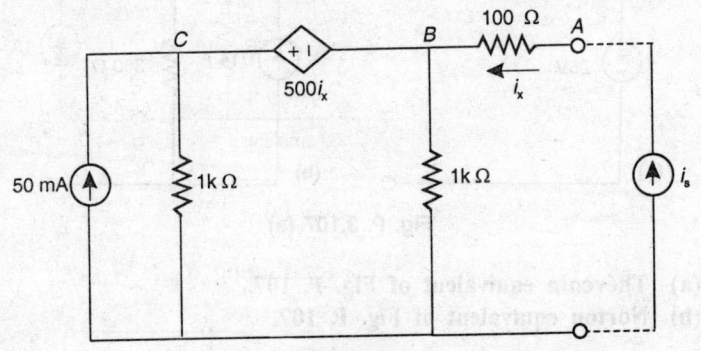

Fig. 3.107

Arbitrary network for finding Thévenin equivalent via excitation by a fictitious external source i_s.

Solution:

At node A, we have,

$$0.01v_A - 0.01v_B = i_s$$

At the supernode that encloses the floating voltage source, we get

$$0.05 = 0.01(v_A - v_B) + 0.001v_B + 0.001v_C$$

Finally, the constraint equation on v_C and v_B is

$$v_C - v_B = 500 \times i_x = 500 \times i_s$$

Writing the node equation in matrix form, we have

$$\begin{bmatrix} 0.01 & -0.01 & 0 \\ -0.01 & 0.011 & 0.001 \\ 0 & -1 & 1 \end{bmatrix} \begin{bmatrix} v_A \\ v_B \\ v_C \end{bmatrix} = \begin{bmatrix} i_s \\ 0.05 \\ 500 i_s \end{bmatrix}$$

Solving for the node voltage produces

$$\begin{bmatrix} v_A \\ v_B \\ v_C \end{bmatrix} = \begin{bmatrix} 600 & 500 & -0.5 \\ 500 & 500 & -0.5 \\ 500 & 500 & 0.5 \end{bmatrix} \begin{bmatrix} i_s \\ 0.05 \\ 500 i_s \end{bmatrix} \tag{1}$$

From the first row of equation (1) we have,

$$v_A = 350 \times i_s + 25$$

Now, we have $v_{oc} = 25$ V, $R_{th} = 350$ Ω, and $i_{sc} = v_{oc}/R_{th} = 1/14$ A. The Thévenin and Norton equivalent circuits are shown in Fig. P. 107 (a)

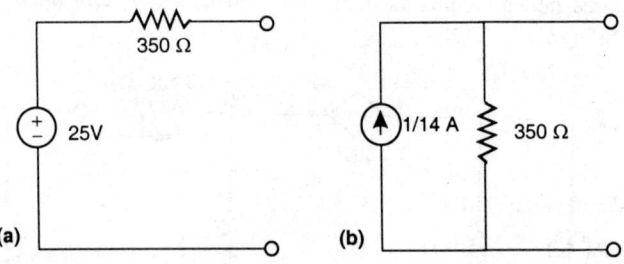

Fig. P. 3.107 (a)

(a) Thévenin equivalent of Fig. P. 107.
(b) Norton equivalent of Fig. P. 107.

Prolem 3.108. Find the Thevenin and Norton equivalent circuits seen at terminals *A–D* in Fig. P. 3.108.

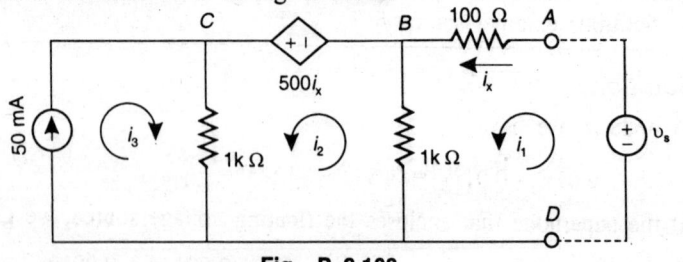

Fig. P. 3.108

Arbitrary network for finding Thévenin equivalent via excitation by a fictitious external source i_S.

For loop 1, We, have $v_s = 1100 i_1 - 1000 i_2$

and, For loop 2, $0 = -500 i_1 + 1000(i_2 + 0.05) + 1000(i_2 - i_1)$

Equivalently, it can be rewritten as,

$$-50 = -1500 i_1 + 2000 i_2$$

Writing the loop equations in matrix form, we have

$$\begin{bmatrix} 1100 & -1000 \\ -1500 & 2000 \end{bmatrix} \begin{bmatrix} i_1 \\ i_2 \end{bmatrix} = \begin{bmatrix} v_s \\ -50 \end{bmatrix}$$

Solving for the loop currents, we get

$$\begin{bmatrix} i_1 \\ i_2 \end{bmatrix} = 10^{-3} \begin{bmatrix} 2.8571 & 1.4286 \\ 2.1429 & 1.5714 \end{bmatrix} \begin{bmatrix} v_s \\ -50 \end{bmatrix} \tag{1}$$

From the first row of equation (1)

$$i_1 = 2.8571 \times 10^{-3} v_s - 0.071429 \tag{2}$$

Now, we have $i_{sc.} = 0.071429 = 1/14$ A, $R_{th} = 10^3 / 2.8571 = 350\,\Omega$, and $v_{oc} = R_{th}\, i_{sc}\ 25$ V.

Problem 3.109.   Consider the circuit of Fig. P. 3.109 (a). Find (a) the value of R_L for maximum power transfer and (b) the corresponding $P_{L,max}$

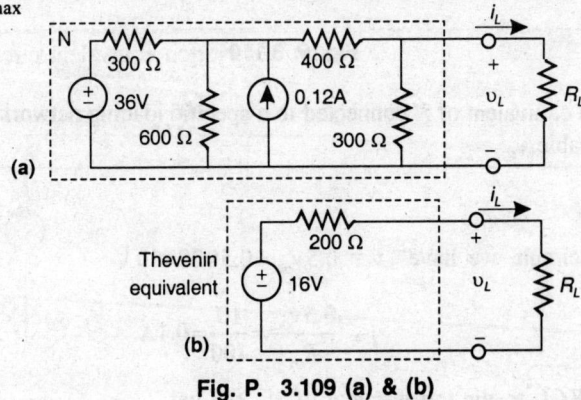

(a)

(b)

Fig. P. 3.109 (a) & (b)

(a) Network N connected to load R_L.

(b) Thévenin equivalent of N connected to R_L.

Solution:

To compute R_{th}, the independent voltage source becomes a short and the independent current source becomes an open. Finding the equivalent resistance seen at the terminals produces $R_{th} = 200\ \Omega$. Hence, maximum power is transferred when $R_L = 200\ \Omega$.

The network N reduces to its Thévenin equivalent shown in Fig. P. 3.109 (b) with $v_{oc} = 16$ V.

Using $v_{oc} = 16$ V and $R_L = R_{th} = 200\ \Omega$ into equation

$$P_{L,\max} = \frac{v^2_{oc}\,\text{V}}{4\,R_{th}}$$

we have, $$P_{L,\max} = \frac{16^2}{800} = 320\,\text{mW}$$

Prolem 3.110.   In the circuit of Fig. P. 3.110, suppose $R = 50\ \Omega$ and v_a is adjustable. Find the value of v_a such that the maximum power is transferred from N to N_L. What is the value of $P_{L,\max}$?

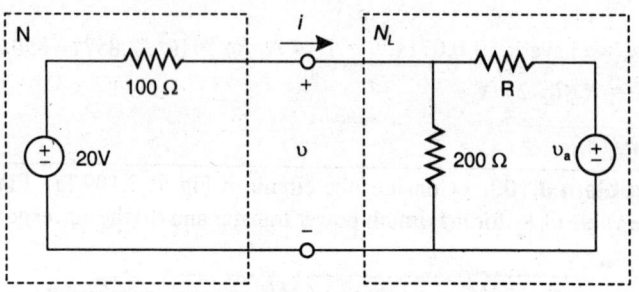

Fig. P. 3.110

Thévenin equivalent of N connected to a specific loading network N_L with variable v_a.

Solution:

From the circuit, we have $v = 0.5\,v_{oc} = 0.5 \times 20 = 10\,\text{V}$

$$i = \frac{0.5\,v_{oc}}{R_{th}} = \frac{10}{100} = 0.1\,\text{A}$$

Applying KCL to the top terminal of N_L, we get

$$0.1 = \frac{10}{200} + \frac{10 - v_a}{50}$$

from which $v_a = 7.5$ V. It is this value which produces maximum power transfer to N_L.

Now, maximum power transfer. $P_{L,\max} = \dfrac{v_{oc}^2}{4 R_{th}} = \dfrac{400}{4 \times 100} = 1 \text{ W}$

Problem 3.111. Compute $v_L(t)$ for the inductor circuit of Fig. P. 3.111 when $i_L(t) = e^{-t^2}$ A.

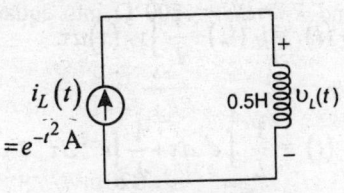

Fig. P. 3.111

Simple inductor driven by a current source.

Solution:

A direct differentiation of the inductor current $i_L(t)$ gives

$$v_L(t) = 0.5 \frac{d\left(e^{-t^2}\right)}{dt} = 0.5(-2t)e^{-t^2} = -te^{-t^2} \text{ V}$$

Problem 3.112. For the circuit of Fig. P. 3.112 (a), determine $i_L(0)$ and $i_L(t)$ for $t \geq 0$ when $v_L(t) = e^{-|t|}$ V. For reference, Fig. P. 3.112 (b) shows a plot of $v_L(t)$.

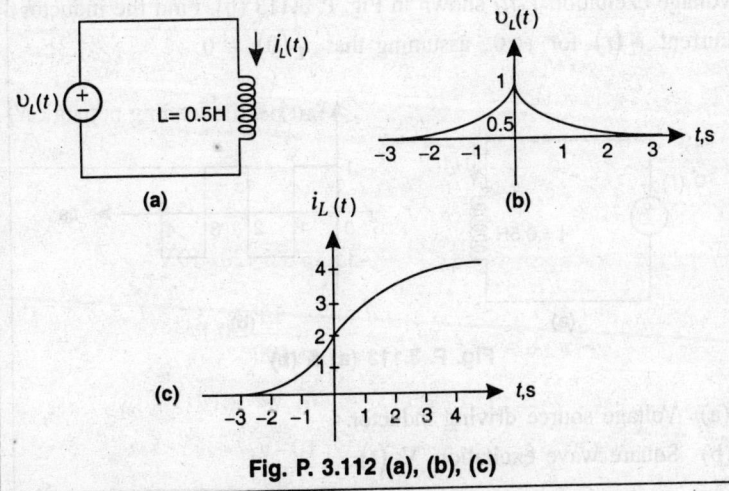

Fig. P. 3.112 (a), (b), (c)

(a) Simple inductor driven by a voltage source.

(b) Source waveform $v_L(t)$.

(c) Resulting inductor current $i_L(t)$.

Solution:

Direct application of equation

$$i_L(t) = i_L(t_o) + \frac{1}{L}\int_{t_o}^{t} v_L(\tau)d\tau.$$

We have,

$$i_L(t) = \frac{1}{L}\int_{-\infty}^{0} e^{\tau}\,d\tau + \frac{1}{L}\int_{0}^{t} e^{-\tau}\,d\tau$$

$$= \frac{1}{L} + \frac{1}{L}\left(1 - e^{-t}\right)A$$

Then,

$$i_L(0) = \frac{1}{L} = 2A$$

$$i_L(t) = \frac{1}{L}\left(2 - e^{-t}\right) = 4 - 2e^{-t} \, A$$

The graph of $i_L(t)$ for all t is given in Fig. P. 3.112 (c).

Problem 3.113. Consider the circuit of Fig. P. 3.113 (a) with the voltage excitation $V_2(t)$ shown in Fig. P. 3.113 (b). Find the inductor current $i_L(t)$ for $t \geq 0$, assuming that $i_L(0) = 0$.

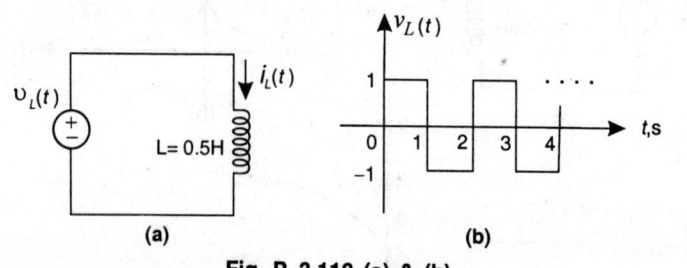

Fig. P. 3.113 (a) & (b)

(a) Voltage source driving inductor.

(b) Square wave excitation $V_L(t)$.

Solution:

From equation $\quad i_L(t) = i_L(t_o) + \dfrac{1}{L}\int\limits_{t_o}^{t} v_L(\tau)d\tau.$

with $t = n$ and n even, we have

$$i_L(t) = i_L(0) + \frac{1}{L}\int\limits_{0}^{n} v_L(\tau)d\tau = \frac{1}{L}\int\limits_{0}^{n} v_L(\tau)d\tau$$

The last integral computes the net area under the $v_L(t)$ curve (see Fig. P. 3.113 (b)) from 0 to n. Since n is even, the net area is zero. Hence $i_L(n) = 0$ for all even values of n.

Now, for $t = n$ and n odd, we have

$$i_L(n) = i_L(0) + \frac{1}{L}\int\limits_{0}^{n-1} v_L(\tau)d\tau + \frac{1}{L}\int\limits_{n-1}^{n} v_L(\tau)d\tau = \frac{1}{L}\int\limits_{n-1}^{n} v_L(\tau)d\tau = 2$$

This follows because n is odd and $n-1$ is even, making the integral of $v_L(t)$ over $[0, n-1]$ equal to zero. From Fig. P. 3.113 (b), the area under the curve $[n, n+1]$ is 1. Multiplying by $1/L$ yields 2.

Now, If n is even, then the value of the inductor current over the interval $[n, n+1]$ is

$$i_L(t) = i_L(n) + \frac{1}{L}\int\limits_{n}^{t} d\tau = i_L(n) + 2(t-n) = 2(t-n)\,\text{A}$$

Observe that $i_L(t) = 2t - 2n$ A is the equation of a straight line having slope 2 and y-intercept $-2n$. Equivalently, the segment over $[n, n+1]$ has slope -2 and intercepts the t-axis at $t = n$ for even values of n.

For n is odd, and for the interval $[n, n+1]$, the inductor current is

$$i_L(t) = i_L(n) - \frac{1}{L}\int\limits_{n}^{t} d\tau = i_L(n) - 2(t-n) = 2 - 2(t-n)\,\text{A}$$

Here $i_L(t) = -2t + 2 + 2n$ is the equation of a straight line, but with slope -2 and y-intercept $2 + 2n$. Equivalently, the segment over $[n, n+1]$ with n odd has slope $+2$ and intercepts the t-axis at $t = n+1$ for odd values of n, i.e., $n+1$ is even. Thus, the segments computed in steps 3 and 4 intercept the t-axis at the same point. Fig. P. 3.113 (c) sketches the response for $t \geq 0$.

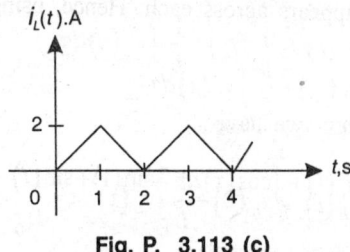

Fig. P. 3.113 (c)

Triangular shape of inductor current for the square wave voltage excitation of Fig. P. 3.113 (b) applied to the circuit of Fig. P. 3.113 (a).

Problem 3.114 For the circuit of Fig. P. 3.114, in which $v_s(t) = \cos(t)$ V for $t \geq 0$ and 0 otherwise, find the input current $i_s(t)$ for $t \geq 0$ and the energy stored in each of the inductors for the intervals $[0, t]$ for $0 \leq t \leq 1$ and for $1 \leq t$.

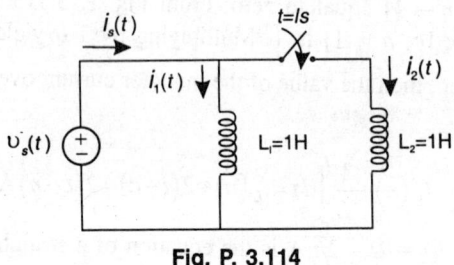

Fig. P. 3.114

Parallel inductive circuit with switch in which $v_s(t) = \cos(t)$ V for $t \geq 0$ and 0 otherwise.

Solution:

Since no voltage is applied to either inductor or $t < 0$, $i_1(0) = 0$, no voltage appears across the second inductor until $t \geq 1$. Hence, $i_2(1) = 0$.
Now, for $0 \leq t < 1$,

$$i_s(t) = i_1(t) = \frac{1}{L_1} \int_0^t v_s(\tau) d\tau$$

$$= \int_0^t \cos(\tau) d\tau = \sin(t) \, \text{A}$$

At $t = 1$, the switch closes. The two inductors are then in parallel, and the source voltage appears across each. Hence, using equation

$$i_L(t) = i_1(t_o) + \frac{1}{L}\int_{t0}^{t} v_L(\tau)d\tau$$

We have, $i_1(t) = i_1(1) + \int_1^t \cos(\tau)d\tau = \sin(1) + \sin(t) - \sin(1) = \sin(t)\,\text{A}$

Using the same equation, we get

$$i_2(t) = i_2(1) + \int_1^t \cos(\tau)d\tau = \sin(t) - \sin(1)\,\text{A}$$

From KCL, the input current

$$i(t) = i_1(t) + i_2(t) = 2\sin(t) - \sin(1)\,\text{A for } t \geq 1$$

Using eqn, for the stored energy during the interval $[t_o, t_1]$,

$$W_L[t_o, t_1] = \frac{1}{2}L\left[i_L^2(t_1) - i_L^2(t_o)\right]\text{ Joule,}$$

we have, for $0 \leq t \leq 1$, $W_{L1}(0, t) = 0.5\sin^2 t\,\text{J}, W_{L2}(0, t) = 0$ and for $t \geq 1$,

$$W_{L1}(o, t) = 0.5\sin^2 t\,\text{J}$$
$$W_{L2}(0, t) = 0.5\left[\sin^2(t) - 2\sin(1)\sin(t) + \sin^2(1)\right]\text{J}$$

Problem 3.115. For the capacitor of Fig. P. 3.115, compute $i_C(t)$ when $v_{in}(t) = e^{-500t}\sin(1000\,t)\,\text{V}$ for $t > 0$.

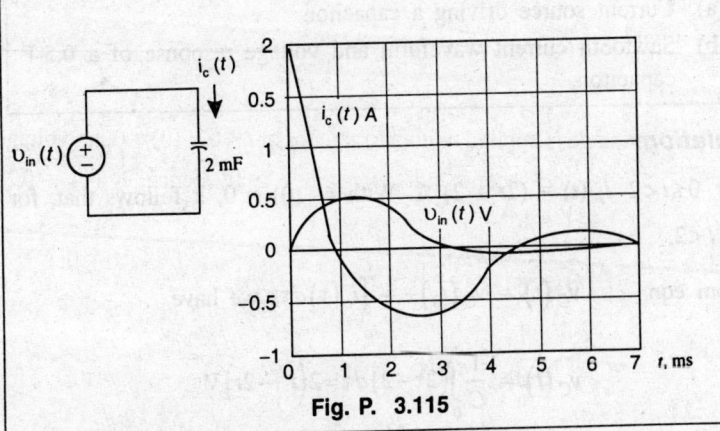

Fig. P. 3.115

(a) A 2-mf capacitor connected to a voltage source;

(b) Plots of capacitor voltage and current waveforms.

Solution:

Using eqn, $i_c(t) = C\dfrac{dv_c(t)}{dt}$, we have

$$i_c(t) = C\frac{dv_c(t)}{dt} = -e^{-500t}\sin(1000t) + 2e^{-500t}\cos(1000t)\,\text{A}$$

Problem 3.116. Suppose the current source with (sawtooth) current given by Fig. P. 3.116 (b) drives a relaxed 0.5-F capacitor (zero initial voltage) as illustrated in Fig. P. 3.116 (a). Compute and plot the voltage across the capacitor.

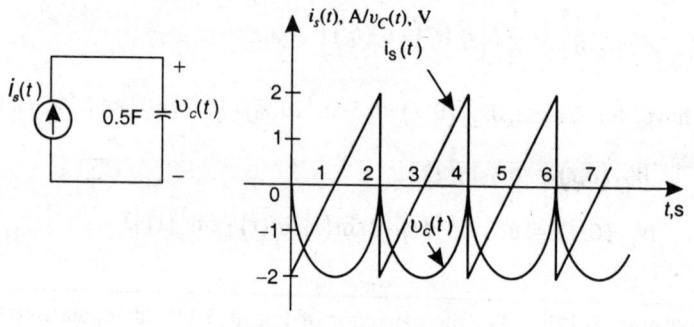

Fig. P. 3.116 (a) & (b)

(a) Current source driving a capacitor.

(b) Sawtooth current waveform and voltage response of a 0.5-F capacitor.

Solution:

For $0 \le t < 2$ $i_s(t) = (2t - 2)$ A. With $v_C(0) = 0$, it follows that, for $0 \le t < 2$,

From eqn $v_C(t) = v_C(t_0) + \dfrac{1}{C}\displaystyle\int_{t_o}^{t} i_c(\tau)\,d\tau$, we have

$$v_C(t) = \frac{1}{C}\int_{0}^{t}(2\tau - 2)\,d\tau = 2(t^2 - 2t)\,\text{V}$$

Observe that $t = 2$, $v_C(2) = 0$. Hence, the capacitor voltage over the interval $2 \le t < 4$ is simply a right-shifted version of the voltage over the first interval. Right-shifting is achieved by replacing t with $t - 2$. In other words, for $2 \le t < 4$, $v_C(t) = 2\left[(t-2)^2 - 2(t-2)\right]$ V.

For the interval $2k \le t \le 2(k+1)$,

$$v_C(t) = 2\left[(t-2k)^2 - 2(t-2k)\right], \quad k = 0, 1, 2.$$

Lastly, observe that the voltage across the capacitor, as illustrated in Fig. P. 3.116 (b), is continuous despite the discontinuity of the capacitor current. Again, this follows because the capacitor voltage is the intergral of the capacitor current supplied by the source.

Problem 3.117. Show that the equivalent inductance of the circuit of Fig. P. 3.117 (a) is given by the reciprocal of the sum-of-reciprocals formula,

$$L_{eq} = \cfrac{1}{\cfrac{1}{L_1} + \cfrac{1}{L_2} + \cfrac{1}{L_3}}$$

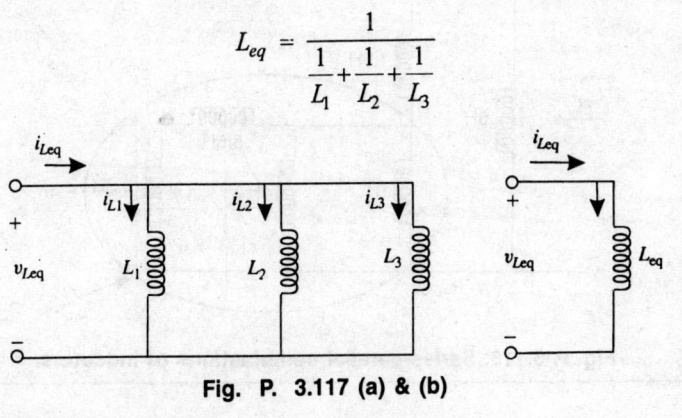

Fig. P. 3.117 (a) & (b)

(a) Parallel connection of three inductors.

(b) Equivalent inductance.

Solution:

Here, Using KCL, we get $i_{Leq} = i_{L1} + i_{L2} + i_{L3}$

Differentiating both sides with respect to time yields

$$\frac{di_{Leq}}{dt} = \frac{di_{L1}}{dt} + \frac{di_{L2}}{dt} + \frac{di_{L3}}{dt} \tag{1}$$

Now for each inductor, $\dfrac{di_{Li}}{dt} = \dfrac{v_{Li}}{L_i}$

Substituting into the eqn (1) and noting that $v_{L1} = v_{L2} = v_{L3} = v_{Leq}$, we have

$$\frac{di_{Leq}}{dt} = \frac{v_{L1}}{dt} + \frac{v_{L2}}{dt} + \frac{v_{L3}}{dt} = \left(\frac{1}{L_1} + \frac{1}{L_2} + \frac{1}{L_3} \right) v_{Leq}$$

Using eqn $v_{Leq}(t) = L_{eq} \dfrac{di_{Leq}}{dt}$

we have, $\quad L_{eq} = \dfrac{1}{1/L_1 + 1/L_2 + 1/L_3}$ **Proved**

Problem 3.118. Find the equivalent inductance L_{eq} of the circuit of Fig. P. 3.118.

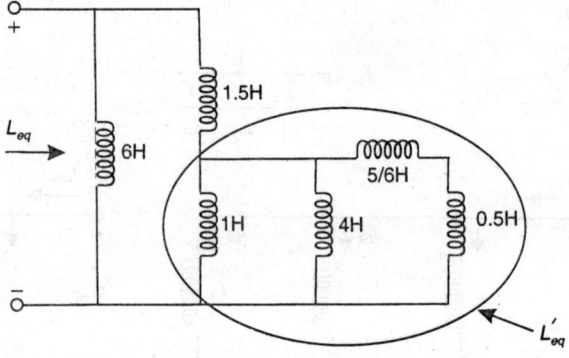

Fig. P. 3.118. Series-parallel combinations of indcutors.

Solution:

Observe that the series inductance of the 5/6-H and 0.5-H inductors equals 4/3 H. This inductance is in parallel with a 1-H and a 4-H inductance. Hence,

$$L'_{eq} = \frac{1}{\dfrac{1}{1} + \dfrac{1}{4} + \dfrac{3}{4}} = \frac{1}{2} H$$

The equivalent circuit at this point is given by Fig. P. 3.118 (a). This figure consists of a series combination of a 1.5-H and a 0.5-H inductor connected in parallel with a 6-H inductor. It follows that

$$L_{eq} = \cfrac{1}{\cfrac{1}{0.5+1.5}+\cfrac{1}{6}} = \frac{6}{4} = 1.5\,\text{H}$$

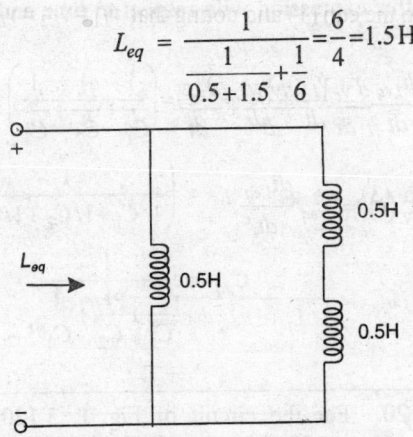

Fig. P. 3.118 (a). Equivalent circuit of Fig. P. 3.118.

Problem 3.119. Compute the equivalent capacitance C_{eq} of the series connection of capacitor of Fig. P. 3.119 (a).

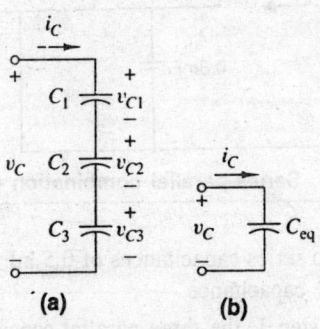

(a)   (b)

Fig. P. 3.119 (a) & (b)

(a) **Series combination of three capacitors.**

(b) **Equivalent capacitance C_{eq}.**

Solution:

Here for each capacitor, $k = 1, 2, 3,$

We have, $i_{Ck} = C_k \dfrac{dv_{Ck}}{dt}$ (1)

But by KCL, $i_C = i_{Ck}$. Hence, $\dfrac{dv_{Ck}}{dt} = \dfrac{i_C}{C_K}$ (2)

From KVL, we get $v_C = v_{C1} + v_{C2} + v_{C3}$

Differentiating this expression with respect to time and using the result of eqn (2),

We have, $\dfrac{dv_C}{dt} = \dfrac{dv_{C1}}{dt} + \dfrac{dv_{C2}}{dt} + \dfrac{dv_{C3}}{dt} = \dfrac{i_C}{C_1} + \dfrac{i_C}{C_2} + \dfrac{i_C}{C_3}$ (3)

Now, using eqn (3), we get $\quad i_C = \left(\dfrac{1}{1/C_1 + 1/C_2 + 1/C_3}\right)\dfrac{dv_C}{dt}$

Hence, $\qquad\qquad C_{eq} = \dfrac{1}{\dfrac{1}{C_1} + \dfrac{1}{C_2} + \dfrac{1}{C_3}}$

Problem 3.120. For the circuit of Fig. P. 3.120, compute the equivalent capacitance C_{eq}.

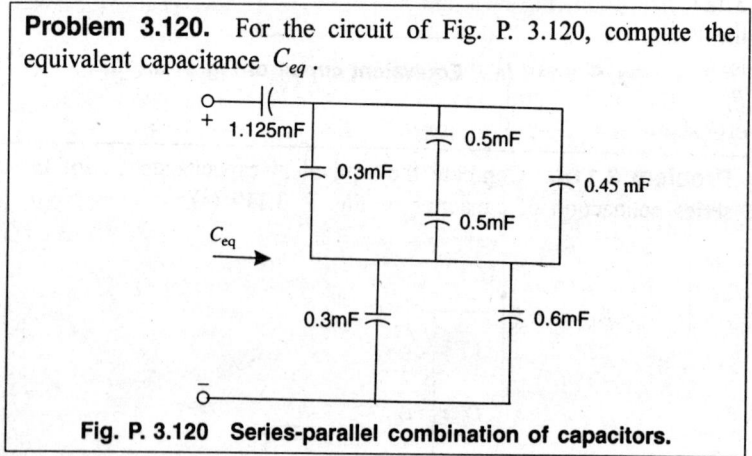

Fig. P. 3.120 Series-parallel combination of capacitors.

Solution:

Observe that the two series capacitances of 0.5 mF and 0.5 mF combine to make a 0.25-mF capacitance.

First, as a result of step 1, the three parallel capacitances 0.3 mF, 0.25 mF, and 0.45 mF add to an equivalent capacitance of 1 mF. Further, the two parallel capacitances 0.3 mF and 0.6 mF at the bottom of the circuit add to make a 0.9-mF capacitance. The new equivalent circuit is shown in Fig. P. 3.120 (a).

Fig. P. 3.120 (a)

Circuit equivalent to that of Fig. P. 3.120.

Using eqn, $C_{eq} = \dfrac{1}{\dfrac{1}{C_1}+\dfrac{1}{C_2}+...+\dfrac{1}{C_n}}$

We have $C_{eq} = \dfrac{1}{1/1.125+1/1+1/0.9} = \dfrac{1}{3}\,\text{mF}$

Problem 3.121. Three identical loads Z having unity power factor are connected to a single-phase power source, as shown in Fig. P. 3.121. The following facts are known: (1) The line voltage across the loads is $V_L = 11{,}000$ V (rms) (2) the resistance per conductor is $R = 0.6\ \Omega$; and (3) the total power delivered to the loads is $P_{tot} = 4000$ kW. Find the power loss in the connecting lines and the efficiency of power transmission.

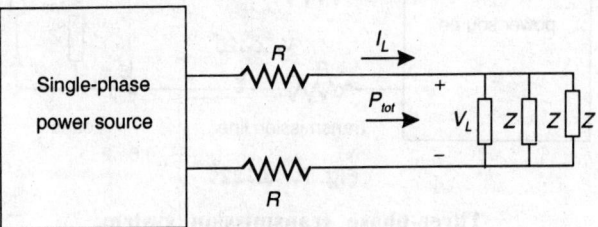

Fig. P. 3.121

Single-phase transmission system.

Solution:

Since, average power absorbed by a load is,

$$P = VI \times pf$$

We have, $\qquad I_L = \dfrac{P_{tot}}{V_L \times pf} = \dfrac{4\times10^6}{11{,}000} = 363.64\,\text{A}$

The heat loss is given by I^2R. Since there are two 0.6-Ω conductors,

$$P_{loss} = 2\times(363.64)^2 \times 0.6 = 158.677\,\text{kW}$$

By definition, efficiency $= \dfrac{P_{tot}}{P_{tot}+P_{loss}} = \dfrac{4000}{4158.7} = 0.962 = 96.2\%$

Problem 3.122. The same three loads of Fig. P. 3.121 are now rearranged into a Δ configuration and connected to a three-phase power source, as shown in Fig. P. 3.122. The same line voltage V_L = 11,000 V appears across each load. Hence, the total average power delivered to the loads is also P_{tot} = 4000 kW. The distances between the loads and the source are the same in both Fig. P. 3.121 and P. 3.122. A smaller cross-sectional area is chosen for each conductor in Fig. P. 3.122 so that both systems use the same amount of copper.

(a) Find the resistance R' in the three-phase system of Fig. P. 3.122.
(b) Find the power loss in the connecting lines.
(c) Find the efficiency of power transmission.

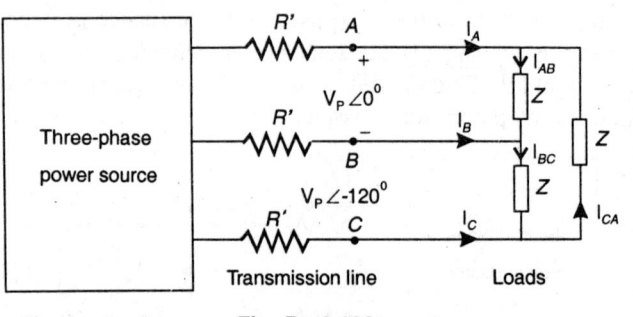

Fig. P. 3.122

Three-phase transmission system.

Solution:

Let the cross-sectional area of each conductor in Fig. P. 3.121 be A_1, and let that in Fig. P. 3.122 be A_3. The line length is L. The volume of copper in Fig. P. 3.121 is $2 \times A_1 \times L$, whereas that of Fig. P. 3.122 is $3 \times A_3 \times L$. Since both systems use the same amount of copper,

$$A_3 = A_1 \times \frac{2}{3}$$

Since the resistance of a conductor is proportional to the length and inversely proportional to the cross-sectional area, therefore R' = 3/2 × R = 0.9 Ω.

Now, line current, $I_L = \dfrac{P_{tot}}{\sqrt{3} V_L \times pf} = \dfrac{4 \times 10^6}{11,000\sqrt{3}} = 209.95$

Since there are three conductors, each with a resistance of 0.9 Ω,

We have $P_{loss} = 3 \times 209.95^2 \times 0.9 = 119 \text{kW}$

By definition, efficiency $= \dfrac{P_{tot}}{P_{tot} + P_{loss}} = \dfrac{4000}{4119} = 0.971 = 97.1\%$

Problem 3.123. Three resistive loads represented by resistances $Ra = 20\ \Omega$, $R_b = 15\Omega$, and $R_c = 20\Omega$ are connected to a 60-Hz, three-phase, four-wire system, as shown in Fig. P. 3.123. Other circuit parameters are: $V_p = 130$ V and $Z_g = 0.2 + j\,0.4\Omega$. The impedances of the lines connecting the loads to the service panel are assumed to be negligible.

(a) Find the total power delivered to the loads.

(b) Find the magnitude of the current in the neutral line.

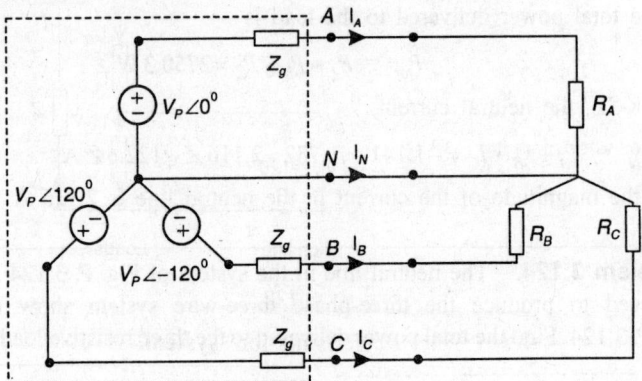

Fig. P. 3.123

Three unequal resistive loads connected to a three-phase four-wire system.

Solution:

R_A and Z_g are in series and connected to the voltage source $130\ \angle 0$ V. By Ohm's law, current in phase A,

$$I_A = \frac{130e^{j0}}{(0.2 + j0.4) + 20} = 6.43 - j0.127 = 6.434e^{-1.134°}\ \text{A}$$

proceeding similarly for phase B and phase C,

We have, $I_B = \dfrac{130e^{-j120°}}{(0.2 + j0.4) + 15} = -4.47 - j7.29 = 8.55e^{-j121.5°}\ \text{A}$

and, $$I_C = \frac{130e^{j120°}}{(0.2+j0.4)+20}$$

$$= -3.106+j5.634 = 6.434e^{-j118.9°} \text{ A}$$

For a resistive load, the absorbed power is given by $|I|^2 R$.

Then absorbed power in phase A, B, and C are,

$$P_A = 6.43^2 \times 20 = 826.9 \text{ W}$$

$$P_B = 8.55^2 \times 15 = 1096.5 \text{ W}$$

$$P_C = 6.43^2 \times 20 = 826.9 \text{ W}$$

and the total power delivered to the load is

$$P_{tot} = P_A + P_B + P_C = 2750.3 \text{ W}$$

From KCL, the neutral current,

$$I_N = I_A + I_B + I_C = -1.141 - j1.782 = 2.116 \angle -122.64° \text{ A}$$

Thus, the magnitude of the current in the neutral line is 2.116 A.

Problem 3.124. The neutral line in the system of Fig. P. 3.124 is removed to produce the three-phase three-wire system show in Fig. P. 3.124. Find the total power delivered to the three resistive loads.

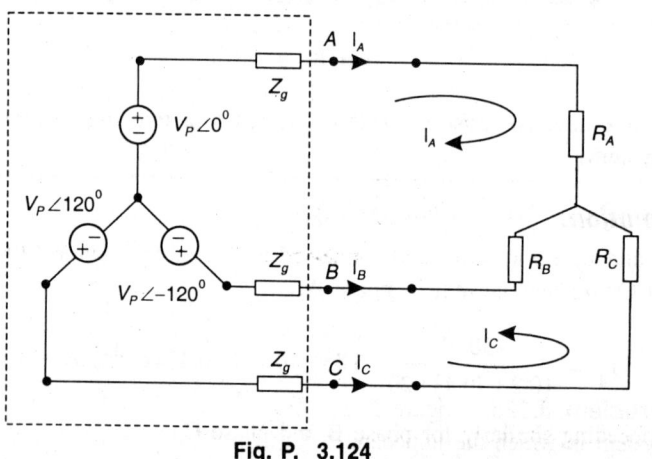

Fig. P. 3.124

Three unequal resistive loads connected to a three-phase three-wire system.

Solution:

The mesh equations in matrix form may be written as,

$$\begin{bmatrix} R_A + R_B + 2Z_g & R_B + Z_g \\ R_B + Z_g & R_C + R_B + 2Z_g \end{bmatrix}\begin{bmatrix} I_A \\ I_C \end{bmatrix} = \begin{bmatrix} V_p - V_p e^{-j120°} \\ V_p e^{j120°} - V_p e^{-j120°} \end{bmatrix}$$

Substituting numerical values $R_A = 20$, $R_B = R_C = 20$, $Z_g = 0.2 + j\,0.4$, and $V_p = 130$ into this matrix equation we have,

$$\begin{bmatrix} 35.4 + j0.8 & 15.2 + j0.4 \\ 15.2 + j0.4 & 35.4 + j0.8 \end{bmatrix}\begin{bmatrix} I_A \\ I_C \end{bmatrix} = \begin{bmatrix} 195 + j112.58 \\ j225.17 \end{bmatrix} \tag{1}$$

Solving the matrix equation (1), we have

$$\begin{bmatrix} I_A \\ I_C \end{bmatrix} = \begin{bmatrix} 6.7746 - j0.4086 \\ -2.7648 + j6.1712 \end{bmatrix} = \begin{bmatrix} 6.787\angle 3.4513° \\ 6.762\angle 114.13° \end{bmatrix} A$$

From KCL, then $\qquad I_B = -I_A - I_C = 7.7053\angle - 121.36° \, A$

Using the formula $P = |I|^2 R$ and the obtained current values, we have

individual powers as, $\qquad P_A = (6.787)^2 \times 20 = 921.2 \, W$

$$P_B = (7.705)^2 \times 15 = 890.6 \, W$$

and $\qquad\qquad\qquad P_C = (6.762)^2 \times 20 = 914.5 \, W$

and the total power delivered to the load is

$$P_{tot} = P_A + P_B + P_C = 2.726 \, W$$

Problem 3.125. Figure P. 3.125 shows a three-phase four-wire system in which the impedances Z_g, Z_l, and Z_N may or may not be zero. Derive an equivalent circuit for phase A only, the analysis of which yields I_A and V_{AN}. The currents and voltages in the other phases may be inferred without further circuit analysis.

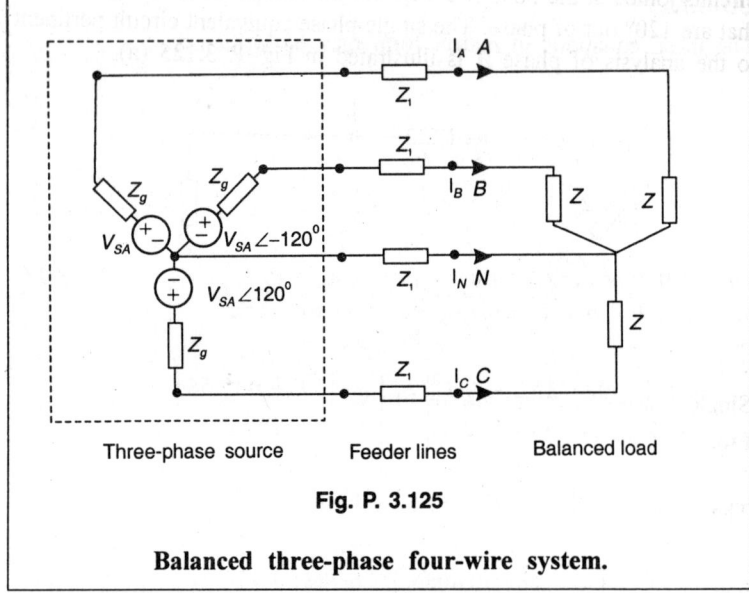

Three-phase source   Feeder lines   Balanced load

Fig. P. 3.125

Balanced three-phase four-wire system.

Solution:

Applying KCL to node N, we get,

$$\frac{V_{N0} - V_{sA}\angle 0°}{Z + Z_1 + Z_g} + \frac{V_{N0} - V_{sA}\angle -120°}{Z + Z_1 + Z_g} + \frac{V_{N0} - V_{sA}\angle 120°}{Z + Z_1 + Z_g} + I_N = 0$$

or, after grouping terms, we have

$$\frac{3\,V_{N0}}{Z + Z_1 + Z_g} + I_N = V_{sA}\left(1\angle 0° + 1\angle -120° + 1\angle 120°\right) = 0 \tag{1}$$

If $Z_N = 0$, then there is a short circuit between nodes N and 0; hence $V_{N0} = 0$ and equation implies $I_N = 0$. If $Z_N = \infty$, then there is an open circuit between nodes N and 0; hence $I_N = 0$ and equation (1) implies $V_{N0} = 0$. If Z_N is finite but nonzero, then substituting $I_N = V_{N0}/Z_N$ into equation (1) leads to

$$\left(\frac{3}{Z + Z_1 + Z_g} + \frac{1}{Z_N}\right) V_{N0} = 0$$

which again shows that $V_{N0} = 0$ and $I_N = 0$.

Since $V_{N0} = 0$, we may replace the impedance Z_N by a short circuit without affecting any voltage or current outside of Z_N. After this replacement, the circuit of Fig. P. 3.125 reduces to three separate identical circuits joined at the node N except for the independent voltage sources that are $120°$ out of phase. The single-phase equivalent circuit pertinent to the analysis of phase A is illustrated in Fig. P. 3.125 (a)

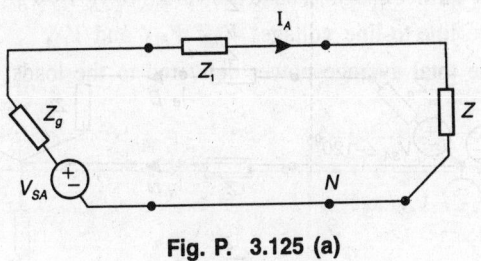

Fig. P. 3.125 (a)

Single-phase equivalent circuit for phase A of Fig. P. 3.125.

From Fig. P. 3.125 (a).

The current,
$$I_A = \frac{V_{sA}}{Z+Z_1+Z_g}$$

and
$$V_{AN} = ZI_A$$

Since the source voltage in phase B is $e^{-j120°}$ times the source voltage in phase A, then, according to the proportionality property, all responses in phase B are $e^{-j120°}$ times the corresponding quantities in phase A. Similarly, all responses in phase C are $e^{-j120°}$ times the corresponding quantities in phase A. In a summary, we can write

$$I_B = I_A e^{-j120°}$$

$$I_C = I_A e^{120°}$$

$$V_{BN} = V_{AN} e^{-j120°}$$

$$V_{CN} = V_{AN} e^{j120°}$$

Problem 3.126. Figure P. 3.126 shows a three-phase system referred to as the Y-Y configuration because both the sources and the loads are connected in the form of a Y. The following parameters are known:

Source voltage $\qquad V_{sA} = 120\angle 0° \, V(\text{rms})$

Source impedance $\qquad Z_g = 0.05 + j0.15 = 0.158\angle 71.56° \, \Omega$

Load impedance $\qquad Z = 4 + j3 = 5\angle 36.87° \, \Omega$

Feeder impedance $Z_1 = 0.1 + j0.2 = 0.224\angle 64.43°\,\Omega$

(a) Compute I_A and V_{AN}.

(b) Without further analysis, state the values of I_B, I_C, V_{BN}, and V_{CN}.

(c) Find the line-to-line voltages V_{AB}, V_{BC}, and V_{CA}.

(d) Find the total average power delivered to the loads.

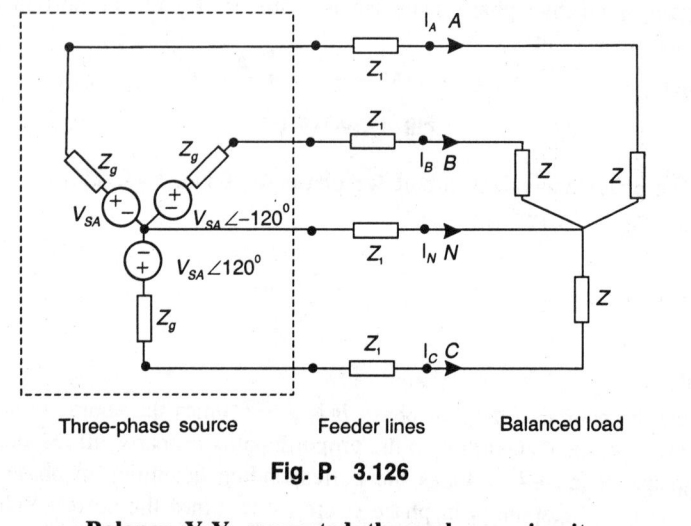

Three-phase source   Feeder lines   Balanced load

Fig. P. 3.126

Balance Y-Y connected three-phases circuit.

Solution:

The single-phase equivalent circuit for Fig. P. 3.126 is shown in Fig. P. 3.126 (a). From Fig. P. 3.126 (a),

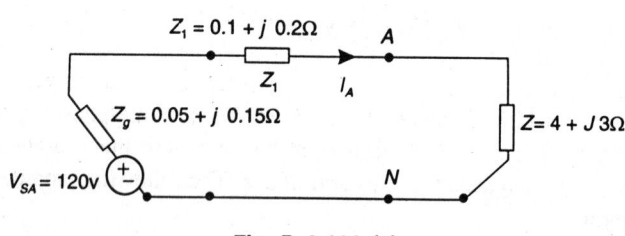

Fig. P. 3.126 (a)

Single-phase equivalent circuit for Fig. P. 3.126.

We have, $I_A = \dfrac{V_{sA}}{Z + Z_1 + Z_g}$

$$\frac{V_{sA}}{(4+j3)+(0.1+j0.2)+(0.05+j0.15)} = \frac{120\angle 0°}{5.333\angle 38.9°}$$

$$= 22.5\angle -38.9° \, A$$

By Ohm's law,

$$V_{AN} = ZI_A = (4+j3)\times 22.5\angle -38.9° = 112.5\angle -2.03° \, V$$

Again, a positive phase sequence is assumed. By the proportionality property, we get

Then, $I_B = 22.5\angle -158.9° \, A$ $I_C = 22.5\angle 81.1° \, A$

$\quad\quad V_{BN} = 112.5\angle -122.03° \, V$ $V_{CN} = 112.5\angle 117.97° \, V$

Now, the line-to-line voltage one given as,

$$V_{AB} = \sqrt{3}\, V_{AN} e^{j30°} = 194.85 e^{j27.97°} \, V$$

$$V_{BC} = V_{AB} e^{-j120°} = 194.85 e^{j92.03°} \, V$$

$$V_{CA} = V_{AB} e^{j120°} = 194.85 e^{j147.97°} \, V$$

The average power absorbed by each load is $P_{av} |I|^2 R_e[Z]$. Hence,

$$P_A = P_B = P_C = 22.5^2 \times 4 = 2025 \, W$$

and the total average power delivered to the loads is

$$P_{tot} = 3 \times 2025 = 6075 \, W$$

Alternatively, the power factor of each load,

$$\text{p.f.} = \cos\left[\tan^{-1}(0.75)\right] = \cos(36.87°) = 0.8$$

Hence, the total power delivered to the load is,

$$P_{tot} = \sqrt{3}\, V_L I_L \times \text{p.f.} = \sqrt{3}\times 194.85 \times 22.5 \times 0.8 = 6075 \, W$$

Problem 3.127. Figure P. 3.127 shows a Δ-Δ-configuration of a Δ-connected three-phase source supplying power to a Δ-connected load. The following parameters are known:

Source voltage $V_p = 180\angle 0° \, V\,(rms)$

Source impedance $Z_g = 0.15 + j0.45 = 0.474\angle 71.56° \, \Omega$

Load impedance $Z_\Delta = 12 + j9 = 15\angle 36.87° \, \Omega$

Feeder impedance $\quad\quad Z_1 = 0.1 + j0.2 = 0.224\angle 64.43° \, \Omega$

(a) Compute the line currents I_A, I_B and I_C.

(b) Find the load currents I_{AB}, I_{BC}, I_{CA}.

(c) Find the line voltages V_{AB}, V_{BC}, and V_{CA}.

(d) Find the total average power delivered to the loads.

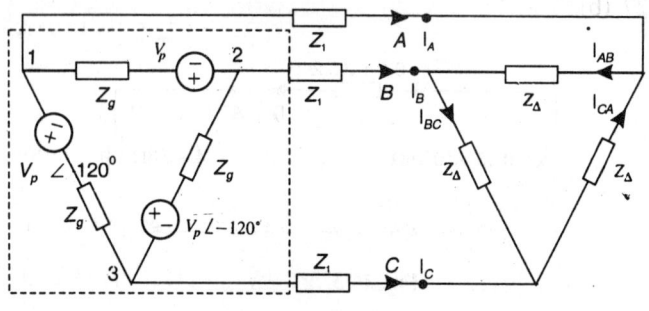

Fig. P. 3.127

Δ-connected sources supplying power to a Δ-connected load.

Solution:

Apply the transformations to the three-phase source and the balanced load to obtain the Y-Y configuration of Fig. P. 3.127 (a). Here

Source phase voltage $\qquad V_{sA} = \dfrac{180}{\sqrt{3}} \angle -30° = 103.92 \angle -30°\,\text{V}$

Source impedance $\qquad \dfrac{Z_g}{3} = 0.05 + j0.15 = 0.158 \angle 71.56° \,\Omega$

Load impedance $\qquad \dfrac{Z_\Delta}{3} = 4 + j3 = 5 \angle 36.87° \,\Omega$

Feeder impedance $\qquad Z_1 = 0.1 + j0.2 = 0.224 \angle 64.43° \,\Omega$

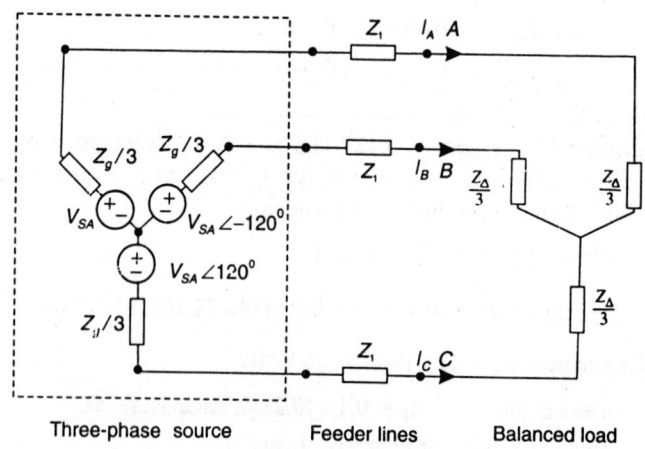

Three-phase source   Feeder lines   Balanced load

Fig. P. 3.127 (a)

Equivalent Y-Y configuration of Fig. P. 3.127.

The single-phase equivalent circuit for Fig. P. 3.127 (a) is given in Fig. P. 3.127 (b).

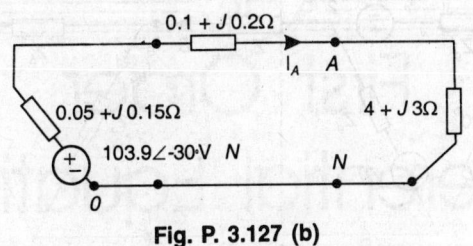

Fig. P. 3.127 (b)

Single-phase equivalent circuit for Fig. P. 3.127 (a).

From Fig. P. 3.127 (a). we get,

$$I_A = \frac{103.92\angle -30°}{(4+j3)+(0.1+j0.2)+(0.05+j0.15)} = 19.49 \angle -68.91° A$$

Using the 120° phase relationship, the other two line currents are given by,

$$I_B = 19.49\angle -188.9° \; A, I_C = 19.49\angle 51.09° A$$

From the current phasor diagram we have line-to-line currents are,

$$I_{AB} = \frac{I_A}{\sqrt{3}} e^{j30°} = 11.25 e^{-j38.9°} \; A$$

$$I_{BC} = I_{AB} e^{-j120°} = 11.25 e^{-j158.9°} A$$

$$I_{CA} = I_{AB} e^{j120°} = 11.25 e^{-j81.09°} A$$

From the original circuit of Fig. P. 3.127. and by Ohm's law,

$$V_{AB} = I_{AB} Z_\Delta = 11.25 e^{-j38.91°} \times e^{j36.87°} = 16.8 e^{-j2.04} \; V$$

By the 120° phase relationships, the other two line voltages are

$$V_{BC} = V_{AB} e^{-j120°}$$

$$= 168.8 e^{-j122.04°} \; V, V_{CA} = V_{AB} e^{j120°} = 168.8 e^{j117.96°} \; V$$

Hence, the total power delivered to the load is given by

$$P_{tot} = \sqrt{3} V_L I_L \times \text{pf} = \sqrt{3} \times 168.8 \times 19.49 \times 0.8 = 4558 \, W$$

First Order
Differential Equations

Problem 4.1. In the network of the following figure, switch K is moved from position 1 to position 2 at $t = 0$, a steady-state current having previously been established in the RL circuit. Find the particular solution for the current $i(t)$.

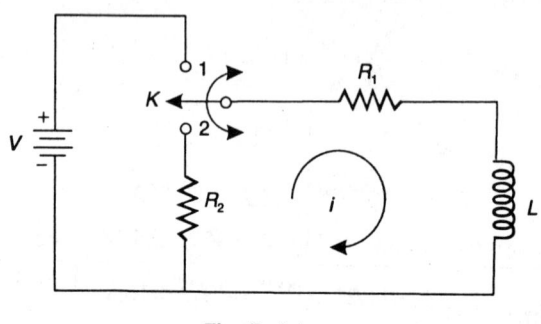

Fig. P. 4.1

Solution:

At $t = 0^-$, inductor is short circuited. So $i(0^-) = i(0^+)$

$$= \frac{V}{R_1}$$

For $t > 0$, the switch K is at positions 2, so applying KVL in this circuit:

$$L\frac{di}{dt} + i(R_1 + R_2) = 0$$

or
$$\frac{di}{dt}+i\left(\frac{R_1+R_2}{L}\right)=0$$

Its solution is given as:

$$i(t)=Ae^{-\left(\frac{R_1+R_2}{L}\right)t}$$

Where A = const and is calculated by initial condition. So at $t = 0^+$

$$i(0) = Ae^{-0} = A = \frac{V}{R_1}$$

So
$$i(t) = \frac{V}{R_1}e^{-\left(\frac{R_1+R_2}{L}\right)t} \text{ Ans.}$$

Problem 4.2. The switch K is moved from position a to b at $t = 0$, having been in position a for a long time before $t = 0$. Capacitor C_2 is uncharged at $t = 0$. (a) Find the particular solution for $i(t)$ for $t > 0$. (b) Find the particular solution for $v_2(t)$ for $t > 0$.

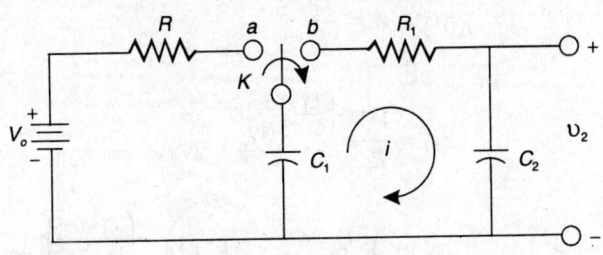

Fig. P. 4.2

Solution:

At time $t = 0^-$, the voltage across capacitors C_1 is equal to V_0.

So $\qquad V_C = V_0 \quad$ at $t = 0^-$

$\therefore \qquad V_C(0^-) = V_C(0^+)$

So $\qquad V_C(0^+) = V_0$

So $\qquad i_c(0^+) = \frac{V_0}{R_1}$

Now, at time $t > 0$, the circuit equation is:

$$iR_1 + \frac{1}{c_2}\int i\,dt + \frac{1}{c_1}\int i\,dt = 0$$

or $$iR_1 + \frac{C_1+C_2}{C_1 C_2}\int i\,dt = 0$$

Now, differentiating the above equation, we get

$$R_1 \frac{di}{dt} + \left(\frac{C_1+C_2}{C_1 C_2}\right) i = 0$$

or $$\frac{di}{dt} + \left(\frac{C_1+C_2}{C_1 C_2 R_1}\right) i = 0$$

Its particular solution is given as

$$i(t) = A e^{-\left(\frac{C_1+C_2}{C_1 C_2 R_1}\right)t}$$

A = const and is given by initial condition. So at $t = 0^+$

$$i(0^+) = A = \frac{V_0}{R_1}$$

∴ $$i(t) = \frac{V_0}{R_1} e^{-\left(\frac{C_1+C_2}{C_1 C_2 R_1}\right)t} \tag{1}$$

and $$v_2(t) = \frac{1}{C_2}\int_0^t i(t)\,dt = \frac{1}{C_2}\int_0^t \frac{V_0}{R_1} e^{-\left(\frac{C_1+C_2}{C_1 C_2 R_1}\right)t} dt$$

$$v_2(t) = \frac{C_1 V_0}{C_1+C_2}\left[1 - e^{-\left(\frac{C_1 C_2}{C_1 C_2 R_1}t\right)}\right] \textbf{ Ans.} \tag{2}$$

Problem 4.3. In the Network of the figure, the switch K is in position a for a long period of time. At $t = 0$, the switch is moved from a to b (by a "make-before-break" mechanism). Find $v_2(t)$ using the numerical values given in the network. Assume that the initial current in the 2-H inductor is zero.

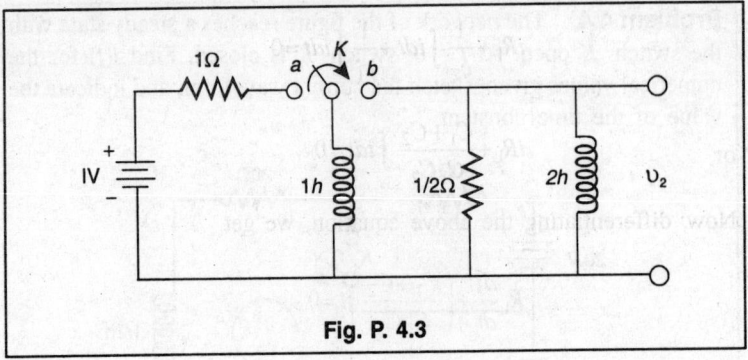

Fig. P. 4.3

Solution:

At $t = (0^-)$ the current across in inductor is equal to:

$$i(0^-) = \frac{1\,\text{V}}{1\Omega} = 1\,\text{amp.}$$

So $\qquad i(0^+) = i(0^-) = 1\,\text{amp., so } v(0^+) = -1 \times \frac{1}{2} = -\frac{1}{2}\,\text{V}$

At time $t > 0$, the circuit equation is

$$\frac{1}{1}\int v_2\, dt + \frac{v_2}{\frac{1}{2}} + \frac{1}{2}\int v_2\, dt = 0$$

or $\qquad 1.5\int v_2\, dt + 2v_2 = 0$

Differentiating the above equation, we get

$$1.5 v_2 + 2\frac{dv_2}{dt} = 0$$

Its solution is given as

$$v_2(t) = Ae^{-\left(\frac{1.5}{2}\right)t}$$

Where A = const and is given by initial condition. So at $t = 0^+$

$$v_2\left(0^+\right) = A = -\frac{1}{2}\,\text{V}.$$

$\therefore \qquad v_2(t) = \left(\frac{-1}{2}\right)e^{-\frac{3}{4}t} \quad \textbf{Ans.}$

Problem 4.4. The network of the figure reaches a steady state with the switch K open. At $t = 0$, switch K is closed. Find $i(t)$ for the numerical values given, sketch the current waveform, and indicate the value of the time constant.

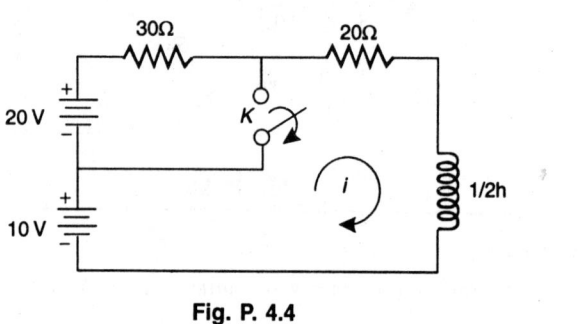

Fig. P. 4.4

Solution:

At time $t = (0^-)$, current in the circuit is given as

$$i(0^-) = \frac{30\,\text{V}}{50\,\Omega}$$

$$= \frac{3}{5}\,\text{amp.}$$

So $i(0^+) = \dfrac{3}{5}\,\text{amp.}$

At time $t > 0$, the circuit equation is given as

$$20i + \frac{1}{2}\frac{di}{dt} = 10$$

or $\dfrac{di}{dt} + 40i = 20$

Its solution is given as

$$i(t) = e^{-40t} \int 20\,e^{40t}\,dt + k\,e^{-40t}$$

$$= 20\,e^{-40t}\left[\frac{e^{40t}}{40}\right] + K e^{-40t}$$

$$i(t) = \frac{1}{2}\left[1 + 2K e^{-40t}\right]$$

at $t = 0^+$

$$i(0^+) = \frac{1}{2} + k = \frac{3}{5}$$

or

$$K = \frac{3}{5} - \frac{1}{2} = \frac{1}{10}$$

∴

$$i(t) = \frac{1}{2} + \frac{1}{10} e^{-40t} \quad \textbf{Ans.}$$

The above equation can also be solved by

$$i(t) = (i(0^+) - i(\infty)) e^{-\frac{t}{\tau}} + i(\infty)$$

Where

$$i(0^+) = \frac{3}{5}, i(\infty) = \frac{1}{2}, \tau = \frac{1}{40} = \frac{L}{R} \quad \textbf{Ans.}$$

Problem 4.5. The network of Prob. 4.4 reaches a steady state in position 2 and at $t = 0$ the switch is moved to position 1. Find $i(t)$ for the numerical values given for the element, sketch the waveform, and show the value of the time constant.

Solution:

At time $t = (0^-)$

$$i(0^-) = \frac{10\,V}{20\,\Omega} = 1/2 \text{ amp.}$$

$$i(0^+) = i(0^-) = 1/2 \text{ amp.}$$

At time $t > 0$, the circuit equation is given as

$$i(50) + \frac{1}{2} \frac{di}{dt} = 30 \text{ V}$$

or

$$\frac{di}{dt} + 100i = 60 \text{ V}$$

Its solution is given as:

$$i(t) = 60 e^{-100t} \int e^{100t} + K e^{-100t}$$

$$i(t) = \frac{60}{100} + K e^{-100t}$$

$$i(t) = \frac{3}{5} + K = e^{-100t}$$

at time $t = 0^+$

$$i(0^+) = \frac{3}{5} + K = 1/2$$

So
$$K = \frac{1}{2} - \frac{3}{5}$$

$$= -\frac{1}{10}$$

∴
$$i(t) = \frac{3}{5} - \frac{1}{10}e^{-100t}$$

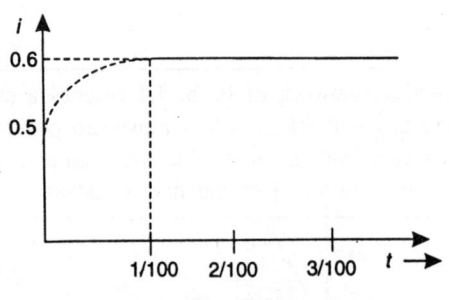

Fig. P. 4.5

Problem 4.6. In the given network, $v_1 = e^{-t}$ for $t \geq 0$ and is zero for all $t < 0$. If the capacitor is initially uncharged, find $v_2(t)$. Let $R_1 = 10$, $R_2 = 20$, and $C = \dfrac{1}{20}$ F, and for these values sketch $v_2(t)$ indentifying the value of the time constant on the sketch.

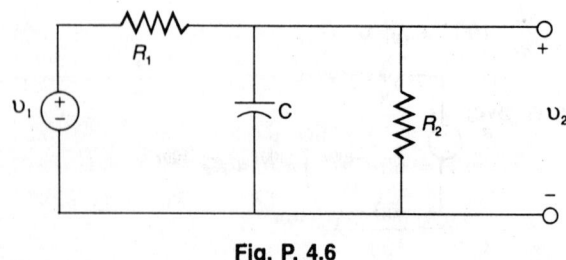

Fig. P. 4.6

Solution:

For, $\qquad t \geq 0, \quad \dfrac{v_2 - v_1}{R_1} + \dfrac{C dv_2}{dt} + \dfrac{v_2}{R_2}$

or $\quad \dfrac{dv_2}{dt} + 3v_2 = 2e^{-t}$

∴ Its solution is given as

$$v_2(t) = e^{-3t} \int 2e^{-t} \cdot e^{+3t} dt + Ke^{-3t} = 2e^{-3t} \left[\frac{e^{2t}}{2} \right] + Ke^{-3t}$$

$\qquad v_2(t) = e^{-t} + Ke^{-3t}$

at time $\quad t = 0^-, v_2(t) = 0$

So $\qquad 0 = 1 + K \quad$ or $\quad K = -1$

So $\quad v_2(t) = e^{-t} + (-1)e^{-3t}$

or $\quad v_2(t) = e^{-t} - e^{-3t}$

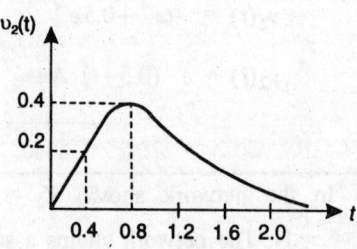

Fig. P. 4.6 (a)

Problem 4.7. In the network shown in the following figure, switch K is closed at $t = 0$ connecting a source e^{-t} to the RC network. At $t = 0$, it is observed that the capacitor voltage has the value $v_c(0) = 0.5$ V. For the element values given, determine $v_2(t)$.

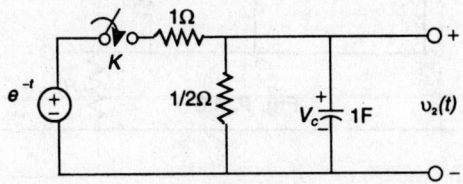

Fig. P. 4.7

Solution:

At time $t > 0$

$$\frac{e^{-t} - v_2}{1} + \frac{v_2}{\frac{1}{2}} + \frac{C dv_2}{dt} = 0$$

or $\qquad e^{-t} - v_2 + 2v_2 + \dfrac{dv_2}{dt} = 0$

or $\qquad v_2 + \dfrac{dv_2}{dt} = -e^{-t}$

$\therefore \qquad v_2(t) = e^{-t} \int (-e^{-t}) \cdot e^{t} \, dt + K e^{-t}$

$$v_2(t) = -t e^{-t} + K e^{-t}$$

At time $t = 0$, $v_2(t) = v_c(0) = 0.5$ V

or $\qquad 0.5 = 0 + K$

$\therefore \qquad K = 0.5$

or $\qquad v_2(t) = -t e^{-t} + 0.5 e^{-t}$

$$v_2(t) = e^{-t}(0.5 - t) \textbf{ Ans.}$$

Problem 4.8. In the network shown $V_0 = 3$V, $R_1 = 10\Omega$, $R_2 = 5\Omega$, and $L = \dfrac{1}{2}$ H. The network attains a steady state, and at $t = 0$ switch K is closed. Find $v_a(t)$ for $t \geq 0$.

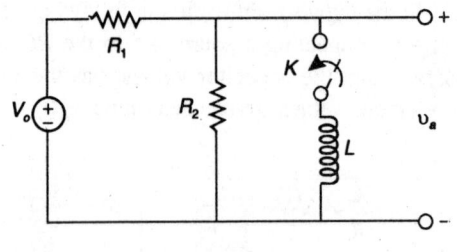

Fig. P. 4.8

Solution:

For $t > 0$, the circuit equation is:

$$\frac{3-V_a}{10}+\frac{v_a}{5}+\frac{1}{\frac{1}{2}}\int v_a\,dt=0$$

or $$0.3-0.1v_a+0.2v_a+2\int v_a\,dt=0$$

or $$0.3+0.1v_a+2\int v_a\,dt=0$$

Differentiating the above equation

$$0.1\frac{dv_a}{dt}+2v_a=0$$

or $$\frac{dv_a}{dt}+20v_a=0$$

or $$v_a(t)=4e^{-20t}$$

At time $t=0$ $$v_a(0)=\frac{3}{5+10}\times5=1\text{V}$$

$\therefore$ $$v_a(t)=1\cdot e^{-20t}$$

or $$v_a(t)=e^{-20t}\ \textbf{Ans.}$$

Problem 4.9. The network of the following figure consists of a current source of value I_0 (a constant), two resistors, and a capacitor. At $t=0$, the switch K is opened. For the element values given in the figure, determine $v_2(t)$ for $t\geq0$.

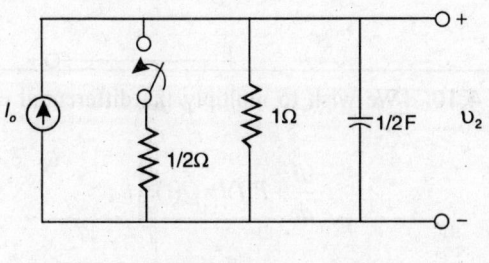

Fig. P. 4.9

Solution:

At time $t<0$, the capacitor acts as open circuit.

So $$v_c(0^-)=\frac{I_0}{3}$$

At time $t > 0$, the circuit equation is given as

$$\frac{1}{2}\frac{dv_2}{dt}+\frac{v_2}{1} = I_0$$

or $\qquad \dfrac{dv_2}{dt}+2v_2 = 2I_0$

Its solution is given as

$$v_2(t) = e^{-2t}\int(2I_0)\cdot e^{2t}\,dt+Ke^{-2t}=2I_0\,e^{-2t}\left[\frac{e^{2t}}{2}\right]+Ke^{-2t}$$

$$v_2(t) = I_0+Ke^{-2t}$$

At time $t = 0$, $v_2(0)=v_c(0^-)=\dfrac{I_0}{3}$

$\therefore \qquad\qquad \dfrac{I_0}{3} = I_0 + K$

or $\qquad\qquad K = \dfrac{-2I_0}{3}$

$\therefore \qquad\qquad v_2(t) = I_0 -\dfrac{2I_0}{3}e^{-2t}$

or $\qquad\qquad v_2(t) = \dfrac{I_0}{3}\left(3-2e^{-2t}\right)$ **Ans.**

Problem 4.10. We wish to multiply the differential equation

$$\frac{di}{dt}+P(t)i=Q(t)$$

by an "integrating factor" R such that the left-hand side of the equation equals the derivative $d(Ri)/dt$. (a) Show that the required integrating factor is $R=e^{\int Pdt}$ (b) Using this integrating factor, find the solution to the differential equation.

Solution:

(a) Multiplying $R = e^{\int Pdt}$ on both the side of the above equation

We get $\qquad R\dfrac{di}{dt} + RP(t)i = Q.R \qquad\qquad (1)$

$\therefore \qquad \dfrac{d(Ri)}{dt} = R\dfrac{di}{dt} + i\dfrac{dR}{dt} \qquad\qquad (2)$

$\therefore$ LHS of eq (1) is equal to $d(Ri)/dt$ according to the question.

So $\quad R\dfrac{di}{dt} + Rp(t)i = \dfrac{d(Ri)}{dt} = i\dfrac{dR}{dt} + R\dfrac{di}{dt}$

On comparison $\qquad \dfrac{dR}{dt} = Rp(t)$

or $\qquad\qquad\qquad \dfrac{dR}{R} = p(t)\,dt$

or $\qquad\qquad\qquad lnR = \int p(t)\,dt$

or $\qquad\qquad\qquad R = e^{\int p(t)\,dt}$

So, the required integrating factor $R = e^{\int p(t)\,dt}$

(b) Now, the equation (1) can be written as:

$$\dfrac{d(iR)}{dt} = Q\cdot R$$

or $\qquad\qquad \dfrac{d(i\cdot e^{pt})}{dt} = Q\cdot e^{pt}$

This equation may be integrated to give

$$i\cdot e^{pt} = \int Q\cdot e^{pt}\,dt + K$$

or $\qquad\qquad i(t) = e^{-pt}\int Q\cdot e^{pt}\,dt + k\,e^{-pt}$ **Ans.**

Problem 4.11. In the network shown in the following figure, the switch K is closed at $t = 0$, a steady-state having previously been attained. Solve for the current in the circuit as a function of time.

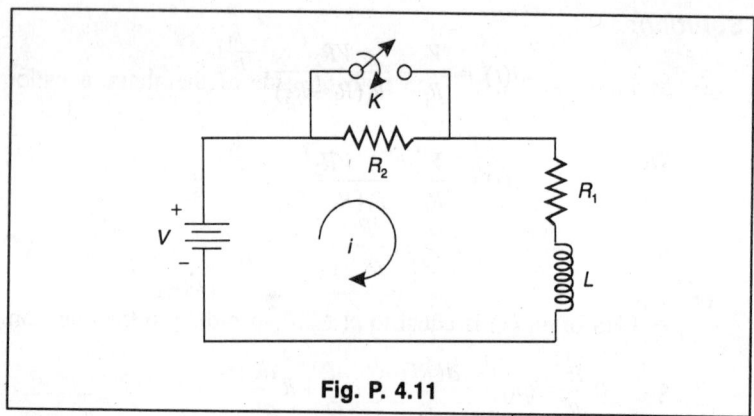

Fig. P. 4.11

Solution:

For $t = (0^-)$ $i(0^-) = \dfrac{V}{R_1 + R_2}$

$\because$ $(i(0^-)) = i(0^+)$

$= \dfrac{V}{R_1 + R_2}$

For $t > 0$, the circuit equation is given as

$$L\frac{di}{dt} + R_1 i = V$$

or $\dfrac{di}{dt} + \dfrac{R_1 i}{L} = \dfrac{V}{L}$

Its solution is given as

$$i(t) = e^{-\frac{R_1}{L}t} \int \frac{V}{L} e^{\frac{R_1}{L}t} \, dt + K e^{\frac{R_1}{L}t}$$

or $i(t) = \dfrac{V}{R_1} + K e^{\frac{-R_1}{L}t}$

At $t = 0,\ i(0) = \dfrac{V}{R_1 + R_2}$

$\therefore$ $\dfrac{V}{R_1 + R_2} = \dfrac{V}{R_1} + K$

or $K = \dfrac{-V R_2}{R_1 (R_1 + R_2)}$

$$\therefore \qquad i(t) = \frac{V}{R_1} - \frac{VR_2}{R_1(R_1+R_2)} e^{\frac{-R_1}{L}t}$$

$$i(t) = \frac{V}{R_1} - \frac{VR_2}{R_1(R_1+R_2)} e^{\frac{-R_1}{L}t}$$

or $\qquad i(t) = \frac{V}{R_1}\left(1 - \frac{R_2}{(R_1+R_2)} e^{\frac{-R_1}{L}t}\right)$ **Ans.**

Problem 4.12. In the network shown in the following figure, the voltage source follows the law $v(t) = Ve^{-\alpha t}$, where α is a constant. The switch is closed at $t = 0$: (a) Solve for the current assuming that $\alpha \neq R/L$. (b) Solve for the current when $\alpha = R/L$.

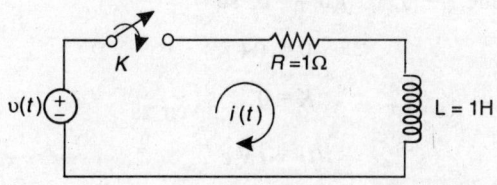

Fig. P. 4.12

Solution:

(a) At time $t > 0$ and $\alpha \neq R/L$

$$\frac{di}{dt} + i = Ve^{-\alpha t}$$

$$\therefore \qquad i(t) = e^{-t}\int Ve^{-\alpha t}\cdot e^t dt + Ke^{-t}$$

$$= V\cdot e^{-t}\left[\frac{e^{t-\alpha t}}{1-\alpha}\right] + Ke^{-t}$$

$$= \frac{V\cdot e^{-t}\cdot e^{(1-\alpha)t}}{(1-\alpha)} + Ke^{-t}$$

$$i(t) = \frac{V\cdot e^{-\alpha t}}{(1-\alpha)} + Ke^{\,t}$$

At time $t = 0^-$, $i(0^-) = 0$

So $0 = \dfrac{V}{(1-\alpha)} + K$

or $K = -\dfrac{V}{1-\alpha}$

∴ $i(t) = \dfrac{V}{1-\alpha}(e^{-\alpha t} - e^{-t})$

(b) When $\alpha = R/L = 1$

    Then $i(t) = e^{-t}\int V \cdot e^{-t} \cdot e^{-t}\, dt + Ke^{-t}$

                                 $i(t) = Ve^{-t}[t] + Ke^{-t}$

At time $t = 0$, $i(0^-) = 0$, so

                                 $0 = 0 + K$

∴ $K = 0$

∴ $i(t) = tVe^{-t}$

                                 $i(t) = V \cdot te^{-t}$ **Ans.**

Problem 4.13.   In the network shown in Fig. P. 4.13, $v(t) = 0$ for $t < 0$, and $v(t) = t$ for $t \geq 0$. Show that $i(t) = t - 1 + e^{-t}$ for $t \geq 0$, and sketch this waveform.

Solution:

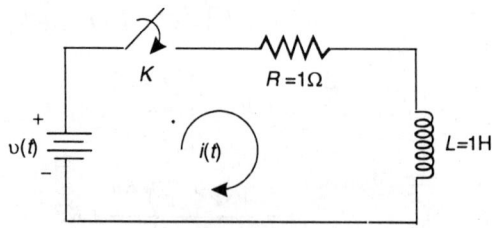

Fig. P. 4.13

For $t \geq 0$, $v(t) = t$

$$\therefore \qquad \frac{di}{dt} + i = t$$

$$\therefore \qquad i(t) = e^{-t} \int t \cdot e^{t} \, dt + K e^{-t}$$

$$= e^{-t} (t-1) e^{t} + K e^{-t}$$

$$i(t) = (t-1) + K e^{-t}$$

$$\therefore \qquad i(0) = 0 \text{ at time } t = 0$$

$$\therefore \qquad i(0) = 0 = -1 + K$$

or $\qquad K = 1$

$$\therefore \qquad i(t) = (t-1) + e^{-t} \text{ Ans.}$$

Fig. P. 4.13 (a)

Problem 4.14. In the network shown in the following figure, the switch is closed at $t = 0$ connecting a voltage source $v(i) = V\sin\omega t$ to a series *RL* circuit. For this system, solve for the response $i(t)$.

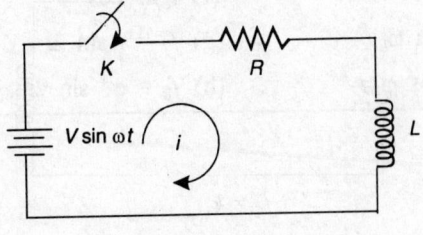

Fig. P. 4.14

Solution:

$$\frac{di}{dt} + i = V \sin \omega t$$

$\therefore$
$$i(t) = e^{-t} \int V \sin \omega t \cdot e^t \, dt + Ke^{-t}$$

$$= V \cdot e^{-t} \left[\frac{e^t \sin \omega t - \omega \cos \omega t \cdot e^t}{\omega^2 + 1} \right] + Ke^{-t}$$

$$= \frac{V \sin \omega t - V \omega \cos \omega t}{\omega^2 + 1} + Ke^{-t}$$

At time $t = 0$, $i(t) = 0$. So at $t = 0$.

$$0 = \frac{-V\omega}{\omega^2 + 1} + K$$

or
$$K = \frac{V\omega}{\omega^2 + 1}$$

$\therefore$
$$i(t) = \frac{V}{\omega^2 + 1} \left[\sin \omega t - \omega \cos \omega t + e^{-t} \omega \right] \textbf{ Ans.}$$

Problem 4.15. Consider the differential equation

$$\frac{di}{dt} + dt = f_k(t)$$

Where a is real and positive. Find the general solution of this equation if all $f_k = 0$ for $t < 0$ and for $t \geq 0$ have the following values:

(a) $f_1 = k_1 t$ (e) $f_5 = \sin^2 t$

(b) $f_2 = te^{-2t}$ (f) $f_6 = \cos^2 t$

(c) $f_3 = \sin \omega_0 t$ (g) $f_7 = t \sin 2t$

(d) $f_4 = \cos \omega_0 t$ (h) $f_8 = e^{-t} \sin 2t$

Solution:

(a)
$$f_1 = k_1 t$$

$\therefore$
$$i(t) = e^{-at} \int k_1 t \cdot e^{at} \, dt + k e^{-at}$$

$$i(t) = \frac{k_1 t}{a} - \frac{k_1}{a^2} + k e^{-at}$$

at $\qquad t = 0, \quad i(0) = 0$

So $\qquad\qquad k = \dfrac{k_1}{a^2}$

$\therefore \qquad\qquad i(t) = \dfrac{k_1 t}{a} - \dfrac{k_1}{a^2} + \dfrac{k_1}{a^2} e^{-at}$

or $\qquad\qquad i(t) = \dfrac{k_1 t}{a} + \dfrac{k_1}{a^2}(e^{-at} - 1)$

(b) When $\qquad\qquad f_2 = te^{-2t}$

$$i(t) = e^{-at} \int t \cdot e^{-2t} \cdot e^{at}\, dt + k e^{-at}$$

$$= \frac{te^{-2t}}{a-2} - \frac{e^{-2t}}{(a-2)^2} + k e^{-at}$$

At time $t = 0$, $\quad i(0) = 0$, so

$$0 = -\frac{1}{(a-2)^2} + k$$

or $\qquad\qquad k = \dfrac{1}{(a-2)^2}$

$\therefore \qquad\qquad i(t) = \dfrac{te^{-2t}}{(a-2)} - \dfrac{e^{-2t}}{(a-2)^2} + \dfrac{1}{(a-2)^2} e^{-at}$

(c) Similar to Problem 4.15 for $f_3 = V_m \sin\omega_0 t$

(d) Similar to Problem 4.15 for $f_3 = V_m \cos\omega_0 t$

(e) $\qquad f_5 = \sin^2 t$

$$i(t) = e^{-at} \int \sin^2 t \cdot e^{-at}\, dt + k e^{-at}$$

$$= \frac{1}{2a} - \frac{a\cos 2t + 2\sin 2t}{2(a^2+4)} + k e^{-at}$$

At time $t = 0$, $\quad i(0) = 0$, so

$$0 = \frac{1}{2a} - \frac{a}{2(a^2+4)} + k$$

or $$k = \frac{a}{2(a^2+4)} \cdot \frac{1}{2a} = \frac{a^2 - a^2 - 4}{2a(a^2+4)}$$

or $$k = \frac{a^2 - 4 - a^2}{2a(a^2+4)} = \frac{-2}{a(a^2+4)}$$

$\therefore$ $$i(t) = \frac{1}{2a} - \frac{a\cos 2t + 2\sin 2t}{2(a^2+4)} - \frac{2e^{-at}}{a(a^2+4)}$$

(f) For $f_6 = \cos^2 t$, Similarly as above

(g) $\qquad f_7 = t\sin 2t$

$$i(t) = e^{-at} \int t\sin 2t \cdot e^{at}\, dt + k e^{-at}$$

Let $u = t\sin 2t$ and $v = e^{at}$

$$i(t) = \left[t\sin 2\frac{e^{at}}{a} - \int \frac{e^{at}}{a}\sin 2t\, dt + \frac{2}{a}\int \frac{e^{at}}{a}\cos 2t \right.$$

$$\left. \frac{-2t\cos 2t}{a^2}e^{at} \right] \frac{e^{-at} \times a^2}{(a^2-4)} + ke^{-at}$$

$$= \left[e^{at}\left(\frac{at\sin 2t - 2t\cos 2t}{a^2} \right) - \frac{1}{a}\left(\frac{a\cos 2t - 2\cos 2t}{a^2+4} \right)e^{at} \right.$$

$$\left. + \frac{2}{a^2}\left(\frac{a\cos 2t - 2\sin 2t}{a^2+4} \right)e^{at} \right] \frac{e^{-at}a^2}{a^2-4} + ke^{-at}$$

$$= \left[\frac{t(a\sin 2t - 2\cos 2t)}{a^2\theta - 4} + \frac{a^2\sin 2t - 4\sin 2t - 4a\cos 2t}{(a^2+4)(a^2-4)} \right] + ke^{-at}$$

$$= \left[\frac{t(a\sin 2t - 2\sin 2t)}{a^2-4} + \frac{(a^2-4)\sin 2t - 4a\cos 2t}{(a^4 - (4)^2)} \right] + ke^{-at}$$

or $\qquad$ at $t = 0$, $i(0) = 0$

So $$k = \frac{4a}{a^4 - 4^4} = \frac{4a}{a^4 - 16}$$

So $$i(t) = \frac{t(a\sin 2t - 2\sin 2t)}{a^2-4} + \frac{(a^2-4)\sin 2t - 4a\cos 2t}{a^4 - 16} + \frac{4ae^{-at}}{a^4 - 16}$$

Ans.

(h) For $f_8 = e^{-t} \sin 2t$; we can find in similar manner as all above.

Problem 4.16. In the network shown, the switch K is closed at $t = 0$. The current waveform is observed with a cathode ray oscilloscope. The initial value of the current is measured to be 0.01 amp. The transient appears to disappear in 0.1 sec. Find (a) the value of R, (b) the value of C, and (c) the equation of $i(t)$.

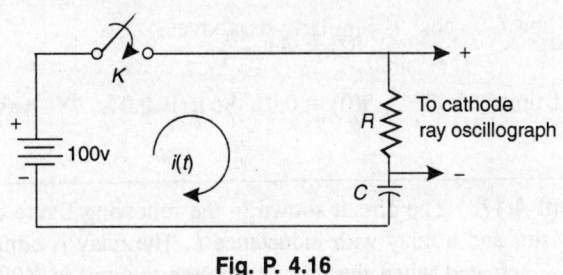

Fig. P. 4.16

Solution:

(a) In steady state, current in the circuit is zero but at time $t = 0^+$, capacitor acts as a short circuited. So, current through

$$i(0) = \frac{100}{R}.$$

$$\because \qquad\qquad i(0) = 0.01 \text{ amp.}$$

So $$R = \frac{100}{0.01}$$

$$\therefore \qquad\qquad R = 10 \text{ k}\Omega$$

(b) Time constant of the circuit is $T = RC = \dfrac{0.1 \text{ sec}}{4}$

So $$C = \frac{0.025}{10 \times 10^3} \text{F}$$

or $$C = 2.5 \times 10^{-6} \text{F}$$

or $$C = 2.5 \text{ }\mu\text{F}$$

(c) For $t > 0$, the equation of $i(t)$ is given as

$$iR + \frac{1}{C} \int i\, dt = 100$$

or differenciating both sides w.r.t. time

$$\frac{Rdi}{dt} + \frac{1}{C}i = 0$$

or

$$\frac{di}{dt} + \frac{1}{RC}i = 0$$

So

$$i(t) = A e^{-\frac{1}{RC}t}$$

So

$$i(t) = A e^{-40t}$$

At time $t = 0$, $\quad i(0) = 0.01$, So $i(t) = 0.01e^{-40t}$ **Ans.**

Problem 4.17. The circuit shown in the following figure consists of a resistor and a relay with inductance L. The relay is adjusted so that it is actuated when the current through the coil is 0.008 amp. The switch K is closed at $t = 0$, and it is observed that the relay is actuated when $t = 0.1$, sec. Find: (a) the inductance L of the coil, (b) the equation of $i(t)$ with all terms evaluated.

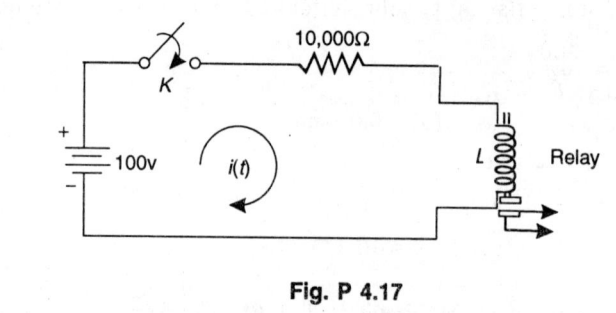

Fig. P 4.17

Solution:

At time $t > 0$ $\quad iR + L\dfrac{di}{dt} = 100$ at time $t = (0^+)$, inductor acts as open

circuit. So, $\quad i(0^+) = 0$

At time $t = (\infty)$, inductor acts as short circuit

So $\quad i(\infty) = \dfrac{100}{R} = \dfrac{100}{10,000} = 10^{-2}$ amp.

So $\quad i(t) = (i(0^+) - i(\infty))e^{-\frac{R}{L}t} + i(\infty) = (0 - 10^{-2})e^{-\frac{10^4 t}{L}} + 10^{-2}$

$$i(t) = \left(10^{-2} 1 - e^{-\frac{10^4 t}{L}} \right)$$

When time $t = 0.1$ sec,

$$i(t) = 0.008 \text{ amp.}$$

So

$$0.008 = \left(10^{-2} 1 - e^{-\frac{10^{4t}}{L} \times 0.1} \right)$$

$$0.8 = \left(1 - e^{-\frac{10^3}{L}} \right)$$

$$e^{-\frac{10^3}{L}} = 0.2$$

Taking antilog we, get

$$-\frac{10^3}{L} = -1.609$$

$$L = 621.33 \text{ H } \textbf{Ans.}$$

Problem 4.18. A switch is closed at $t = 0$, connecting a battery of voltage V with a series RC circuit. (a) Determine the ratio of energy delivered to the capacitor to the total energy supplied by the source as a function of time. (b) Show that this ratio approaches 0.50 as $t \to \infty$.

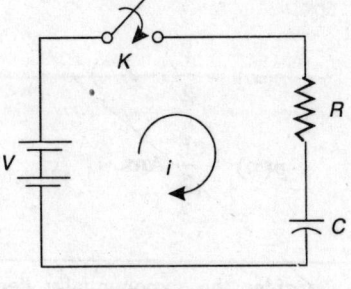

Fig. P. 4.18

Solution:

$$V = iR + \frac{1}{C} \int i \, dt$$

$$\because \qquad i(0) = \frac{V}{R}, \; i(\infty)=0$$

$$\therefore \qquad i(t) = \left(\frac{V}{R}-0\right)e^{-\frac{t}{RC}} +0$$

$$i(t) = \frac{V}{R}e^{-\frac{t}{RC}}$$

$$q=\int_0^t idt = \frac{V}{R}\left[\frac{e^{-\frac{t}{RC}}}{-1/RC}\right]_0^t =CV\left[1-e^{-\frac{t}{RC}}\right]$$

or

$$q = CV\left[1-e^{-\frac{t}{RC}}\right]$$

$$W_c =\frac{1}{2}CV^2 = \frac{q^2}{2C}=\frac{CV^2}{2}\left[1-e^{-\frac{t}{RC}}\right]^2$$

$$W_T =qV = CV^2\left[1-e^{-\frac{t}{RC}}\right]$$

(a)

$$p(t)=\frac{W_c}{W_T} = \frac{1}{2}\left(1-e^{-\frac{t}{RC}}\right)$$

(b)

$$p(\infty) = \frac{1}{2}(1-e^{-\infty})$$

$$p(\infty) = \frac{1}{2} \; \textbf{Ans.}$$

Problem 4.19. Consider the exponentially decreasing function $i = Ke^{-t/\tau}$ where T is the time constant. Let the tangent drawn from the curve at $t = t_1$ intersect the line $i = 0$ at t_2. Show that for any such point, $i(t_1)$, $t_2 - t_1 = T$.

Solution:

Given $i = K\,e^{-t/T}$ [where T = time constant]

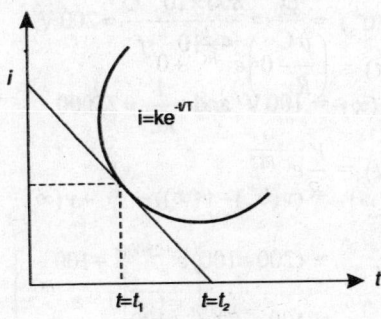

Fig. P. 4.19

$\therefore$ $\qquad\qquad i = K\,e^{-t/T}$

Differentiating both sides to get a tangent:

$$\frac{di}{dt} = \frac{-K}{T}e^{-t/T}$$

or

$$\frac{di}{dt} = \frac{i-0}{t_1-t_2} = \frac{-K}{T}e^{-t/T}$$

$$\frac{-Ke^{-t/T}}{t_1-t_2} = \frac{-Ke^{-t/T}}{T}$$

$$t_2 - t_1 = T \quad \textbf{Ans.}$$

Problem 4.20. The capacitor in the circuit shown in Fig. P. 4.20 has initial charge $Q_0 = 800\ \mu c$, with polarity as indicated. If the switch is closed at $t = 0$, obtain the current and charge for $t > 0$.

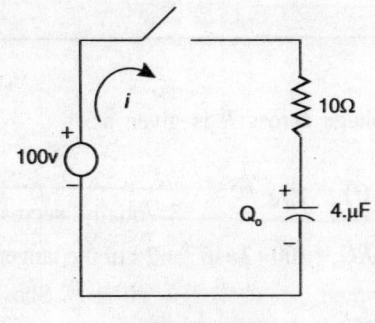

Fig. P. 4.20

Solution:

$$v_c(0^+) = \frac{Q_0}{C} = \frac{800 \times 10^{-6} C}{4 \times 10^{-6} f} = 200 \text{ V}$$

At $\quad t = \infty$, $v_c(\infty) = 100 \text{ V}$ and $\dfrac{1}{RC} = 25000$

$\therefore \qquad\qquad v_c = (v(0^+) - v(\infty)) e^{-\frac{t}{RC}} + v(\infty)$

$$= (200 - 100) e^{-25000t} + 100$$

$$= 100 e^{-25000t} + 100$$

$\therefore \qquad\qquad v_c = 100(1 + e^{-25000t})$

Now, charge $Q = Cv_c$ for $t > 0$

$\therefore \qquad Q = 4 \times 10^{-6} \times 100(1 + e^{-25000t}) = 4 \times 10^{-4}(1 + e^{-25000t})$

Current i is given by:

$$i = \frac{dq}{dt} = 4 \times 10^{-4} \cdot e^{-25000t} \cdot (-25000)$$

$$= -4 \times 10^{-4} \times 25000 \cdot e^{-25000t}$$

$$i = -10 \cdot e^{-25000t} \textbf{ Ans.}$$

Problem 4.21. A 2 μF capacitor, with initial charge $Q_0 = 100$ μc, is connected across a 100 Ω resistor at $t = 0$. Calculate the time in which the transient voltage across the resistor drops from 40 to 10 volts.

Solution:

In a RC circuit voltage across R is given as.

$$v_R(t) = V_0 e^{-\frac{t}{RC}}$$

Here, $\qquad\qquad RC = 100 \times 2 \times 10^{-6} = 2 \times 10^{-4}$

$\therefore \qquad\qquad \dfrac{1}{RC} = \dfrac{1}{2 \times 10^{-4}} = 5000, \; v_R(40) = V_0 \, e^{-\frac{t_1}{2 \times 10^{-4}}} \qquad (1)$

$$v_R(10) = V_0 e^{-\frac{t_2}{2\times10^{-4}}} \qquad (2)$$

$$\therefore \qquad \frac{40}{10} = \frac{e^{-5000 t_1}}{e^{-5000 t_2}}$$

$$= e^{-5000(t_1 - t_2)}$$

$$4 = e^{-5000(t_1 - t_2)}$$

Taking ln both sides.

$$\ln 4 = \ln e^{-5000(t_1 - t_2)}$$

$$1.386 = -5000(t_1 - t_2)$$

or $$(t_2 - t_1) = \frac{1.386}{5000}$$

$$(t_2 - t_1) = 0.277 \text{ ms } \textbf{Ans.}$$

Problem 4.22. In the RC circuit shown in Fig. P. 4.22, the switch is closed on position 1 at $t = 0$ and then moved to 2 after the passage of one time constant. Obtain the current transient for (a) $0 < t < \tau$ (b) $t > \tau$.

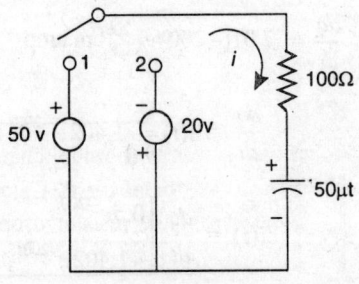

Fig. P. 4.22

Solution:

(a) For $0 < t < \tau$, $q = Ae^{-t/\tau} + B$

Where $A = -B = -2.5$ mc

and $\tau = RC = \dfrac{1}{200}$

$$\therefore \qquad q = 2.5\left(1 - e^{-\frac{t}{1/200}}\right) \text{mc}$$

$$q = 2.5\left(1 - e^{-200t}\right)\text{mc}$$

$$\therefore \qquad i = \frac{dq}{dt} = 2.5(-(-200)e^{-200t})\text{m amp}$$

$$i = 500 \cdot e^{-200t} \text{ m amp}$$

or $\qquad i = 0 \cdot 5 e^{-200t}$ amp

(b) For $t \geq \tau \qquad q = [q(\tau) - q(\infty)]e^{-(t-\tau)/\tau} + q(\infty)$

Where $\qquad q(\tau) = 2.5(1 - e^{-1})\text{mc}$

$$q(\infty) = -(50 \times 10^{-6})(20)c$$

$$= -(1000 \times 10^{-6})c = -1\text{mc}$$

$$\therefore \qquad q = [2.5(1 - e^{-1}) + 1]e^{-t/\tau} \cdot e^1 - 1\text{mc}$$

$$q = [7.01]e^{-200t} - 1 \text{ mc}$$

$$\therefore \qquad \frac{dq}{dt} = 7.01(-200)e^{-200t} \text{ m amp}$$

or $\qquad \dfrac{dq}{dt} = i(t) = -1.402e^{-200t}$ amp

$$\therefore \qquad i(t) = \frac{dq}{dt}\begin{cases} 0.5e^{-200t}, & (0 < t < \tau) \\ -1.402e^{-200t}, & (t \geq \tau) \end{cases} \textbf{Ans.}$$

Problem 4.23. A 10 μf capacitor, with initial charge Q_0 is connected across a resistor at $t = 0$. Given that the power transient for the capacitor is $800 \, e^{-4000t}$ (W), Find R, Q_0, and the initial stored energy in the capacitor.

Solution:

Given $\qquad P_R = P_C = 800 \, e^{-4000t}$ \hfill (1)

$$P_R = v_R \cdot i_R = (v_0 \, e^{-t/\tau}) \cdot \left(\frac{v_0}{R} e^{-t/\tau} \right)$$

$$P_R = \frac{v_0^2}{R} e^{-2t/\tau} \qquad (2)$$

Comparing eq (1) and eq (2)

$$\therefore \qquad \frac{2}{\tau} = 4000$$

or $$\tau = \frac{1}{2000} = RC = R \times 10 \times 10^{-6}$$

$$\therefore \qquad R = \frac{10^5}{2000} = 50\,\Omega \text{ and } \frac{v_0^2}{R} = 800$$

$$v_0^2 = 800\,R = 800 \times 50 = 40,000$$

$$\therefore \qquad v_0 = 200 \text{ V}$$

$$\therefore \qquad q_0 = Cv_0 = 200 \times 10 \times 10^{-6}, q_0 = 2000\,\mu C$$

Initial stored energy is given as:

$$W_R = \int_0^t P_R \, dt = \int_0^t 800 \, e^{-4000t} \, dt$$

$$W_R = -0.2 \, e^{-4000t} + c$$

at $t = 0, W(0) = -0.2 + c = 0$

$$\therefore \qquad W_R = 0.2(1 - e^{-4000t})\,\text{J}$$

at $t = 0$, W_R (initial stored energy)

$$W_R = \frac{1}{2} C v_0^2 = \frac{1}{2} 10 \times 10^{-6} \times (200)^2 = 0.2\,\text{J} \quad \textbf{Ans.}$$

Problem 4.24. A series *RL* circuit, with $R = 10\,\Omega$ and $L = 1H$, has a 100 V source applied at $t = 0$. Find the current for $t > 0$.

Solution:

$$i(t) = v_{0/R} (1 - e^{-\frac{R}{L}t}) \text{ for } t > 0 = \frac{100}{10} (1 - e^{-\frac{10}{1}t})\,\text{amp}$$

$$i(t) = 10(1 - e^{-10t})\,\text{amp} \quad \textbf{Ans.}$$

Problem 4.25. In Fig. P. 4.25, the switch is closed on position 1 at $t = 0$, then moved to 2 at $t = 1$ ms. Find the time at which the voltage across the resistor is zero, reversing polarity.

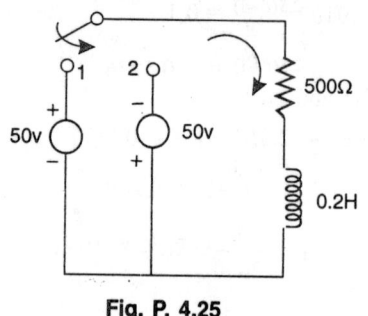

Fig. P. 4.25

Solution:

When switch is on position 1.

Then

$$i(t) = i_0 \left(1 - e^{-\frac{R}{L}t}\right)$$

$\therefore$

$$i(1\,\text{ms}) = \frac{50}{500}(1 - e^{-2.5}) \qquad\qquad \left[\because \frac{R}{L} = 2500\right]$$

$$= \frac{1}{10}(1 - e^{-2.5})$$

$$i(1\text{ms}) = 0.091$$

When switch is thrown on position 2 then $i(1\text{ms})$ will acts as $i(0)$. Now after time $t' = 1$ms current in inductor is:

$$i(t) = (i(0) - i(\infty))\, e^{-\frac{(t - t')}{RC}} + i(\infty)$$

Where

$$i(\infty) = -\frac{50}{500} = -0.1 \text{ amp.} \qquad \text{(because polarity is reversed)}$$

$\therefore$

$$i(t) = (0.091 - (-0.1))\, e^{-2500(t-1)\times 10^{-3}} + (-0.1)$$

$$\text{(where } t \text{ in m sec)}$$

$\therefore$

$$i(t) = 0.191 e^{-2.5(t-1)} - 0.1$$

$\therefore$

$$v_R = i(t) \times R = (0.191 e^{-2.5(t-1)} - 0.1)\, R$$

For v_R to be zero. Let the time required is t'. Hence.

$$v_R = (0.191)e^{-2.5(t'-1)} - 0.1)R = 0$$

or $$0.191e^{-2.5(t'-1)} = 0.1$$

or $$e^{-2.5(t'-1)} = -0.5236$$

Taking ln both side $\quad -2.5(t'-1) = -0.6471$

$$(t'-1) = 0.260$$

or $$t' = 1+0.260$$

or $$t' = 1.260 \, \text{ms} \quad \textbf{Ans.}$$

Problem 4.26. A series RL circuit with $R = 100 \, \Omega$ and $L = 0.2$ H, has a 100 V source applied at $t = 0$; then a second sources of 50 V with the same polarity is switched in at $t = t'$, replacing the first source. Find t' such that the current is constant at 0.5 A for $t > t'$.

Solution:

$\therefore$ Time const $\left(\dfrac{L}{R}\right) = \dfrac{0.2}{100}$

$\therefore \qquad \dfrac{R}{L} = 500$

When switch is at position 1 for time t'. Then current is given as:

$$i(t') = \frac{100 \, \text{V}}{100 \, \Omega}(1-e^{-500t'})$$

$$i(t') = (1-e^{-500t'})$$

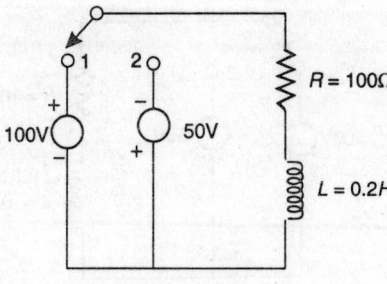

Fig. P 4.26

This $i(t')$ will acts as $i(0)$ when switch is thrown at position 2. Now the current in the circuit t at any time t is given as:

$$i(t) = (i(0) - i(\infty))e^{-\frac{(t-t')R}{L}} + i(\infty)$$

Where $i(\infty) = \dfrac{50\,V}{100} = 0.5\,amp$

$\therefore$ $i(t) = ((1-e^{-500t'}) - 0.5)e^{-(t-t')500} + 0.5$

For constant current of 0.5 A, $i(t) = 0.5$ A

$\therefore$ $0.5 = (0.5 - e^{-500t'})e^{-(t-t')500} + 0.5$

$\qquad\qquad = (0.5\,e^{-500(t-t')} - e^{-500t}$

or $e^{-500t} = 0.5\,e^{-500t} \cdot e^{500t'}$

or $e^{500t'} = \dfrac{1}{0.5} = 2$

Taking ln both sides. $500t' = 0.693$

or $t' = 1.386 \times 10^{-3}$

$\therefore$ $t' = 1.386\,m$ **Ans.** sec

Problem 4.27. The circuit of Problem 4.26 has a 50 V source of opposite polarity switched in at $t = 0.50$ ms, replacing the first source. Obtain the current for (a) $0 < t < 0.5$ ms (b) $t > 0.5$ ms.

Solution:

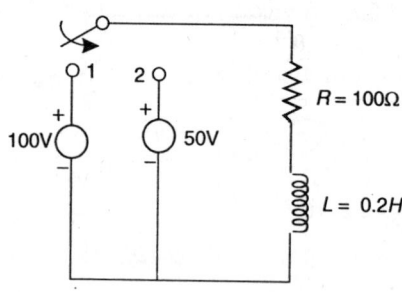

Fig. P. 4.27

Time constant of the circuit, $\dfrac{L}{R} = \dfrac{0.2}{100}$

When switch is at position 1, then current is given as:

$$i(t) = i_0 \left(1 - e^{-\frac{R}{L}t}\right)$$

Where $\qquad i_0 = \dfrac{100}{100} = 1 \text{ amp}$

$\therefore \qquad i(t) = 1(1 - e^{-500t}), 0 < t < 0.5 \text{ ms}$

or $\qquad i(t) = (1 - e^{-500t})$, for $0 < t < 0.5 \text{ ms}$

at $\qquad 0.5 \text{ ms} \cdot i(0.5) = (1 - e^{-500 \times 0.5 \times 10^{-3}}) = (1 - 0.7788)$

$$i(0.5) = 0.2211$$

This $i(0.5)$ will acts as $i(0)$ when switch is thrown at position 2. Then the current is given as:

$$i(t) = (i(0) - i(\infty)) e^{-\left(\frac{t - t'}{RC}\right)} + i(\infty)$$

Where $\qquad i(\infty) = -\dfrac{50}{100} = -0.5 \text{ amp}$

$\therefore \qquad i(t) = (0.2211 + 0.5) e^{-500(t - 0.0005)} - 0.5$

$$i(t) = (0.721) e^{-500(t - 0.0005)} - 0.5$$

For $t > 0.0005$ sec or $t > 0.5$ ms. **Ans.**

Problem 4.28. A voltage transient, $35 e^{-500t}$ (V), has the value 25 V at $t_1 = 6.73 \times 10^{-4}$ 5. Show that at $t = t_1 + \tau$ the function has a value 36.8% of that at t_1.

Solution:

Given $\tau = \dfrac{1}{500} = 2 \times 10^{-3}$

$\therefore \qquad v(t) = 35 e^{-500t} = 35 e^{-t/\tau}$

at $\qquad v(\tau) = 35 e^{-1} = 35 \times 0.3678 = 12.87 \text{ V}$

at $t=t_1+\tau = 6.73\times10^{-4}+20\times10^{-4}$

                              $= 26.73\times10^{-4}$

                     $v(t) = 35\,e^{-500}\,(26.73\times10^{-4})=35\,e^{-1.336}=9.196$

∴ $\dfrac{v(t)}{v(t_1)} = \dfrac{9.196}{25}=0.3678$

or $\dfrac{v(t)}{v(t_1)} = 36.78\%$

Hence the function at $(t = t_1 + \tau)$ has a value of 36.8% at t_1.

Problem 4.29. The circuit shown in Fig. P. 4.29 is switched to position 1 at $t = 0$, then to position 2 at $t = 3\tau$. Find the transient current i for (a) $0 < t < 3\tau$ (b) $t > 3\tau$.

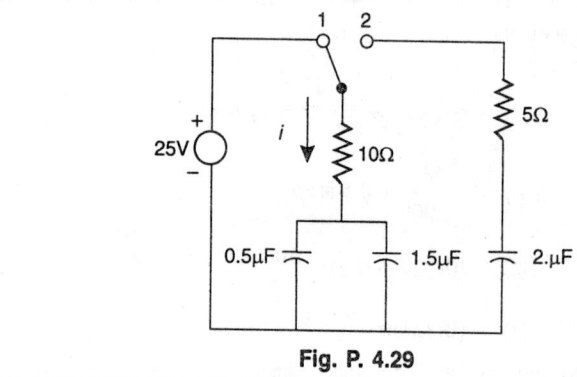

Fig. P. 4.29

Solution:

For $0 < t < 3\tau$, $q = A e^{-t/\tau} + B$

Where $A = -B = -25\times(0.5+1.5)\,\mu C = -50\,\mu C$

and $\tau = RC = \dfrac{1}{50{,}000}$ $[\because R=10\,\Omega, C=-2\mu F]$

∴ $q = 50\left(1-e^{-t(50000)}\right)\mu C$

∴ $i = \dfrac{dq}{dt}=50(-(-50{,}000))\,e^{-50{,}000t}\ \mu\ \text{amp}$

$$i = 2.5e^{-50,000t} \text{ amp}$$

For $t > 3\tau$ $\quad q(3\tau) = 50(1-e^{-3}) = 50(0.95) = 47.51\mu C$

$\therefore$ $\qquad$ at $t = 3\tau$

$$v_c = v_0(1-e^{-t/\tau}) = 25(1-e^{-3}) = 23.75$$

$\therefore$ $\quad q(\infty) = (+23.75 \text{ V}) \times (1.\mu F) = +23.75\mu C$

$\therefore$ $\quad q(t) = (q(0) - q(\infty))e^{-\left(\frac{t-\tau}{RC'}\right)} + q(\infty)$

$\therefore$ $\quad q(t) = (47.51 - 23.75)e^{-\frac{(t-\tau)}{RC'}} + 23.75\mu C$

Where $\qquad RC' = (10+5)(1\mu f) = 15 \times 10^{-6}$

$\therefore$ $\quad i(t) = \dfrac{dq(t)}{dt} = (23.75)\left(-\dfrac{1}{RC'}\right)e^{-\left(\frac{t-\tau}{RC'}\right)}$

$$= \frac{-23.75}{15 \times 10^{-6}}e^{-66666(t-0.00006)} \mu \text{ amp}$$

$$i(t) = 1.58e^{-66666(t-0.00006)} \text{ amp } \textbf{Ans.}$$

Problem 4.30. An *RL* circuit with $R = 300\ \Omega$ and $L = 1H$, has voltage $v = 100\cos(100t + 45°)$ (V) applied by closing a switch at $t = 0$. Obtain the resulting current for $t > 0$.

Solution:

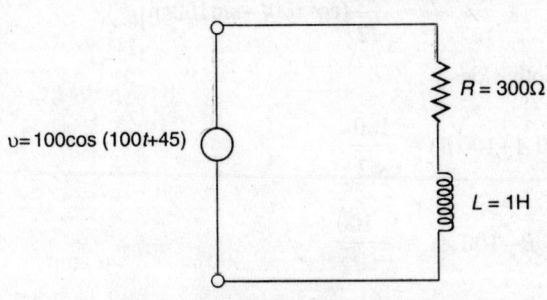

$v = 100\cos(100t+45)$
$R = 300\Omega$
$L = 1H$

Fig. P. 4.30

Time const $= \dfrac{L}{R} = \dfrac{1}{300}$

The circuit equation for $t > 0$ is

$$\frac{di}{dt} + \frac{R}{L} i = 100 \cos(100t + 45°)$$

or $\qquad \dfrac{di}{dt} + 300 i = 100 \cos(100t + 45°)$ \hfill (1)

Its complimentry function (i_c) is given as

$$i_c = K e^{-m_1 t}, \text{ where } m_1 = 300$$

$\therefore \qquad i_c = K e^{-300t}$

Its particular solution is assumed as

$$i_p = A \cos 100t + B \sin 100t \hfill (a)$$

$$\frac{d_{ip}}{dt} = -100 A \sin 100t + 100 B \cos 100t \hfill (b)$$

putting eq. (a) and (b) in eq. (1) we get:

$-100\,A \sin 100t + 100\,B \cos 100t + 300\,A \cos 100t + 300\,B \sin 100t$

$$= \frac{100}{\sqrt{2}} (\cos 100t - \sin 100t)$$

or $\quad -100\,A \sin 100t + 100\,B \cos 100t + 300\,A \cos 100t + 300\,B \sin 100t$

$$= \frac{100}{\sqrt{2}} (\cos 100t - \sin 100t)$$

or $\quad \cos 100t\,(300\,A + 100\,B) + \sin 100t\,(300\,B - 100\,A)$

$$= \frac{100}{\sqrt{2}} (\cos 100t - \sin 100t)$$

Equating both sides

$$(300\,A + 100\,B) = \frac{100}{\sqrt{2}}$$

$$(300\,B - 100\,A) = -\frac{100}{\sqrt{2}}$$

or $$3A+B = \frac{1}{\sqrt{2}} \qquad (1)$$

$$3B+A = -\frac{1}{\sqrt{2}} \qquad (2)$$

Adding equations, (1) and (2)

or $$9A+3B = \frac{3}{\sqrt{2}}$$

or $$3B-A = -\frac{1}{-\sqrt{2}}$$

$$\underline{\quad - \quad + \quad - \quad}$$

$$-10\,A = \frac{3}{\sqrt{2}}+\frac{1}{\sqrt{2}} = \frac{1}{\sqrt{2}}$$

$$A = \frac{\sqrt{2}}{5}=0.2828, \; 3B-A=-\frac{1}{\sqrt{2}}$$

or $$3B-\frac{\sqrt{2}}{5} = -\frac{1}{\sqrt{2}}$$

or $$3B = \frac{\sqrt{2}}{5}-\frac{1}{\sqrt{2}}=0.2828-0.707, B=-0.1414$$

∴ $$i_p = 0.2828\cos100t-0.1414\sin100t$$

∴ $$A\cos\theta = 0.282, \quad A\sin\theta=-0.1414$$

$$A = \sqrt{(0.2828)^2+(-0.1414)^2}, \; A=0.3162$$

$$A\cos\theta = 0.2828$$

$$A\cos\theta = \frac{0.2828}{0.3162}=0.89 \; \therefore \; \theta=26.57°$$

∴ $$i_p = 0.3162\cos(100t+26.57°)$$

∴ Total output current $i = i_c + i_p$

or $$i = Ke^{-300t}+0.3162\cos(100t+26.57°)$$

at $t = 0, \; i = 0$

∴ $$i(o) = K+0.3162\cdot\cos(26.57°)=0$$

or $k = -0.2827$

∴ $i = -0.2827\,e^{-300t} + 0.3162\cos(100t + 26.57°)$ **Ans.**

Problem 4.31. Write simultaneous differential equations for the circuit shown in Fig. P. 4.31 and solve for i_l and i_2. The switch is closed at $t = 0$ after having been open for an extended period of time.

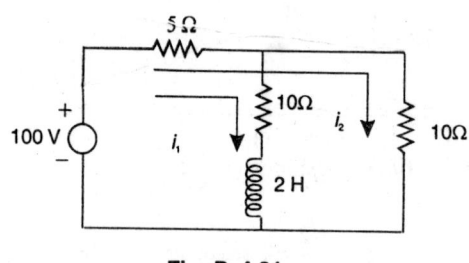

Fig. P. 4.31

Solution:

From KVL $5i_1 + 10i_1 + 5i_2 + 2\dfrac{di_1}{dt} = 100$ V   (1)

$$5i_1 + 5i_2 + 10i_2 = 100 \text{ V}$$ (2)

From eq (2), $i_2 = \dfrac{100 - 5i_1}{15}$

Putting it in eq (1), we get

$$15i_1 + \dfrac{100 - 5i_1}{15} \times 5 + 2\dfrac{di_1}{dt} = 100\text{V}$$

or $\dfrac{di_1}{dt} + 6.66\,i_1 = \dfrac{100}{3}$

$$i_1(t) = (i_1(0) - i_1(\infty))e^{-\frac{t}{RC}} + i_1(\infty)$$

Where $i_1(0) = \dfrac{100}{15}$

$$i_1(\infty) = 5 \text{ amp}$$

$$RC = \dfrac{1}{6.66}$$

$$\therefore \qquad i_1(t) = 1.67e^{-6.66t} + 5 \text{ amp.}$$

$$i_2(t) = \frac{100 - 5(1.67e^{-6.66t} + 5)}{15}$$

$$i_2(t) = 6.66 - \left(\frac{1.67e^{-6.66t} + 5}{3}\right)$$

or $$i_2(t) = -0.55e^{-6.66t} + 5$$

Problem 4.32. For the *RL* circuit t shown in Fig 4.32, find the current i_L at the following times (a) –1 ms (b) 0^+ (c) 0.3 ms (d) ∞.

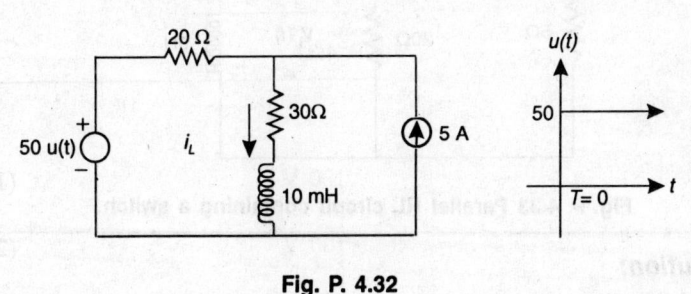

Fig. P. 4.32

Solution:

Time Const

$$\frac{L}{R} = \frac{10 \times 10^{-3}}{50} = \frac{1}{5000}$$

For $t < 0$, $50u(t)$ source gives 0 (i_L) current but 5A current gives

$$i_L(0^-) = \frac{5}{20+30} \times 20 = 2 \text{ amp.}$$

(b) $\therefore i_L(0^+) = i_L(0^-) = 2 \text{ amp.}$

(a) At $t = -1$ ms, $i_L(-1)$ is also 2 amp.

Now, $$i(t) = (i(0) - i(\infty))e^{-Rt/L} + i(\infty)$$

Now, $$i(\infty) = 1(\text{amp}) + 2(\text{amp})$$

[∵ 1 amp due to $50u(t)$ source and 2 amp due to 5 A source]

$$= 3 \text{ amp.}$$

∴ $$i(t) = (2-3)e^{-5000t} + 3 \text{ amp}$$

(c) For $t = 0.3 \text{ ms, } i(t) = -1e^{-1.5} + 3 = -0.223 + 3 = 2.776$

∴ $i(0.3 \text{ ms}) = 2.776$ **Ans.**

Problem 4.33.   For the circuit of Fig. P. 4.33, find $i_L(t)$ and $v_L(t)$ for $t > 0$ given that $i_L(0^-) = 10$ A and the switch S closes at $t = 0.4$ s. Then compute the energy dissipated in the 5-Ω resistor over the time interval [0.4, ∞).

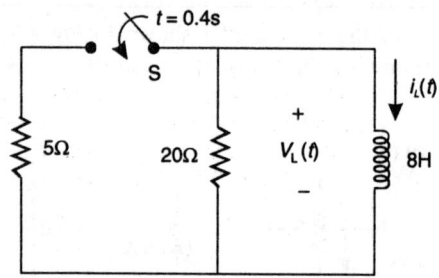

Fig. P. 4.33 Parallel RL circuit containing a switch.

Solution:

From the continuity property of the inductor current, $i_L(0^+) = i_L(0^-)$ = 10 A.

Then $$i_L(t) = e^{-\frac{R_{th}}{L}t} \, i_L(0^+) = 10e^{-2.5t} \text{ A}$$

For this time interval the Thévenin equivalent resistance seen by the inductor is $R_{th} = 20 \parallel 5 = 4\,\Omega$, i.e. the equivalent resistance of a parallel 20-Ω and 5-Ω combination. Then, the response for $t \geq t_0 = 0.4$ s is, given by,

$$i_L(t) = e^{-\frac{R_{th}}{L}(t-t_0)} \, i_L(t_0) = i_2(0.4)\, e^{-0.5(t-0.4)} = 3.679\, e^{-0.5(t-0.4)} \text{ A}$$

Now, Using step functions,

$$i_L(t) = 10\, e^{-2.5t}[u(t) - u(t-0.4)] + 3.679\, e^{-0.5(t-0.4)} \, u(t-0.4) \text{ A}$$

As we know, $v_L(t) = R_{th}\, i_L(t)$

In particular, $v_L(0^+) = -200$ V. Hence for $0 \leq t < 0.4$,

$$v_L(t) = e^{-\frac{R_{th}}{L}t} v_L(0^+) = 200 e^{-2.5t} \text{ V}$$

Fot $t \geq 0.4$, $\dot{v}_L(t) = e^{-\frac{R_{th}}{L}(t-0.4)} v_L(0.4) = -73.576 e^{-0.5(t-0.4)} \text{ V}$

The power absorbed by the 5-Ω resistor for $0.4 \leq t$ is

$$p_{5\Omega}(t) = \frac{v_L^2(t)}{5} = \frac{[-73.576 e^{-0.5(t-0.4)}]^2}{5} = 1082.7 e^{-(t-0.4)} \text{ W}$$

The energy dissipated over $(0, \infty)$ is expressed by,

$$W_{5\Omega}(0.4, \infty) = \int_{0.4}^{\infty} p_{5\Omega}(q) dq = 1082.7 \int_{0.4}^{\infty} e^{-(q-0.4)} dq = 1082.7 \text{ J}$$

Problem 4.34. Find $v_c(t)$ for $t \geq 0$ for the circuit of Fig. P. 4.34 given that $v_c(0) = 9$ V

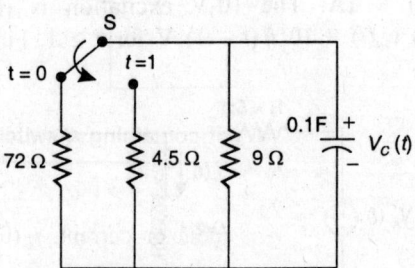

Fig. P. 4.34 Parallel *RC* circuit with switch to effect different time constants.

Solution:

By the continuity of the capacitor voltage, $v_c(0^+) = v_c(0^-) = 9$ V. Then the capacitor voltage is given by,

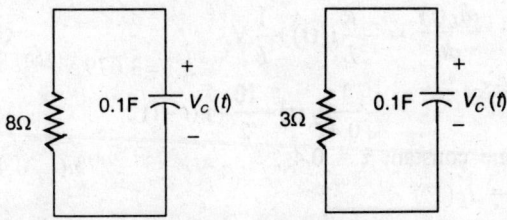

Fig. P. 4.34 (a) & (b) Equivalent circuits for Fig. P. 4.34
(a) $0 \leq t < 1$. (b) $1 \leq t$.

$$v_c(t) = e^{-\frac{1}{R_{th}C}t} \quad v_c(0^+) = 9\,e^{-1.25t} \text{ V} \tag{1}$$

Figure P. 4.34 (b) depicts the pertinent equivalent circuit. Observe that $R_{th} = 3\Omega$.

Then, for $t > 1$,

$$v_c(t) = e^{-\frac{1}{R_{th}C}(t-t_0)} \quad v_c(t_0^+) = 2.58\,e^{-\left(\frac{t-1}{0.3}\right)} \text{ V} \tag{2}$$

Using the shifting unit step function, the two eqns A and eqns B can be combined as,

$$v_c(t) = 9\,e^{-1.25t}\,[u(t) - u(t-1)] + 2.58\,e^{-\left(\frac{t-1}{0.3}\right)}\,u(t-1)\,\text{V} \textbf{ Ans.}$$

Problem 4.35. For the circuit of Fig. P. 4.35, suppose a 10-V unit step excitation is applied at $t = 1$ when it is found that the inductor current is $i_L(1^-) = 1$A. The 10-V excitation is represented mathematically as $v_{in}(t) = 10\ u(t-1)$ V for $t \geq 1$. Find $i_L(t)$ for $t \geq 1$.

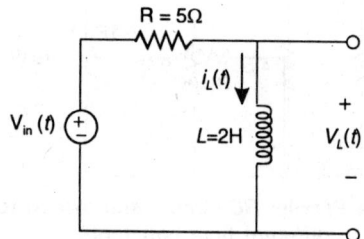

Fig. P. 4.35 Driven series *RL* circuit with $i_L(1^-)$ = 1 A

Solution:

The differential equation for the given circuit is given by

$$\frac{di_L(t)}{dt} = -\frac{R}{L}i_L(t) + \frac{1}{L}V_s$$

$$= -\frac{1}{0.4}i_L(t) + \frac{10}{2} + u(t-1)$$

Where the time constant $\tau = 0.4$s.

Since $i_L(1^-) = i_L(1^+)$

Then, $i_L(t) = i_L(\infty) + [i_L(1^+) - i_L(\infty)]e^{-\left(\frac{t-1}{\tau}\right)}u(t-1)\,\text{A}$

Here, the presence of $u(t-1)$ emphasizes that the response is only valid for $t \geq 1$.

Since $\tau = 0.4 > 0$, we replace the inductor in the circuit of Fig. P. 4.35 by a short circuit to compute $I_{sc} = i_L(\infty) = 2$ A. It follows that:

$$i_L(t) = [2+(1-2)e^{-2.5(t-1)}]$$

$$= [2 - e^{-2.5(t-1)}]u(t-1)\,\text{A} \tag{1}$$

Using eqn (1), we have, for $t > 1$,

$$v_L(t) = L\frac{di_L(t)}{dt} = -\frac{L}{\tau}[i_L(1^+)-i_L(\infty)]e^{-\left(\frac{t-1}{\tau}\right)}u(t-1^-)$$

$$= 5e^{-2.5(t-1)}u(t-1^+)\,\text{V}$$

Problem 4.36. The source in the circuit of Fig. P. 4.36 furnishes a 12-V excitation for $t < 0$ and a 24-V excitation for $t \geq 0$, denoted by $v_{in}(t) = 12u(-t) + 24\,u(t)$ V. The switch in the circuit closes at $t = 10$ s. First determine the value of the capacitor voltage at $t = 0^-$, which by continuity equals $v_c(0^+)$. Next determine $v_c(t)$ for all $t \geq 0$.

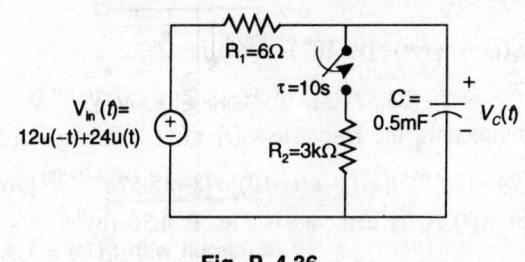

Fig. P. 4.36

Solution:

For $t < 0$, the 12-V excitation has been applied for a long time. Therefore, at $t = 0^-$, the capacitor has reached its final value and looks like an open circuit to the source. Hence the entire source voltage of 12 V appears across the capacitor at $t = 0^-$, i.e. $v_c(0^-) = v_c(0^+) = 12$ V by the continuity property of the capacitor voltage.

For $0 \leq t \leq 10$ s, $\tau = R_1C = 3$ It is important to realize here that for $0 \leq t \leq 10$ the circuit behaves as if the switch were not present. Hence, the computation of $v_c(\infty)$ proceeds as if no switching would take place at $t = 10$ s. Here $V_{oc} = v_c(\infty) = 24$ V. Hence, for $0 \leq t \leq 10$, the capacitor voltage is given by,

$$v_c(t) = v_c(\infty)+[v_c(0^+)-v_c(\infty)]e^{-\left(\frac{t}{R_{th}C}\right)}$$

$$= 24+(12-24)e^{-t/3} = 24-12e^{-t/3}\,\text{V} \qquad (1)$$

Using eqn (1) and using the continuity property of the capacitor voltage,

we have, $v_c(10^-) = v_c(10^+)=24-12e^{-10/3}=23.57\,\text{V}$

For $t > 10$, the resistive part of the circuit can be replaced by its Tévenin equivalent, and resultant circuit illustrated is Fig. P. 4.36 (a).

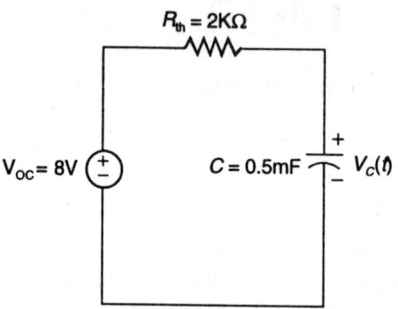

Fig. P. 4.36 (a) Circuit equivalent to that of Fig. P. 4.36 for $t \geq 10$.

The value for $v_c(\infty)$, however, is now 8 V and the new time constant is now $R_{th}C = 1$ s. Hence, for $t > 10$.

$$v_c(t) = v_c(\infty)+[v_c(10^+)-v_c(\infty)]e^{-\frac{t-10}{R_{th}C}}$$

$$= 8+(23.57-8)e^{-(t-10)} = 8+15.57\,e^{-(t-10)}\,\text{V}$$

Using step function, the response $v_c(t)$ for $t \geq 0$ is given by,

$$v_c(t)=(24-12e^{-t/3})[u(t)-u(t-10)]+[8+15.57\,e^{-(t-10)}]u(t-10)\,\text{V}$$

The plot of $v_c(t)$ A, is depicted is Fig. P. 4.36 (b).

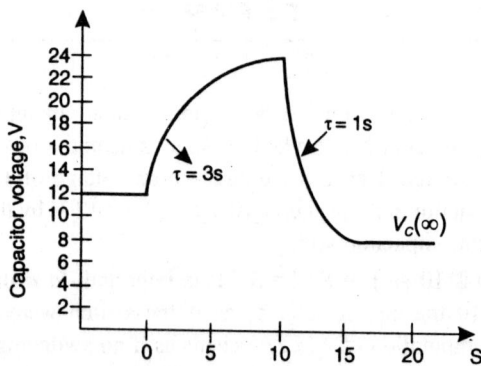

Fig. P. 4.36 (b) Capacitor voltage vc(t) for $t \geq 0$.

5

Initial Conditions in Network

Problem 5.1. In the network of the following figure, switch K is closed at $t = 0$ with the capacitor uncharged. Find values for i, di/dt and d^2i/dt^2 at $t = 0^+$, for element values as follows: $V = 100$ V, $R = 1000$ Ω, and $C = 1$ μF.

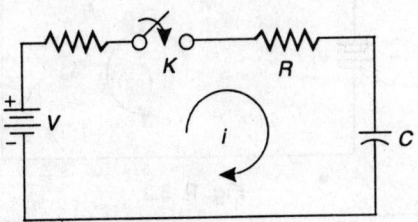

Fig. P. 5.1

Solution:

Applying KVL in the above circuit

$$iR + \frac{1}{C} \int i \, dt = V$$

or

$$iRC + \int i \, dt = VC$$

Differentiating both sides we get w.r.t. time t

$$\frac{di}{dt} + \frac{1}{RC} i = 0$$

or

$$i(t) = A e^{-\frac{t}{RC}}$$

At time $t = 0^+$, capacitor is short circuited.

So
$$i(0^+) = \frac{V}{R} = \frac{100}{1000} = 0.1$$

and
$$RC = 10^3 \times 10^{-6} = 10^{-3}$$

∴
$$i(t) = 0.1e^{-10^3 t} \text{ or, } i(0^+) = 0.1$$

or
$$\frac{di(t)}{di} = -100\, e^{-10^3 t} \text{ or } \frac{di(0^+)}{di} = -100$$

or
$$\frac{d^2i(t)}{dt^2} = 10^5\, e^{-10^3 t} \text{ or } \frac{d^2i(0^+)}{dt^2} = 10^5 \text{ Ans.}$$

Problem 5.2.   In the given network, K is closed at $t = 0$ with zero current in the inductor. Find the values of i, di/dt, and d^2i/dt^2 at $t = 0^+$ if $R = 10\,\Omega$, $L = 1$ H, and $V = 100$ V.

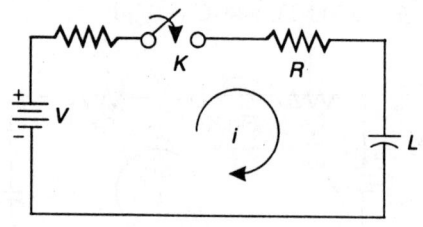

Fig. P. 5.2

Solution:

$$i(0^+) = 0 \text{ (Inductor is open circuited).}$$

$$i(\infty) = \frac{V}{R} \text{ (Inductor is short circuited)} = \frac{100}{10} = 10.$$

∴
$$i(t) = (i(0^+) - i(\infty))e^{-\frac{Rt}{L}} + i(\infty) = (0-10)e^{-10t} + 10$$

$$i(t) = 10(1 - e^{-10t})$$

∴
$$i(0^+) = 0$$

$$\frac{di(t)}{dt} = 100\, e^{-10t}$$

or $$\frac{di(0^+)}{dt} = 100$$

$$\frac{d^2i(t)}{dt^2} = -10^3 e^{-10t}$$

or $$\frac{d^2i(0^+)}{dt} = -10^3 \text{ Ans.}$$

Problem 5.3. In the network of Fig. P. 5.3, K is changed from position a to b at $t = 0$. Solve for i, di/dt, and d^2i/dt^2 at $t = 0^+$ if $R = 1000\ \Omega$, $L = 1$ H, and $C = 0.1\ \mu$F, and $V = 100$ V.

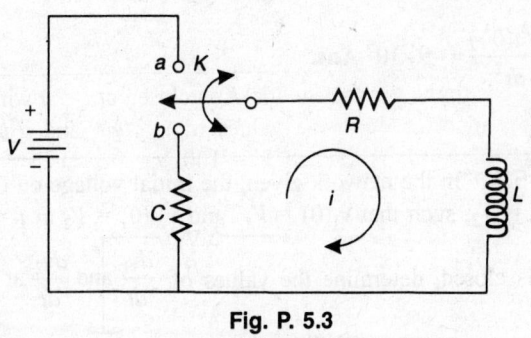

Fig. P. 5.3

Solution:

$$i(0^-) = \frac{V}{R}$$

$$= \frac{100}{1000} = 0.1 \text{amp.,}$$

at time $t > 0$ $$= iR + L\frac{di}{dt} + \frac{1}{C}\int i\,dt = 0 \qquad (1)$$

or at time $t = 0^+$, voltage across capacitor is zero, and $i(0^+)$ in inductor $\leq i(0^-)$ so inductor is not open circuited. So, $\dfrac{1}{C}\int i(0^+)dt = 0$

$\therefore$ $$i(0^+)R + L\frac{di(0^+)}{dt} = 0$$

or $$10^3 \cdot (0.1) + 1\frac{di(0^+)}{dt} = 0$$

or
$$\frac{di(0^+)}{dt} = -10^2 \text{ Amp/sec.}$$

Now, differentiating eq (1), we get:

$$L\frac{d^2i}{dt^2} + R\frac{di}{dt} + \frac{1}{C}i = 0$$

or
$$\frac{d^2i(0^+)}{dt^2} + 10^3\frac{di(0^+)}{dt} + 10^7\,i\,(0.1) = 0$$

or
$$\frac{d^2i(0^+)}{dt^2} + (-10^3 \times 10^{+2}) + 10^7 \cdot (0.1) = 0$$

or
$$\frac{d^2i(0^+)}{dt^2} = -9 \times 10^5 \text{ Ans.}$$

Problem 5.4. In the network given, the initial voltage on C_1 is V_1 and on C_2 is V_2, such that $v_1(0) = V_1$ and $v_2(0) = V_2$ at $t = 0$, the switch K is closed, determine the values of $\dfrac{dv_1}{dt}$ and $\dfrac{dv_2}{dt}$ at $t = 0^+$.

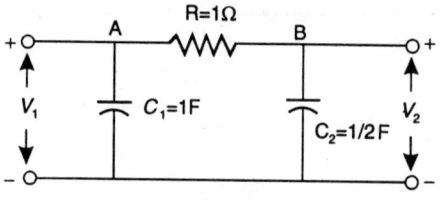

Fig. P. 5.4

Solution:

$\therefore$
$$v_1(0^+) = V_1$$

and
$$v_2(0) = V_2$$

At node A:
$$C_1\frac{dV_1}{dt} + \frac{V_1 - V_2}{R} = 0$$

At $t = 0^+$,
$$C_1\frac{dv_1(0)}{dt} = \frac{V_2 - V_1}{R}$$

or $$\frac{dv_1(0)}{dt} = \frac{V_2 - V_1}{RC_1}$$

$$= \frac{1-2}{1 \times 1} = -1 \text{V/sec}$$

$\therefore$ $$\frac{dv_1(0)}{dt} = -1 \text{V/sec}$$

At note B, $$C_1 \frac{dV_2}{dt} + \frac{V_2 - V_1}{R} = 0$$

or at $t = 0^+$, $$\frac{dv_2(0^+)}{dt} = \frac{V_1 - V_2}{RC_2} = \frac{2-1}{1 \times 1/2} = 2 \text{V/s}$$

$\therefore$ $$\frac{dv_2(0^+)}{dt} = 2 \text{V/s} \quad \textbf{Ans.}$$

Problem 5.5. In the given network $v_1 = e^{-t}$ for $t \geq 0$ and is zero for all $t < 0$, if the capacitor is initially uncharged, find the value of $\dfrac{d^2 v_2}{dt^2}$ and $\dfrac{d^3 v_2}{dt^3}$ at $t = 0^+$, Let $R_1 = 10 \ \Omega$, $R_2 = 20 \Omega$ and $C = \dfrac{1}{20} \text{F}$.

Fig. P. 5.5

Solution:

At time $t > 0$

$$\frac{-v_1 + v_2}{R_1} + \frac{C dv_2}{dt} + \frac{v_2}{R_2} = 0$$

or $$\frac{-e^{-t} + v_2}{10} + \frac{1}{20} \frac{dv_2}{dt} + \frac{v_2}{20} = 0$$

or $$-2e^{-t} + 2v_2 + v_2 + \frac{dv_2}{dt} = 0$$

or $\qquad \dfrac{dv_2}{dt} + 3v_2 = +2e^{-t}$

or $\qquad v_2(t) = e^{-3t} \int +2e^{-t} \cdot e^{+3t}\, dt + k e^{-3t}$

$$= +2e^{-3t}\, \dfrac{e^{+2t}}{+2} + k e^{-3t}$$

$$v_2(t) = e^{-t} + k e^{-3t}$$

At time $t = 0,$ $\qquad v_2(t) = 0,$ so $k = -1$

$\therefore \qquad\qquad\qquad v_2(t) = e^{-t} - e^{-3t},$ or $v_2(t) = e^{-t} - e^{-3t}$

or $\qquad\qquad \dfrac{dv_2}{dt} = 3e^{-3t} - e^{-t}$ or $\dfrac{d^2 v_2}{dt^2} = e^{-t} - 9e^{-3t}$

or $\qquad\qquad \dfrac{d^2 v_2(0^+)}{dt^2} = 1 - 9 = -8$

or $\qquad\qquad \dfrac{d^2 v_2(0^+)}{dt^2} = -8\,\text{V}^2/\text{s}^2$

or $\qquad\qquad \dfrac{d^3 v_2}{dt^3} = 27 e^{-3t} - e^{-t}$

or $\qquad\qquad \dfrac{d^3 v_2(0^+)}{dt^3} = 27 - 1 = 26$

or $\qquad\qquad \dfrac{d^3 v_2(0^+)}{dt^3} = 26\ \text{V}^3/\text{s}^3$ **Ans.**

Problem 5.6. The network shown in Fig. P. 5.6 is in the steady state with the switch k closed. At $t = 0$, the switch is opened. Determine the voltage across the switch, v_k, and dv_k/dt at $t = 0^+$.

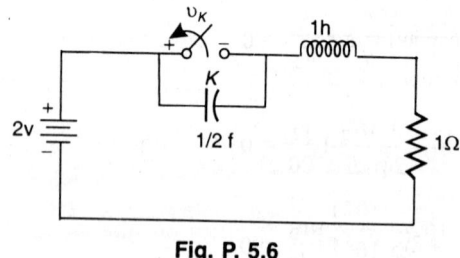

Fig. P. 5.6

Solution:

At time $t = 0^-$, inductor is short circuited and capacitor is open circuited.

So $$i(0^-) = \frac{2}{1} = 2 \text{ amp}$$

$\therefore$ $$i(0^+) = i(0^-) = 2 \text{ amp}$$

$\because$ $$v_k = \frac{1}{c} \int i \, dt \quad \text{for } t > 0$$

$\therefore$ $$v_k(0^+) = \frac{1}{c} \int i(0^+) \, dt$$

or $$v_k(0^+) = 0 \text{ V}$$

or $$\frac{dv_k(0^+)}{dt} = \frac{1}{C} i(0^+) = \frac{2}{\frac{1}{2}} = 4 \text{ V/sec}$$

$\therefore$ $$\frac{dv_k(0^+)}{dt} = 4 \text{ V/sec} \quad \textbf{Ans.}$$

Problem 5.7. In the given network, the switch K is opened at $t = 0$. At $t = 0^+$, solve for the values of v, dv/dt, and d^2v/dt^2 if $I = 10$ amp, $R = 1000 \ \Omega$, and $C = 1 \mu\text{F}$.

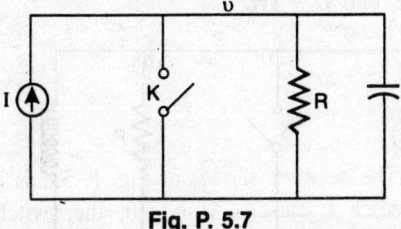

Fig. P. 5.7

Solution:

At time $t = 0$, capacitor acts as short circuited

So, $v(0^+) = 0$ at time $t > 0$

$$10 = \frac{v}{10^3} + 10^{-6} \frac{dv}{dt} \tag{1}$$

or $$10 = \frac{v(0^+)}{10^3} + 10^{-6} \frac{dv(0^+)}{dt} \quad \text{at time } t = 0^+$$

so $\qquad 10 = 0 + 10^{-6}\dfrac{dv(0^+)}{dt}$

or $\qquad \dfrac{dv(0^+)}{dt} = 10^7 \text{ V/sec}$

Differentiating eq (1), we get

$$10^{-6}\frac{d^2v}{dt^2} + \frac{dv}{10^3\,dt} = 0$$

or $\qquad 10^{-6}\dfrac{d^2v(0^+)}{dt^2} + \dfrac{dv}{10^3\,dt} = 0$

or $\qquad 10^{-6}\dfrac{d^2v(0^+)}{dt^2} = -\dfrac{dv(0^+)}{10^3\,dt}$

$$\frac{d^2v(0^+)}{dt^2} = -\frac{10^7}{10^3}\times10^6$$

$$\frac{d^2v(0^+)}{dt^2} = -10^{10}\,V^2/s^2 \text{ \textbf{Ans.}}$$

Problem 5.8. The network shown in Fig. P. 5.8 has the switch K opened at $t = 0$. Solve for v, dv/dt, and d^2v/dt^2 at $t = 0^+$ if $I = 1$ amp, $R = 100\ \Omega$, and $L = 1$H.

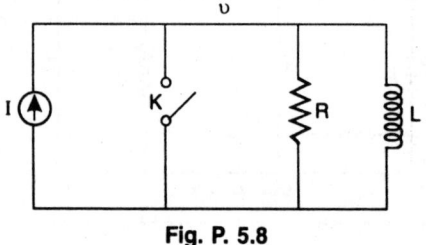

Fig. P. 5.8

Solution:

At time $t = 0^+$, inductor acts as open circuited.

So $v = I.R$

$\therefore\ v = 1\cdot100 = 100\,\text{V} \quad \therefore\ v(0^+) = 100\,\text{V}$

At time $t > 0$ $\qquad I = \dfrac{v}{R} + \dfrac{1}{L}\displaystyle\int v\,dt$ $\qquad\qquad$ (1)

Differentiating both sides w.r.t. time t

$$0 = \frac{1}{R}\frac{dv}{dt} + \frac{1}{L}v \qquad (2)$$

or

$$0 = \frac{1}{R}\frac{dv(0^+)}{dt} + \frac{1}{L}v(0^+)$$

$$\frac{dv(0^+)}{dt} = -\frac{R}{L}v(0^+) = -\frac{100}{1}\cdot 100$$

$$\frac{dv(0^+)}{dt} = -10^4 \text{ V/sec}$$

Now, again differentiating eq (2) w.r.t. time t

$$0 = \frac{1}{R}\frac{d^2v(0^+)}{dt^2} + \frac{1}{L}\frac{dv(0^+)}{dt}$$

or

$$\frac{d^2v(0^+)}{dt^2} = -\frac{R}{L}\frac{dv(0^+)}{dt} = -\frac{100}{1}(-10^4)\text{ V}^2/\text{s}^2$$

$$\frac{d^2v(0^+)}{dt^2} = 10^6 \text{ V}^2/\text{s}^2 \text{ Ans.}$$

Problem 5.9. In the network shown in the following figure, a steady state is reached with the switch K open. At $t = 0$, the switch is closed. For the element values given, determine the value of $v_a(0^-)$ and $v_a(0^+)$.

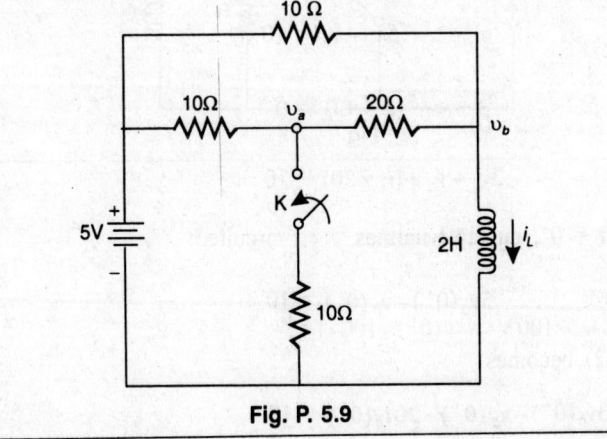

Fig. P. 5.9

Solution:

At time $t = 0^-$, inductor acts as a short circuited.

So
$$v_a = \frac{5}{30} \times 20 = \frac{10}{3} \text{ V}$$

$\therefore$
$$v_a(0^-) = \frac{10}{3} \text{ V}, \quad \therefore \quad i_L(0^-) = \frac{5}{30} + \frac{5}{10}$$

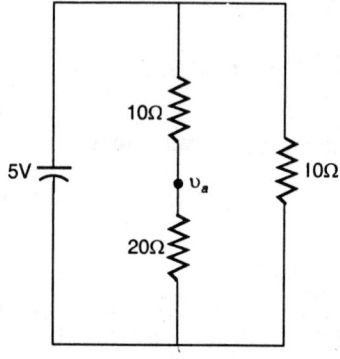

Fig. P. 5.9 (a)

or
$$i_L(0^-) = 2/3 \quad \text{amp}$$

At time $t > 0$.

At node a:
$$\frac{v_a - 5}{10} + \frac{v_a}{10} + \frac{v_a - v_b}{20} = 0$$

or
$$v_a(5) - v_b = 10$$

or
$$5v_a - v_b = 10 \tag{1}$$

At node b:
$$\frac{v_b - 5}{10} + \frac{v_b - v_a}{20} + i_L = 0$$

or
$$3v_b - v_a + (i_L \times 20) = 10 \tag{2}$$

At time $t = 0^+$, eq (1) becomes

$$5v_a(0^+) - v_b(0^+) = 10 \tag{3}$$

and eq (2) becomes

$$3v_b(0^+) - v_a(0^+) + 20i_L(0^+) = 10 \tag{4}$$

Solving eq (3) and eq (4), we get

$v_a(0^+) = 40/21\text{V}$ **Ans.**

Problem 5.10. In the following figure is shown a network in which a steady state is reached with switch K open. At $t = 0$, the switch is closed. For the element given, determine the values of $v_a(0^-)$ and $v_a(0^+)$.

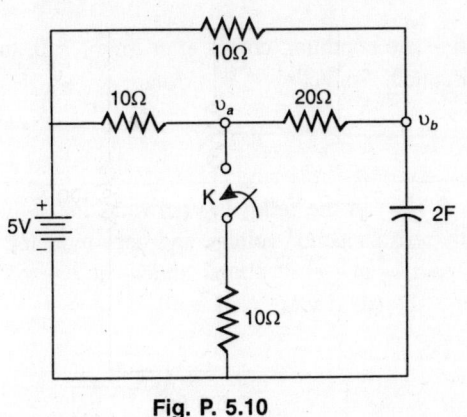

Fig. P. 5.10

Solution:

At time $t = 0^-$, capacitor acts as a open circuited.

So $v_a(0^-) = v_b(0^-) = 5\text{V}$

At $t > 0$, $\dfrac{v_a - 5}{10} + \dfrac{v_a - v_b}{20} + \dfrac{v_a}{10} = 0$

or $v_a\left(\dfrac{1}{10} + \dfrac{1}{20} + \dfrac{1}{10}\right) - \dfrac{5}{10} - \dfrac{v_b}{20} = 0$

or $5v_a - v_b = 10$

At $t = 0^+$, $5v_a(0^+) - v_b(0^+) = 10$ $\qquad\qquad$ (1)

Since voltage across capacitor does not change instantaneously, so

$$v_b(0^-) = v_b(0^+) = 5\text{V}$$

∴ eq (1) becomes $5v_a(0^+) - 5 = 10$

or $v_a(0^+) = \dfrac{15}{5} = 3\text{V}$

∴ $v_a(0^+) = 3\text{V}$ **Ans.**

Problem 5.11. In the network given in Fig. P. 5.10, determine $v_b(0^+)$ and $v_b(\infty)$ for the condition stated in Problem 5.10.

Solution:

$\therefore$ $$v_b(0^-) = 5V$$

$$= v_b(0^+)$$

When steady-state condition comes after time $t > 0$, the capacitor acts as short circuited. So $v_b(\infty) = 0$ **Ans.**

Problem 5.12. In the following network, the switch is closed at $t = 0$, with zero capacitor voltage and zero inductor current. Solve for (a) v_1 and v_2 at $t = 0^+$ (b) v_1 and v_2 at $t = \infty$, (c) dv_1/dt and dv_2/dt at $t = 0^+$ (d) d^2v_2/dt^2 at $t = 0^+$.

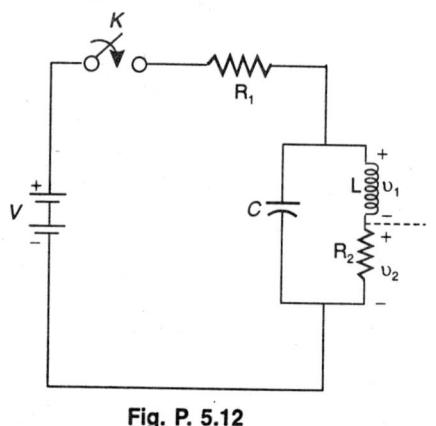

Fig. P. 5.12

Solution:

(a) At time $t = 0^+$, L acts as open circuited and C acts as short circuited.

So $$V_C(0^+) = 0 \text{ V}$$

$\therefore$ $$v_C(0^+) = v_1(0^+) + v_2(0^+)$$

So $$v_1(0^+) = 0$$

$$v_2(0^+) = 0$$

(b) At time $t = \infty$ capacitor acts as open circuited and inductor acts as short circuited.

So $\qquad v_C(\infty) = \dfrac{vR_2}{R_1 + R_2}$ volts

and $\qquad v_L = 0, \therefore v_1 = 0$

$\therefore \qquad v_2(\infty) = v_C(\infty) - v_1(\infty)$

$\qquad\qquad v_2(\infty) = \dfrac{vR_2}{R_1 + R_2}$ volts

(c) Let i_1 and i_2 be the loop current
Applying KVL in mesh 1

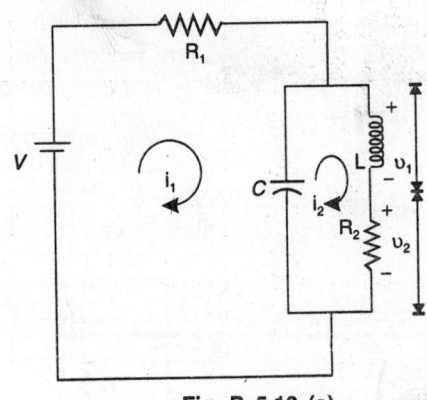

Fig. P. 5.12 (a)

$$i_1 R_1 + \frac{1}{c}\int (i_1 - i_2)dt = v \qquad\qquad (1)$$

or $\qquad i_1 R_1 + i_2 R_2 + L\dfrac{di_2}{dt} = v$

or $\qquad \dfrac{di_2}{dt} + \dfrac{i_2 R_2}{L} + \dfrac{i_1 R_1}{L} = \dfrac{v}{L} \qquad\qquad (2)$

or $\qquad \dfrac{di_2(0^+)}{dt} + i_2(0^+)\dfrac{R_2}{L} + i_1(0^+)\dfrac{R_1}{L} = \dfrac{v}{L}$

$\because \qquad i_1(0^+) = \dfrac{v}{R_1}$

and $\qquad i_2(0^+) = 0$

$$\therefore \quad \frac{di_2\left(0^+\right)}{dt} + 0 + \frac{v}{L} = \frac{v}{L} \quad \therefore \quad \frac{di_2(0^+)}{dt} = 0 \tag{3}$$

Differentiating eq (2) w.r.t. time t, we get

$$\frac{d^2 i_2(0^+)}{dt^2} + \frac{di_2(0^+)}{dt} \frac{R_2}{L} + \frac{di_1(0^+) R_1}{L} = 0$$

or $\qquad \dfrac{d^2 i_2(0^+)}{dt^2} + 0 + \left(-\dfrac{v}{R_1^2 C} \dfrac{R_1}{L} \right) = 0$ [From eq 1]

or $\qquad \dfrac{d^2 i_2(0^+)}{dt^2} = \dfrac{v}{R_1 L C} \tag{4}$

$\therefore \qquad\qquad v_1 = L\dfrac{di_2}{dt}$

or $\qquad \dfrac{dv_1(0^+)}{dt} = L\dfrac{d^2 i_2(0^+)}{dt^2} = L\dfrac{v}{R_1 L C} = \dfrac{v}{R_1 L C}$

$\therefore \qquad \dfrac{dv_1(0^+)}{dt} = \dfrac{v}{R_1 C}$ and $v_2 = i_2 R_2$

or $\qquad \dfrac{dv_2(0^+)}{dt} = R_2 \dfrac{di_2(0^+)}{dt} = R_2 \times 0 = 0 \quad \therefore \quad \dfrac{dv_2(0^+)}{dt} = 0$

(d) Differentiating $v_2 = i_2 R_2$ twice, we get

$$\frac{d^2 v_2}{dt^2} = R_2 \frac{d^2 i_2}{dt^2}$$

or $\qquad \dfrac{d^2 v_2(0^+)}{dt^2} = R_2 \dfrac{d^2 i_2(0^+)}{dt^2} = R_2 \dfrac{v}{R_1 L C}$

or $\qquad \dfrac{d^2 v_2(0^+)}{dt^2} = \dfrac{v R_2}{R_1 L C}$ **Ans.**

Problem 5.13. The switch K in the network of Fig. P. 5.13 is closed at $t = 0$ connecting the battery to an unenergized network. (a) Determine i, di/dt, and d^2i/dt^2 at $t = 0^+$. (b) Determine v_1, dv_1/dt, and d^2v_1/dt^2 at $= 0^+$.

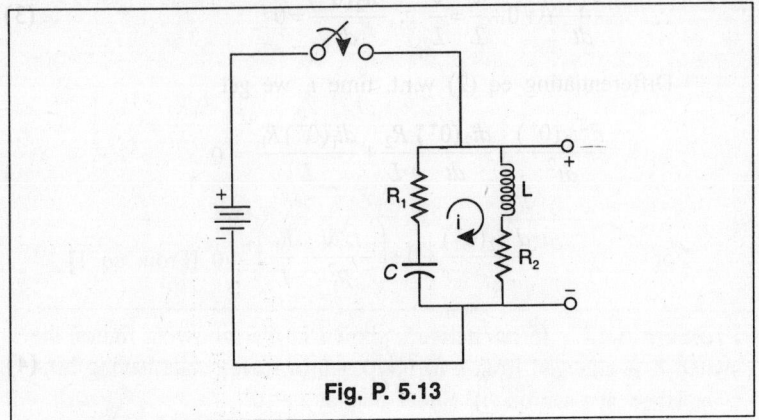

Fig. P. 5.13

Solution:

At time $t = 0^+$, $i(0^+) = 0$

At time $t > 0$.
$$V_0 = L\frac{di}{dt} + R_2 i \qquad (1)$$

or
$$V_0(0^+) = L\frac{di(0^+)}{dt} + R_2 i(0^+)$$

or
$$\frac{V_0}{L} = \frac{di(0^+)}{dt} \qquad (2)$$

Differentiating eq (1), w.r.t. time t is given as:

$$0 = L\frac{d^2i(0^+)}{dt^2} + R_2 \frac{di(0^+)}{dt}$$

or
$$\frac{d^2i(0^+)}{dt^2} = -\frac{R_2}{L^2}V_0 \qquad (3)$$

$$\therefore \qquad v_1 = iR_1 - \frac{1}{c}\int i\,dt \qquad (4)$$

or
$$\frac{dv_1(0^+)}{dt} = \frac{di(0^+)}{dt}R_1 - \frac{1}{C}i(0^+) = \frac{V_0 R_1}{L} - \frac{1}{C}\times 0$$

$$\frac{dv_1(0^+)}{dt} = \frac{V_0 R_1}{L}$$

Again, differentiating eq (4), we get

$$\frac{d^2 v_1(0^+)}{dt^2} = \frac{d^2 i(0^+)}{dt^2} R_1 - \frac{1}{C} \frac{di(0^+)}{dt}$$

$$= \frac{-R_1 R_2 V_0}{L^2} - \frac{1}{C} \frac{V_0}{L} = -V_0 \left(\frac{R_1 R_2}{L^2} + \frac{1}{LC} \right)$$

or $\qquad \dfrac{d^2 v_1(0^+)}{dt^2} = -V_0 \left(\dfrac{R_1 R_2 C + L}{L^2 C} \right)$ **Ans.**

Problem 5.14. In the network shown in the following figure, the switch K is changed from *a* to *b* at *t* = 0 (a steady state having been established at position *a*). Show that at *t* = 0+

$$i_1 = i_2 = \frac{V}{R_1 + R_2 + R_3}, i_3 = 0$$

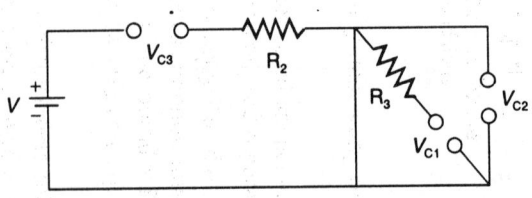

Fig. P. 5.14

Solution:

For time *t* = 0⁻ or steady state conditions. So Inductor is short circuited and capacitor is open circuited. So the equivalent circuit at *t* = 0⁻ is:

∴ $\qquad V_{c3} = V, \ V_{c1} = V_{c2} = 0, \ V_{L1} = V_{L2} = 0$

Fig. P. 5.14 (a)

For time $t > 0$, in mesh (1), (2) and (3), the current in inductor and voltage across capacitor does not change instantaneously.

$\because$

$$V_{L1} = L_1 \frac{d(i_1 - i_2)}{dt} = 0, \therefore i_1 = i_2$$

and

$$V_{L2} = L_2 \frac{di_3}{dt} = 0, \therefore i_3 = 0$$

So, the equivalent circuit at $t = 0^+$ is given as

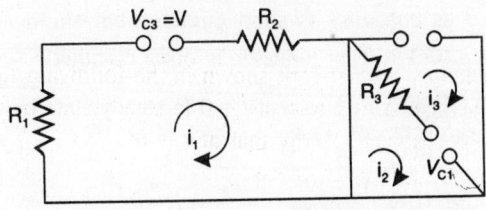

Fig. P. 5.14 (b)

So, Applying KVL:

$$i_1 = i_2 = -\frac{V}{R_1 + R_2 + R_3} \text{ and } i_3 = 0 \text{ Ans.}$$

Problem 5.15. In the circuit shown in the following figure, the switch K is closed at $t = 0$ connecting a voltage, $V_0 \sin \omega t$, to the parallel *RL-RC* circuit. Find (a) di_1/dt and (b) di_2/dt at $t = 0^+$.

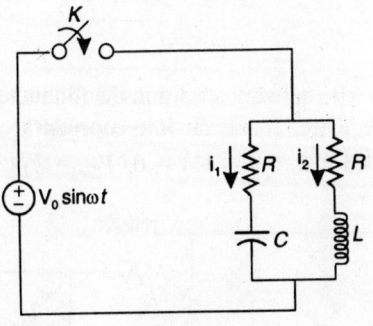

Fig. P. 5.15

Solution:

From the given circuit, applying KVL

(a)
$$i_1 R + \frac{1}{c} \int i_1 dt = V_0 \sin \omega t$$

or
$$R \frac{di_1}{dt} + \frac{1}{c} i_1 = V_0 \, \omega \cos \omega t$$

or
$$R \frac{di_1(0^+)}{dt} + \frac{1}{c} i_1(0^+) = V_0 \omega$$

∵ $i_1(0^+) = 0$ as capacitor is short circuited but $\sin \omega t = 0$
Similarly $i_2(0^+) = 0$ as inductor is open circuited.

∴
$$R \frac{di_1(0^+)}{dt} = V_0 \, \omega$$

or
$$\frac{di_1(0^+)}{dt} = \frac{V_0 \omega}{R} \qquad\qquad (1)$$

(b)
$$i_2 R + L \frac{di_2}{dt} = V_0 \sin \omega t$$

or
$$i_2(0^+) R + L \frac{di_2(0^+)}{dt} = 0$$

or
$$0 + L \frac{di_2(0^+)}{dt} = 0$$

or
$$\frac{di_2(0^+)}{dt} = 0 \quad \textbf{Ans.} \qquad\qquad (2)$$

Problem 5.16. The network shown in the following figure has two independent node pairs. If switch K is opened at $t = 0$, find the following quantities at $t = 0^+$: (a) v_1, (b) v_2, (c) dv_1/dt, (d) dv_2/dt.

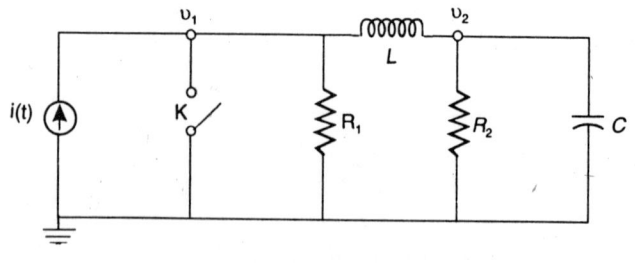

Fig P. 5.16

Solution:

(a) At time $t = 0^+$, inductor acts as open circuited and capacitor acts as short circuited. So $v_1 = R_1 i$

or $\qquad v_1(0^+) = R_1 i(0^+)$

(b) $\qquad v_2(0^+) = 0$

(c) At time $t > 0$,

At node A, $\qquad \dfrac{v_1}{R_1} + \dfrac{1}{L} \int (v_1 - v_2) dt = i(t)$ \hfill (1)

At node B, $\dfrac{v_2}{R_2} + C \dfrac{dv_2}{dt} + \dfrac{1}{L} \int (v_2 - v_1) dt = 0$ \hfill (2)

Differentiating eq (1) and at time $t = 0^+$ can be written as:

$$\frac{dv_1(0^+)}{dt(R_1)} + \frac{1}{L} v_1(0^+) = \frac{di(0^+)}{dt}$$

or $\qquad \dfrac{1}{R_1} \dfrac{dv_1(0^+)}{dt} = \dfrac{di(0^+)}{dt} - \dfrac{R_1 i(0^+)}{L}$

or $\qquad \dfrac{dv_1(0^+)}{dt} = R_1 \left(\dfrac{di_1(0^+)}{dt} - \dfrac{R_1 i(0^+)}{L} \right)$

(d) Adding eq (1) and eq (2), we get

$$\frac{v_1}{R_1} + \frac{v_2}{R_2} + C \frac{dv_2}{dt} = i(t)$$

or $\qquad \dfrac{v_1(0^+)}{R_1} + \dfrac{v_2(0^+)}{R_2} + C \dfrac{dv_2(0^+)}{dt} = i(0^+)$

or $\qquad \dfrac{v_1(0^+)}{R_1} + 0 + C \dfrac{dv_2(0^+)}{dt} = i(0^+)$

or $\qquad v_1(0^+) + R_1 C \dfrac{dv_2(0^+)}{dt} = i(0^+) R_1$

So $\qquad C \dfrac{dv_2}{dt}(0^+) = 0$ **Ans.**

Problem 5.17. The network of the following figure shows that if K is closed at $t = 0$,

$$\frac{d^2 i_2}{dt^2}(0^+) = -\frac{1}{R_1}\left\{\frac{-1}{R_1 C}\left[\frac{v(0)}{R_1 C} - \frac{dv}{dt}(0)\right] - \frac{d^2 v}{dt^2}(0)\right\}$$

Fig. P. 5.17

Solution:

$$i_1 R_1 + \frac{1}{C}\int (i_1 - i_2)\,dt = v\,(t) \tag{1}$$

$$R_2 i_2 + L\frac{di_2}{dt} + \frac{1}{C}\int (i_2 - i_1)\,dt = 0 \tag{2}$$

Adding eq (1) and eq (2), we get

$$i_1 R_1 + i_2 R_2 + L\frac{di_2}{dt} = v\,(t) \tag{3}$$

At time $t = 0$, capacitor acts as short circuited and inductor acts as open circuited.

So $i_1(0^+) = \dfrac{v(0^+)}{R_1}$ and $i_2(0^+) = 0$

Putting the value of $i_1(0^+)$ and $i_2(0^+)$ in eq (3), we get

$$\frac{di_2(0^+)}{dt} = 0$$

and $$\frac{di_1(0^+)}{dt} = \frac{1}{R_1}\frac{dv(0^+)}{dt}$$

Now, differentiating eq (1) twice and at time $t = 0^+$,

$$R_1\frac{d^2 i_1(0^+)}{dt^2} + \frac{1}{C}\frac{di_1(0^+)}{dt} - \frac{1}{C}\frac{di_2(0^+)}{dt} = -\frac{d^2 v(0^+)}{dt^2}$$

or $$\frac{d^2 i_1(0^+)}{dt^2} = -\frac{1}{R_1}\left\{\frac{-1}{R_1 C}\left[\frac{v(0^+)}{R_1 C} - \frac{dv(0^+)}{dt}\right] - \frac{d^2 v(0^+)}{dt^2}\right\}$$ **Ans.**

Problem 5.18. The given network consists of two coupled coils and a capacitor. At $t = 0$, the switch K is closed connecting a generator of voltage, $v(t) = V \sin (t/\sqrt{MC})$. Show that

$$v_a(0^+)=0, \quad \frac{dv_a}{dt}(0^+)=(V/L)\sqrt{MC}, \quad \text{and} \quad \frac{d^2v_a}{dt^2}(0^+) = 0$$

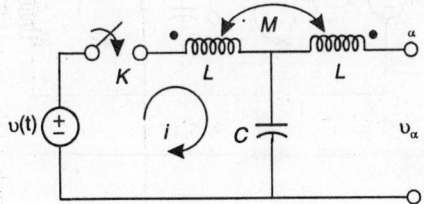

Fig. P. 5.18

Solution:

At $t = 0^+$, $i(0^+) = 0$ [as inductor is open circuited]

So
$$V_L = L\frac{di(0^+)}{dt} = V\sin 0° = 0 \therefore \frac{di(0^+)}{dt} = 0.$$

At time $t > 0$, $L\frac{di}{dt} + \frac{1}{c}\int i\,dt = V\sin(t/\sqrt{MC})$ (1)

or $L\frac{d^2i(0^+)}{dt^2} + \frac{1}{C}i(0^+) = V\cos(0)\times\frac{1}{\sqrt{MC}}$

or $\frac{d^2i(0^+)}{dt^2} + 0 = \frac{V}{L\sqrt{MC}}$

or $\frac{d^2i(0^+)}{dt^2} = \frac{V}{L\sqrt{MC}}$ (2)

$$V_a = M\frac{di}{dt} + \frac{1}{C}\int i\,dt \quad\quad (3)$$

or $\frac{dv_a(0^+)}{dt} = M\frac{d^2i(0^+)}{dt^2} + \frac{1}{C}i(0^+)$

$$= M\cdot\frac{V}{L\sqrt{MC}} = \frac{V}{L}\sqrt{M/C}$$

or $\dfrac{dv_a(0^+)}{dt} = \dfrac{V}{L}\sqrt{\dfrac{M}{C}}$ (4)

$$\dfrac{d^2v_a(0^+)}{dt^2} = M\dfrac{d^3i(0^+)}{dt^3} + \dfrac{1}{c}\dfrac{di(0^+)}{dt}$$ (5)

From, eq (2) $\dfrac{d^3i(0^+)}{dt^3} = 0$

So $\dfrac{d^2v_a(0^+)}{dt^2} = 0$ **Ans.**

Problem 5.19.   In the network of the following figure, the switch K is opened at $t = 0$ after the network has attained a steady state with the switch closed. (a) Find an expression for the voltage across the switch at $t = 0^+$. (b) If the parameters are adjusted such that $i(0^+) = 1$ and $di/dt\,(0^+) = -1$, what is the value of the derivative of the voltage across the switch, $dv_k/dt\,(0^+)$?

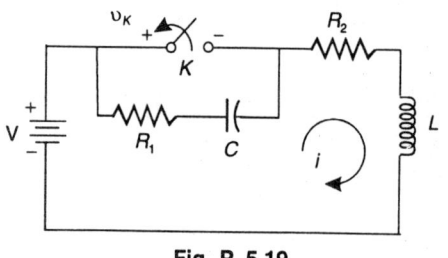

Fig. P. 5.19

Solution:

At time $t = 0^-$, $i(0^-) = \dfrac{V}{R_2}$ [as inductor is short circuited].

At time $t > 0$ capacitor acts as short circuited.

(a) So, $v_k = i(0^+)\,R_1,\; v_k = \dfrac{VR_1}{R_2}$

(b) Applying KVL, we get:

$$iR_1 + \dfrac{1}{C}\int idt + iR_2 + L\dfrac{di}{dt} = V$$

and
$$V_k = iR_1 + \frac{1}{C}\int idt$$

$$\therefore \qquad \frac{dv_k}{dt} = \frac{di}{dt}R_1 + \frac{1}{C}i$$

At $t = 0^+$, $\quad \dfrac{dv_k(0^+)}{dt} = R_1\dfrac{di(0^+)}{dt} + \dfrac{1}{C}i(0^+) = R_1(-1) + \dfrac{1}{C} \quad (1)$

$$\frac{dv_k(0^+)}{dt} = \frac{1 - R_1 C}{C} \quad \textbf{Ans.}$$

Problem 5.20. In the network shown in the following figure, the switch K is closed at $t = 0$ connecting the battery with an unenergized system.

(a) Find the voltage v_a at $t = 0^+$.
(b) Find the voltage across capacitor C_1 at $t = \infty$.

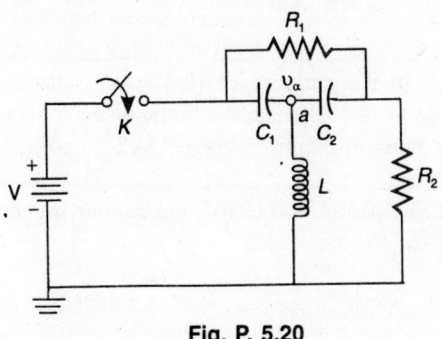

Fig. P. 5.20

Solution:

(a) At time $t = 0^+$, capacitor acts as short circuited and inductor acts as open circuited. So voltage v_a is equal to V.

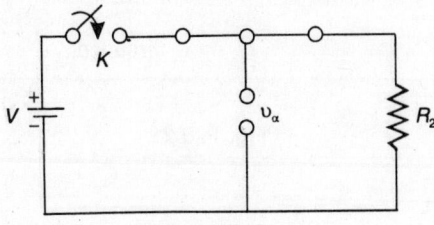

Fig. P. 5.20 (a)

$\therefore \quad v_a(0^+) = V$ volts

(b) At time $t = \infty$, capacitor acts as open circuit and inductor acts as short circuit. So equivalent circuit is drawn as
So voltage across capacitor C_1 is equal to voltage across R_1 and R_2.

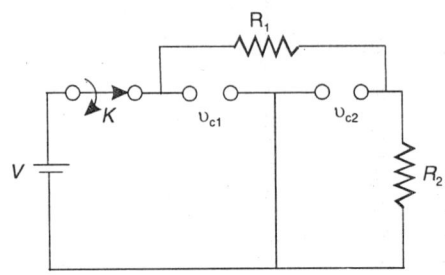

Fig. P. 5.20 (b)

$\therefore \quad v_{c1}(\infty) = V$ Volts **Ans.**

Problem 5.21. In the network of Fig. P. 5.21, the switch K is closed at $t = 0$. At $t = 0^-$, all capacitor voltages are inductor currents are zero. Three node-to-datum voltages and identified as v_1, v_2 and v_3.
(a) Find v_1 and dv_1/dt at $t = 0^+$. (b) Find v_2 and dv_2/dt at $t = 0^+$.
(c) Find v_3 and dv_3/dt at $t = 0^+$.

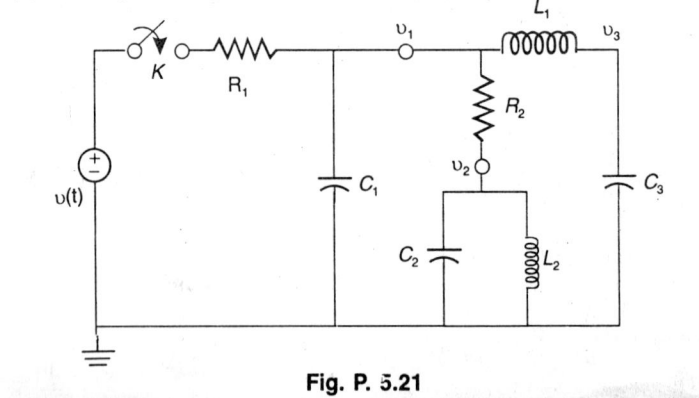

Fig. P. 5.21

Solution:

At time $t = 0^+$, $i(0^+) = \dfrac{v(0^+)}{R_1}$ [as capacitor is short circuited].

(a) Given, $\qquad v_1(0^+)=v_2(0^+)=v_3(0^+)=0$.

For time $t > 0$ $\qquad \dfrac{v_1-v}{R_1}+C_1\dfrac{dv_1}{dt}+\dfrac{1}{L}\int(v_1-v_3)dt=0$ $\qquad\qquad$ (1)

At time $t = 0^+$, current through inductor is zero, because at $t = 0^-$, current through inductor is zero.

$\therefore \qquad\qquad \dfrac{1}{L}\int(v_1-v_3)dt = 0$

$\therefore \qquad \dfrac{v_1(0^+)-v(0^+)}{R_1}+C_1\dfrac{dv_1(0^+)}{dt} = 0$

$\because \qquad\qquad v_1(0^+) = 0$

$\therefore \qquad \dfrac{dv_1(0^+)}{dt} = \dfrac{v(0^+)}{R_1C_1}=\dfrac{i(0^+)}{R_1C_1}$

(b) $\quad C_2\dfrac{dv_2}{dt}+\dfrac{1}{L_2}\int v_2\,dt+\dfrac{v_2-v_1}{R_2}=0$

At time $t = 0^+$, $v_1(0^+) = 0$, $v_2(0^+) = 0$

So, $\qquad C_2\dfrac{dv_2}{dt} = 0$ or $\dfrac{dv_2}{dt}=0$

(c) $\quad C_3\dfrac{dv_3}{dt}+\dfrac{1}{L}\int(v_3-v_1)dt=0$

At time $t=0^+$, $v_1(0^+)=0$, $v_3(0^+)=0$

$\therefore C_3\dfrac{dv_3}{dt}=0 \therefore \dfrac{dv_3}{dt}=$ **Ans.**

Problem 5.22. In the network of the following figure, a steady state is reached, and at $t = 0$, the switch K is opened. (a) Find the voltage across the switch, v_k at $t = 0^+$. (b) Find dv_k/dt at $t = 0^+$.

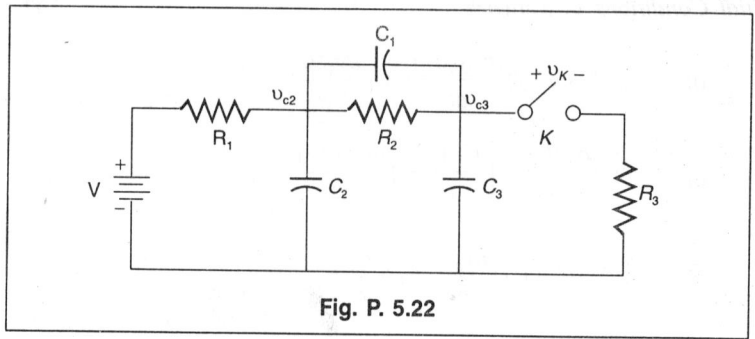

Fig. P. 5.22

Solution:

(a) At time $t = 0^-$, all capacitors are open circuited.

So $$i = \frac{V}{R_1 + R_2 + R_3}$$

$\therefore$ $$v_{c3}(0^-) = i \cdot R_3$$

$$= \frac{V \cdot R_3}{R_1 + R_2 + R_3}$$

At time $t = 0^+$, $v_k(0^+) = v_{c3}(0^+) = v_{c3}(0^-)$

$\therefore$ $$v_k(0^+) = \frac{V \cdot R_3}{R_1 + R_2 + R_3} \quad \textbf{Ans.}$$

(b) For time $t > 0$

$$\frac{v_{c2} - V}{R_1} + \frac{v_{c2} - v_{c3}}{R_2} + C_1 \frac{d}{dt}(v_{c2} - v_{c3}) + C_2 \frac{dv_{c2}}{dt} = 0$$

or $$v_{c2}(0^+) = \frac{V(R_2 + R_3)}{(R_1 + R_2 + R_3)}$$

and $$v_{c3}(0^+) = v_k(0^+)$$

$$= \frac{V \cdot R_3}{R_1 + R_2 + R_3}$$

$\therefore$ $$\frac{-V}{(R_1 + R_2 + R_3)} + \frac{V}{(R_1 + R_2 + R_3)} + C_1 \frac{d}{dt}\left(v_{c2}(0^+) - v_k(0^+)\right)$$

$$+ c_2 \frac{dv_{c2}(0^+)}{dt} = 0$$

or $$(C_1+C_2)\frac{dv_{c2}(0^+)}{dt}-C_1\frac{dv_k(0^+)}{dt}=0$$

or $$\frac{dv_k(0^+)}{dt}=\left(\frac{C_1+C_2}{C_1}\right)\frac{dv_{c2}(0^+)}{dt} \qquad (1)$$

At node 3 $$(C_1+C_3)\frac{dv_{c3}}{dt}-C_1\frac{dv_{c2}}{dt}=\frac{V}{R_1+R_2+R_3}$$

or $$(C_1+C_3)\frac{dv_k(0^+)}{dt}-C_1\frac{dv_{c2}(0^+)}{dt}=\frac{V}{R_1+R_2+R_3} \qquad (2)$$

Putting the value of $\dfrac{dv_{c2}(0^+)}{dt}$ from eq (1) to eq (2) we get

$$(C_1+C_3)\frac{dv_k(0^+)}{dt}-\frac{C_1^2}{(C_1+C_2)}\frac{dv_k(0^+)}{dt}=\frac{V}{(R_1+R_2+R_3)}$$

or $$\frac{dv_k(0^+)}{dt}\frac{(C_1^2+C_1C_2+C_1C_3+C_2C_3-C_1^2)}{(C_1+C_2)}=\frac{V}{(R_1+R_2+R_3)}$$

or $$\frac{dv_k(0^+)}{dt}=\frac{V(C_1+C_2)}{(R_1+R_2+R_3)(C_1C_2+C_1C_3+C_2C_3)} \qquad \textbf{Ans.}$$

Problem 5.23. In the network of the following figure, a steady state is reached with the switch K closed and with $i = I_0$, a constant. At $t = 0$, switch K is opened. Find: (a) $v_2(0^-)$, (b) $v_2(0^+)$, and (c) $dv_2/dt)(0^+)$.

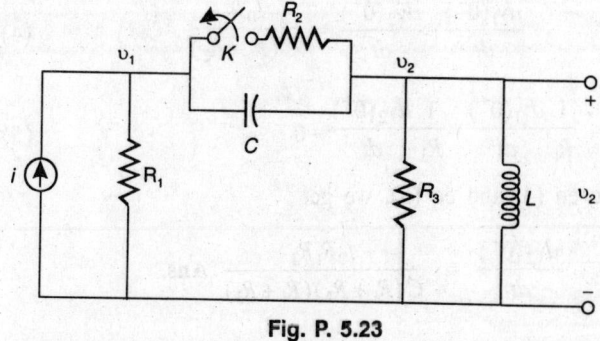

Fig. P. 5.23

Solution:

(a) At time $t = 0^-$, capacitor is open circuited and inductor is short circuited.

So $v_2(0^-) = 0$ volts, $i_L(0^-) = i = I_0$

(b) At time $t > 0$, $v_2 = L\dfrac{di_L}{dt}$

or $v_2(0^+) = L\dfrac{dI_L(0^+)}{dt} = L\dfrac{di_L(0^+)}{dt} = L\dfrac{dI_0}{dt}$

$\therefore$ $v_2(0^+) = 0$ as I_0 is a constant

(c) At node 1 $i = I_0 = \dfrac{v_1}{R_1} + \dfrac{cd(v_1 - v_2)}{dt}$ (1)

and at node 2 $C\dfrac{d(v_2 - v_1)}{dt} + \dfrac{v_2}{R_3} + \dfrac{1}{L}\displaystyle\int v_2\, dt = 0$ (2)

Adding eq (1) and eq (2) we get

$$\dfrac{v_1}{R_1} + \dfrac{v_2}{R_3} + \dfrac{1}{L}\int v_2\, dt = 0$$

differentiating the above equation

$$\dfrac{1}{R_1}\dfrac{dv_1}{dt} + \dfrac{1}{R_3}\dfrac{dv_2}{dt} + \dfrac{1}{L}v_2 = 0 \qquad (3)$$

at $t = 0^+$, eq (1) and eq (3) becomes equal to

$$C\dfrac{dv_1(0^+)}{dt} - C\dfrac{dv_2(0^+)}{dt} = I_0 - \dfrac{v_1(0^+)}{R_1} = I_0 - \dfrac{I_0 R_1 R_2}{R_1(R_1 + R_2)}$$

or $\dfrac{dv_1(0^+)}{dt} - \dfrac{dv_2(0^+)}{dt} = \dfrac{I_0 R_1}{C(R_1 + R_2)}$ (4)

and $\dfrac{1}{R_1}\dfrac{dv_1(0^+)}{dt} + \dfrac{1}{R_3}\dfrac{dv_2(0^+)}{dt} = 0$ (5)

Solving eq (4) and eq (5), we get

$$\dfrac{dv_2(0^+)}{dt} = \dfrac{I_0 R_1 R_3}{C(R_1 + R_2)(R_1 + R_3)} \quad \textbf{Ans.}$$

6

Second Order Linear Circuits

Problem 6.1. Suppose the switch S in Fig. P. 6.1 (a) has remained in position A for a long time and moves to position B at $t = 0$. The 1-µF capacitor is assumed to be ideal, whereas the practical inductor is modeled by a 10-mH ideal inductor in series with a 20-Ω resistor. Find and plot $vc(t)$ for $t \geq 0$ for the following three cases: cases 1: $R_2 = 405\ \Omega$; case 2 : $R_2 = 0$; and case 3 : $R_2 = 180\ R_2$. Each of these cases represents a different type of response.

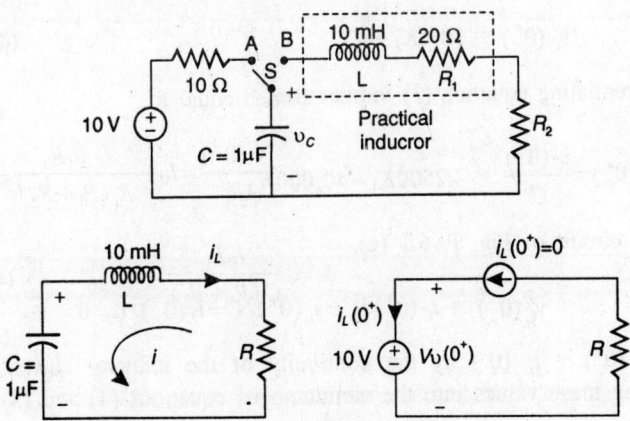

Fig. P. 6.1 (a) Discharge of a capacitor through a practical inductor in series with a resistance R_2. **(b)** Equivalent circuit for $t \geq 0$. **(c)** Equivalent circuit at $t = 0^+$ for calculating $i_C(0^+)$, in which the inductor has been replaced by an independent current source of value $i_L(0^+)$ and the capacitor by an independent voltage source of value $v_C(0^+)$.

Solution:

For the series *RLC*, the differential equation is, given by

$$\frac{d^2v_c}{dt^2}+\frac{R}{L}\frac{dv_c}{dt}+\frac{1}{LC}v_c=0 \tag{1}$$

With *L* and *C* fixed, the series *RLC* characteristic equation is

$$s^2+\frac{R}{L}s+\frac{1}{LC}=s^2+(20+R_2)10^2s+10^8=0 \tag{2}$$

Case 1. $R_2=405\,\Omega, R=425\,\Omega$ [Since for $t>0$, $R=R_1+R_2$].

If $R_2 = 405\ \Omega$, the characteristic equation (2) is $s^2 + 42,500s + 10^8 = 0$. Solving for the roots by the quadratic formula yields

$$S_{1,2} = -21,250\pm18,750=-2500,-40,000$$

Real distinct roots imply an overdamped response of the form

$$v_C(t) = K_1 e^{-2500t}+K_2 e^{-40,000t} \tag{3}$$

valid for $t > 0$.

Evaluating at $t = 0^+$ implies

$$v_C(0^+) = 10=K_1+K_2 \tag{4}$$

and differentiating equation (3) implies that,

$$v_C'(0^+)=\frac{i_C(0^+)}{C} = -2500K_1-40,000K_2 \tag{5}$$

From the circuit of Fig. P. 6.1 (c),

$$v_C'(0^+) = i_C(0^+)/C=i_L(0^+)/C=i_L(0^-)/C=0$$

where $i_L(0^+) = i_L(0^-)$ by the continuity of the inductor current. Substituting these values into the simultaneous equations (4) and (5), we get

$$K_1 = 10.667 \text{ and } K_2 = -0.667$$

Now, for $t > 0$ $v_C(t) = 10.667e^{-2500t}-0.667e^{-40,000t}$ V

The plot of $v_C(t)$ is depicted in Fig. P. 6.1 (d).

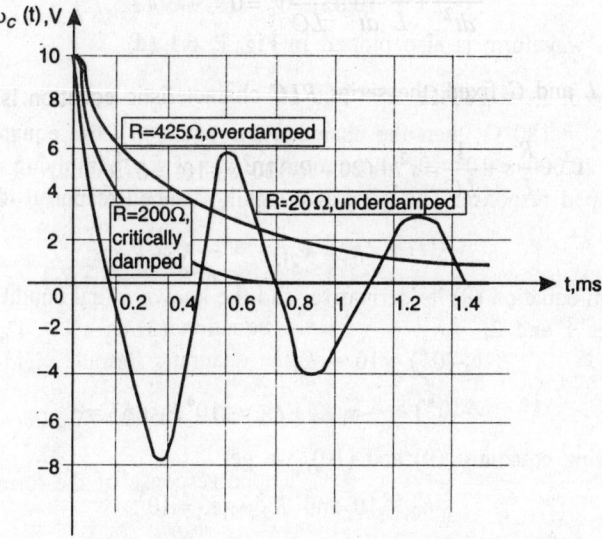

Fig. P. 6.1 (d) Waveforms of $v_C(t)$ in Problem 6.1 for three different degrees of damping. Critical damping represents the boundary between overdamped and the oscillatory behaviour of underdamped.

Case 2. $R_2 = 0$, $R = 20\ \Omega$

If $R_2 = 0$, then from equation (2), the characteristic equation is $s^2 + 2000s + 10^8 = 0$. Since $b^2 - 4c = -396{,}000{,}000 < 0$, the roots are complex. From the quadratic root formula

$$s_1 = 1000 + j9950 = -\sigma + j\omega_d \quad \text{and} \quad s_2 = -1000 - j9950 = -\sigma + j\omega_d$$

The underdamped response form is expressed by,

$$v_C(t) = e^{-\sigma t}[A\cos(\omega_d t) + B\sin(\omega_d t)]$$

$$= e^{-1000t}[A\cos(9950t) + B\sin 9950t)]\,\text{V} \qquad (6)$$

It remains to compute A and B in equation (6). From equation (6) and its derivative, $v_C(0^+) = 10 = A$ \hfill (7)

$$v'_C(0^+) = -\sigma A + \omega_d B = -1000A + 9950B \qquad (8)$$

As in case 1, $v'_C(0^+) = i_C(0^+)/C = 0$. Substituting into equation (7) and

(8) and solving yields $A = 10$ and $B = \dfrac{\sigma A}{\omega_d} = 1.005$

Now, for $t > 0$, the capacitor voltage is expressed by,

$$v_C(t) = e^{-1000t}[10\cos(9950t)+1.005\sin(9950t)]$$

$$= 10.05\,e^{-1000t}\cos(9950t+5.7°)\,V$$

This waveform is also plotted in Fig. P. 6.1 (d).

Case 3. $R_2 = 180\ \Omega$, $R = 200\ \Omega$.

If $R_2 = 180\ \Omega$, then the characteristic equation from equation (2) is $s^2 + 20,000s + 10^8 = 0$, whose roots are $s_1 = s_2 = 10^4$, implying a critically damped response. The general critically damped response form is

$$v_C(t) = (K_1+K_2t)e^{-s_1t}=(K_1+K_2t)e^{-10^4t} \tag{8}$$

From equation (8) its derivative, and the known initial conditions from cases 1 and 2,

$$v_C(0^+) = 10 = K_1 \tag{9}$$

$$v_C'(0^+) = -s_1K_1+K_2=-10^4K_1+K_2=0 \tag{10}$$

Solving equations (9) and (10), we get

$$K_1 = 10 \text{ and } K_2=s_1K_1=10^5$$

Hence, for $t > 0$

$$v_C(t) = 10(1+10^4t)e^{-10^4t}\,V$$

This waveform is also plotted in Fig. P. 6.1 (d).

Problem 6.2. This example illustrates the analysis of the second-order *RC* circuit shown in Fig. P. 6.2. The objective is to find $v_{C1}(t)$ and $v_{C2}(t)$ for $t > 0$ given the initial conditions $v_{C1}(0) = 10$ V and $v_{C2}(0) = 0$.

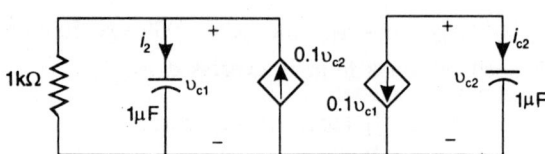

Fig 6.2 Oscillatory second-order *RC* circuit with controlled sources

Solution:

From the properties of a capacitor and KCL at the left node,

$$10^{-6}\frac{dv_{C1}}{dt} = i_{C1}=0.1v_{C2}-10^3v_{C1}$$

Multiplying through by 10^6, we get

$$\frac{dv_{C1}}{dt} = 10^6 \, i_{C1} = 10^5 \, v_{C2} - 10^3 \, v_{C1} \tag{1}$$

Considering the right node, we have

$$10^6 \frac{dv_{C2}}{dt} = i_{C1} = -0.1 v_{C1}$$

Rewriting as, $\dfrac{dv_{C2}}{dt} = -10^5 v_{C1}$ $\tag{2}$

Differentiating both sides of equation (1) will give an expression for the derivative of v_{C2}. Hence, we differentiate both sides of equation (1) and substitute equation (2) into the result of this differentiation to obtain a differential equation in only $v_{C1}(t)$

$$\frac{d^2 v_{C2}}{dt^2} = 10^5 \frac{dv_{C2}}{dt} - 10^3 \frac{dv_{c1}}{dt}$$

$$= -10^{10} v_{c1} - 10^3 \frac{dv_{C1}}{dt}$$

After rearranging terms, we get

$$\frac{d^2 v_{C1}}{dt^2} + 10^3 \frac{dv_{C1}}{dt} + 10^{10} \, v_{C1} = 0 \tag{3}$$

The differential equation (3) has the characteristic equation

$$s^2 + 10^3 s + 10^{10} = 0$$

From the quadratic formula, the roots are complex:

$$s_{1,2} = -500 \pm j99,998.75 = -\sigma \pm j\omega_d$$

As we know, complex roots imply an underdamped response of the form

$$v_{C1}(t) = e^{-\sigma t} [A \cos(\omega_d t) + B \sin(\omega_d t)]$$

$$= e^{-500t} [A \cos(99,998.75t) + B \sin(99,998.75t)] \tag{4}$$

Now, at $t = 0$, $v_{C1}(0) = 10 = A$. Also, from equation (1) and the initial conditions,

$$\frac{dv_{C1}(0)}{dt} = 10^5 \, v_{C2}(0) - 10^3 \, v_{C1}(0) = -10^4 \tag{5}$$

Differentiating equations (4) evaluating at $t = 0$, and equating the result with equation (5) produces

$$-10^4 = \frac{dv_{C1}(0)}{dt} = B\omega_d$$

in which case $B = 5.0001 \times 10^{-2}$.

Hence, the final form of the response is expressed as,

$$v_{C1}(t) = e^{-500t}[10\cos(99{,}998.75t) + 5.0001 \times 10^{-2}\sin(99{,}998.75t)]\,V$$

A plot of the (underdamped) response is illustrated in Fig. P. 6.2 (a).

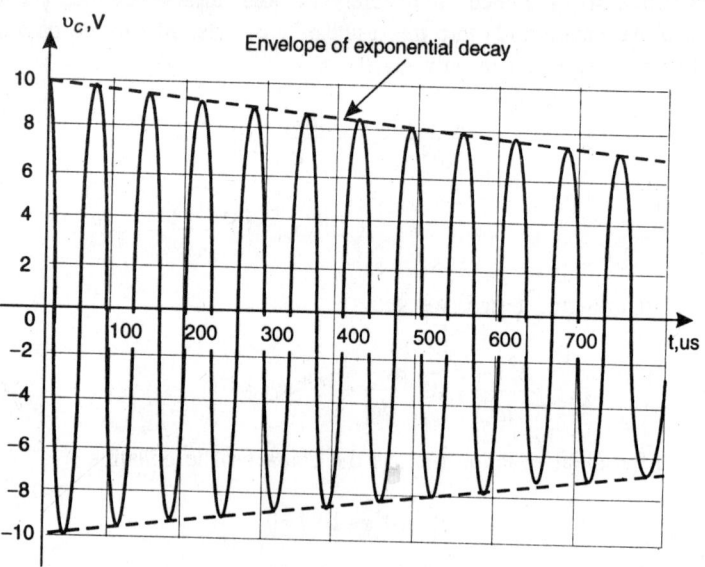

Fig. P. 6.2. (a) Underdamped (oscillatory) response showing envelope of exponential decay.

Problem 6.3. In the *RLC* circuit of Fig. P. 6.3, the capacitor is initially charged to $v_0 = 200$ V. Find the current transient after the switch is closed at $t = 0$.

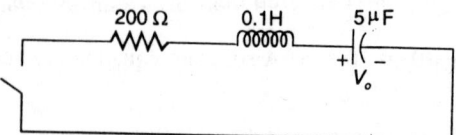

Fig. P. 6.3

Solution:

$$\alpha = \frac{R}{2L} = \frac{200}{2 \times 0.1} = 1000 \, s^{-1}$$

$$\omega_0^2 = \frac{1}{LC} = \frac{1}{0.1 \times 5 \times 10^{-6}} = 2 \times 10^6 \, s^{-2}$$

$$\therefore \qquad \omega_0 = 1414 \, 21 \, s^{-2}$$

$$j\beta = \sqrt{\alpha^2 - \omega_0^2} = \sqrt{10^6 - 2 \times 10^6} = 10^3 \sqrt{-1}$$

$$j\beta = 10^3 j \therefore B = 10^3$$

$$\because \quad \omega_0 > \alpha$$

$$\therefore \qquad i(t) = e^{-\alpha t} (A_3 \cos \beta t + A_4 \sin \beta t)$$

or $\qquad i(t) = e^{-1000 t} (A_3 \cos 1000 t + A_4 \sin 1000 t)$

Where, A_3 and A_4 are constant and can be obtained from the initial condition.

$$i(0^+) = 0 \text{ and } v_C(0^+) = 200 \text{ V}$$

$$i(0) = (A_3) = 0 \therefore A_3 = 0$$

$$\left. \frac{di}{dt} \right|_{t=0} = A_4 \beta = \frac{v_0}{L}$$

$$\therefore \qquad A_4 \beta = \pm \frac{200}{0.1}, A_4 = \pm \frac{2000}{10^3} = \pm 2$$

$$\therefore \qquad i(t) = e^{-1000 t} (0 + (\pm) 2 \sin 100 t)$$

$$i(t) = \pm 2 e^{-1000 t} \sin 1000 t \text{ A.} \quad \textbf{Ans.}$$

Problem 6.4. A series *RLC* circuit with $R = 200 \, \Omega$, $L = 0.1$ H and $C = 1000 \, \mu F$, has a voltage source of 200 V applied at $t = 0$. Find the current transient, assuming zero initial charge on the capacitor.

Solution:

$$\alpha = \frac{R}{2L} = \frac{200}{2 \times 0.1} = 1000 \, s^{-1}$$

$$\omega_0^2 = \frac{1}{LC} = \frac{1}{0.1 \times 100 \times 10^{-6}} = 10^5$$

$$\omega_0 = 316.227 \, s^{-2}$$

$\therefore$ $\omega_0 < \alpha$.   Hence $i(t)$ is given as

$$i(t) = e^{-\alpha t} \left(A_1 e^{\beta t} + A_2 e^{-\beta t} \right)$$

Where $\beta = \sqrt{\alpha^2 - \omega_0^2} = 948.68$

$\therefore$ $i(t) = A_1 \left(e^{-\alpha t + \beta t} \right) + A_2 \left(e^{-\alpha t - \beta t} \right)$

$$i(t) = A_1 e^{(-\alpha + \beta)t} + A_2 e^{-(\alpha + \beta)t}$$

$$i(t) = A_1 e^{-52t} + A_2 e^{-1948t}$$

Here A_1 and A_2 are constant and can be calculated by zero initial condition.

$$i(0) = A_1 + A_2 = 0 \quad \therefore \quad A_1 = -A_2$$

$$\left. \frac{di}{dt} \right|_{t=0} = A_1 (-\alpha + \beta) + A_2 (-\alpha - \beta) = \frac{200}{0.1}$$

or $-A_2 (-\alpha + \beta) + A_2 (-\alpha - \beta) = 2000$

$$+ A_2 \alpha - A_2 \beta - A_2 \alpha - A_2 \beta = 2000$$

$$-2A_2 \beta = 2000, \, A_2 \beta = -1000$$

or $A_2 = \dfrac{-1000}{\beta} = -1.05 \quad \therefore \quad A_1 = 1.05$

$\therefore$ $i(t) = 1.05 \left(e^{-52t} - e^{-1948t} \right)$ **Ans.**

Problem 6.5.   What value of capacitor is place of the 100 μF in Problem 6.4 results in the critically damped case?

Solution:

$$\alpha = \frac{R}{2L} = \frac{200}{2 \times 0.1} = 1000$$

$$\omega_0^2 = \frac{1}{LC} = \frac{1}{0.1C} = \frac{10}{C}$$

For critically damped case $\alpha = \omega_0$ ∴ $1000 = \sqrt{\dfrac{10}{C}}$

or

$$(10^3)^2 = \frac{10}{C} \text{ or } 10^6 = \frac{10}{C}$$

or

$$C = \frac{10}{10^6} = 10\,\mu F, \therefore C = 10\,\mu F \text{ Ans.}$$

Problem 6.6. Find the natural resonant frequency, $|\beta|$ of a series *RLC* circuit with $R = 200\ \Omega$, $L = 0.1$ H and $C = 5\ \mu f$.

Solution:

$$\alpha = \frac{R}{2L} = \frac{200}{2 \times 0.1} = 1000 = 10^3$$

$$\omega_0^2 = \frac{1}{LC} = \frac{1}{0.1 \times 5 \times 10^{-6}} = 2 \times 10^6$$

$$|\beta| = \left|\sqrt{\alpha^2 - \omega_0^2}\right| = \left|\sqrt{(10^3)^2 - 2 \times 10^6}\right| = \left|\sqrt{10^6 - 2 \times 10^6}\right| = \left|10^3\sqrt{-1}\right|$$

$$|\beta| = 10^3 \text{ rad/sec Ans.}$$

Problem 6.7. A voltage of 10 V is applied at $t = 0$ to a series *RLC* circuit with $R = 5\Omega$, $L = 0.1$ H, $C = 500\ \mu F$. Find the transient voltage across the resistance.

Solution:

$$\alpha = \frac{R}{2L} = \frac{5 \times 10}{2 \times 0.1} = 25\,s^{-1}$$

$$\omega_0^2 = \frac{1}{LC} = \frac{10}{0.1 \times 500 \times 10^{-6}} = \frac{10^6}{50}\,s^{-2} = 2 \times 10^4\,s^{-2}$$

∴

$$\omega_0 = 141.42\,s^{-2}$$

$$|\beta| = \left|\sqrt{\alpha^2 - \omega_0^2}\right| = \left|\sqrt{625 - 2 \times 10^4}\right|$$

$$\beta = \left| \sqrt{-19375} \right|, \ \beta = 139.19$$

$$i(t) = e^{-\alpha t} (A_3 \cos \beta t + A_4 \sin \beta t)$$

From initial condition. $A_3 = 0$

and $\qquad A_4 \beta = \dfrac{100}{1.0} \ \therefore \ A_4 = \dfrac{100}{139.19} = 0.719$

$\therefore \qquad i(t) = 0.719 e^{-25t} + \sin 139 t$

$\therefore \qquad v(t) = R \cdot i(t) = 5 \times i(t) = 5 \times 0.719 \, e^{-25t} \sin 139 t$

or $\qquad v(t) = 3 \cdot 6 e^{-25t} \sin 139 t$ **Ans.**

Problem 6.8. In the two-mesh circuit shown in Fig. P. 6.8, the switch is closed at $t = 0$. Find i_1 and i_2, for $t > 0$.

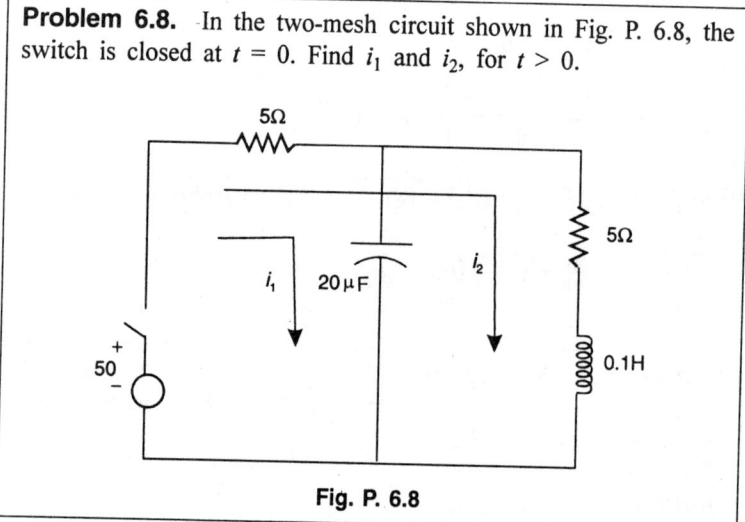

Fig. P. 6.8

Solution:

Applying KVL in both the mesh in first order differential form:

$$5i_1 + 5i_2 + \frac{1}{20 \times 10^{-6}} \int i_1 \, dt = 50 \text{ V} \qquad (1)$$

$$5i_1 + 5i_2 + 5i_2 + 0.1 \frac{di_2}{dt} = 50 \text{ V} \qquad (2)$$

Taking the time derivative of eq (1)

$$5 \frac{di_1}{dt} + 5 \frac{di_2}{dt} + \frac{10^6}{20} i_1 = 0$$

$$5\frac{di_2}{dt} = -5\times10^4\,i_1 - 5\frac{di_1}{dt}$$

or
$$\frac{di_2}{dt} = -10^4\,i_1 - \frac{di_1}{dt} \qquad (3)$$

From eq (2)

$$5i_1 + 10i_2 + 0.1\left(-10^4 i_1 - \frac{di_1}{dt}\right) = 50$$

or
$$10i_2 = 50 + \left(10^3 i_1 + 0.1\frac{di_1}{dt}\right) - 5i_1$$

or
$$i_2 = 5 + 100i_1 + 0.01\frac{di_1}{dt} - 0.5i_1 \qquad (4)$$

Putting eq (4) in eq (1), we get

$$5i_1 + 5(5 + 100i_1 + 0.01\frac{di_1}{dt} - 0.5i_1) + 5\times10^4 \int i_1 dt = 50V$$

or
$$5i_1 + 25 + 500i_1 + 0.05\frac{di_1}{dt} - 2.5i_1 + 5\times10^4 \int i_1 dt = 50\,\text{V}$$

Differentiating it once again, we get

$$2.5\frac{di_1}{dt} + 0 + 500\frac{di_1}{di} + 0.05\frac{d^2 i_1}{dt^2} + 5\times10^4\,i_1 = 0$$

or
$$\frac{d^2 i_1}{dt^2} + 10050\frac{di_1}{dt} + 10^6\,i_1 = 0$$

or
$$(s^2 + 10050\,s + 10^6\,s)i_1 = 0$$

$$s_{1,2} = \frac{-10050 \pm \sqrt{(10050)^2 - 4(10^6)}}{2}$$

$$= \frac{-10050 \pm 9848.98}{2} \quad \textbf{Ans.}$$

or
$$s_1 = -100.5, \; s_2 = -9949.49$$

$$\therefore \qquad i_1(t) = A_1\,e^{s_1 t} + A_2\,e^{s_2 t}$$

$$i_1(t) = A_1\,e^{-100t} + A_2\,e^{-9950t}$$

Where A_1 and A_2 are constant and is given by initial value.

$$\therefore \qquad i_1(0) = A_1 + A_2 = 10$$

$$v(0) = \frac{1}{c} \int i_1(t)\, dt$$

$$50 = \frac{1}{C} \left[\frac{A_1 e^{-100t}}{-100} + \frac{A_2 e^{-9950t}}{-9950} \right]_{t=0}$$

or
$$50 = \frac{1}{C} \left[\frac{A_1}{-100} - \frac{A_2}{9950} \right]$$

or
$$A_2 = 9.899 \quad \text{and} \quad A_1 = 0.101$$

$$\therefore \qquad i_1(t) = 0.101 e^{-100t} + 9.899 e^{-9950t} \quad \textbf{Ans.}$$

Similarly, $i_2 = 5 + 100 i_1 + 0.01 \dfrac{di_1}{dt} - 0.5 i_1$

Putting the value of $i_1(t)$ and $\dfrac{di_1(t)}{dt}$

We, get $i_1 = 5 - 5.05 e^{-100t} + 0.05 e^{-9950t}$ **Ans.**

Problem 6.9. A voltage has the S-domain representation $100 \angle 30°$ V. Express the time function for (a) $s = -2$ Np/s (b) $s = (-1 + j5)\ s^{-1}$

Solution:

  (a) For $\qquad s = -2$ Np/s

$$v(t) = 100 \cos(0.t + 30°)\, e^{-2t} = 100 \cos 30° \cdot e^{-2t}$$

  (b) For $\qquad s = (-1 + j5) = \alpha + jB$

$$v(t) = 100 \cos(5 \cdot t + 30°)\, e^{-t} \quad \textbf{Ans.}$$

Problem 6.10. Give the complex frequencies associated with the current $i(t) = 5.0 + 10\ e^{-3t} \cos(50t + 90°)$

Solution:

For transient current $i'(t) = 10 e^{-3t} \cos(50t + 90°)$, comparing it with standard form $= A_1 e^{\alpha t} \cos(\beta t + \phi)$

$$\therefore \qquad \alpha = -3, \quad \beta = 50$$

$\therefore$ In s-domain $\qquad s = \alpha \pm jB$

or $\qquad\qquad\qquad s = -3 \pm j50 \ s^{-1}$

For, constant current of 5 A, $s = 0$ **Ans.**

Problem 6.11. A phasor current $25 \angle 40°$ A has complex frequency $s = (-2 + j3)s^{-1}$ What is the magnitude of $i(t)$ at $t = 0.2$s?

Solution:

$$i(t) = 25\cos(3t+40°)e^{-2t}$$

$$i(0.2) = 25\cos(3 \times 0.2 + 40°)e^{-2 \times 0.2}$$

$$3 \times 0.2 = 0.6 \, \text{radian} = 0.6 \times 57.3°$$

$$= 34.38°$$

$$\therefore \qquad i(0.2) = 25\cos(34.38+40°)e^{-0.4} \, \text{A}$$

$$= 25 \times 0.269 \times 0.67 \, \text{A}$$

$$i(0.2) = 4.51 \, \text{A} \, \textbf{Ans.}$$

Problem 6.12. Calculate the impedance $Z(s)$ for the circuit shown in Fig. P. 6.12. at (a) $s = 0$ (b) s = $j1$ rad/s (c) $s = j2$ rad/s (d) $|S| = \infty$.

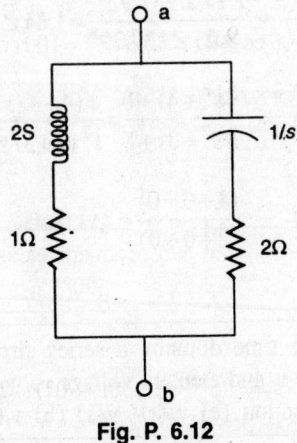

Fig. P. 6.12

Solution:

$$z(s) = \frac{\left(\dfrac{1}{s}+2\right)(2s+1)}{(2s+1)+\left(\dfrac{1}{s}+2\right)} = \frac{(1+2s)(1+2s)}{2s^2+3s+1}$$

or

$$z(s) = \frac{(1+2s)^2}{2s^2+3s+1} = \frac{4s^2+4s+1}{2s^2+3s+1}$$

(a)
$$s=0, \ z(0) = \frac{(1+0)^2}{0+0+1} = 1\,\Omega$$

$$S=j1 \ \ z(j1) = \frac{4(j1)^2+4(j1)+1}{2(j1)^2+3(j1)+1} = \frac{-4+j4+1}{-2+j3+1}$$

$$= \frac{-3+j4}{-1+j3} = \frac{5\angle126.86°}{3.16\angle108.43°}$$

$$z(j1) = 1.58\angle18.43°$$

(c) $\ \ s=j2\,\text{rad/s} \ \ z(s) = \dfrac{4(j2)^2+4(j2)+1}{2(j2)^2+3(j2)+1}$

$$= \frac{-16+j8+1}{-8+j6+1} = \frac{-15+j8}{-7+j6}$$

or
$$z(s) = \frac{17\angle151.92°}{9.21\angle139.39°} = 1.84\angle12.53°$$

(d)
$$|s|=\infty, z(s) = \frac{4s^2+4s+1}{2s^2+3s+1} = \frac{s(4+4/s+1/s^2)}{s^2(2+3/s+1/s^2)}$$

or
$$z(\infty) = \frac{(4+0+0)}{(2+0+0)} = 2\,\Omega \ \ \textbf{Ans.}$$

Problem 6.13. In the time domain, a series circuit of R, L and C has an applied voltage v_i and element voltage v_R, v_L and v_c. Obtain the voltage transfer function (a) $v_R(s)/\, v_i(s)$ (b) $v_c(s)/v_i(s)$.

Solution:

Fig. P. 6.13

$$i(s) = \frac{v_1(s)}{R + sL + 1/sC}$$

$\therefore$
$$v_R(s) = R \cdot i(s) = \frac{R v_i(s)}{R + sL + 1/sC}$$

or
$$\frac{v_R(s)}{v_i(s)} = \frac{R}{R + sL + 1/sC}$$

or
$$\frac{v_R(s)}{v_i(s)} = \frac{Rs/L}{s^2 + \dfrac{R}{L}s + \dfrac{1}{LC}} \quad \textbf{Ans.}$$

(b)
$$v_c(s) = \frac{i(s)}{Cs}$$

$\therefore$
$$v_c(s) = \frac{v_i(s)}{Cs\left(R + SL + \dfrac{1}{sC}\right)}$$

or
$$\frac{v_c(s)}{v_i(s)} = \frac{sC}{Cs\left(RCs + s^2 LC + 1\right)}$$

or
$$\frac{v_c(s)}{v_i(s)} = \frac{1/LC}{\left(s^2 + \dfrac{R}{L}s + \dfrac{1}{LC}\right)} \quad \textbf{Ans.}$$

Problem 6.14. Obtain the network function $H(s)$ for the circuit shown in Fig. P. 6.14. The response is the voltage $v_i(s)$.

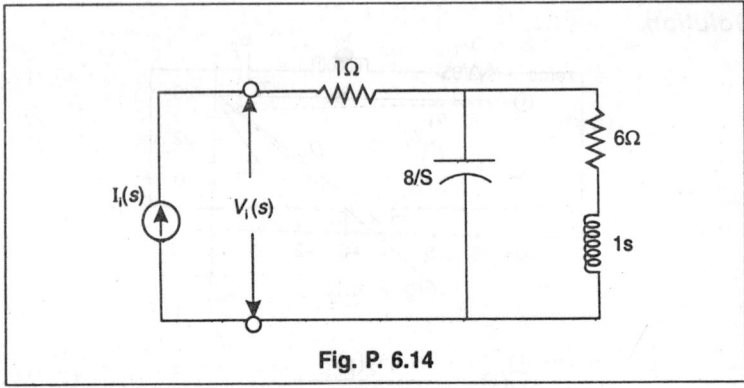

Fig. P. 6.14

Solution:

$H(s)$ response is equal to:

$$H(s) = \frac{v_i(s)}{I_i(s)} = z(s)$$

$$z(s) = \frac{(s+6)\dfrac{8}{s}}{s+6+\dfrac{8}{s}} + 1 = \frac{s^2 + 14s + 56}{s^2 + 6s + 8}$$

$$z(s) = \frac{(s+7-j2.65)(s+7+j2.65)}{(s+2)(s+4)} \quad \textbf{Ans.}$$

Problem 6.15. Construct the s-plane plot for the transfer function of Problem 6.15. Evaluate $H(j3)$ from the plot.

Solution:

$$H(s) = z(s) = \frac{(s+7-j2.65)(s+7+j2.65)}{(s+2)(s+4)}$$

$Z(s)$ has two poles and two zeros. Poles lie on -ve half of s-plane at $s = -2$ and $s = -4$, and zeroes lie on $s_1 = -7 + j\,2.65$ and $s_2 = -7 - j\,2.65$.

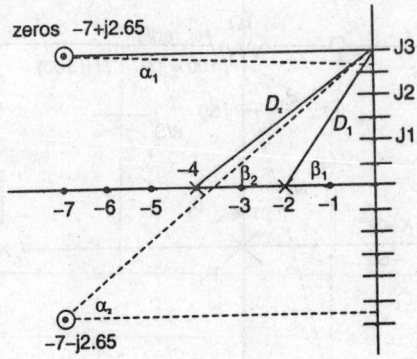

Fig. P. 6.15

From above

$$H(s) = \frac{(s+7-j2.65)(s+7+j2.65}{(s+2)(s+4)}$$

$$H(j3) = \frac{(j3+7-j2.65)(j3+7+j2.65)}{(j3+2)(j3+4)}$$

$$= \frac{(7+j0.35)(7+j5.65)}{(2+j3)(4+j3)}$$

$$= \frac{(7.008\angle2.86)(9.01\angle38.91)}{(3.61\angle56.31°)(5.01\angle36.87°)}$$

$$= \frac{(7.008\times9.01)(\angle2.86°+38.91°)}{(3.61\times5.01)(\angle56.31°+36.87°)}$$

$$= \frac{63.18\angle41.77°}{18.08(\angle93.18°)}=3.50\angle-51.41° \text{ Ans.}$$

Problem 6.16. Write the transfer function $H(s)$ whose pole-zero plot is given in Fig. P. 6.16. and find $H(j100)$.

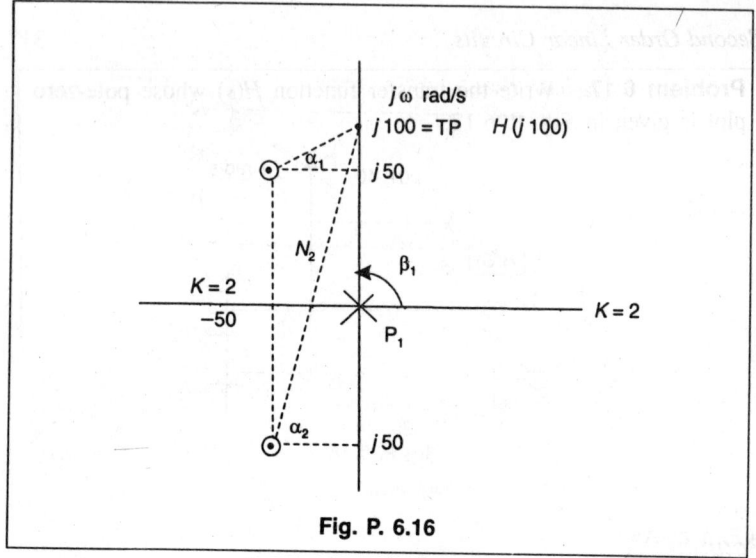

Fig. P. 6.16

Solution:

Given figure has one pole at origin and two zeros at $s = -50 + j50$ and $-50 - j50$

Transfer function

$$H(s) = K\frac{(s+z_1)(s+z_2)}{s} = \frac{(s-50-j50)(s-50+j50)}{s}$$

$$H(s) = K = \frac{(s^2 + 100s + 2500)}{s}$$

$$H(j100) = K = \frac{(j100-50-j50)(j100-50+j50)}{j100}$$

$$= K\frac{(-50+j50)(-50+j150)}{j100}$$

$$= K\frac{(70.8\angle135.43)(158\angle108.56)}{100\angle+90}$$

$$= K\frac{11060}{100}\angle135.43+108.56-90° \quad \because K=2$$

$$\therefore \qquad H(j100) = 223.6\angle153.99° \textbf{ Ans.}$$

or This is equivalent to $H(j\omega) = 222.6\angle26.57°$.

Problem 6.17. Write the transfer function $H(s)$ whose **pole-zero** plot is given in Fig. P. 6.17.

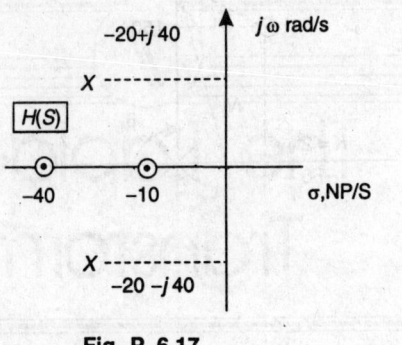

Fig. P. 6.17

Solution:

The given figure has two poles $P_1 = -20 + j40$ and $P_2 = -20 - j40$ and two zeroes, $z_1 = -10$ and $z_2 = -40$.

∴ Transfer function $H(s)$ =

$$H(s) = K \frac{(s+z_1)(s+z_2)}{(s+p_1)(s+p_2)}$$

∴

$$H(s) = K \frac{(s+10)(s+40)}{(s+(-20+j40))(S+(-20-j40))}$$

$$H(s) = K \frac{(s^2+50s+400)}{s^2+40s+2000} \quad \textbf{Ans.}$$

The Laplace Transform

Problem 7.1. Find the laplace transform form of each of the following functions.

(a) $f(t) = At$ (b) $f(t) = te^{-at}$

(c) $f(t) = e^{-at} \sin \omega t$ (d) $f(t) = \sin h\omega t$

(e) $f(t) = \cos h\omega t$ (f) $f(t) = e^{-at} \sin h\omega t$

Solution:

(a)
$$f(t) = At$$

$\therefore$
$$Lf(t) = AL t = \frac{A}{S^2}$$

$\therefore$
$$Lf(t) = L At = \frac{A}{S^2}$$

(b)
$$f(t) = te^{-at}$$

$$Lf(t) = L te^{-at}$$

$$L te^{-at} = \frac{1}{(s+a)^2}$$

(c)
$$f(t) = e^{-at} \sin \omega t$$

$$Lf(t) = L e^{-at} \sin \omega t$$

$$\mathcal{L}e^{-at}\sin\omega t = \frac{\omega}{(s+a)^2 + \omega^2}$$

(d)
$$f(t) = \sinh\omega t$$
$$\mathcal{L}f(t) = \mathcal{L}\sin h\omega t$$
$$\mathcal{L}\sin h\omega t = \frac{\omega}{s^2 - \omega^2}$$

(e)
$$f(t) = \cosh\omega t$$
$$\mathcal{L}\cosh\omega t = \frac{s}{s^2 - \omega^2}$$

(f)
$$f(t) = \mathcal{L}e^{-at}\sin h\omega t$$

$$\mathcal{L}e^{-at}\sin h\omega t = \frac{\omega}{(s+a)^2 - \omega^2}$$

$$\therefore \quad \mathcal{L}e^{-at}\sin h\omega t = \frac{\omega}{(s+a)^2 - \omega^2} \quad \textbf{Ans.}$$

Problem 7.2. Find the inverse laplace transform of each of the following functions.

(a) $$F(s) = \frac{s}{(s+2)(s+1)}$$

(b) $$F(s) = \frac{1}{s^2 + 7s + 12}$$

(c) $$F(s) = \frac{5s}{s^2 + 3s + 2}$$

(d) $$F(s) = \frac{3}{s(s^2 + 6s + 9)}$$

(e) $$F(s) = \frac{s+5}{s^2 + 2s + 5}$$

(f) $$F(s) = \frac{2s+4}{s^2 + 4s + 13}$$

(g) $$F(s) = \frac{2s}{(s^2 + 4)(s+5)}$$

Solution:

(a) $F(s) = \dfrac{s}{(s+2)(s+1)}$

$F(s) = -\dfrac{s}{s+2} + \dfrac{s}{s+1} = \dfrac{s+1-1}{s+1} - \dfrac{s+2-2}{s+2}$

$= 1 - \dfrac{1}{s+1} - 1 + \dfrac{2}{s+2}$

$F(s) = \dfrac{2}{s+2} - \dfrac{1}{s+1}$

Taking laplace inverse of both sides

$\mathcal{L}^{-1} F(s) = 2 \cdot e^{-2t} - e^{-t}$ **Ans.**

(b) $F(s) = \dfrac{1}{s^2+7s+12} = \dfrac{1}{s^2+4s+3s+12}$

$= \dfrac{1}{s(s+4)+3(s+4)} = \dfrac{1}{(s+3)(s+4)}$

$F(s) = \dfrac{1}{s+3} - \dfrac{1}{s+4}$

Taking laplace inverse of both sides

$\mathcal{L}^{-1} F(s) = e^{-3t} - e^{-4t}.$ **Ans.**

(c) $F(s) = \dfrac{5s}{s^2+3s+2} = \dfrac{5s}{(s+1)(s+2)}$

or, $F(s) = 5\left[\dfrac{s}{s+1} - \dfrac{s}{s+2}\right] = 5\left[\dfrac{s+1-1}{s+1} - \dfrac{s+2-2}{s+2}\right]$

$= 5\left[1 - \dfrac{1}{s+1} - 1 + \dfrac{2}{s+2}\right] = 5\left[\dfrac{2}{s+2} - \dfrac{1}{s+1}\right]$

Taking laplace inverse of both sides

$\mathcal{L}^{-1} F(s) = 5\left[2e^{-2t} - e^{-t}\right]$ **Ans.**

(d) $F(s) = \dfrac{3}{s(s^2+6s+9)} = \dfrac{3}{s(s+3)^2}$

let $\qquad F'(s) = \dfrac{1}{(s+3)^2}$

$$\frac{F'(s)}{s} = \int\limits_0^t F(t)\,dt.$$

$\therefore \qquad f'(t) = 3t\,e^{-3t}$

$\therefore \qquad \mathcal{L}^{-1} F(s) = F(t) = \int\limits_0^t 3t\,e^{-3t}\,dt$

or, $\qquad F(t) = 3\left(t\int\limits_0^t e^{-3t}\,dt - \int \frac{dt}{dt}\cdot\int e^{-3t}\,dt \right)$

$$= 3\left(\frac{t}{-3}\Big[e^{-3t} - 1 \Big]_0^t \int -\frac{1}{9}\Big[e^{-3t} - 1 \Big] \right)$$

$$= \left(\frac{-t\,e^{-3t}}{3} - \frac{1}{9}e^{-3t} + \frac{1}{9} \right)\times 3$$

$\therefore \qquad F(t) = \dfrac{1}{3} - \dfrac{1}{3}e^{-3t} - t\,e^{-3t}$ **Ans.**

(e) $\qquad F(s) = \dfrac{s+5}{s^2+2s+5}$

$$= \frac{s+5}{s^2+2s+4+1} = \frac{s+5}{(s+1)^2+2^2}$$

$$= \frac{s+1+4}{(s+1)^2+2^2} = \frac{s+1}{(s+1)^2+2^2} + \frac{2\times 2}{(s+1)^2+2^2}$$

Taking laplace inverse of both sides

$$\mathcal{L}F(s) = e^{-t}\left(\cos 2t + 2\sin 2t \right)\ \textbf{Ans.}$$

(f) $\qquad F(s) = \dfrac{2s+4}{s^2+4s+13} = \dfrac{2(s+2)}{s^2+4s+4+9}$

$$F(s) = \frac{2(s+2)}{(s+2)^2+3^2}$$

Taking laplace inverse of both sides

$$\mathcal{L}F(s) = 2e^{-2t}\cos 3t \quad \textbf{Ans.}$$

(g) $$F(s) = \frac{2s}{\left(s^2+4\right)(s+5)}$$

$$= \frac{1}{29}\frac{10s^2+50s+8s+40-10s^2-40}{\left(s^2+4\right)(s+5)}$$

$$= \frac{1}{29}\frac{10S(s+5)+8(s+5)-10\left(s^2+4\right)}{\left(s^2+4\right)(s+5)}$$

$$F(s) = \frac{1}{29}\left[\frac{10s}{s^2+4}+\frac{8}{s^2+4}-\frac{10}{s+5}\right]$$

Taking laplace inverse of both sides

$$\mathcal{L}^{-1}F(s) = \frac{1}{29}\left[10\cos 2t+4\sin 2t-10e^{-5t}\right]$$

$$\therefore \quad \mathcal{L}^{-1}F(s) = \left[\frac{10}{29}\cos 2t+\frac{4}{29}\sin 2t-\frac{10}{29}e^{-5t}\right]\textbf{Ans.}$$

Problem 7.3. A series *RL* circuit, with $R = 10\ \Omega$ and $L = 0.2$ H, has a constant voltage $V = 50$ V applied at $t = 0$. Find the resulting current using the laplace transform method.

Solution:

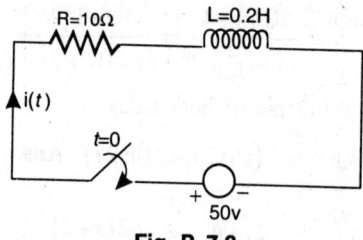

Fig. P. 7.3

Applying KVL in the single loop:

$$Ri(t)+L\frac{di(t)}{dt} = 50 \text{ V}$$

The Laplace Transform

Taking laplace on both sides:

$$i(s)R + Lsi(s) = \frac{50}{s}$$ [Initial condition is 0]

or, $$i(s)(R + Ls) = \frac{50}{s}$$

or, $$i(s)(10 + 0.2s) = \frac{50}{s}$$

or, $$i(s) = \frac{50}{s(10+0.2s)}$$

or, $$i(s) = \frac{50}{0.2s(s+50)}$$

or, $$i(s) = \frac{250}{s(s+50)}$$

$$i(s) = \frac{250}{50}\left[\frac{1}{s} - \frac{1}{s+50}\right]$$

$$i(s) = 5\left[\frac{1}{s} - \frac{1}{s+50}\right]$$

Taking laplace invers of both sides:

$$i(t) = 5[1-e^{-50t}] \text{ Ans.}$$

Problem 7.4. In the series *RL* circuit of Fig. P. 7.4, the switch is in position 1 long enough to establish the steady-state and is switched to position 2 at $t = 0$. Find the current.

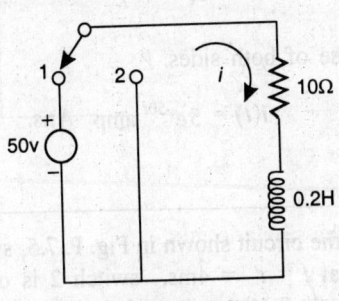

Fig. P. 7.4

Solution:

Since the switch is on position 1 for a long time to establish steady state, Hence, current

$$i\left(0^-\right) = \frac{50\,V}{10\,\Omega} = 5\,\text{amp}$$

Now the switch is thrown on position 2, the current at any time is given as:

$$10i(t) + 0.2\frac{di(t)}{dt} = 0$$

Taking laplace transform on both sides:

$$10i(s) + 0.2si(s) - 0.2i(0^+) = 0$$

Since $i\left(0^-\right) = i\left(0^+\right)$, because current in the inductor can not change instantaneously.

$\therefore$ $\qquad 10i(s) + 0.25i(s) = 0.2 \times 5$

or, $\qquad i(s)\,(10 + 0.2s) = 1$

or, $\qquad i(s) = \dfrac{1}{10 + 0.25}$

or, $\qquad i(s) = \dfrac{1}{0.2(s + 50)}$

or, $\qquad i(s) = \dfrac{5}{(s + 50)}$

Taking laplace inverse of both sides.

$$i(t) = 5e^{-50t} \text{ amp } \textbf{Ans.}$$

Problem 7.5. In the circuit shown in Fig. P. 7.5, switch 1 is closed at $t = 0$ and then, at $t = t' = 4\text{ms}$, switch 2 is opened. Find the current in the intervals $0 < t < t'$ and the $t > t'$.

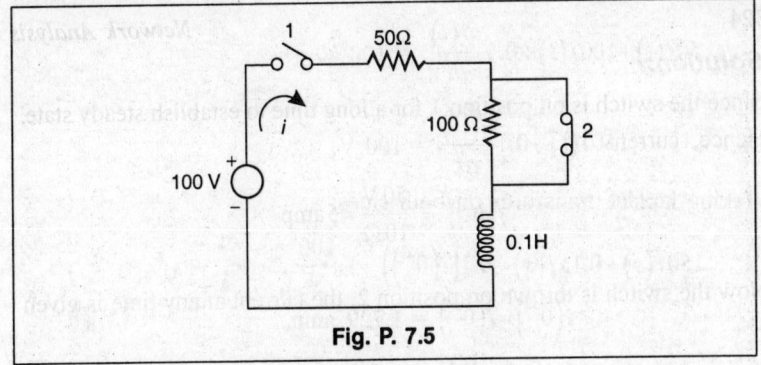

Fig. P. 7.5

Solution:

At time $t = 0$, circuit element is 50 Ω in series with 0.1 H inductance with zero initial condition. Hence, current $i(t)$ for $t > 0$ is:

$$50i(t) + 0.1\frac{di(t)}{dt} = 100 \text{ V}$$

Taking laplace transform on both sides

$$(50 + 0.2\ s)\ I(s) = \frac{100}{s}$$

or,

$$I(s) = \frac{100}{s(50 + 0.1s)} = \frac{1000}{s(s + 500)} = \frac{1000}{500}\left[\frac{1}{s} - \frac{1}{s + 500}\right]$$

$$I(s) = 2\left[\frac{1}{s} - \frac{1}{s + 500}\right]$$

Taking laplace inverse of both sides

$$I(t) = 2\left(1 - e^{-500t}\right)$$

Hence, in the interval $0 < t < t'$, the current

$$I(t) = 2\left(1 - e^{-500t}\right)$$

at time $t = t' = 4$ ms

$$I(4\ \text{ms}) = 2\left(1 - e^{\frac{-500 \times 4}{1000}}\right) = 2\left(1 - e^{-2}\right) = 1.729 \text{ amp}$$

At time $t = 4$ ms, circuit parameter is changed with initial current $i(0^-) = 1.729$ amp. Now, the current equation for $t > t'$ is:

$$50i(t)+100i(t)+0.1\frac{di(t)}{dt} = 100 \text{ V}$$

or, $$150i(t)+0.1\frac{di(t)}{dt} = 100 \text{ V}$$

Taking laplace transform on both sides:

$$150i(s)+0.1si(s)-0.1\left(i\left(0^+\right)\right) = \frac{100}{s}$$

∵ $$i\left(0^+\right)=i\left(0^-\right) = 1.729 \text{ amp}$$

or $$(150+0.1s)i(s) = \frac{100}{s}+0.1729$$

or, $$i(s) = \frac{100}{s\,(150+0.15)}+\frac{0.1729}{(150+0.15)}$$

or, $$i(s) = \frac{2}{3}\left[\frac{1}{s}-\frac{1}{s+500}\right]+\frac{0.1729}{0.1(s+1500)}$$

Taking laplace inverse both sides

$$i(t) = \frac{2}{3}\left(1-e^{-1500t_1}\right)+1.729e^{-1500t_1}$$

or, $$i(t) = 1.06e^{-1500t_1}+0.667$$

Where $$t_1 = t-t'$$

∴ $$i(t) = 1.06e^{-1,500(t-t')}+0.667 \text{ amp} \quad \textbf{Ans.}$$

Problem 7.6.   In the series $R\,L$ circuit shown in Fig. P. 7.6, the switch is closed on position 1 at $t = 0$ and then at $t = t' = 50\ \mu s$, it is moved to position 2. Find the current in the interval $0 < t < t'$. and $t > t'$.

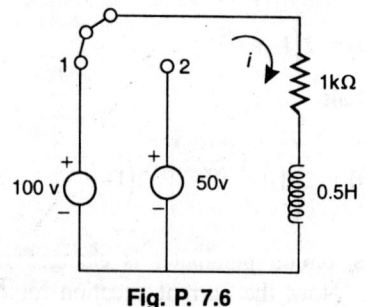

Fig. P. 7.6

Solution:

When switch is on position 1, with zero initial condition, the current at any time t is given as:

$$1000i(t)+0.5\frac{di(t)}{dt} = 100V$$

or, Taking laplace transform on both sides

$$1000i(s)+0.5si(s) = \frac{100}{s}$$

or, $$i(s) = \frac{100}{s(1000+0.5s)}$$

or, $$i(s) = \frac{200}{s(s+2000)}$$

or, $$i(s) = \frac{200}{2000}\left[\frac{1}{s}-\frac{1}{s+2000}\right]$$

Taking laplace inverse on both sides

$$i(t) = 0.1\left(1-e^{-2000t}\right)amp$$

This is the current for $0 < t < t'$, because at $t > t'$, switch is thrown at position 2. At $t = t' = 50\ \mu s$, $i(t)$ is equal to:

$$i(50\mu s) = 0.1\left[1-e^{-2000\times50\times10^{-6}}\right]$$

$$= 0.1\left[1-e^{-0.1}\right]=0.0095\,amp$$

Now, the current equation at $t > t'$ is given as:

$$1000i(t)+0.5\frac{di(t)}{dt} = -50\ V$$

or, $$I(s)(1000+0.5s)-0.5i(0^+) = \frac{-50}{s}$$

$$i(0^+) = i(0^-)=0.0095\,amp$$

or, $$I(s) = \frac{-50}{s(0.5s+1000)}+\frac{4.75\times10^{-3}}{(0.5s+1000)}$$

or, $$I(s) = \frac{-100}{s(s+2000)} + \frac{4.75 \times 10^{-3}}{0.5(s+2000)}$$

or, $$I(s) = \frac{-100}{2000}\left[\frac{1}{s} - \frac{1}{s+2000}\right] + \frac{0.0095}{s+2000}$$

Taking laplace inverse on both sides

$$I(t) = -0.05\left[1 - e^{-2000t_1}\right] + 0.0095\,e^{-2000t_1}$$

$$I(t) = 0.0595\,e^{-2000t_1} - 0.05\,\text{amp}\quad\textbf{Ans.}$$

where $t_1 = t - t'$

Now, $$I(t) = 0.0595\,e^{-2000(t-t')} - 0.05\,\text{amp}\,\textbf{Ans.}$$

Problem 7.7. A series *RC* circuit with $R = 10\ \Omega$ and $C = 4\,\mu\text{F}$, has an initial charge $Q_0 = 800\ \mu\text{c}$ on the capacitor at the time the switch in closed, applying a constant voltage source $V = 100$ V. Find the resulting current transient if the charge is (a) of the same polarity as that deposited by the source and (b) of the opposite polarity.

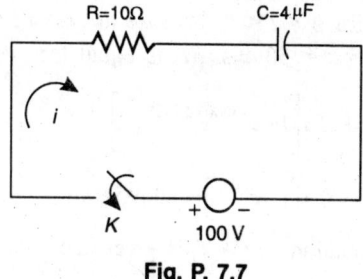

Fig. P. 7.7

Solution:

When switch is closed, then current through the circuit with initial charge $Q_0 = 800\ \mu\text{c}$ is given as

$$iR + \frac{1}{c}\int i\,dt = V$$

or, $$10i(t) + \frac{1}{4\times10^{-6}}\int i(t)\,dt = 100\ \text{V}$$

Taking laplace transform

$$10i(s)+\frac{1}{4\times10^{-6}}\left[\frac{I(s)}{s}+\frac{v(0)}{s}\right]=\frac{100}{s}$$

(a) When the charge has same polarities as that deposited by the source, then

$$10i(t)+\frac{I(s)}{4\times10^{-6}s}+\frac{800\times10^{-6}}{4\times10^{-6}s}=\frac{100}{s}$$

or, $$\left(10+\frac{10^6}{4s}\right)i(s)+\frac{200}{s}=\frac{-100}{s}$$

$$i(s)\left(10+\frac{10^6}{4s}\right)=\frac{-100}{s}$$

or, $$i(s)\left(\frac{s40+10^6}{4s}\right)=\frac{-100}{s}$$

or, $$i(s)=\frac{-100\times4}{40\left(s+\frac{10^6}{40}\right)}$$

or, $$i(s)=\frac{-10}{\left(s+2s\times10^3\right)}$$

Taking laplace inverse on both sides

$$i(t)=-10e^{-25\times10^3}t\ amp$$

(b) When the charge has opposite polarities as that deposited by the source, then

$$10i(s)+\frac{i(s)}{4\times10^{-6}s}-\frac{800\times10^{-6}}{4\times10^{-6}s}=\frac{100}{s}$$

or, $$i(s)\frac{10s+25\times10^4}{s}=\frac{300}{s}$$

or, $$i(s)\left(s+25\times10^3\right)=0$$

or, $\qquad i(s) = \dfrac{30}{s+25\times10^3}$

Taking laplace inverse on both sides

$$i(t) = -30e^{-25\times10^3}t \text{ amp } \textbf{Ans.}$$

Problem 7.8. In the *RC* circuit shown in Fig. P. 7.8, the switch is closed on position 1 at $t = 0$ and then at $t = t' = \tau$ (the time constant) is moved to position 2. Find the transient current in the intervals $0 < t < t'$ and $t > t'$

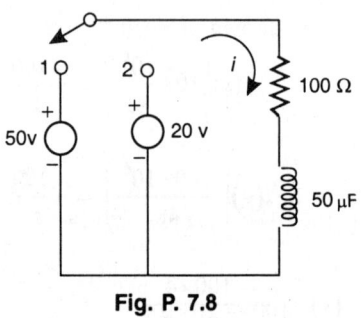

Fig. P. 7.8

Solution:

When switch is on position 1, then current in the circuit is given as

$$Ri(t) + \frac{1}{C}\int i(t)dt = V$$

or, $\quad 100i(t) + \dfrac{1}{50\times10^{-6}}\int i(t)dt = 50 \text{ V}$

In laplace form

$$10i(s) + \frac{i(s)}{50\times10^{-6}s} - \frac{v(0)}{50\times10^{-6}s} = \frac{50}{s}$$

Since $\qquad\qquad\qquad v(0) = 0.$

$\therefore \qquad 100i(s) + \dfrac{i(s)}{50\times10^{-6}s} = \dfrac{50}{s}$

or, $\qquad i(s) + \left(100 + \dfrac{20\times10^3}{s}\right) = \dfrac{50}{s}$

or, $$i(s) = \frac{50}{\left(100s + 20\times10^3\right)}$$

or, $$i(s) = \frac{0.5}{s+200}$$

Taking laplace inverse

$$i(t) = 0.5e^{-200t} \text{ amp}$$

This is the current for time interval $0 < t < t'$. At $t = t' = $ Time constant.

$$i(\tau) = 0.5e^{-1} = 0.1830 = i(0)$$

Now, when switch is thrown at position 2, then current in the circuit is given as

$$100i(t) + 20\times10^3 \int i(t)\,dt = -20V$$

or, $$100i(s) + \frac{20\times10^3}{s}\left[i(s)+i(0)\right] = \frac{-20}{s}V$$

or, $$\left(\frac{100s+20\times10^3}{s}\right)i(s) = \frac{-20}{s} - \frac{20\times10^3\,i(0)}{s}$$

or, $$\left(100s+20\times10^3\right)i(s) = -20 - 20\times10^3\times0.1830$$

or, $$\left(100s+20\times10^3\right)i(s) = -20 - 36.6$$

$$i(s) = \frac{-56.6}{100s+20\times10^3}$$

or, $$i(s) = \frac{-56.6}{100(s+200)}$$

or, $$i(s) = \frac{-0.566}{s+200}$$

Taking laplace inverse we get

$$i(t) = -0.566\,e^{-200t_1}$$

Where $t_1 = t - t'$

$\therefore$ $i(t) = -0.566 e^{-200(t-t')}$ **Ans.**

Problem 7.9. A series *RLC* circuit with $R = 5\ \Omega$, $L = 0.1\ H$ and $C = 500\ \mu F$, has a constant voltage $V = 10V$ applied at $t = 0$. Find the resulting current.

Solution:

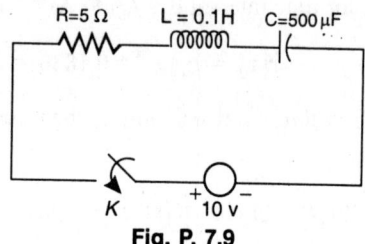

Fig. P. 7.9

The time–domain equation of the given circuit is

$$Ri(t) + L\frac{di(t)}{dt} + \frac{1}{C}\int_0^t i(t)\,dt = V$$

In laplace form

$$Ri(s) + Lsi(s) + \frac{1}{Cs} + i(s) = \frac{V}{s}$$

or, $$\left(5 + 0.1s + \frac{1}{500\times10^{-6}s}\right)i(s) = \frac{10}{s}$$

or, $$\left(s^2 + 50s + 20\times10^3\right)i(s) = \frac{10}{0.1}$$

or, $$i(s) = \frac{100}{s^2 + 50s + 20\times10^3}$$

or, $$= \frac{100}{s^2 + 2\times25s + 625 + 19375} = \frac{100}{(s+25)^2 + (139.194)^2}$$

$$= \frac{100\times139.194}{139.194\left((s+25)^2 + (139.194)^2\right)}$$

Taking laplace inverse $= \dfrac{100}{139.194}\left[e^{-25t}\sin 139.194\right]$

or, $\qquad\qquad i(t) = 0.718 e^{25t}\sin 139.194 t$ **Ans.**

Problem 7.10. A pair of coupled inductors is connected in two different ways, as shown in Fig. P. 7.10. In each case, find the differential equation relating v and i, and then find the equivalent inductance "seem" at the two terminals of each box.

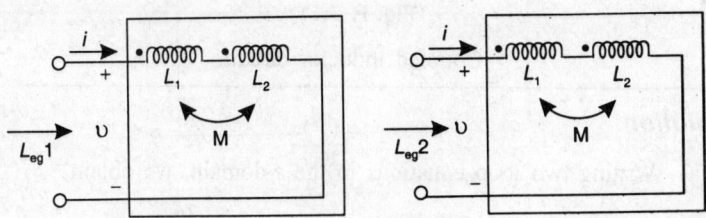

Fig. P. 7.10

Equivalent inductance of two series-connected inductors.

(a) $L_{eq1} = L_1 + L_2 + 2M$.

(b) $L_{eq2} = L_1 + L_2 - 2M$.

Solution:

Using KVL in Fig. P. 7.10 (a)

We get, $v(t) = \left(L_1\dfrac{di}{dt} + M\dfrac{di}{dt}\right) + \left(M\dfrac{di}{dt} + L_2\dfrac{di}{dt}\right) = (L_1 + L_2 + 2M)\dfrac{di}{dt}$

Comparing this relationship with

$$v(t) = L_{eq1}\dfrac{di}{dt}$$

We get $\qquad\qquad L_{eq1} = L_1 + L_2 + 2M$

Similarly, for Fig. P. 7.10 (b), using KVL, we obtain,

$$v(t) = \left(L_1\dfrac{di}{dt} - M\dfrac{di}{dt}\right) + \left(-M\dfrac{di}{dt} + L_2\dfrac{di}{dt}\right) = (L_1 + L_2 - 2M)\dfrac{di}{dt}$$

Hence, $\qquad\qquad L_{eq2} = L_1 + L_2 - 2M$

Problem 7.11. In the circuit of Fig. P. 7.11, $R_1 = 4\ \Omega$, $R_2 = 2\Omega$, $L_1 = 8H$, $L_2 = 6$ H, and $M = 4$ H. If $E = 36$ V, $i_2\ (0^-)$ and S is closed at $t = 0$:

(a) Find $i_1\ (t)$ and $i_2\ (t)$.

(b) Show that $v_2\ (t)$ and di_1/dt are positive for all $t > 0$.

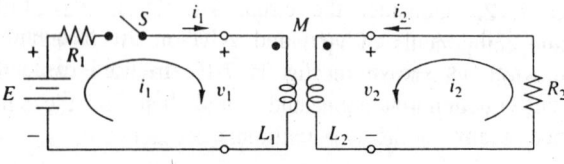

Fig. P. 7.11

Coupled inductor circuit.

Solution:

(a) Writing two loop equations in the s-domain, we obtain,

$$4I_1 + \left(82s\,I_1 + 4s\,I_2\right) = \left(8s + 4\right)I_1 + 4sI_2 = \frac{36}{s} \tag{1}$$

$$\left(4I_1 + 6s\,I_2\right) + 2I_2 = 4s\,I_1 + \left(6s + 2\right)I_2 = 0 \tag{2}$$

Writing eq^n (1) and eq^n (2) in matrix notation, we have

$$\begin{bmatrix} 8s+4 & 4s \\ 4s & 6s+2 \end{bmatrix} \begin{bmatrix} I_1 \\ I_2 \end{bmatrix} = \begin{bmatrix} 36/s \\ 0 \end{bmatrix}$$

Solving for I_1 and I_2 by Cramer's rule, we get

$$I_1(s) = \frac{6.75\left(s + \dfrac{1}{3}\right)}{s(s+0.25)(s+1)} = \frac{9}{s} - \frac{3}{s+0.25} - \frac{6}{s+1}$$

$$I_2(s) = \frac{-4.5}{(s+0.25)(s+1)} = \frac{-6}{s+0.25} + \frac{6}{s+1}$$

Using inverse Laplace transforming $I_1(s)$ and $I_2(s)$, gives,

$$i_1(t) = \left(9 - 3e^{-0.25t} - 6e^{-t}\right)u(t)\,\text{A}$$

$$i_2(t) = 6\left(e^{-t} - e^{-0.25t}\right)u(t)\,\text{A}$$

(b) Using the expressions found in part (a) implies

$$v_2(t) = -R_2\,i_2(t) = 12\left(e^{-0.25t} - e^{-t}\right)u(t)\,\text{V} \tag{3}$$

and $\qquad \dfrac{di_1}{dt} = 0.75e^{-0.25t} + 6e^{-t}$, for $t > 0$ $\qquad\qquad$ (4)

From equations (3) and (4), $v_2(t) > 0$ and $di_1/dt > 0$ for all $t > 0$.

Problem 7.12. Consider the circuit of Fig. P. 7.12 Find the steady-state components of $v_1(t)$ and $v_2(t)$ at the frequency of 1 rad/s for the following two cases: (a) The 2-F capacitor is disconnected. (b) The 2-F capacitor is connected.

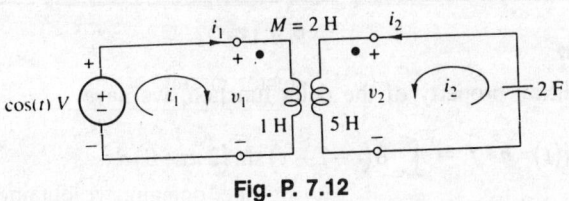

Fig. P. 7.12

Solution:

First we note that $\omega = 1$ rad/s that the phasor for the voltage source is $1\angle 0$ V.

(a) With the capacitor disconnected, $I_2 = 0$. Hence I_2 induces no component in V_1, and the equation for V_1 is $V_1 = 1\angle 0 = j1 \times I_1$

From which it follows that $\qquad I_1 = 1\angle -90°$

and $\qquad\qquad\qquad\qquad\quad V_2 = j\omega M I_1 = 2\angle 0$

Observe that $\qquad\qquad\qquad v_1(t) = \cos(t)$ V

and $\qquad v_2(t) = 2\cos(t)$ V $\,$ are in phase

(b) With the 2-F capacitor connected, the two loop equations are given as,

$$V_1 = 1 = j\omega L_1 I_1 + j\omega M I_2 = j I_1 + j2 I_2 \qquad (1)$$

and, since $V_2 = -I_2 / j\omega C$,

$$\left(j\omega M I_1 + j\omega L_2 I_2\right) + \dfrac{I_2}{j\omega C} = j2 I_1 + j4.5 I_2 = 0 \qquad (2)$$

Solving eqns (1) and eqns (2), we have

$$I_1 = -j9 \text{ and } I_2 = j4$$

Using equation $\bar{V}_2 = \pm j\omega M \bar{I}_1 + j\omega L_2 \bar{I}_2$, we obtain,

$$V_2 = j\omega M I_1 + j\omega L_2 I_2 = j2(-j9) + j5(j4) = -2$$

in which case $v_2(t) = 2\cos(t + 180°)$ V. Clearly, $v_1(t) = \cos(t)$ V and $v_2(t) = 2\cos(t + 180°)$ V are 180° out of phase!

Problem 7.13. Compute the convolution of $h(t) = \delta(t - T)$ with $f(t) = \sin(2\pi t + \theta)$.

Solution:

Using shifting property of the delta function, we have

$$y(t) = h * f = \int_{-\infty}^{\infty} \delta(t - T - \tau)\sin(2\pi\tau + \theta)\,d\tau$$

$$= \sin\left[(2\pi\tau + \theta)\right]_{\tau = t - T} = \sin\left[2\pi(t - T) + \theta\right]$$

Problem 7.14. Compute the convolution of $h(t) = -2\delta(t - 3)$ with the function $f(t)$ sketched in Fig. P. 7.14.

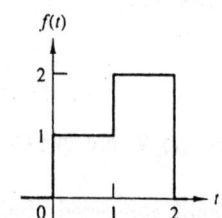

Fig. P. 7.14. Function $f(t)$

Solution:

From the sifting property of the delta function we have,

$$y(t) = h * (t)\, f(t) = \int_{-\infty}^{\infty} -2\delta(\tau - 3)\, f(t - \tau)\,d\tau = -2f(t - 3)$$

In other words, $f(t)$ is shifted to the right by three units and multiplied by -2, as depicted in Fig. P. 7.14 (a).

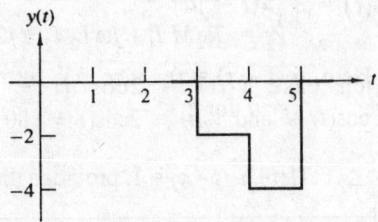

Fig 7.14 (a) Resulting convolution, $y = h*f$

Problem 7.15. Find $y(t) = h*f$ when $h(t) = u(t)$ and $f(t) = u(t + 1) - u(t - 1)$. $h(t)$, $f(t)$, and the solution, $y(t) = h*f$, are all plotted in Fig. P. 16.5.

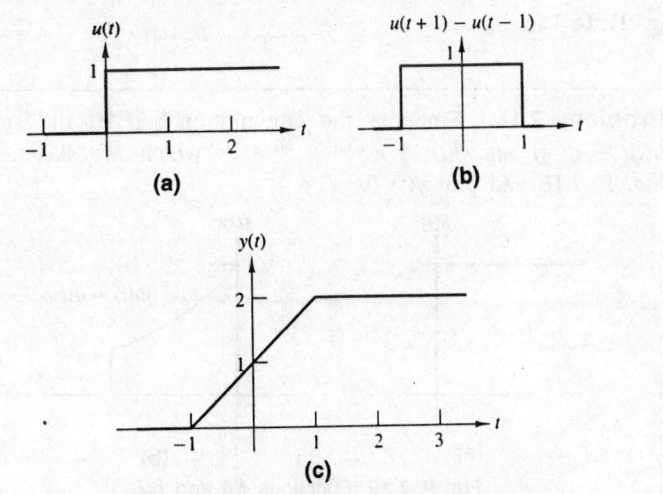

Fig 7.15 (a) $h(t) = u(t)$, the step function.
(b) $f(t) = u(t+1) - u(t-1)$.
(c) Resulting $y(t)$ for the convolution $h*f$.

Solution:

From the definition of the convolution integral, we have

$$y(t) = h*f = \int_{-\infty}^{\infty} h(t-\tau) f(\tau) d\tau$$

$$= \int_{-\infty}^{\infty} u(t-\tau)\left[u(\tau+1) - u(\tau-1)\right] d\tau$$

Observing that $u(\tau+1) - u(\tau-1)$ is nonzero only for $-1 \le \tau \le 1$ yields

$$y(t) = \int_{-1}^{1} u(t-\tau)\,d\tau \tag{1}$$

Case 1. $t < -1$: Here, $u\,(t-\tau) = 0$, since τ is restricted to the region $[-1.1]$. Therefore, $y\,(t) = 0$ for $t < -1$.

Case 2. $-1 \le t \le 1$: Here, $u\,(t-\tau) = 1$, provided that $\tau \le t$. Therefore, for $-1 \le t \le 1$

$$y(t) = \int_{-1}^{1} u(t-\tau)\,d\tau = \int_{-1}^{t} d\tau = t+1$$

Case 3. $-1 \le t$: Here, $u\,(t-\tau) = 1$, for all τ in $[-1.1]$ and evaluation of the integnal over $[-1, 1]$ yields $y\,(t) = 2$ for $t \ge 1$.

Combining these three cases leads to the fuction illustrated in Fig. P. 16.15c.

Problem 7.16. Compute the convolution $y(t)$ of the signals $h(t) = u(-t)$ and $f(t) = e^{-t[u(t) - u(t-T)]}$ which are sketched in Fig. P. 7.16 and plot y(t) for $T = 1$.

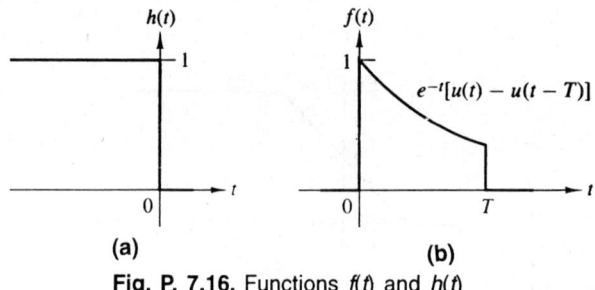

Fig. P. 7.16. Functions *f(t)* and *h(t)*

Solution:

The convolution of the signal $h(t)$ and $f(t)$ is expressed by,

$$y(t)=h*f = \int_{-\infty}^{\infty} h(t-\tau)\,f(\tau)\,d\tau = \int_{-\infty}^{\infty} u(\tau-t)e^{-\tau}\big[u(\tau)-u(\tau-T)\big]d\tau$$

Since $u\,(\tau) - u\,(\tau - T)$ is nonzero only for $0 \le \tau \le T$, the lower upper limits of intergration become 0 and T, respectively:

$$y(t)=h*f = \int_{0}^{T} e^{-\tau} u(\tau-t)\,d\tau \tag{1}$$

Case 1. $t < 0$: Here $t < 0$ implies that $\tau - t \ge 0$ 0ver $0 \le \tau \le T$. Hence $u\,(\tau - t) = 1$ over $0 \le \tau \le T$., and

$$y(t) = \int_0^T e^{-\tau} d\tau = 1 - e^{-T}$$

Case 2. $0 \le \tau \le T$. For this case, $u(\tau - t)$ in equation (1) is nonzero only when $\tau \ge t$. Hence, in the region $0 \le \tau \le T$, it must also be true that $0 \le \tau \le T$ for the integral of equation (1) to be nonzero. Thus, the lower and upper limits of integration with respect to the variable τ become t and T, respectively:

$$y(t) = \int_t^T e^{-\tau} u(\tau - t)\, d\tau = \left[-e^{-\tau}\right]_t^T = e^{-t} - e^{-T}$$

Case 3. $t \ge T$: A simple calculation shows that $y(t) = 0$ in this region.

Now, Plot $y(t)$ for $T = 1$. Combining the results obtain so far with $T = 1$ implies that $y(t)$ has the graph depicted in Fig. P. 7.16 (a)

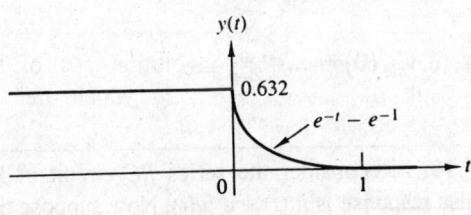

Fig. P. 7.16 (a)

Plot of the resulting function $y(t)$

Problem 7.17. Consider the *RC* circuit of Fig. P. 7.17, whose impulse response is

$$h(t) = e^{-t} u(t)$$

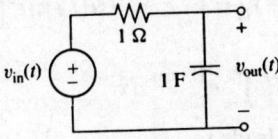

Fig. P. 7.17 *RC* circuit

If the input $v_{in}(t)\, e^{-at}$ V has been applied for a long time (theoretically, from $t = -\infty$), find $v_{out}(0)$ when $a = 0, 0.5,$ and 2.

Solution:

Since the capacitor voltage $v_{out}(t)$ is the convolution of the input with the impulse response, we have,

$$v_{out}(0) = \int_{-\infty}^{\infty} e^{-(0-\tau)} u(0-\tau) e^{-a\tau} d\tau$$

Because $u(-\tau)$ is in the integral, the upper limit of integration becomes 0. Hence,

$$v_{out}(0) = \int_{-\infty}^{0} e^{-(0-\tau)} e^{-a\tau} d\tau = \int_{-\infty}^{0} e^{(1-a)\tau} d\tau$$

For $a = 0$, $v_{out}(0) = e^{\tau} \big]_{-\infty}^{0} = 1 \text{V}$

for $a = 0.5$, $v_{out}(0) = \dfrac{e^{0.5\tau}}{0.5} \bigg]_{-\infty}^{0} = 2 \text{V}$

for $a = 2$, $v_{out}(0) = -e^{-\tau} \big]_{-\infty}^{0} = \infty$

Problem 7.18. Reconsider the series RC circuit of Fig. P. 7.18, whose impulse response is $h(t) = e^{-t}u(t)$. Now suppose that the input excitation is $v_{in}(t) = e^{-a|t|}$ V, where $a \neq 1$ and $a > 0$. Determine the response $v_{out}(t)$ for all t.

Solution:

Here, the input is expressed by,

$$v_{in}(t) = \begin{cases} e^{at}u(-t), & t < 0 \\ e^{-at}u(t), & t \geq 0 \end{cases}$$

Now, for $t < 0$,

$$v_{out}(t) = h * v_{in} = \int_{-\infty}^{0} e^{-(t-\tau)} u(t-\tau) e^{a\tau} u(-\tau) d\tau$$

$$= e^{-t} \int_{-\infty}^{0} e^{(1+a)\tau} d\tau \tag{1}$$

Evaluating equation (1) leads to

$$v_{out}(t) = e^{-t} \dfrac{e(1+a)\tau}{1+a} \bigg]_{-\infty}^{\tau} = \dfrac{e^{at}}{1+a} \text{V} \tag{2}$$

which is valid for $t < 0$.

For the case when $t \geq 0$,

$$v_{out}(t) = h*v_{in} = \int_{-\infty}^{0} e^{-(t-\tau)} e^{-a\tau} d\tau + \int_{0}^{t} e^{(t-\tau)} e^{-a\tau} d\tau \tag{3}$$

The first integral is simply equation (2) evaluated at $t = 0$ and multiplied by e^{-t}, i.e. $e^{-t}/(1 + a)$. Hence, equation (3) reduces to

$$v_{out}(t) = \frac{e^{-t}}{a+1} + e^{-t} \frac{e^{(1-a)\tau}}{1-a}\Bigg]_{0}^{t} = \frac{e^{-t}}{a+1} + \frac{e^{-at}}{1-a} - \frac{e^{-t}}{1-a} \text{ V} \tag{4}$$

A plot of the waveform $v_{in}(t)$ with $a = 2$ appears in Fig. P. 7.18 (a) and of the impulse response in Fig. P. 7.18 (b). A plot of the response, equations (2) and (4), with $a = 2$, appears in Fig. P. 7.18 (c).

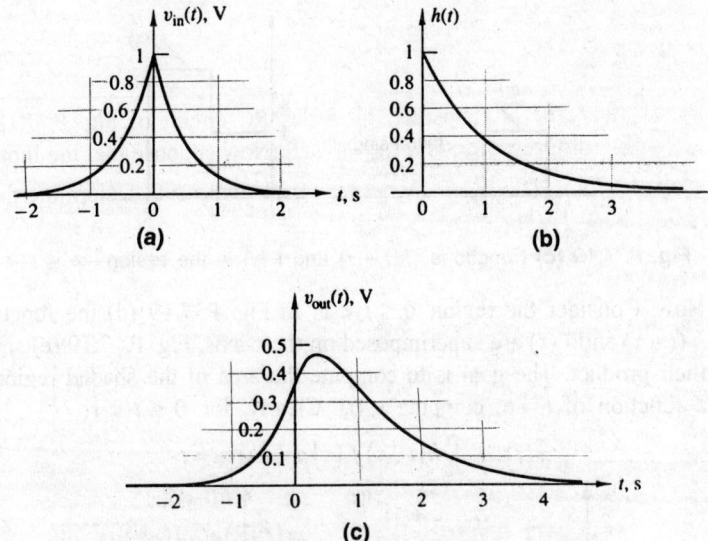

Fig. P. 7.18 (a) Input waveform $v_{in}(t)$ with $a = 2$.
(b) Impulse response $h(t)$. **(c)** $v_{out}(t)$ for series
RC circuit excited by $v_{in}(t) = e^{-2|t|}$ V.

Problem 7.19. Graphically compute the convolution $y = h*f$ of the two waveforms $h(t)$ and $f(t)$ sketched in Fig. P. 7.19 (a) & (b).

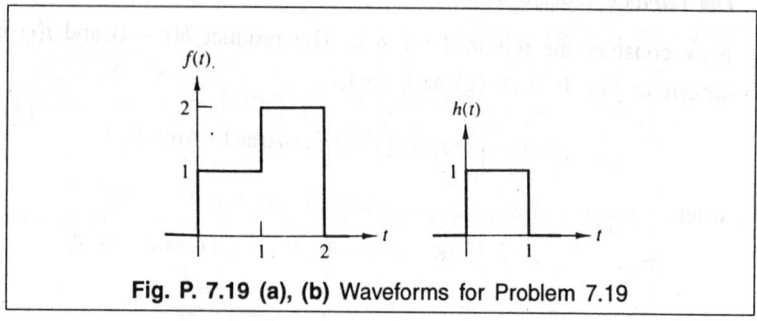

Fig. P. 7.19 (a), (b) Waveforms for Problem 7.19

Solution:

There are five regions to consider:

    (1) $-\infty \le t < 0$ (2) $0 \le t < 1$ (3) $1 \le t < 2$,

    (4) $2 \le t < 3$ and   (5) $3 \le t$

Now, Consider the region $-\infty \le t < 0$. Figure P. **7.19** (c). shows $h(t - \tau)$ and $f(\tau)$ on the same τ-axis. Their product, $h(t - \tau) f(\tau)$, is clearly zero. Hence, in the first t-region, $y(t) = \int_{-\infty}^{\infty} h(t-\tau) f(\tau) d\tau = 0$

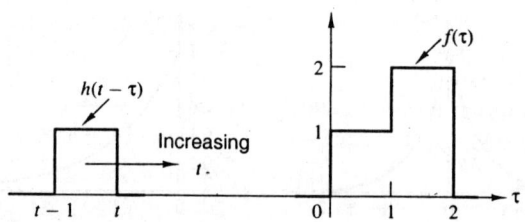

Fig. P. 7.19 (c) Functions $h(t - \tau)$ and f (τ) in the resion $-\infty \le t < 0$.

Now, Consider the region $0 \le t < 1$. In Fig. P. 7.19 (d) the functions $h(t - \tau)$ and $f(\tau)$ are superimposed on the τ-axis. Fig. P. 7.19 (e) shows their product. The goal is to compute the area of the shaded region as a function of t, i.e. compute $y(t)$. Clearly, for $0 \le t < 1$,

$$y(t) = \int_0^t h(t-\tau) f(\tau) d\tau = \text{Area} = t$$

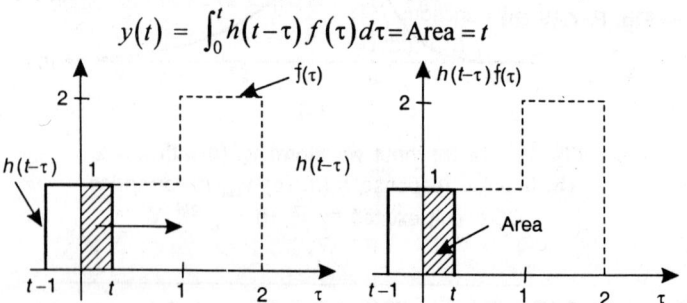

Fig. P. 7.19 (d) Functions $h(t - t)$ and $f(t)$ in the region $0 \le t < 1$. **(e)** Product of $h(t - t)$ and $f(t)$.

Now consider the region $1 \le t < 2$. The product $h(t - t)$ and $f(t)$ is shown in Fig. P. 7.19 (g) and, area,

$$y(t) = \int_0^2 h(t-\tau)f(\tau)d\tau = \text{Area}1 + \text{Area}2 = t$$

where Area 1 and Area 2 are shown in the Fig. P. 7.19(f).

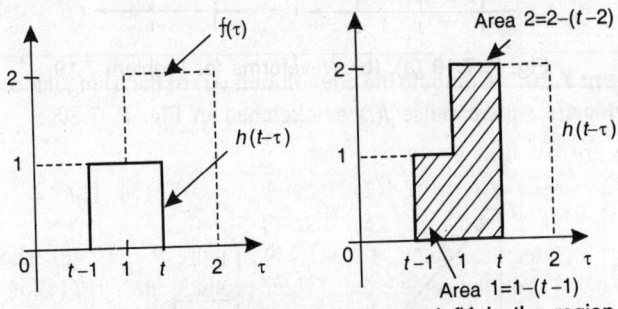

Fig. P. 7.19 (f) Functions $h(t - \tau)$ and $f(t)$ in the region $1 \le t < 2$. **(g)** Product of $h(t - t)$ and $f(t)$.

Now consider the region $2 \le t < 3$. From the fig one computes the area as follows:

The product of $h(t - \tau)$ and $f(\tau)$ is shown in Fig. P. 7.19 (h) Then required area, $y(t) = \text{Area} = 2[2-(t-1)] = 2(3-t) = 6 - 2t$

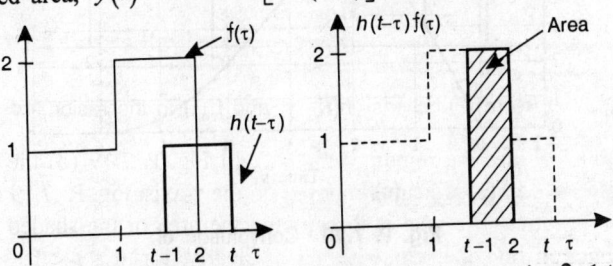

Fig. P. 7.19 (h) Functions $h(t - \tau)$ and $f(\tau)$ in the region $2 \le t < 3$. **(i)** Product of $h(t - \tau)$ and $f(\tau)$.

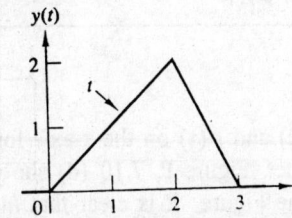

Fig. P. 7.19 (j) Plot of resulting function y (t).

Finally, consider the region $3 \le t$. Clearly, $h\,(t-\tau)\,f(t) = 0$ in this region, in which case $y(t) = 0$ for $t \ge 3$.

Hence, Fig. P. 17.19 (j) sketches the function $y(t)$ resulting from the convolution.

Problem 7.20.   Compute the convolution $y(t)$ of the triangular pulse $h(t)$ with the square pulse $f(t)$ as sketched in Fig. P. 7.20.

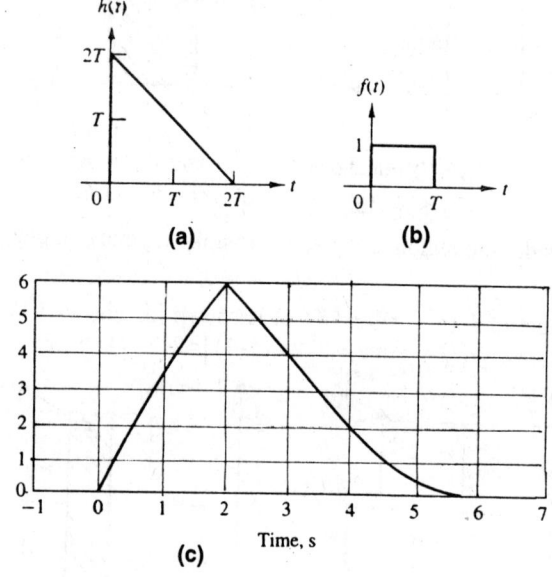

Fig. P. 7.20   Convolution of

(a) triangular signal with

(b) square pulse to produce the signal sketched in

(c) for $T = 2$

Solution:

First Draw $h\,(t - \tau)$ and $f(\tau)$ on the τ-axis for $t < 0$ and compute the area of their product. Figure P. 7.10 (d) shows $h(t - \tau)$ and $f(\tau)$ on the τ-axis. From the Figure it is clear that $h(t - \tau)$ and $f(\tau) = 0$ for $t < 0$; hence, $y(t) = 0$ in the first region.

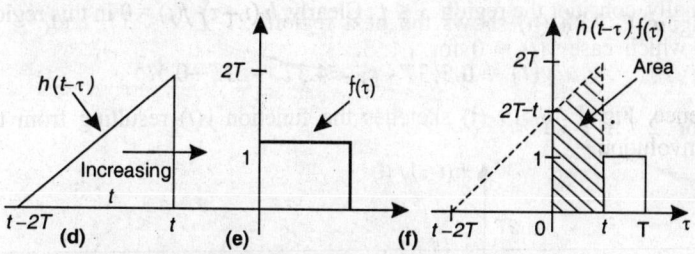

Fig. P. 7.20 (d), (e) & (f) Graph of $h(t - \tau)$ and   (e) $f(t)$
on the t-axis for $t < 0$.
Graph of $h(t - \tau)$ and $f(\tau)$ on τ-axis for $0 \leq t < T$.

Now, Consider the region $0 \leq t < T$, as illustrated in Fig. P. 7.20 (e)
The shaded area of the fig is the difference between the area of the large
triangle, expressed as,

$$\text{Area } A = 0.5(2T)^2$$

and the area of the smaller triangle to the left of the vertical axis, is
expressed by,

$$\text{Area } B = 0.5(2T - t)^2$$

Hence, $y(t) = \text{Area} = \text{Area } A - \text{Area } B = 2Tt - 0.5t^2$

for $0 \leq t < T$.

Now, consider the region $T \leq t < 2\,T$, as depicted in Fig. P. 7.20 (f)

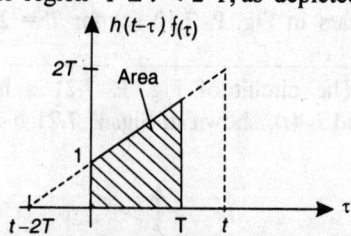

Fig. P. 7.20 (g) Graph of $h(t - \tau)$ and
$f(\tau)$ on τ-axis for $T \leq t < 2T$.

In this figure the shaded area, which determines $y\,(t)$, is again the
difference of two triangular areas; specifically,

$$y(t) = 0.5[T - (t - 2T)]^2 - 0.5(2T - t)^2 = 2.5T^2 - Tt$$

For $T\mathcal{L} < 2T$.

Figure P. 7.2h (h) shows the next region, 2T $\mathcal{L}$ $t < 3T$. and,

$$y(t) = 0.5(3T-t)^2 = 4.5T^2 - 3tT + 0.5t^2$$

For $2\,T \le t\;3\,T$.

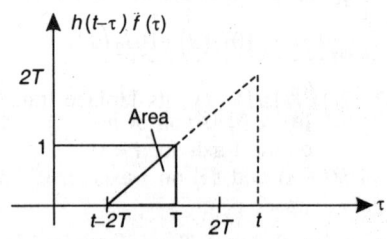

Fig. P. 7.20 (h). Graph of $h(t-\tau)$ and
$f(\tau)$ on τ-axis for $2T \le t < 3T$.

Finally, Consider the region $3T \le t$. Here the product
$h(t - \tau)\ f(t) = 0$, in which case $y\ (t) = 0$ for $t > 3T$. Hence, expressing
$y\ (t)$ for different regions of interval, we have

we have, $y\ (t) = \begin{cases} 0, & t < 0 \\ 2Tt - 0.5t^2, & 0 \le t < T \\ 2.5T^2 - Tt, & T \le t < 2T \\ 4.5T^2 - 3Tt + 0.5t^2, & 2T \le t < 3T \\ 0, & 3T \le t \end{cases}$

A plot of $y\ (t)$ appears in Fig. P. 7.10 (c) for $T = 2$.

Problem 7.21. The circuit of Fig. P. 7.21 a has two source
excitations, $i_1\ (t)$ and $i_2\ (t)$, shown in Fig. P. 7.21 b and fig 7.21(c).
Compute $V_{out}\ (s)$.

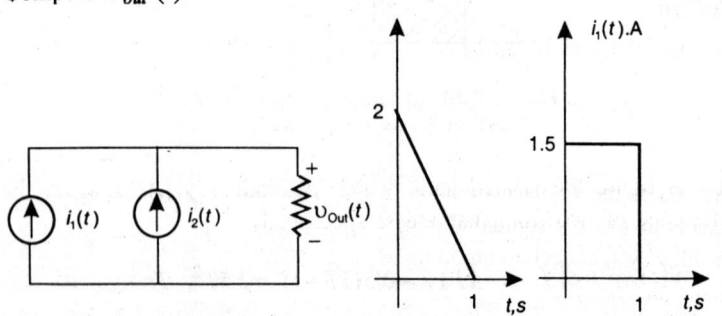

Fig. P. 7.21 (a) Resistive circuit driven by two current sources.
(b) T riangular signal $i_1\ (t)$. **(c)** Pulse signal, $i_2\ (t)$

Solution:

By superposition and Ohm's law,

$$v_{out}(t) = 10i_1(t) + 10i_2(t)$$

From the linearity of the Laplace transform, we have,

$$v_{out}(s) = 10I_1(s) + 10I_2(s)$$

Now, since, $i_1(t) = (2-2t)u(t)u(1-t)$, its laplace transform is given by,

$$I_1(s) = \int_{0^-}^1 (2-2t)e^{-st}\, dt = 2\int_{0^-}^1 e^{-st}\, dt - 2\int_{0^-}^1 te^{-st}\, dt$$

or, $I_1(s) = 2\dfrac{1-e^{-s}}{s} + 2\dfrac{-e^{-s}}{s} + 2\dfrac{e^{-st}}{s^2}\bigg]_{0^-}^1 = 2\left(\dfrac{1}{s} - \dfrac{1}{s^2} + \dfrac{e^{-s}}{s^2}\right)$ In a similar

manner, the laplace transform for $i_2(t)$ is,

$$I_2(s) = 1.5\dfrac{1-e^{-s}}{s}$$

Since $V_{out}(s) = 10I_1(s) + 10I_2(s)$, it follows that

$$V_{out}(s) = \dfrac{35}{s} - \dfrac{20}{s^2} + e^{-s}\left(\dfrac{20}{s^2} - \dfrac{15}{s}\right)$$

Problem 7.22. Find $f(t)$ when $F(s) = \dfrac{1}{s(s+a)}$

Solution:

Using partial fraction, we have

$$F(s) = \dfrac{1}{s(s+a)} = \dfrac{A}{s} + \dfrac{B}{s+a}$$

where A/s is the Laplace transform of a weighted step, $Au(t)$, and $B/(s+a)$ is that of a weighted exponential, $Be^{-at}u(t)$. To find A, multiply both sides by s, cancel common numerator and denominator factors, and evaluate the result at $s = 0$, producing $A = 1/a$. Similarly, to find B, multiply both sides by $s+a$, cancel common numerator and denominator factors, and evaluate the result at $s = -a$, obtaining $B = -1/a$. Recall that, by linearity, $\mathcal{L}^{-1}\big[aF(s)\big]=a\mathcal{L}^{-1}\big[F(s)\big]$. Hence,

$$f(t) = \frac{1}{a}u(t)-\frac{1}{a}e^{-at}u(t)=\frac{1}{a}\left(1-e^{-at}\right)u(t)$$

Problem 7.23. Find $f(t)$ when

$$F(s) = \frac{cs^2+b}{s(s+a)}$$

Solution:

Using partial fraction, $\quad F(s) = \dfrac{cs^2+b}{s(s+a)}=K+\dfrac{A}{s}+\dfrac{B}{s+a}$ $\qquad\qquad$ (1)

The value of K is $\qquad K=\lim\limits_{s\to\infty}\dfrac{cs^2+b}{s(s+a)}=c$

A is found by multiplying equation (1) by s and evaluating the result at $s = 0$.

$$A = \frac{cs^2+b}{s+a}\bigg]_{s=0}=\left[Ks+A+\frac{Bs}{s+a}\right]_{s=0}=\frac{b}{a}$$

B is found in a similar way: multiply both sides by $s + a$ and evaluate the result at $s + - a$ to obtain

$$B = -\frac{b+a^2c}{a}$$

Now the inverse Laplace transform of the right-hand side of equation (1) with the values of K, A and B properly inserted, can be given by,

$$f(t) = c\delta(t)+\frac{b}{a}u(t)-\frac{b+a^2c}{a}e^{-at}u(t)$$

Problem 7.24. Find $f(t)$ when

$$F(s) = \frac{3s^2+s+3}{(s+1)(s^2+4)}=\frac{D}{s+1}+\frac{A+jB}{s+j2}+\frac{A-jB}{s-j2}=\frac{D}{s+1}+\frac{C_1s+C_2}{s^2+4}$$

Solution:

Here,
$$D = \frac{3s^2+s+3}{s^2+4}\bigg|_{s=-1} = 1$$

Given that $D = 1$, to find C_2, we evaluate $F(s)$ at $s = 0$, in which case $0.75 = 1 + 0.25\,C_2$, or $C_2 = -1$. With $D = 1$ and $C_2 = -1$, we evaluate $F(s)$ at $s = 1$ to obtain $0.7 = 0.5 + 0.2\,(C_1 - 1)$ or, equivalently, $C_1 = 2$. Thus,

$$F(s) = \frac{1}{s+1}+2\frac{s}{s^2+4}-0.5\frac{2}{s^2+4}$$

Hence, the inverse transform is,

$$f(t) = \left[e^{-t}+2\cos(2t)-0.5\sin(2t)\right]u(t)$$

Problem 7.25. Using the time shift property, find $F(s)$ for $f(t)$ sketched in Fig. P. 7.25.

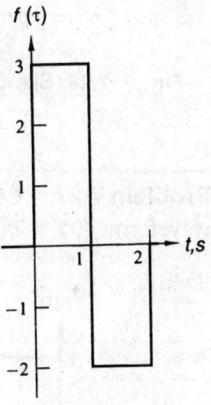

Solution:

Using step functions and shifted step functions, we get

$$f(t) = 3u(t)-5u(t-1)+2u(t-2)$$

Then using the shift property, we have,

$$F(s) = \frac{3}{s}-\frac{5e^{-s}}{s}+\frac{2e^{-2s}}{s}$$

Fig. P. 7.25

Problem 7.26. Compute the inverse transform of the function

$$F(s) = \frac{1-e^{-s}}{s(s+1)}=\frac{1}{s(s+1)}-\frac{e^{-s}}{s(s+1)}$$

Solution:

Here,
$$\mathcal{L}^{-1}\left[\frac{1}{s(s+1)}\right] = \left(1-e^{-t}\right)u(t)$$

Now, Using the shift theorem, we have,

$$\mathcal{L}^{-1}\left[\frac{e^{-s}}{s(s+1)}\right] = \left(1-e^{-(t-1)}\right)u(t-1)$$

By the linearity of the inverse Laplace transform,

$$f(t) = \left(1-e^{-t}\right)u(t)-\left[1-e^{-(t-1)}\right]u(t-1)$$

A sketch of $f(t)$ appears in Fig. P. 7.26.

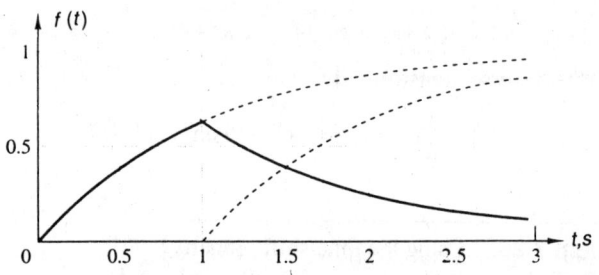

Fig. P. 7.26. Sketch of $f(t)=\left(1-e^{-t}\right)u(t)-\left[1-e^{-(t-1)}\right]u(t-1)$.

Problem 7.27. Compute $F(s)$ for the pulse $f(t)$ and the triangular waveform $g(t) = tf(t)$ sketched in Fig. P. 7.27.

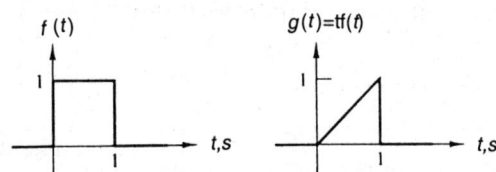

Fig. P. 7.27 Signals illustrating the multiplication-by-t property.

Solution:

By inspection, $f(t) = u(t) - (t - 1)$. Then $F(s) = (1 - e^{-s})/s$. Since $g(t) = tf(t)$, the multiplication-by property implies that

$$G(s) = -\frac{d}{ds}\left(\frac{1-e^{-s}}{s}\right) = -\frac{e^{-s}}{s}+\frac{1-e^{-s}}{s^2}$$

Problem 7.28. Compute the response, denoted here by the input current $i_{in}(t)$, to the input voltage excitation $v_{in}(t) = \delta(t)$, given the series RLC circuit of Fig. P. 7.28. Suppose the inital conditions are $i_L(0^-) = 1$ A and $v_c(0^-) = -2$ V.

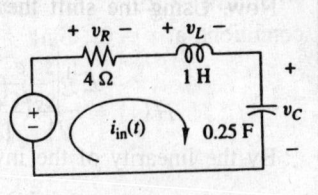

Fig. P. 7.28
Series RLC circuit for

Solution:

Taking sum of the voltages around the loop, $v_R + v_L + v_C = v_{in}$. Substituting for each of the element voltages using the mesh current $i_{in}(t)$, we have, integrodifferential equation, expressed by,

$$Ri_{in}(t) + L\frac{di_{in}}{dt}(t) + \frac{1}{C}\int_{-\infty}^{t} i_{in}(q)dq = v_{in}(t) \qquad (1)$$

With the aid of the differentiation and integration formulas, taking the Laplace transform of both sides of equation, (1) we get,

$$Ri_{in}(s) + Ls\, I_{in}(s) Li_L\left(0^-\right) + \frac{1}{Cs} I_{in}(s) + \frac{v_C(0-)}{s} = v_{in}(s)$$

or, $$L\frac{s^2 + \dfrac{R}{L}s + \dfrac{1}{LC}}{s} I_{in}(s) = V_{in}(s) + Li_L\left(0^-\right) - \frac{V_C\left(0^-\right)}{s} \qquad (2)$$

Putting the values in equation (2), we get

$$I_{in}(s) = \frac{s}{s^2 + 4s + 4} + \frac{1}{s+2} = \frac{2}{s+2} - \frac{2}{(s+2)^2} \qquad (3)$$

Taking the inverse Laplace transform of eqn (3), we get,

$$i_{in}(t) = (2 - 2t)e^{-2t} u(t)\,\text{A}$$

A plot of this response appears in Fig. P. 7.28 (a)

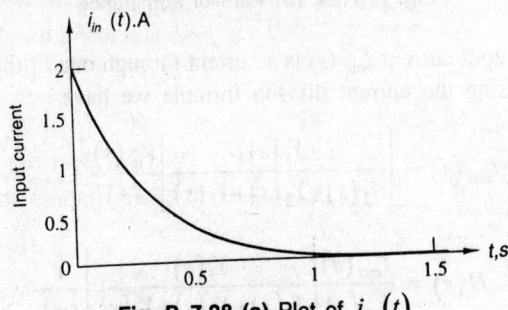

Fig. P. 7.28 (a) Plot of $i_{in}(t)$

Problem 7.29. The circuit of Fig. P. 7.29. has zero initial conditions at $t = 0^-$. Find

$$H(s) = \frac{\mathcal{L}\left[i_{out}(t)\right]}{\mathcal{L}\left[v_{in}(t)\right]} = \frac{I_{out}(s)}{V_{in}(s)}$$

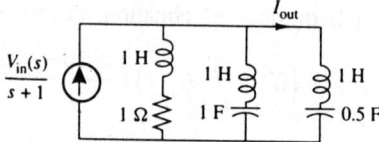

Fig. P. 7.29

Solution:

A source transformation to obtain three parallel branches, as given in Fig. P. 7.29 (a)

This circuit has the parallel structure of Fig. P. 7.29.

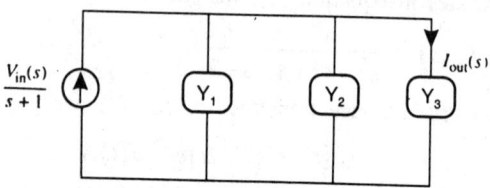

Fig. P. 7.29 (a) Circuit after source transformation.

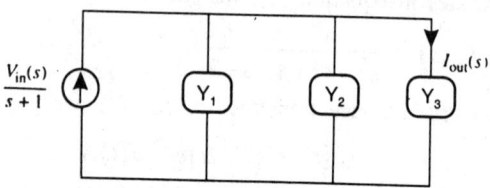

Fig. P. 7.29. (b) Parallel admittance

Since the output current $I_{out}(s)$ is a current through one of three parallel branches, using the current divison formula we have,

$$I_{out}(s) = \left[\frac{Y_3(s)}{Y_1(s) + Y_2(s) + Y_3(s)}\right]\frac{V_{in}(s)}{s+1}$$

Hence, $$H(s) = \frac{I_{out}(s)}{V_{in}(s)}\left[\frac{Y_3(s)}{Y_1(s) + Y_2(s) + Y_3(s)}\right]\frac{1}{s+1} \qquad (1)$$

As impedances in series add, and as admittance is the inverse of impedance we have

$$Y_1(s) = \frac{1}{s+1}, Y_2(s) = \frac{1}{s+\dfrac{1}{s}} = \frac{s}{s^2+1}$$

$$Y_3(s) = \frac{1}{s+\dfrac{2}{s}} = \frac{s}{s^2+2}$$

Using, $Y_1(s)$, $Y_2(s)$ *and* $Y_3(s)$ in equation (1), we have

$$H(s) = \frac{\dfrac{s}{s^2+2}}{\dfrac{1}{s+1}+\dfrac{s}{s^2+1}+\dfrac{s}{s^2+2}}\left(\dfrac{1}{s+1}\right) = \frac{s\left(s^2+1\right)}{3s^4+2s^3+6s^2+3s+2}$$

Problem 7.30. Construct the transfer function of the ideal op amp circuit of Fig. P. 7.30, where $Z_f(s)$ and $I_f(s)$ denote a feedback impedance and feedback current, respectively.

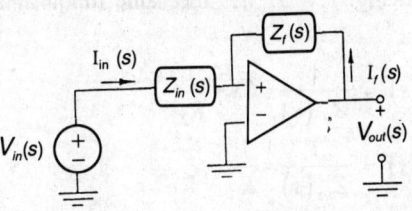

Fig. P. 7.30. Simple ideal op amp circuit

Solution:

Since no current enters the inputs of an ideal op amp, $I_{in}(s) = -I_f(s)$. Further, the voltage at the negative op amp terminal is driven to virtual ground. Hence, $V_{in}(s) = Z_{in}(s)I_{in}(s)$ and $V_{out}(s) = Z_f(s)I_f(s)$. Combining these relationships with $I_{in}(s) = -I_f(s)$, we have transfer equation given by,

$$H(s) = \frac{V_{out}(s)}{V_{in}(s)} = -\frac{Z_f(s)}{Z_{in}(s)} = -\frac{Y_{in}(s)}{Y_f(s)}$$

Problem 7.31. The ideal op amp circuit of Fig. P. 14.15. is called a leaky integrator. If the input to the leaky integrator circuit is $v_{in}(t) = e^{-t} u(t)$, find the values of R_1, R_2, and C leading to an output response $v_{out}(t) = 2te^{-t} u(t)$, assuming that $v_C(0^-) = 0$.

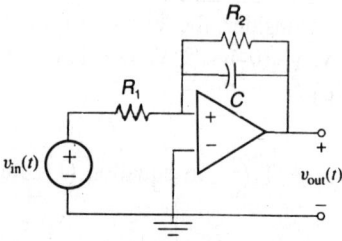

Fig. P. 7.31. Ideal op amp circuit known as leaky integrator.

Solution:

By definition of the transfer function. we have

$$H(s) = \frac{\mathcal{L}[response]}{\mathcal{L}[input]} = \frac{V_{out}(s)}{V_{in}(s)} = -\frac{\dfrac{2}{(s+1)^2}}{\dfrac{1}{s+1}} = -\frac{2}{s+1} \tag{1}$$

Here observe that Fig. P. 7.31 has the same topology as Fig. P. 7.30, where

$$Y_f(s) = \frac{1}{Z_f(s)} = Cs + \frac{1}{R_2}$$

and

$$Y_{in}(s) = \frac{1}{Z_{in}(s)} = \frac{1}{R_1}$$

Now,

$$H(s) = -\frac{Y_{in}(s)}{Y_f(s)} = -\frac{\dfrac{1}{R_1}}{Cs + \dfrac{1}{R_2}} \tag{2}$$

Equating the coefficients, we have

$$\frac{\dfrac{1}{R_1}}{Cs + \dfrac{1}{R_2}} = \frac{2}{s+1}$$

Here, One possible solution is $R_1 = 0.5\Omega$, $R_2 = 1\Omega$, and $C = 1$ F. If we rewrite equation (2) as,

$$H(s) = -\frac{Y_{in}(s)}{Y_f(s)} = -\frac{\dfrac{1}{CR_1}}{s + \dfrac{1}{CR_2}}$$

Problem 7.32. Consider the circuit of Fig. P. 7.32, in which $v_{in}(t) = 5\, u(t)$ V, $v_{C1}(0^-) = 3$ V, $v_{C2}(0^-) = 0$ V, and $i_L(0^-) = 2$ A. Find $v_{out}(t)$.

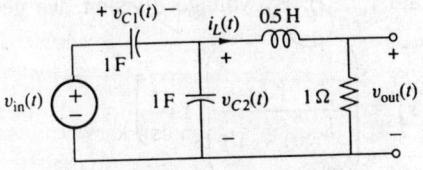

Fig. P. 7.32. *RLC* circuit

Solution:

Figure 7.32 (a). Illustrates the equivalent complex-frequency domain circuit of Fig. P. 7.32.

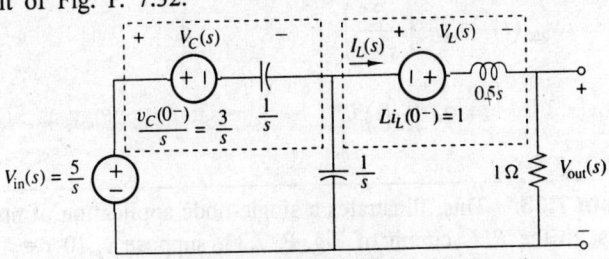

Fig. P. 7.32 (a) Complex-frequency domain equivalent of circuit of Fig. P. 7.32

Using source transformation and combine parallel capacitors, we have, Now, using second source transformation and combine series voltage source to produce a series circuit, we have resulting Fig. P. 7.32.

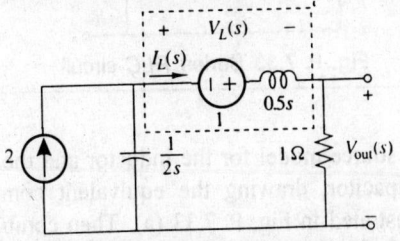

Fig. P. 7.32 (b) Circuit equivalent of Fig. P. 7.32 after source transformation.

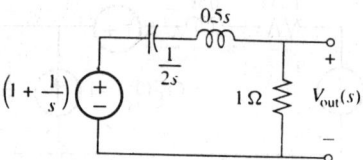

Fig. P. 7.32 (c) Simplified version of Fig. P. 7.32

Now, using voltage division to find V_{out} *(s)* and take the inverse transform to obtain v_{out} *(t)*. By voltage division, we get

$$V_{out}(s) = \left(\frac{1}{\dfrac{1}{2s}+0.5s+1} \right)\left(1+\frac{1}{s}\right)$$

$$= \frac{2}{s+1}$$

Taking the inverse transform produces, the o/p voltage,

$$v_{out}(t) = \mathcal{L}^{-1}\left(\frac{2}{s+1}\right)$$

$$= 2e^{-t}u(t)\,\text{V}$$

Problem 7.33. This illustrates a single-node application of nodal analysis. In the *RLC* circuit of Fig. P. 7.33, suppose $v_c\,(0^-) = 1$ V, $i_L\,(0^-) = 2$A, and $v_{in}\,(t) = u\,(t)$ V. Find $v_c\,(t)$ for t $\geq$ 0.

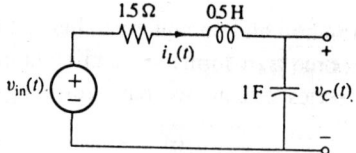

Fig. P. 7.33 Series *RLC* circuit

Solution:

Using the voltage source model for the inductor and the current source model for the capacitor, drawing the equivalent complex-frequency domain circuit illustrated in Fig. P. 7.33 (a). Then combine the voltage sources and the series impedance into single terms, as shown in Fig. P. 14.25b.

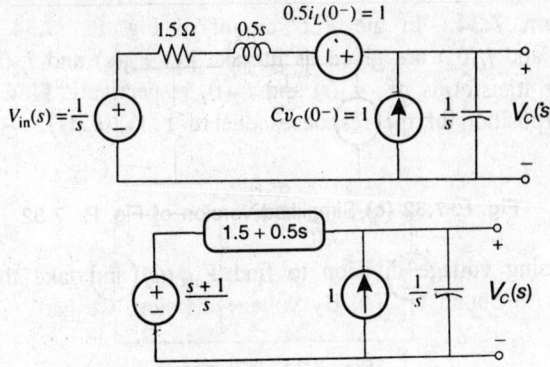

Fig. P. 7.33 **(a)** Complex-frequency domain equivalent accounting for initial conditions of the circuit of Fig. P. 7.33

(b) Circuit equivalent to (a) where voltage sources have been combined.

Now, summing the currents leaving the top node of $V_C(s)$, we have,

$$\frac{1}{1.5+0.5s}\left(V_c(s)-\frac{s+1}{s}\right)-1+sV_C(s)=0$$

Grouping terms produces

$$\left(\frac{1}{1.5+0.5s}+s\right)V_c(s)=\frac{s+1}{s(0.5s+1.5)}+1$$

Doing the required algebra leads to

$$V_C(s)=\frac{s^2+5s+2}{s(s+1)(s+2)}$$

Using the expression for $V_C(s)$, we get

$$V_C(s)=\frac{s^2+5s+2}{s(s+1)(s+2)}=\frac{1}{s}+\frac{2}{s+1}+\frac{-2}{s+2}$$

Inverting this transform, we have the true response,

$$v_C(t)=\left(1+2e^{-t}-2e^{-2t}\right)u(t)\text{ V}$$

Problem 7.34. In the *RLC* circuit of Fig. P. 7.34, suppose $v_c (0^-)$ and $i_L(0^-)$ are given as literals. Let $V_{s1}(s)$ and $I_{s2}(s)$ be the Laplace transforms of $v_{s1}(t)$ and $i_{s2}(t)$, respectively. Find $V_C(s)$ as a superposition of the responses due to $V_{s1}(s)$, $I_{s2}(s)$, $v_C(0-)$, and $i_L(0^-)$.

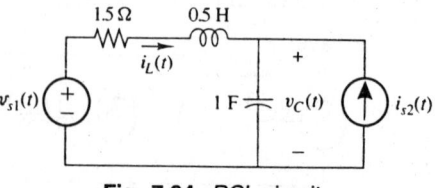

Fig. 7.34. RCL circuit

Solution:

Using the voltage source model for the inductor and the current source model for the capacitor, draw the equivalent complex-frequency domain circuit as shown in Fig. P. 7.34 (a). Combining voltage sources, combining current, sources, and writing the series impedance as a single expression yields the circuit of Fig. P. 7.34 (b).

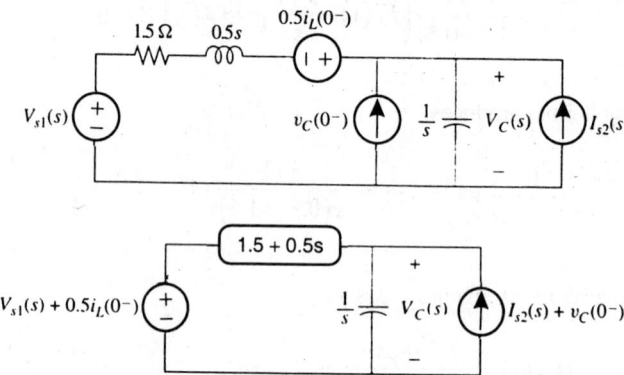

Fig. 7.34 (a) Complex-frequency domain equivalent accounting for initial conditions of the circuit of Fig. P. 7.34.

   (b) Circuit equivalent to

   (a) where voltage sources have been combined.

Summing the currents leaving the top node of $V_c (s)$, we have,

$$\frac{1}{1.5+0.5s}\left[V_C(s)-V_{S1}(s)-0.5i_L\left(0^-\right)\right]-\left[I_{s2}(s)+v_C\left(0^-\right)\right]$$
$$+sV_C(s)=0 \qquad (1)$$

Rewriting the eqn (1), we get

$$\left[V_C(s) - V_{s1}(s) - 0.5\,i_L\left(0^-\right) \right]$$

$$-\left(0.5s + 1.5\right)\left[I_{s2}(s) + v_C\left(0^-\right) \right] + s\left(0.5s + 1.5\right)V_C(s) = 0$$

By regrouping terms, we have

$$\left(\frac{s^2 + 3s + 2}{2} \right) V_C(s) = V_{s1}(s) + \left(\frac{s+3}{2} \right) I_{s2}(s) + \left(\frac{s+3}{2} \right) v_C\left(0^-\right) + 0.5\,i_L\left(0^-\right)$$

or,

$$V_C(s) = \frac{2}{s^2 + 3s + 2} V_{s1}(s) + \left(\frac{(s+3)}{s^2 + 3s + 2} \right) I_{s2}(s)$$

$$+ \left(\frac{s+3}{s^2 + 3s + 2} \right) V_c(0^-) + \frac{1}{s^2 + 3s + 2} i_L(0^-)$$

Problem 7.35. In the circuit of Fig. P. 7.35, suppose $i_{in}(t) = \delta(t)$, $i_L(0^-) = 1$ A, $v_C(0^-) = 1$ V. Find the voltages $v_C(t)$ and $v_L(t)$.

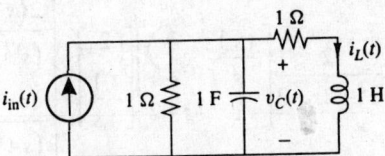

Fig. P. 7.35 Two-node RLC circuit for Fig. P. 7.35. Given the indicated direction of current $i_L(t)$ what is the implied voltage polarity for $v_L(t)$?

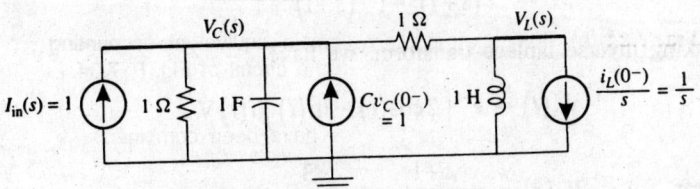

Fig. P. 7.35 (a) Complex-frequency domain equivalent of circuit of Fig. P. 7.35

First draw the complex-frequency domain equivalent circuit, as shown in Fig. P. 7.35 (a)

Solution:

At the node labeled $V_C(s)$ (identified with a bold line), using KCL gives,

$$(1+s)V_C(s)+[V_C(s)-V_L(s)] = 2 \tag{1}$$

Simplifying node equation (1), we get

$$(s+2)V_C(s)-V_L(s) = 2$$

At the node labeled $V_L(s)$, $[V_L(s) - V_c(s)] + (1/s) V_L(s) = -1/s$ or, equivalently,

$$-V_C(s)+\frac{s+1}{s}V_L(s) = -\frac{1}{s} \tag{2}$$

The matrix form of these two node equations (1) and (2) is expressed by,

$$\begin{bmatrix} s+2 & -1 \\ -1 & \dfrac{s+1}{s} \end{bmatrix}\begin{bmatrix} V_C(s) \\ V_L(s) \end{bmatrix} = \begin{bmatrix} 2 \\ -\dfrac{1}{s} \end{bmatrix} \tag{3}$$

Solving the matrix equation (c). Using Cramer's rule, we have

$$\begin{bmatrix} V_C(s) \\ V_L(s) \end{bmatrix} = \frac{2}{s^2+2s+2}\begin{bmatrix} \dfrac{s+1}{s} & 1 \\ 1 & s+2 \end{bmatrix}\begin{bmatrix} 2 \\ -\dfrac{1}{s} \end{bmatrix} = \begin{bmatrix} \dfrac{2(s+1)-1}{(s+1)^2+1} \\ \dfrac{(s+1)-3}{(s+1)^2+1} \end{bmatrix} \tag{4}$$

Breaking up equation (4) into its components, we get

$$V_C(s) = \frac{2(s+1)}{(s+1)^2+1}-\frac{1}{(s+1)^2+1}$$

Taking inverse laplace transform, we have

$$v_C(t) = e^{-t}[2\cos(t)-\sin(t)]u(t)\,\text{V}$$

Also, $$V_L(s) = \frac{s+1}{(s+1)^2+1} - \frac{3}{(s+1)^2+1}$$

By inverse laplace transformer we have

$$v_L(t) = e^{-t}[\cos(t)-3\sin(t)]u(t)\,\text{V}$$

Figure P. 7.35 (b) presents plots of $v_C(t)$ and $v_L(t)$

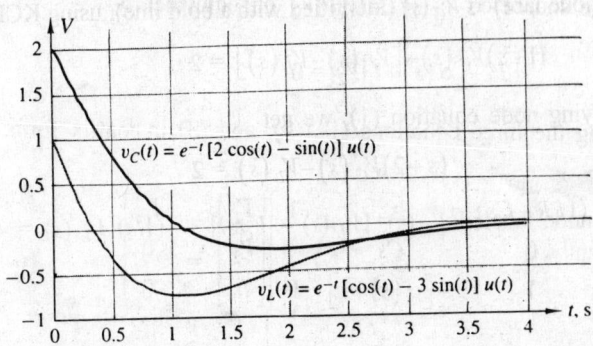

$v_C(t) = e^{-t}[2\cos(t) - \sin(t)] u(t)$

$v_L(t) = e^{-t}[\cos(t) - 3\sin(t)] u(t)$

Fig. P. 7.35 (b) Plots of capacitor and inductor voltages

Problem 7.36. Find the transfer function
$$H(s) = \frac{V_2(s)}{I_{in}(s)} \quad \text{for the circuit in Fig. P. 7.36}$$
Given, both capacitors are initially relaxed, $C = 0.1$ μF, $R = 10$ kΩ, a $(s) = 10^{-10}$ s^2, $g_m = 1/R = 10^{-4}$ S, and $i_{in}(t) = 10^{-6}$ $\delta(t)$, find v_2 (t) for $t > 0$.

The term $i_{in}(t) = 10^{-6}$ d (t) represents a small noise excitation to the circuit.

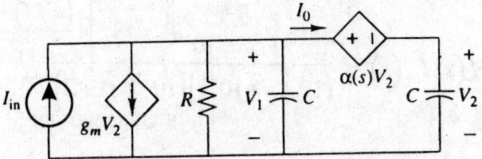

Fig. P 7.36 Two-node circuit with floating
dependent voltage source

Solution:

Summing all currents leaving node 1 and grouping like terms, we get

$$\left(\frac{1}{R} + Cs\right)V_1 + g_m V_2 + I_0 = I_{in} \qquad (1)$$

Summing all currents leaving node 2 and grouping like terms for node 2 produces

$$CsV_2 - I_0 = 0 \qquad (2)$$

The constraint equation introduced by the VCVS (voltage controlled voltage source) is $V_1 - V_2 = \alpha(s)V_2$ or, equivalently,

$$V_1 - [\alpha(s) + 1]V_2 = 0 \tag{3}$$

Writing the three equations (1), (2), and (3) in matrix form, we get,

$$\begin{bmatrix} (1/R + Cs) & g_m & 1 \\ 0 & Cs & -1 \\ 1 & -[\alpha(s) + 1] & 0 \end{bmatrix} \begin{bmatrix} V_1 \\ V_2 \\ I_0 \end{bmatrix} = \begin{bmatrix} I_{in} \\ 0 \\ 0 \end{bmatrix}$$

By Cramer's rule, recalling that $g_m = 1/R$, the transfer function is

$$H(s) = \frac{V_2}{I_{in}} = \frac{\det \begin{bmatrix} (1/R + Cs) & 1 & 1 \\ 0 & 0 & -1 \\ 1 & 0 & 0 \end{bmatrix}}{\det \begin{bmatrix} (1/R + Cs) & 1/R & 1 \\ 0 & Cs & -1 \\ 1 & -[\alpha(s) + 1] & 0 \end{bmatrix}} = \frac{1}{(Cs + 1/R)[\alpha(s) + 2]} \tag{4}$$

By eqn (4),

$$V_2(s) = H(s)I_{in}(s) = \frac{10^{-6}}{\left(10^{-7} s + 10^{-4}\right)\left(10^{-10} s^2 + 2\right)}$$

$$= \frac{10^{11}}{\left(s + 10^3\right)\left(s^2 + 2 \times 10^{10}\right)}$$

$$V_2(s) = \frac{10^{11}}{\left(s + 10^3\right)\left(s^2 + 2 \times 10^{10}\right)}$$

$$= \frac{4.9998}{s + 10^3} + \frac{-2.4999 - j1.7677 \times 10^{-2}}{s - j\sqrt{2} \times 10^5} +$$

$$\frac{-2.4999 + j1.7677 \times 10^{-2}}{s + j\sqrt{2} \times 10^5}$$

Taking inverse laplace transform, we get,

$$v_2(t) = 4.9998e^{-10t}u(t) + 4.9999\cos(\sqrt{2}\times10^5 t - 179.6°)u(t)\,\text{V}\quad\textbf{Ans.}$$

Problem 7.37. Use loop analysis to compute the input impedance of the bridged-T network illustrated in fig 7.37.

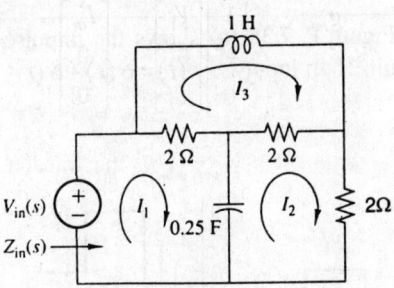

Fig. P. 7.37 Bridged-T network

Solution:

Dropping the specific s-dependence for convenience, using KVL in Loop (1), we get

$$2(I_1 - I_3) + \frac{4}{s}(I_1 - I_2) = 2\frac{s+2}{s}I_1 - \frac{4}{s}I_2 - 2I_3 = V_{in} \qquad (1)$$

and, Sum of the voltages around loop 2, is given by

$$\frac{4}{s}(I_2 - I_1) + 2(I_2 - I_3) + 2I_2 = -\frac{4}{s}I_1 + 4\frac{s+1}{s}I_2 - 2I_3 = 0 \quad (2)$$

Sum of the voltage around loop 3, gives

$$2(I_3 - I_1) + sI_3 + 2(I_3 - I_2) = -2I_1 - 2I_2 + (s+4)I_3 = 0 \qquad (3)$$

Puting the three loop equations in matrix form and solving, we have

$$\begin{bmatrix} 2\dfrac{s+2}{s} & -\dfrac{4}{s} & -2 \\[2mm] -\dfrac{4}{s} & 4\dfrac{s+1}{s} & -2 \\[2mm] -2 & -2 & s+4 \end{bmatrix} \begin{bmatrix} I_1 \\ I_2 \\ I_3 \end{bmatrix} = \begin{bmatrix} V_{in} \\ 0 \\ 0 \end{bmatrix} \qquad (4)$$

Using Cramer's rule to solve this equation for $V_{in}(s)$ in terms of $I_{in}(s) = I_1(s)$, we get

$$Z_{in}(s) = \frac{V_{in}(s)}{I_{in}(s)} = \frac{V_{in}(s)}{I_1(s)} = \frac{8\dfrac{s^2+s+4}{s}}{4\dfrac{(s+2)^2}{s}} = 2\Omega$$

Problem 7.38. Figure P. 7.38 (a) shows the impulse response of a hypothetical circuit. If an input is $f(t) = \delta(t) + \delta(t-1)$, compute the response $y(t)$.

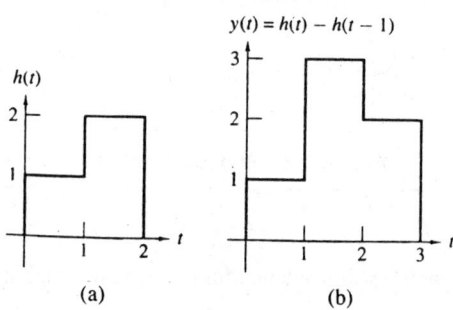

(a)   (b)

Fig. P. 7.38   (a) Impulse response of hypothetical circuit.
                 (b) Response to $\delta(t) + \delta(t-1)$

Solution:

Since $L[\delta(t) + \delta(t-1)] = 1 + e^{-s}$ the response $y(t)$ is simply the sum of $h(t)$ and $h(t-1)u(t-1)$. Doing the addition graphically, we have the resulting waveform as illustrated in Fig. P. 7.38b.

Problem 7.39. The response of a relaxed linear time-invariant circuit to a scaled ramp, $f(t) = 8tu(t)$, is given by $y(t) = (-6 + 4t + 8e^{-t} - 2e^{-2t})u(t)$. Compute the impulse response $h(t)$.

Solution:

If the step function is the integral of the delta function and the ramp the integral of the step, then the delta function equals the second derivative of the ramp. Hence, $\delta(t) = 0.125 f''(t)$. By the linearity and the time-invariance of the circuit, the impulse response $h(t) = 0.125 y''(t)$ and some straightforward calculations gives

$$y'(t) = (4 - 8e^{-t} + 4e^{-2t})u(t) + (-6 + 4t + 8e^{-t} - 2e^{-2t})\delta(t)$$

But the right-hand term in zero. Hence $y''(t) = (8e^{-t} - 8e^{-2t})u(t)$ and $h(t) = (e^{-1} - e^{-2t})u(t)$ **Ans.**

Problem 7.40. Compute the step response of the *RLC* circuit of Fig. P. 7.40.

Fig. P. 7.40. RLC circuit

Solution:

From voltage division, we get

$$H(s) = \frac{\dfrac{1}{Cs}}{R + Ls + \dfrac{1}{Cs}} = \frac{1}{LC} \frac{1}{s^2 + \dfrac{R}{L}s + \dfrac{1}{LC}} = \frac{5}{(s+2)^2 + 1} \qquad (1)$$

Hence, the Laplace transform of the step response is given by,

$$V_{out}(s) = \frac{5}{s[(s+2)^2 + 1]} = \frac{1}{s} + \frac{-s-4}{(s+2)^2 + 1}$$

Rearranging terms, we have

$$V_{out}(s) = \frac{1}{s} - \frac{s+2}{(s+2)^2 + 1} - 2\frac{1}{(s+2)^2 + 1}$$

Taking the inverse transform produces, we obtain the step response,

$$v_{out}(t) = [1 - e^{-2t}\cos(t) - 2e^{-2t}\sin(t)]u(t) \, \text{V}$$

Now, $\dfrac{d}{dt}v_{out}(t) = 2e^{-2t}[\cos(t) + 2\sin(t)] - e^{-2t}[-\sin(t) + 2\cos(t)]$

$$= 5e^{-2t}\sin(t)$$

With *H(s)* given by equation (1), the impulse response is

$$h(t) = L^{-1}[H(s)] = 5e^{-2t}\sin(t)u(t)$$

Problem 7.41. The Laplace transform of a capacitor voltage is given by

$$V_c(s) = \frac{2}{s} - \frac{1}{5s+2}$$

Find the initial capacitor voltage $v_c(0^+)$.

Solution:

Using the initial-value theorem, we have

$$v_c(0^+) = \lim_{x \to \infty} sV_c(s) = \lim_{x \to \infty}\left[2 - \frac{s}{5s+2}\right] = 2 - \frac{1}{5} = 1.8V$$

Problem 7.42. Let the Laplace transform of the velocity of a certain projectile be given by

$$V(s) = \frac{500s + 20}{s(5s+20)(10s+1)}$$

Find the initial velocity $v(0^+)$ and the initial acceleration $a(0^+)$

Solution:

Applying the initial-value theorem directly, we get initial velocity,

$$v(0^+) = \lim_{x \to \infty} sV(s) = \lim_{x \to \infty}\left[\frac{500s+20}{(5s+20)(10s+1)}\right] = 0$$

Since acceleration is the derivative of velocity, and the initial velocity is zero, from the time differentiation property, assuming the velocity is continuous at $t = 0$, we have

$$A(s) = sV(s) - v(0^+) = \frac{500s+20}{(5s+20)(10s+1)}$$

From the initial-value theorem, initial acceleration is,

$$a(0^+) = \lim_{s \to \infty}\left[\frac{500s^2 + 20s}{(5s+20)(10s+1)}\right]$$

$$= \lim_{s \to \infty} \left[\frac{500 + 20/s}{(5s + 20/s)(10s + 1/s)} \right] = 10$$

Problem 7.43. As in Problem 7.42 suppose the velocity $v(t)$ of a certain projectile has the Laplace transform

$$V(s) = \frac{500s + 20}{s(5s + 20)(10s + 1)}$$

with the Laplace transform of the acceleration $a(t)$ given by

$$A(s) = sV(s) - v(0^+) = \frac{500s + 20}{(5s + 20)(10s + 1)}$$

Find the final values of $v(t)$ and $a(t)$ if possible.

Solution:

Since, $V(s)$ and $A(s)$ have poles, therefore both satisfying conditions of final-value theorem, therefore

$$\lim_{t \to \infty} v(t) = \lim_{s \to 0} sV(s) = \lim_{s \to 0} \left[\frac{500s + 20}{(5s + 20)(10s + 1)} \right] = 1$$

and $\quad \lim_{t \to \infty} a(t) = \lim_{s \to 0} sA(s) = \lim_{s \to 0} \left[\frac{500s^2 + 20s}{(5s + 20)(10s + 1)} \right] = 0$

Impedance Functions and Network Theory

Problem 8.1. In the network of (a) of Fig. P. 8.1 $v_1 = V_0 e^{-2t} \cos tu(t)$, and for the network of (b), $i_1 = I_0 e^{-t} \sin 3tu(t)$. The impedance of the passive network N is found to be

$$Z(s) = \frac{(s+2)(s+3)}{(s+1)(s+4)}$$

(a) With N connected to the voltage source as in (a) of the Fig. P. 8.1, what will be the complex frequencies in the current $i_1(t)$?

(b) With N connected to the current source as in (b) of the Fig. P. 8.1, what will be the complex frequencies in the voltage $v_1(t)$?

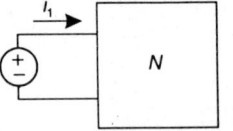 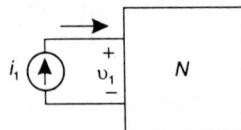

Fig. P. 8.1

Solution:

Given, (a) $\quad v_1 = V_0 e^{-2t} \cos u(t)$

∴ $\quad v_1(s) = \dfrac{V_0(s+2)}{(s+2)^2 + 1^2}$

and $Z(s) = \dfrac{(s+2)(s+3)}{(s+1)(s+4)}$

$\therefore \quad I_1(s) = \dfrac{v_1(s)}{Z(s)} = \dfrac{V_0(s+1)(s+4)}{\left((s+2)^2+1^2\right)(s+3)} = \dfrac{V_0(s+1)(s+4)}{\left(s^2+4s+5\right)(s+3)}$

Therefore, complex frequency of I_1 is equal to:

$(S = -3)$ and $S = -2 \pm j\,1$

(b) $\qquad\qquad i_1 = I_0\,e^{-t}\sin 3t\,u(t)$

$\therefore \qquad\qquad i_1(s) = \dfrac{3I_0}{(s+1)^2+3^2} = \dfrac{3I_0}{s^2+2s+10}$

$\therefore \qquad\qquad V_1(s) = i_1(s)Z(s)$

$\qquad\qquad\qquad = \dfrac{3I_0(s+2)(s+3)}{\left(s^2+2s+10\right)(s+1)(s+4)}$

Therefore, complex frequency of v_1 is equal to:

$(s = -4)$, $(s = -1)$, and $(s = -1 \pm j\,3)$ **Ans.**

Problem 8.2. Repeat Prob. 8.1 if

$$Z(s) = \dfrac{2s^4+3s^3+5s^2+5s+1}{\left(s^2+1\right)\left(2s^2+2s+4\right)}$$

Solve part (b) only.

Solution:

There complex frequency are:

$(s=-1\pm j3),\ (s=\pm j)$ and $s=\left(-1/2\pm j\sqrt{1.75}\right)$ **Ans.**

Problem 8.3. Consider the two series circuits shown in Fig. P. 8.3. Given that

$v_1(t)=\sin 10^3\,t,\ v_2(t)=e^{-1000t}$ for $t>0$, and $C=1\,\mu\text{F}$.

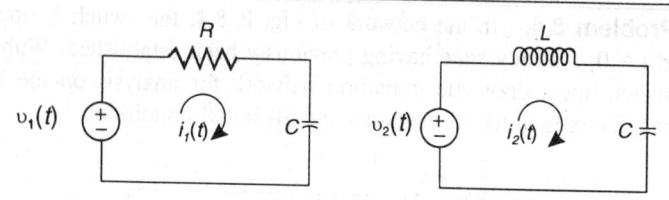

Fig. P. 8.3

(a) Show that it is possible to have $i_1(t) = i_2(t)$ for all $t > 0$.
(b) Determine the required values of R and L for (a) to hold.
(c) Discuss the physical meaning of this problem in terms of the complex frequencies of the two series circuits.

Solution:

$$I_1(s) = \frac{10^3}{\left(s^2+10^6\right)\left(R+\dfrac{1}{sC}\right)} = \frac{10^3\,S}{R\left(s^2+10^6\right)\left(s+\dfrac{1}{RC}\right)}$$

$$I_2(s) = \frac{1}{\left(s+10^3\right)\left(sL+\dfrac{1}{sC}\right)} = \frac{S}{L\left(s^2+\dfrac{1}{LC}\right)\left(s+10^3\right)}$$

(a) So, for $I_1(s) = I_2(s)$

$$\frac{1}{LC} = 10^6 \text{ and} \frac{1}{RC}$$

and $= 10^3 \text{ and} \dfrac{10^3}{R} = \dfrac{1}{L}$

(b) $\therefore$ $C = 1\,\mu\,\text{F}.$

$\therefore$ $\dfrac{1}{LC} = 10^6 \text{ or,} \dfrac{1}{L\times10^{-6}} = 10^6$ or, L = 1 H

and, $\dfrac{1}{RC} = 10^3$ or, $\dfrac{1}{R\times10^{-6}} = 10^3$

$\therefore$ $R = 10^3$

(c) $I_1(s) = \dfrac{10^3\,s}{10^3\left(s^2+10^6\right)\left(s+10^3\right)} = I_2(s)$

$\therefore$ $S = \pm\,j10^3 \text{ and } s = -\,10^3$

Problem 8.4. In the network of Fig. P. 8.4, the switch is opened at $t = 0$, a steady state having previously been established. With the switch open, draw the transform network for analysis on the loop basis, representing all elements and all initial conditions.

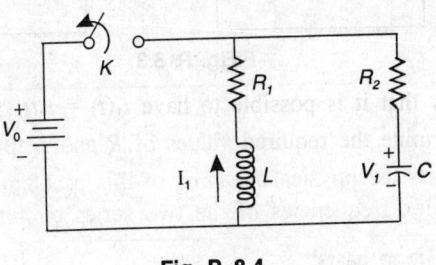

Fig. P. 8.4

Solution:

$$i_1\left(0^-\right) = \frac{V_0}{R_1} \text{ and } V_1\left(0^-\right) = V_0$$

At time $t > 0$.

$$i_1\left(0^+\right) = i_1\left(0^-\right) = \frac{V_0}{R} \text{ and } v_1\left(0^+\right) = v_1\left(0^-\right) = V_0.$$

So, the circuit is drawn as:

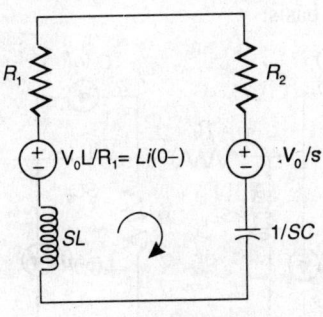

Fig. P. 8.4 (a)

Problem 8.5. This problem is similar to Prob. 8.4, except that the transform network required should be prepared for analysis on the (a) loop basis, and (b) node basis. In this network, initial currents and voltages are a consequence of active elements removed at $t = 0$.

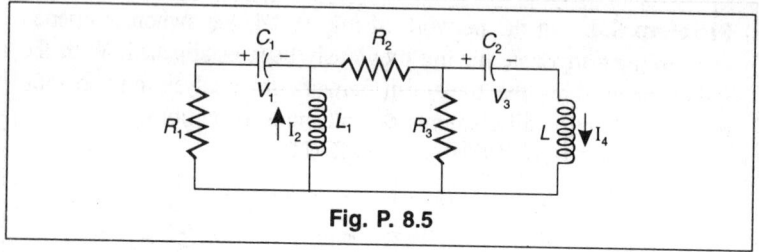

Fig. P. 8.5

Solution:

(a) On the loop basis:

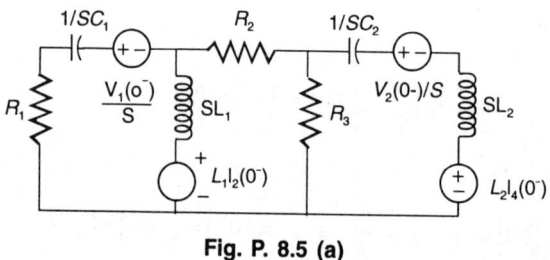

Fig. P. 8.5 (a)

(b) On the node basis:

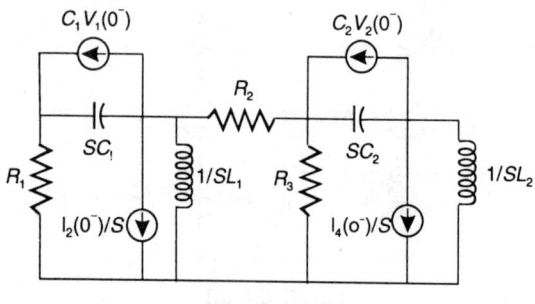

Fig. P. 8.5 (b)

Problem 8.6.   In the network of the Fig. P. 8.6, the switch K is closed at $t = 0$ and at $t = 0^-$ the indicated voltages are on the two capacitors. Repeat Prob. 8.4 for this network.

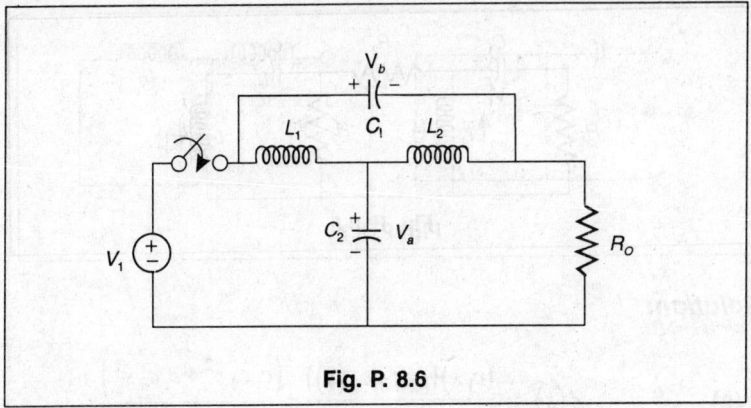

Fig. P. 8.6

Solution:

(a) On the loop basis:

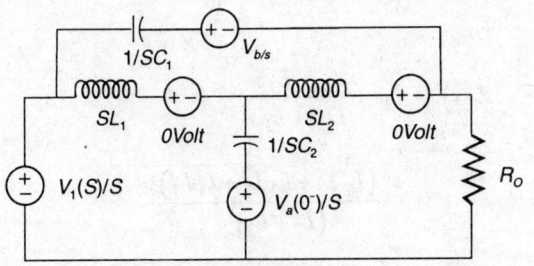

Fig. P. 8.6 (a)

(b) On the node basis: **Ans.**

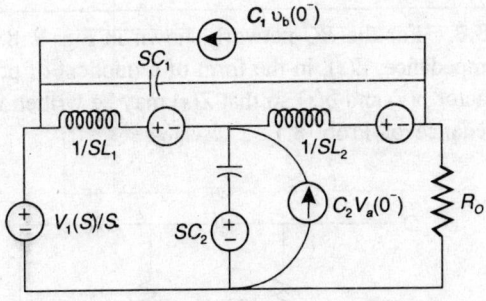

Fig. P. 8.6 (b)

Problem 8.7. Determine the transform impedances for the two networks shown in Fig. P. 8.7.

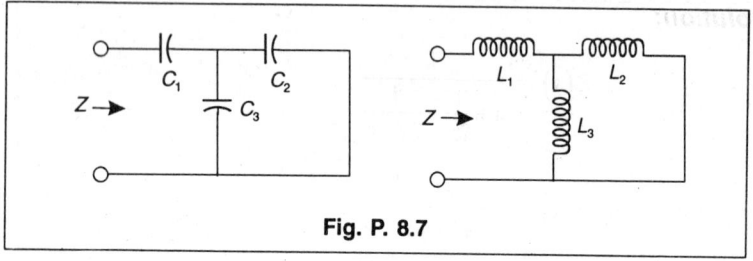

Fig. P. 8.7

Solution:

(a)
$$Z(s) = \frac{(c_1 s)(c_2 s + c_3(s))}{c_1 s + (c_2 s + c_3 s)} = \frac{(c_1 c_2 s^2 + c_1 c_3 s^2)}{s(c_1 + c_2 + c_3)}$$

$$Z(s) = \frac{s c_1 (c_2 + c_3)}{(c_1 + c_2 + c_3)}$$

(b)
$$Z(s) = L_1 s + \frac{L_2 s \cdot L_3 s}{L_2 s + L_3 s}$$

$$= \frac{s^2 (L_1 L_2 + L_2 L_3 + L_1 L_3)}{(L_2 + L_3)}$$

$\therefore$
$$Z(s) = \frac{s(L_1 L_2 + L_2 L_3 + L_1 L_3)}{(L_2 + L_3)} \quad \textbf{Ans.}$$

Problem 8.8. For the *RC* network shown in Fig. P. 8.8, find the transform impedance, $Z(s)$, in the form of a quotient of polynomials, $p(s)/q(s)$. Factor $p(s)$ and $q(s)$ so that $Z(s)$ may be written in the form of the impedance of Prob. 8.1.

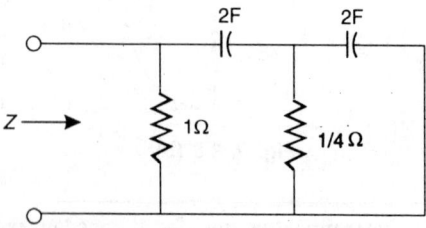

Fig. P. 8.8

Solution:

$$Z(s) = \frac{1}{1+\dfrac{1}{2s}+\dfrac{1}{2s+4}}$$

$$Z(s) = \frac{s+1}{s^2+3s+1}$$

so, $$s = -2.65, -0.37$$

$$\therefore \quad Z(s) = \frac{s+1}{(s+2.65)(s+0.37)} \quad \textbf{Ans.}$$

Problem 8.9. Repeat Prob. 8.8. for the *LC* network of the accompanying Fig. P. 8.9.

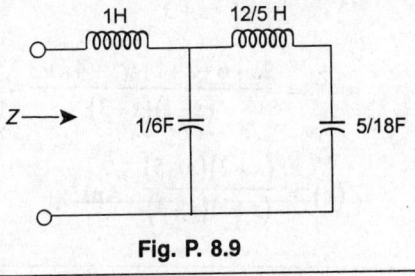

Fig. P. 8.9

Solution:

$$Z(s) = s+\frac{1}{\dfrac{s}{6}+\dfrac{1}{\dfrac{12s}{5}++\dfrac{18}{5s}}}$$

$$Z(s) = \frac{s^4+10s^2+9}{s(s^2+4)}$$

$$\therefore \quad s = 0,\ s = \pm 2j.$$

$$\therefore \quad Z(s) = \frac{s^4+10s^2+9}{s(s+j2)(s-j2)} \quad \textbf{Ans.}$$

Problem 8.10. Repeat Prob. 8.8. for the *RC* network shown in Fig. P. 8.10.

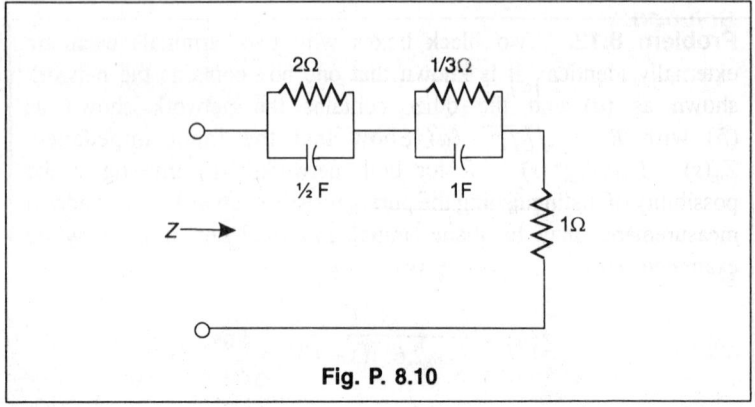

Fig. P. 8.10

Solution:

$$Z(s) = \frac{2}{s+1} + \frac{1}{s+3} + 1$$

$$= \frac{2s+6+s+1+s^2+4s+3}{(s+1)(s+3)} = \frac{s^2+7s+10}{(s+1)(s+3)}$$

$$Z(s) = \frac{(s+2)(s+5)}{(s+1)(s+3)} \quad \textbf{Ans.}$$

Problem 8.11. Repeat Prob. 8.8 for the *RLC* network of the Fig. P. 8.11, except that in this case determine $Y(s)$ rather than $Z(s)$.

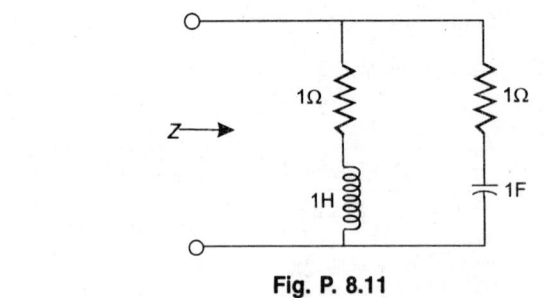

Fig. P. 8.11

Solution:

$$Y(s) = \frac{1}{s+1} + \frac{1}{1+1/s} = \frac{s+1}{s+1}$$

$\therefore$ $Y(s) = 1$ **Ans.**

Problem 8.12. Two black boxes with two terminals each are externally identical. It is known that one box contains the network shown as (a) and the other contains the network shown as (b) with $R = \sqrt{L/C}$. (a) Show that the input impedance, $Z_{in}(s) = V_{in}(s)/I_{in}(s) = R$ for both networks. (b) Investigate the possibility of distinguishing the purely resistive network. Any external measurements may be made, initial and final conditions may be examined, etc.

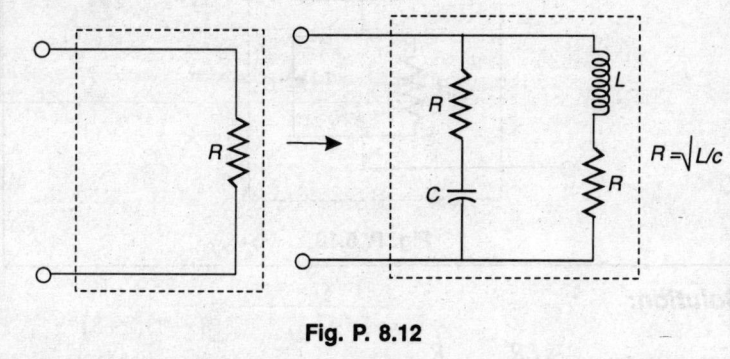

Fig. P. 8.12

Solution:

From Fig. P. 8.12 (b)
$$Y(s) = \frac{1}{R + 1/sC} + \frac{1}{sL + R}$$

$$Y(s) = \frac{sC}{sRC + 1} + \frac{1}{(sL + R)} \tag{1}$$

$\because$
$$R = \sqrt{L/C}$$

$\therefore$ $L = R^2 C$, So, Putting the value of L in eq (1) we get.

$$Y(s) = \frac{sC}{sRC + 1} + \frac{1}{\left(s R^2 C + R \right)} = \frac{sC}{(sRC + 1)} + \frac{1}{R(s RC + 1)}$$

$$Y(s) = \frac{(s RC + 1)}{R(s RC + 1)}$$

$$= \frac{1}{R}$$

$\therefore$
$$Z(s) = \frac{1}{Y(s)} = R$$

So, both circuits are equivalent.

Problem 8.13. Repeat Prob. 8.12 by comparing the network shown in Fig. P. 8.13 to that given in (a) of Fig. P. 8.12.

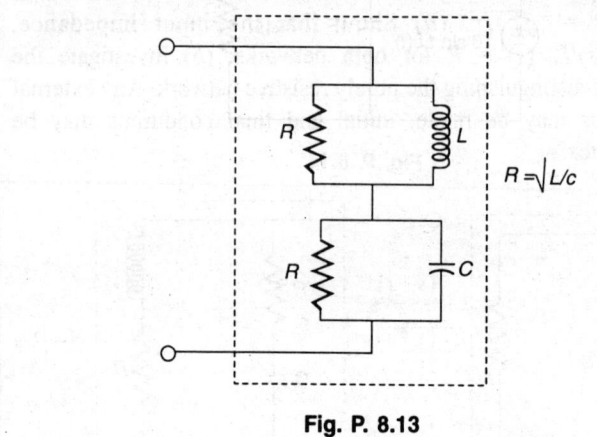

$R = \sqrt{L/c}$

Fig. P. 8.13

Solution:

$$Z(s) = \frac{sLR}{R+sL} + \frac{R}{sRC+1}$$

$$\because \qquad R = \sqrt{\frac{L}{C}} \quad \therefore L = R^2 C$$

$$\therefore \qquad Z(s) = \frac{sR^2CR}{\left(R+sR^2C\right)} + \frac{R}{\left(sRC+1\right)}$$

$$= \frac{sR^2C}{\left(sRC+1\right)} + \frac{R}{\left(sRC+1\right)} = \frac{R(sRC+1)}{\left(sRC+1\right)}$$

$$Z(s) = R$$

So, both circuits are equivalent.

Problem 8.14. If the capacitors are uncharged and the inductor current zero at $t = 0^-$, in the given network, show that the transform of the generator current is

$$I_1(s) = \frac{10\left(s^2+s+1\right)}{\left(s^2+1\right)\left(s^2+2s+2\right)}$$

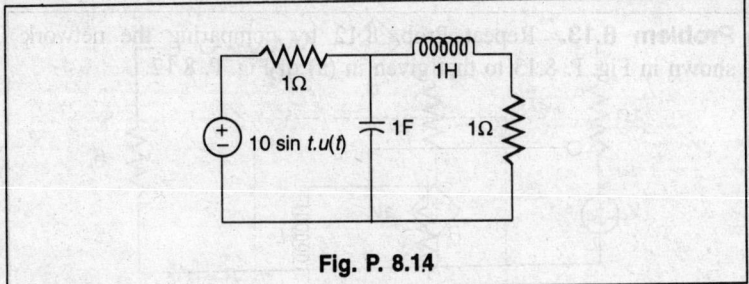

Fig. P. 8.14

Solution:

$$Z(s) = 1 + \frac{(s+1)1/s}{(s+1)+\dfrac{1}{s}}$$

$$Z(s) = \frac{s^2+2s+2}{s^2+s+1}$$

$$v_1(t) = 10 \ \sin t.u(t)$$

$\therefore \qquad v_1(s) = \dfrac{10}{\left(s^2+1\right)}$

$\therefore \qquad I_1(s) = \dfrac{v_1(s)}{Z(s)} = \dfrac{10\left(s^2+s+1\right)}{\left(s^2+1\right)\left(s^2+2s+2\right)}$

$\therefore \qquad I_1(s) = \dfrac{10\left(s^2+s+1\right)}{\left(s^2+1\right)\left(s^2+2s+2\right)}$ **Ans.**

Problem 8.15. For the network of the Fig. P. 8.15, show that **the** equivalent Thevenin network is represented by

$$V_\theta = \frac{V_1}{2}(1+a+b-ab)$$

and

$$Z_\theta = \frac{3-b}{2}$$

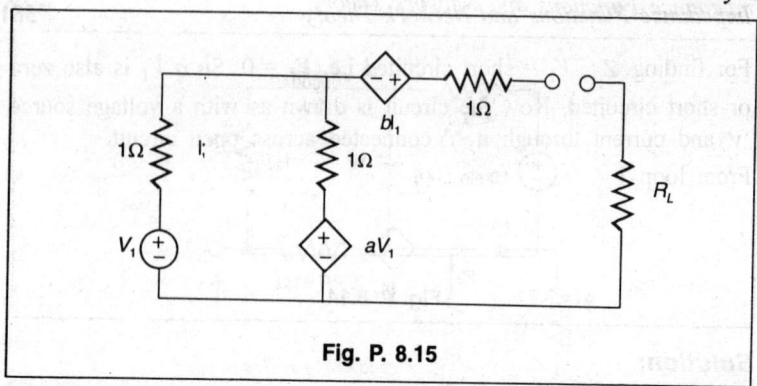

Fig. P. 8.15

Solution:

In loop 1 applying KVL:

$$1 \cdot I_1 + I_1 \cdot 1 + aV_1 = V_1$$

or, $$2I_1 = V_1(1-a) \quad \text{or} \quad I_1 = \frac{V_1(1-a)}{2}$$

The above circuit can be drawn as:

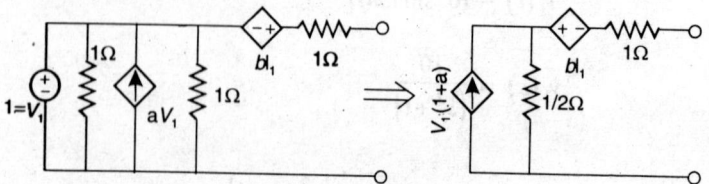

Fig. P. 8.15 (a)

$$\therefore \; v_\theta = \frac{V_1(1+a)}{2} + bI_1, = \frac{V_1(1+a)}{2} + b\left(\frac{V_1(1-a)}{2}\right) = \frac{V_1 + V_1 a + bV_1 - abV_1}{2}$$

$$v_\theta = \frac{V_1(1+a+b-ab)}{2}$$

Fig. P. 8.15 (b)

For finding Z_θ, V_1 = short circuited i,e. $V_1 = 0$. So $a\, V_1$ is also zero or short circuited. Now the circuit is drawn as with a voltage source 'v' and current through it 'i' connected across open circuit.

From loop 1

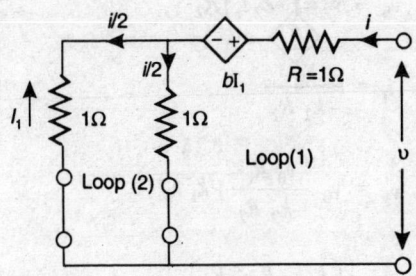

Fig. P. 8.15 (c)

$$i \cdot 1 + b\,I_1 + \frac{i}{2} \cdot 1 = V \quad \because \quad I_1 = \frac{-i}{2}$$

∴ The above equation can be written as

$$i - \frac{i}{2}b + \frac{i}{2} = V \quad \text{or,} \quad \frac{3i}{2} - \frac{i}{2}b = V$$

or, $$\frac{(3-b)i}{2} = V$$

or, $$\frac{V}{i} = \frac{(3-b)}{2} \quad \text{or} \quad Z_\theta = \frac{3-b}{2} \textbf{ Ans.}$$

Problem 8.16. The accompanying network consists of resistors and controlled sources in addition to the independent voltage source v_s. For this network, find the Thevenin equivalent network by determining an expression for the voltage v_θ and the Thevenin equivalent resistance.

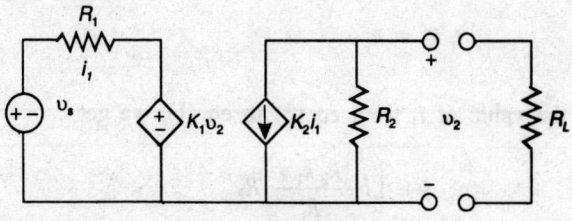

Fig. P. 8.16

Solution:

$$V_s = i_1 R_1 + k_1 v_2$$

or,

$$V_s = i_1 R_1 + k_1 v_\theta$$

Where

$$v_\theta = v_2 = \left(-k_2 i_1\right) R_2$$

∴

$$i_1 = \frac{v_\theta}{-k_2 R_2}$$

or,

$$v_s = v_\theta \frac{v_\theta R_1}{-k_2 R_2} + k_1 v_\theta$$

$$v_s = \left(\frac{k_1 k_2 R_2 - R_1}{k_2 R_2}\right)$$

or,

$$v_\theta = v_s \left(\frac{k_2 R_2}{k_1 k_2 R_2 - R_1}\right)$$

For $z_0,$ $v_s = 0$

$$v_2 = v$$

and current i at terminal v.

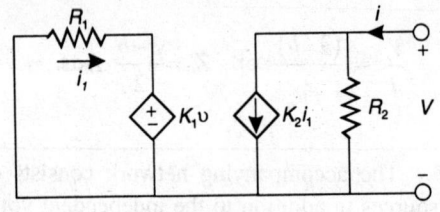

Fig. P. 8.16 (a)

Now,

$$v = \left(i - k_2 i_1\right) R_2 \tag{1}$$

and

$$i_1 R_1 = k_1 v \tag{2}$$

putting the value of i_1 from eq (2) to eq (1), we get

$$V = \left(i - \frac{k_2 k_1 v}{R_1}\right) R_2$$

or, $\quad \dfrac{v}{R_2} + \dfrac{k_1 k_2 v}{R_1} = i$

or, $\quad \dfrac{v(R_1 + k_1 k_2 R_2)}{R_1 R_2} = i$

or, $\quad \dfrac{v}{i} = \dfrac{R_1 R_2}{R_1 + k_1 k_2 R_2}$

or, $\quad Z_0 = \dfrac{R_1 R_2}{R_1 + k_1 k_2 R_2}$ **Ans.**

Problem 8.17. The network of Fig. P. 8.17 contains three resistors and one controlled current source, in addition to independent sources. For this network, determine the Thevenin equivalent network at terminals 1-1′.

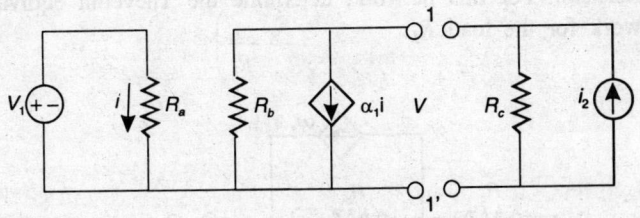

Fig. P. 8.17

Solution:

$$v_1 = i R_a \tag{1}$$

$\therefore \qquad i = \dfrac{v_1}{R_a}$

Thevenin voltage $\quad v_0 = -\alpha_i \, i \, R_b \tag{2}$

$\therefore \qquad v_0 = -\alpha_i \dfrac{v_1}{R_a} R_b$

$\therefore \qquad v_0 = \dfrac{-\alpha_i R_0}{R_a} v_1$

For Thevenin resistance

$v_1 = 0$, So $i = 0$, let $v_{1-1'} = v$

$\therefore \qquad Z_0 = \dfrac{v}{i} = R_b$

$\therefore$ $\qquad\qquad Z_0 = R_b$

So, circuit is

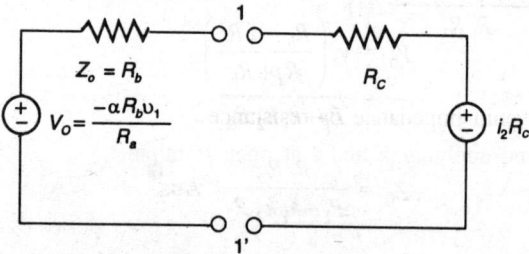

Fig. P. 8.17 (a)

Problem 8.18. The network shown is a simple representation of a transistor. For this network, determine the Thevenin equivalent network for the load R_L.

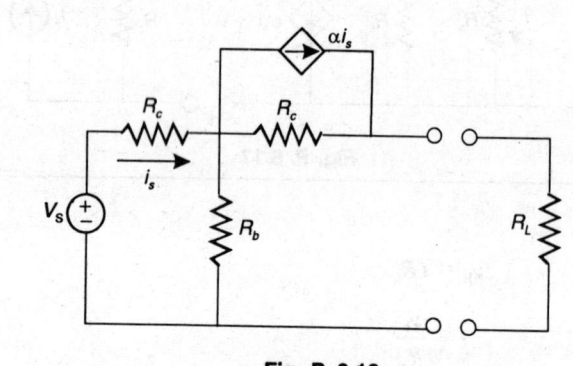

Fig. P. 8.18

Solution:

Let V_0 be the Thevenin voltage. So

$$V_0 = v_s - i_s R_e + (-\alpha i_s) R_c$$

$$V_0 = v_s - (R_e + \alpha R_c) i_s$$

$$V_0 = v_s - (R_e + \alpha R_c) i_s \qquad (1)$$

and $\qquad v_s = (R_e + R_b) i_s$

$$\therefore \qquad i_s = \frac{v_s}{R_e + R_b} \tag{2}$$

$$\therefore \qquad V_0 = v_s - \left(R_e + \alpha R_c \right) \frac{v_s}{R_e + R_b}$$

$$V_0 = v_s \left(\frac{R_b - \alpha R_c}{R_e + R_b} \right)$$

For Thevenin impedance or resistance
$v_s = 0$ and applying V and i at open terminal.

$$\therefore \qquad i_s = \frac{-i \times R_b}{R_e + R_b}$$

So, $$(i_s \cdot R_e) + i' R_c = V,$$

where $$i' = i + \alpha i_s$$

or, $$\frac{-i \cdot R_b R_e}{R_e + R_b} + i' R_c = V$$

or $$\frac{-i \cdot R_b R_e}{R_e + R_b} + (i + \alpha i_s) R_c = V$$

or, $$\frac{-i \cdot R_b R_e}{R_e + R_b} + \left(i + \frac{\alpha(-i R_b)}{R_e + R_b} \right) R_c = V$$

or, $$\frac{V}{i} = R_c - \frac{(\alpha R_c - R_e) R_b}{R_e + R_b}$$

$$\therefore \qquad Z_0 = R_e - \frac{(\alpha R_c - R_e) R_b}{R_e + R_b} \qquad \textbf{Ans.}$$

Problem 8.19. The network in Fig. P. 8.19 contains a resistor and a capacitor in addition to various sources. With respect to the load consisting of R_L in series with L, determine the Thevenin equivalent network.

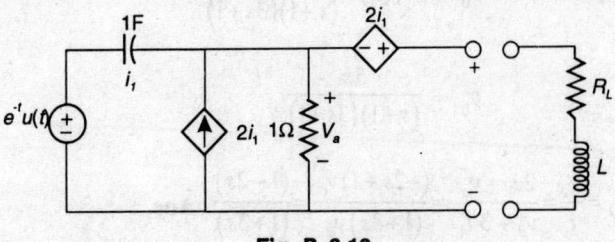

Fig. P. 8.19

Solution:

$$v_a = (i_1 + 2i_1) \times 1$$

$$v_a = 3i_1 \text{ and Thevenin voltage}$$

$$V_0 = v_a + 2i_1 = v_a + \frac{2v_a}{3} = \frac{5v_a}{3}$$

or, $$V_0 = 5i_1$$

$$e^{-t}u(t) - v_a(t) = \frac{1}{C}\int i_1 \, dt.$$

In the laplace form

$$e^{-t} - 3i_1(t) = \frac{1}{C}\int i_1 \, dt.$$

or, $$\frac{1}{s+1} - 3i_1(s) = \frac{1}{Cs}i_1(s)$$

or, $$\frac{s}{(s+1)} - 3s\,i_1(s) = \frac{1}{C}i_1(s)$$

$$\frac{s}{s+1} = \frac{1}{1}i_1(s) + 3s\,i_1(s)$$

or, $$\frac{s}{s+1} = i_1(s)(1+3s)$$

or, $$i_1(s) = \frac{s}{(s+1)(1+3s)}$$

$\therefore$ $$V_0 = 5i_1(s) = \frac{5s}{(s+1)(3s+1)}$$

$\therefore$ $$V_0 = \frac{5s}{(s+1)(3s+1)}$$

and $$Z_0 = \frac{V}{i} = \frac{2i_1 + v_a}{v_a - 3I_1} = \frac{(-2s+1)v_a}{(1+3s)v_a} = \frac{(1-2s)}{(1+3s)} \quad \textbf{Ans.}$$

Problem 8.20. In the given network, the switch is in position *a* until a seady state is reached. At $t = 0$, the switch is moved to position *b*. Under that condition, determine the transform of the voltage across the 0.5-F capacitor using (a) Thevenin's theorem, and (b) Norton's theorem.

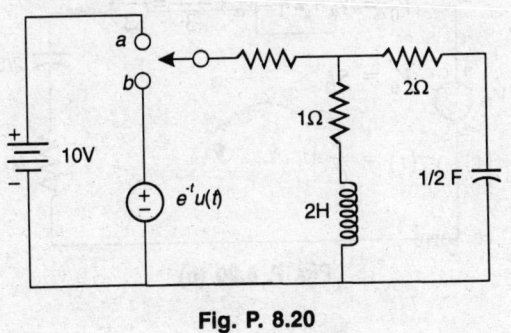

Fig. P. 8.20

Solution:

At time $t = 0^-$

Current through inductor $= \dfrac{10}{2} = 5$ amp.

and voltage through capacitor $= 5 \times 1 = 5$ volt.

At time $t = 0^+$, the circuit can be drawn as:

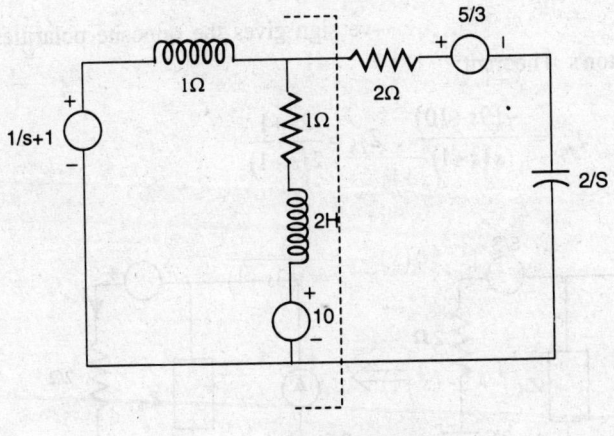

Fig. P. 8.20 (a)

(a) Thevenin Theorem:

$\therefore \qquad v_{Th} = \dfrac{-(9s+10)(2s+1)}{2s(s+1)^2}$ and $Z_{Th} = \dfrac{2s+1}{2(s+1)}$

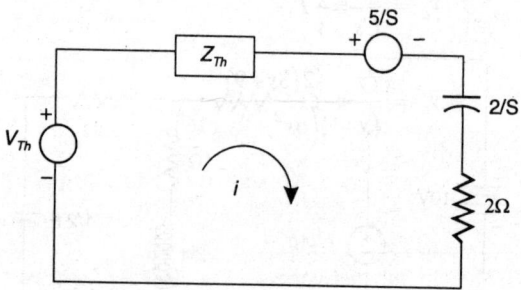

Fig. P. 8.20 (b)

So, $\quad i(s) = \dfrac{v_{Th} + 5/s}{Z_{Th} + \left(2 + \dfrac{2}{s}\right)} = \dfrac{-s(8s+9)}{(s+1)(6s^2+9s+4)}$

$\therefore \qquad V_c = \dfrac{i(s)}{Cs} = \dfrac{2}{s} i(s) = \dfrac{-2(8s+9)}{(s+1)(6s^2+9s+4)}$

$\therefore \qquad V_c = \dfrac{2(8s+9)}{(s+1)(6s^2+9s+4)}$

–ve sign gives the opposite polarities.

(b) Norton's Theorem:

$$I_{sc} = \dfrac{-(9s+10)}{s(s+1)}, \ Z_{Th} = \dfrac{2s+1}{2(s+1)}$$

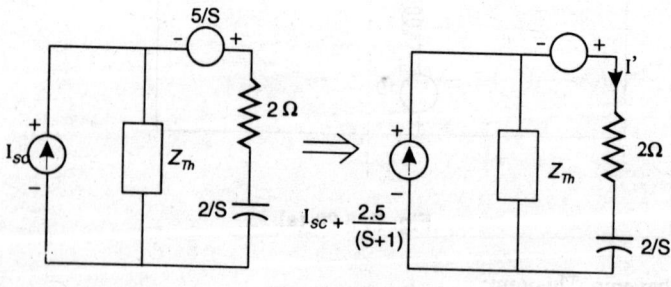

Fig. P. 8.20 (c)

$$\therefore \qquad I' = \frac{-s(8s+9)\cdot}{(s+1)(6s^2+9s+16)}$$

$$\therefore \qquad v_c = \frac{I'}{Cs} = \frac{2}{s}I'$$

$$\therefore \qquad v_c = \frac{-2(8s+9)}{(s+1)(6s^2+9s+16)}$$

–ve sign for opposite polarities. **Ans.**

Problem 8.21. In the network of Fig. P. 8.21, the switch K is closed at $t = 0$, a steady state having previously existed. Find the current in resistor R_3 using (a) Thevenin's theorem, and (b) Norton's theorem.

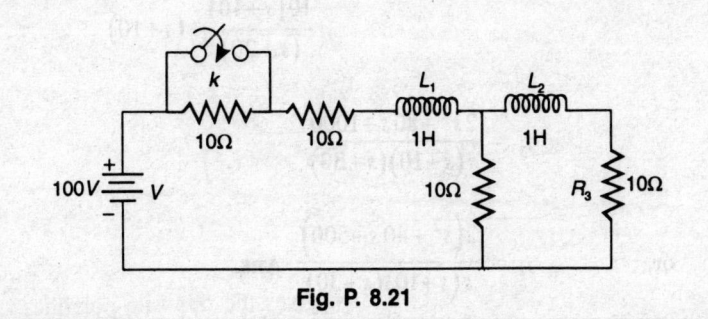

Fig. P. 8.21

Solution:

At time $t = 0^-$, current through inductor $L_1 = 4A$ and through $L_2 = 2A$.

(a) At time $t = 0^+$, the circuit is drawn as:

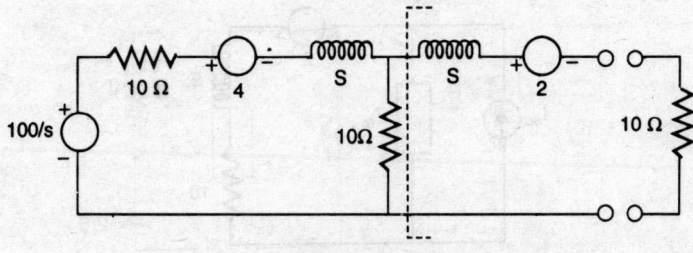

Fig. P. 8.21 (a)

$$v_{Th} = \frac{40(s+25)}{s(s+20)}, \quad Z_{Th} = \frac{10(s+10)}{(s+20)}$$

So, the equivalent circuit is

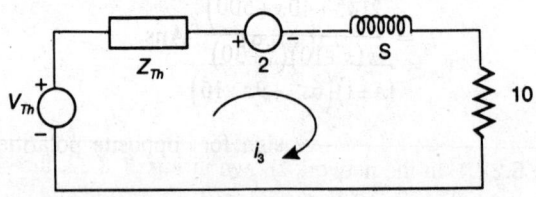

Fig. P. 8.21 (b)

$$\therefore \qquad i_3 = \frac{v_{Th}+2}{Z_{Th}+(s+10)} = \frac{\dfrac{40s+1000}{s(s+20)}+2}{\dfrac{10(s+10)}{(s+20)}+(s+10)}$$

$$i_3 = \frac{2s^2+80s+1000}{s(s+10)(s+30)}$$

or, $$\qquad i_3 = \frac{2(s^2+40s+500)}{s(s+10)(s+30)} \quad \textbf{Ans.}$$

(b) Norton's Theorem:

$$I_{s \cdot c} = \frac{4(s+25)}{s(s+10)}, \quad Z_{Th} = \frac{10(s+10)}{(s+20)}$$

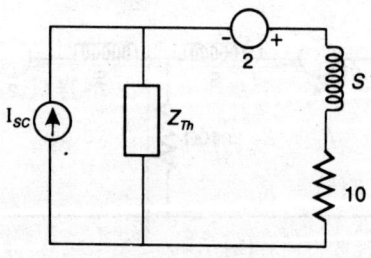

Fig. P. 8.21 (c)

Current in 10 Ω resistor is:

$$i_3 = \frac{I_s C(Z_{Th})}{Z_{Th}+(s+10)} + \frac{2}{Z_{Th}+(s+10)}$$

or, $\qquad i_3 = \dfrac{2\left(s^2+40s+500\right)}{s(s+10)(s+30)}$ **Ans.**

Problem 8.22. In the network shown in Fig. P. 8.22, the elements are chosen such that $L = CR_1^2$ and $R_1 = R_2$. If $v_1(t)$ is a voltage pulse of 1-V amplitude and T-sec duration, shown that $v_2(t)$ is also a pulse and find its amplitude and time duration.

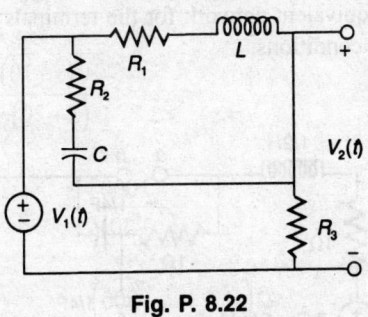

Fig. P. 8.22

Solution:

Impedance of the parallel circuit is:

$$Z(s) = \frac{(SL+R_1)\left(R_2+\dfrac{1}{sC}\right)}{SL+R_1+R_2+\dfrac{1}{sC}}$$

$$Z(s) = \frac{(SL+R_1)(S R_2 C+1)}{S^2 LC + sC(R_1+R_2)+1}$$

Putting $\qquad R_1 = R_2$ and $L = CR_1^2$, we get

$$Z(s) = R_1$$

∴ $\qquad v_2(s) = \dfrac{v_1(s) R_3}{R_1+R_3}$

$\therefore$ $\qquad$ $v_1(s) = 1$ V

$\therefore$ $\qquad$ $v_2(s) = \dfrac{R_3}{R_1 + R_3}$ volt

$\therefore$ $\qquad$ $v_2(t) = \dfrac{R_3}{(R_1 + R_3)} \left[u(t) - u(t - T) \right]$

$\therefore$ Amplitude of $v_2(t) = \dfrac{R_3}{R_1 + R_3}$ and pulse duration or time duration is

similar as $v_1(t)$ [due to resistive load].

So, time duration of $v_2(t) = T$ sec. **Ans.**

Problem 8.23. Using either Thevenin's or Norton's theorem, determine an equivalent network for the terminals *a-b* in Fig. P. 8.23 for zero initial conditions.

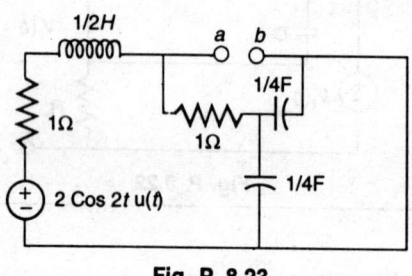

Fig. P. 8.23

Solution:

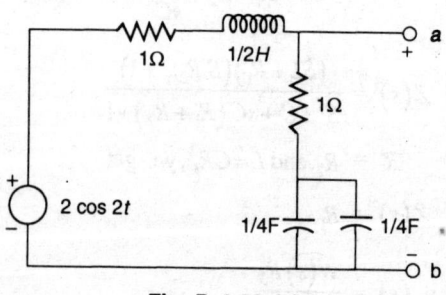

Fig. P. 8.23 (a)

In the laplace form:

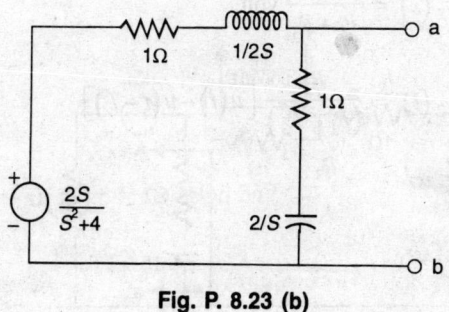

Fig. P. 8.23 (b)

∴ v_0 (Thevenin voltage)

$$= \frac{\dfrac{2s}{s^2+4}\left(1+\dfrac{2}{s}\right)}{\left(2+\dfrac{s}{2}+\dfrac{2}{s}\right)}$$

$$= \frac{2(s+2)}{\left(s^2+4\right)\left(\dfrac{4s+s^2+4}{2s}\right)} = \frac{2(s+2)(2s)}{\left(s^2+4\right)(s+2)^2}$$

$$v_0 = \frac{4s}{(s+2)\left(s^2+4\right)} \text{ Volt}$$

Thevenin impedance or resistance.

$$Z_0 = \frac{\left(1+\dfrac{s}{2}\right)\left(1+\dfrac{2}{s}\right)}{\left(1+\dfrac{s}{2}+1+\dfrac{2}{s}\right)} = 1$$

∴ $Z_0 = 1 \ \Omega$ **Ans.**

Problem 8.24. The network given contains a controlled source. For the element values given, with $v_1(t) = u(t)$, and for zero initial

conditions: (a) determine the equivalent Thevenin network at $a-a'$.
(b) Determine the equivalent Thevenin network at $b-b'$.

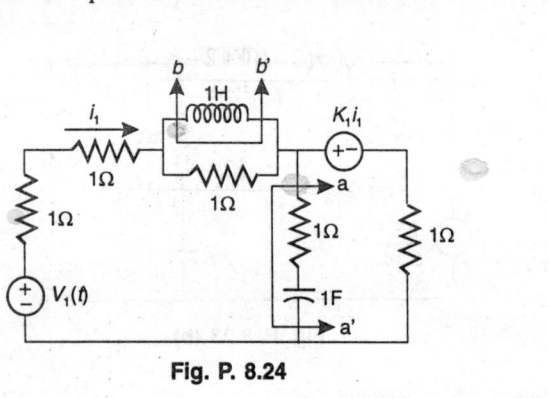

Fig. P. 8.24

Solution:

(a) For $a-a'$ terminals:

$$I_1(s) = \frac{s+1}{s[3s - k_1 s + 2 - k_1]}$$

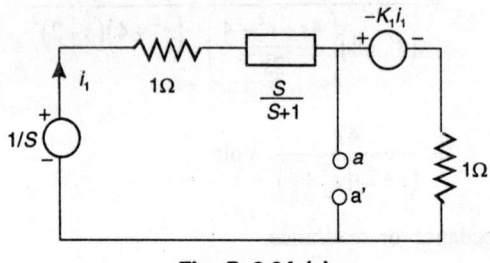

Fig. P. 8.24 (a)

and so, $v_{Th} = (I_1/s)\cdot(1-k_1)$

$$= \frac{(s+1)(1-k_1)}{\cdot\; s[3s - k_1 s + 2 - k_1]}$$

$$i = v\left(1 + \frac{(1-k_1)(s+1)}{2s+1}\right)$$

$$i = v\left(\frac{2s+1+s+1-k_1s-k_1}{2s+1}\right)$$

$$i = v\left(\frac{s(3-k_1)+2-k_1}{(2s+1)}\right)$$

or,

$$\frac{v}{i} = Z_{Th} = \frac{(2s+1)}{s(3-k_1)+(2-k_1)}$$

∴

$$Z_{Th} = \frac{2s+1}{s(3-k_1)+(2-k_1)}$$

(b) For $b-b'$ terminals:

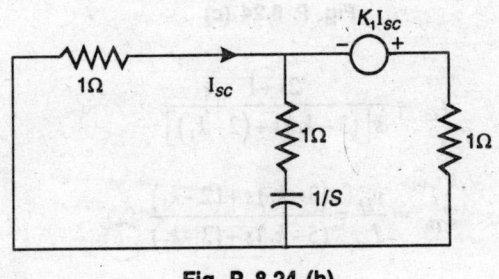

Fig. P. 8.24 (b)

$$2i_1 + \frac{i_1-i_2}{s} + i_1 - i_2 = \frac{1}{s}$$

or,

$$3i_1 s + i_1 - i_2 s - i_2 = 1$$

or,

$$i_1(3s+1) - i_2(s+1) = 1 \qquad (1)$$

From loop 2,

$$I_1(k_1 s + s + 1) - I_2(2s+1) = 0 \qquad (2)$$

After solving eq (1) and eq (2), we get

$$I_1 = \frac{2s+1}{s\left[(5-k_1)s+(3-k_1)\right]}$$

$$\therefore \qquad v_{Th} = I_1 \cdot 1 = \frac{2s+1}{s\left[(5-k_1)s+(3-k_1)\right]}$$

For, Z_{Th}, let I_{sc} be the current across 1Ω resistance. So.

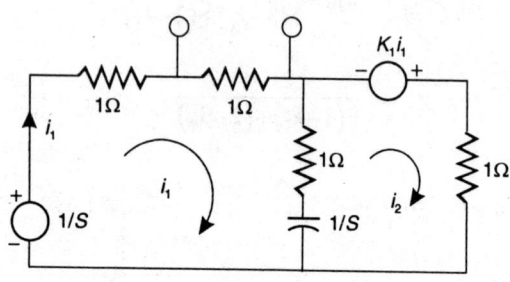

Fig. P. 8.24 (c)

$$\therefore \qquad I_{sc} = \frac{2s+1}{s\left[(3-k_1)s+(2-k_1)\right]}$$

$$\therefore \qquad Z_{Th} = \frac{v_{Th}}{I_{sc}} = \frac{(3-k_1)s+(2-k_1)}{(5-k_1)s+(3-k_1)}$$

$$\therefore \qquad Z_{Th} = \frac{(3-k_1)s+(2-k_1)}{(5-k_1)s+(3-k_1)} \quad \textbf{Ans.}$$

Problem 8.25. Assuming zero initial voltage on the capacitor, determine the equivalent Norton network for the resistor R_k.

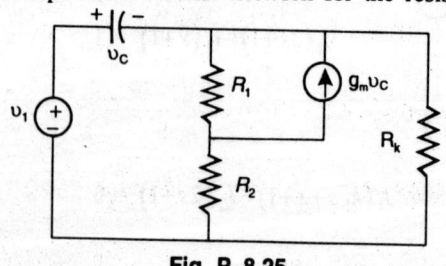

Fig. P. 8.25

Solution:

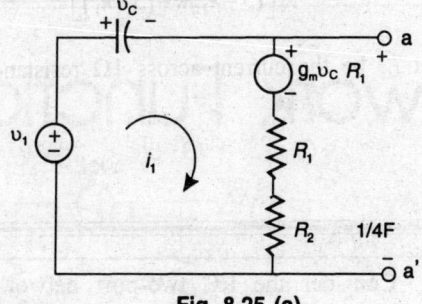

Fig. 8.25 (a)

$$-v_1 + v_c + g_m v_c R_1 + i_1 (R_1 + R_2) = 0$$

or, $\qquad i_1 (R_1 + R_2) = v_1 - v_c (1 + g_m R_1)$

or, $\qquad\qquad i_1 = \dfrac{v_1 - v_c (1 + g_m R_1)}{(R_1 + R_2)}$

$$Z_{Th} = \frac{R_1 + R_2}{1 + sc(R_1 + R_2) + g_m R_1}$$

or $\qquad I_{s \cdot c} = v_1(s) \dfrac{\left[sc(R_1 + R_2) + g_m R_1 \right]}{R_1 + R_2}$ **Ans.**

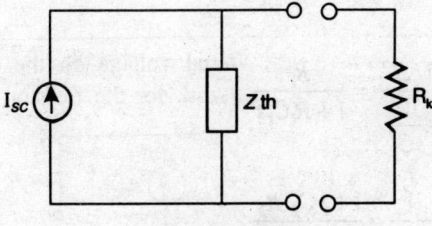

Fig. P. 8.25 (b)

Network Functions

Problem 9.1. Consider the RC two-port network shown in Fig. P. 9.1. For this network show that

$$G_{12} = \left[\frac{s^2 + (R_1C_1 + R_2C_2)s/R_1R_2C_1C_2 + 1/R_1R_2C_1C_2}{s^2 + (R_1C_1 + R_1C_2 + R_2C_2)s/R_1R_2C_1C_2 + 1/R_1R_2C_1C_2} \right]$$

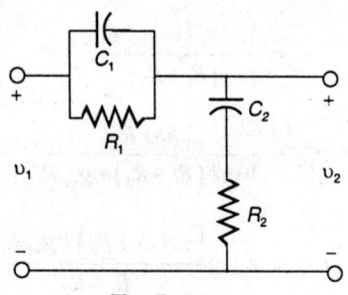

Fig. P. 9.1

Solution:

Let $\quad Z_1 = \dfrac{R_1 - 1/sC_1}{R_1 + \dfrac{1}{sC_1}} = \dfrac{R_1}{1 + R_1C_1s}$

$\qquad Z_2 = R_2 + \dfrac{1}{sC_2} = \dfrac{1 + sC_2\,R_2}{sC_2}$

$\therefore \quad G_{12} = \dfrac{v_2}{v_1} = \dfrac{1 + sR_2C_2}{\left[\dfrac{sC_2R_1}{1 + sR_1C_1} + \dfrac{sC_2(1 + R_2C_2s)}{C_2s} \right]}$

$\therefore \; G_{12} = \left[\dfrac{s^2 + (R_1C_1 + R_2C_2)s/R_1R_2C_1C_2 + 1/R_1R_2C_1C_2}{s^2 + (R_1C_1 + R_1C_2 + R_2C_2)s/R_1R_2C_1C_2 + 1/R_1R_2C_1C_2} \right]$ **Ans.**

Problem 9.2. *(a)* For the given network in Fig. P. 9.2, show that with port 2 open, the input impedance at port 1 is 1 Ω. *(b)* Find the voltage-ratio transfer function, G_{12} for the two-port network.

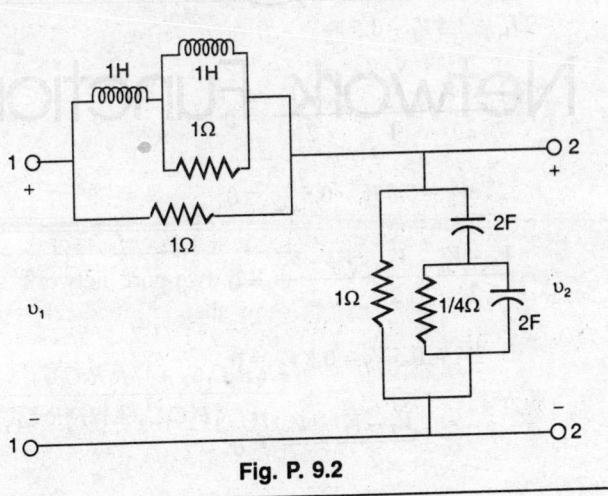

Fig. P. 9.2

Solution:

$$Z_1 = \frac{s(s+2)}{s^2+3s+1} \text{ and } Z_2 = \frac{s+1}{s^2+3s+1}$$

$$\therefore \qquad G_{12} = \frac{V_2}{V_1} = \frac{Z_2}{Z_1+Z_2} = \frac{s+1}{s^2+3s+1} \quad \textbf{Ans.}$$

Problem 9.3. For the resistive two-port network of Fig. P. 9.3, determine the numerical value for *(a)* G_{12}, *(b)* Z_{12}, *(c)* Y_{12}, and α_{12}.

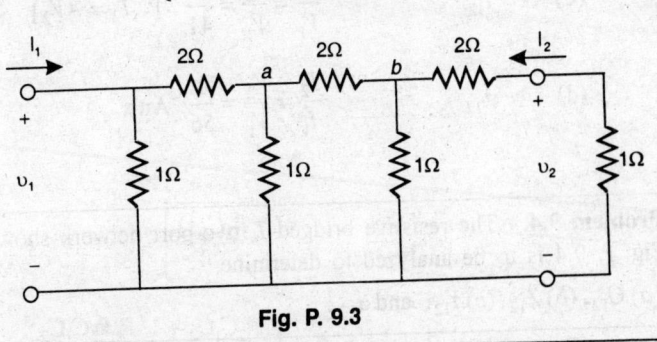

Fig. P. 9.3

Solution:

$$I_1 = \frac{V_1}{1} + \frac{V_1 - V_a}{2} = 1.5\,V_1 - 0.5V_a$$

or, $$I_1 = 1.5\,V_1 - 0.5\,V_a \tag{1}$$

$$\frac{V_a - V_1}{2} + \frac{V_a}{1} + \frac{V_a - V_b}{2} = 0$$

or, $$2V_a - 0.5\,V_1 - 0.5\,V_2 = 0 \tag{2}$$

$$\frac{V_b - V_a}{2} + \frac{V_b}{2} + \frac{V_b - V_2}{2} = 0$$

or $$2V_b - 0.5\,V_a - 0.5\,V_2 = 0 \tag{3}$$

$$\frac{V_2 - V_b}{2} + \frac{V_2}{1} = 0$$

or $$1.5\,V_2 - 0.5\,V_b = 0 \tag{4}$$

Now, Solving the above four equations, we get

$$V_a = 11V_2, V_b = 3V_2, V_1 = 41V_2 \text{ and } I_1 = 56\,V_2$$

$\therefore$ (a) G_{12} $=$ $\dfrac{V_2}{V_1} = \dfrac{1}{41}$

$\therefore$ (b) Z_{12} $=$ $\dfrac{V_2}{I_1} = \dfrac{1}{56}$

$\therefore$ (c) Y_{12} $=$ $\dfrac{I_2}{V_1} = \dfrac{V_2}{V_1} = \dfrac{1}{41}$ $[\because I_2 = -V_2]$

       (d) α_{12} $=$ $\dfrac{I_2}{I_1} = \dfrac{V_2}{V_1} = \dfrac{1}{56}$ **Ans.**

Problem 9.4. The resistive bridged-*T,* two-port network shown in Fig. P. 9.4 is to be analyzed to determine
(a) G_{12}, (b) Z_{12}, (c) Y_{12}, and α_{12}.

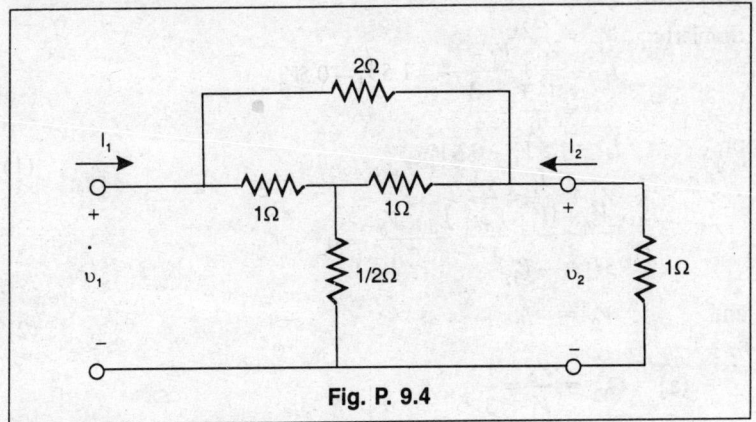

Fig. P. 9.4

Solution:

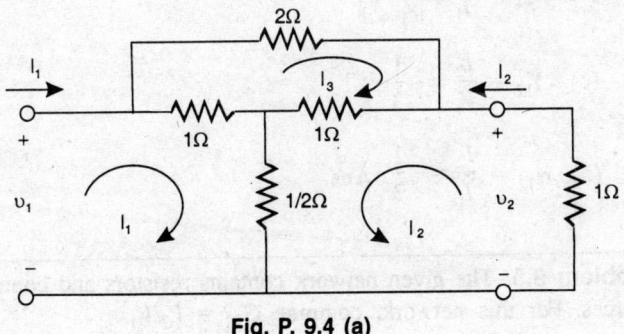

Fig. P. 9.4 (a)

In matrix terms:

$$\begin{bmatrix} 1.5 & 0.5 & -1 \\ 0.5 & 2.5 & 1 \\ -1 & 1 & 4 \end{bmatrix} \begin{bmatrix} I_1 \\ I_2 \\ I_3 \end{bmatrix} = \begin{bmatrix} 1 \\ 0 \\ 0 \end{bmatrix}$$

$$\therefore \quad Y_{11} = \frac{\Delta_{11}}{\Delta} = \frac{\begin{vmatrix} 1 & 0.5 & -1 \\ 0 & 2.5 & 1 \\ 0 & 1 & 4 \end{vmatrix}}{\begin{vmatrix} 1.5 & 0.5 & -1 \\ 0.5 & 2.5 & 1 \\ -1 & 1 & 4 \end{vmatrix}} = \frac{9}{9} = 1$$

$$\therefore \quad Y_{11} = \frac{I_1}{V_1} = 1 \quad \therefore I_1 = V_1$$

Similarly $Y_{12} = \dfrac{\Delta_{21}}{\Delta}$

$\qquad\qquad = \dfrac{-3}{9} = \dfrac{-1}{3}$

$\because\qquad Y_{12} = \dfrac{I_2}{V_1} = \dfrac{-1}{3}$

$\therefore\qquad 3I_2 = -V_1$

and $V_2 = -I_2$

$\therefore$ (a) $G_{12} = \dfrac{V_2}{V_1} = \dfrac{1}{3}$

$\quad$ (b) $Z_{12} = \dfrac{V_2}{I_1} = \dfrac{V_2}{V_1} = \dfrac{1}{3}$

$\quad$ (c) $Y_{12} = \dfrac{I_2}{V_1} = -\dfrac{1}{3}$

$\quad$ (d) $\alpha_{12} = \dfrac{I_2}{I_1} = -\dfrac{1}{3}$ **Ans.**

Problem 9.5. The given network contains resistors and controlled sources. For this network, compute $G_{12} = V_2/V_1$.

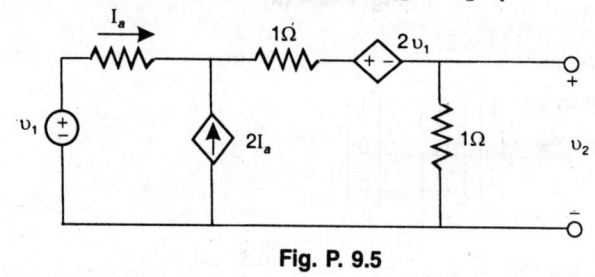

Fig. P. 9.5

Solution:

From the given circuit:

$$V_2 + 2V_1 + (3I_a).1 + I_a \times 1 = V_1 \qquad\qquad (1)$$

and $$V_2 = 3I_a.1 \qquad\qquad (2)$$

Solving eq (1) and (2), we get $G_{12} = \dfrac{V_2}{V_1} = \dfrac{-3}{7}$ **Ans.**

Problem 9.6. For the network of the following figure and the element values specified, determine $\alpha_{12} = I_2 / I_1$.

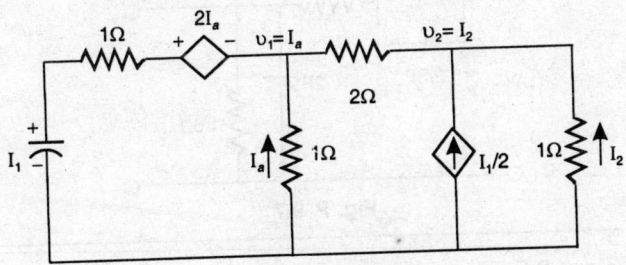

Fig. P. 9.6

Solution:

The above circuit can be drawn as

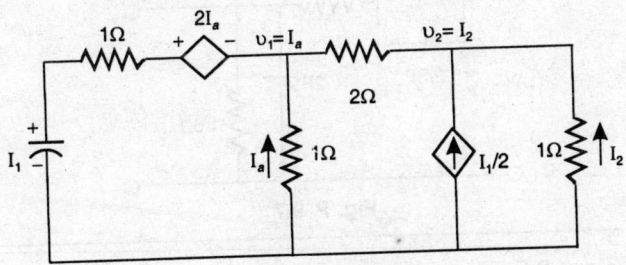

Fig. P. 9.6 (a)

$$\frac{v_1 - (I_1 + 2I_a)}{1} + \frac{v_1}{1} + \frac{v_1 - v_2}{2} = 0$$

or

$$\frac{I_a - (I_1 + 2I_a)}{1} + \frac{I_a}{1} + \frac{I_a - I_2}{2} = 0$$

or

$$0.5 I_a - 0.5 I_2 - I_1 = 0 \qquad (1)$$

and

$$\frac{v_2 - v_1}{2} + \frac{I_1}{2} + \frac{v_2}{1} = 0$$

or

$$\frac{I_2 - I_a}{2} + \frac{I_1}{2} + I_2 = 0$$

or

$$1.5 I_2 - 0.5 I_a + 0.5 I_1 = 0 \qquad (2)$$

Adding eq (1) eq (2), we get

$$I_2 = 0.5 I_1$$

or $\qquad \dfrac{I_2}{I_1} = 0.5$

$\therefore \qquad \alpha_{12} = 0.5$ **Ans.**

Problem 9.7. For the given network, show that

$$Y_{12} = \dfrac{K(s+1)}{(s+2)(s+4)}$$

and determine the value and sign of K.

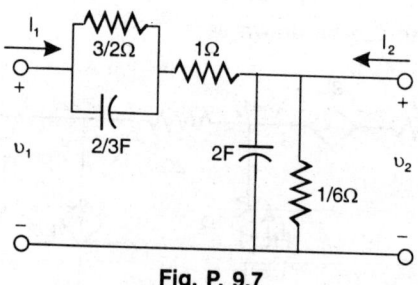

Fig. P. 9.7

$Z_1 = \dfrac{3}{2(s+1)} + 1$

$Z_2 = \dfrac{1}{2(s+3)}$

$V_2 = \dfrac{V_1 Z_2}{Z_1 + Z_2} = -\dfrac{I_2}{6} = \dfrac{6 Z_2}{Z_1 + Z_2} = -\dfrac{I_2}{V_1}$

or $Y_{12} = \dfrac{I_2}{V_1} = -\dfrac{6 Z_2}{Z_1 + Z_2}$

$Y_{12} = \dfrac{-6 \times \dfrac{1}{2(s+3)}}{4 \dfrac{(s^2+6s+8)}{(2s+2)(2s+6)}}$

$= \dfrac{-3(s+1)(s+3) \times 4/(s+3)}{4(s^2+6s+8)} = \dfrac{-3(s+1)}{(s^2+6s+8)}$

$$Y_{12} = \frac{-3(s+1)}{(s+2)(s+4)}$$

$\therefore$ $K = -3$ **Ans.**

Problem 9.8. For the network shown in Fig. P. 9.8, show that the voltage-ratio transfer function is

$$G_{12} = \frac{(s^2+1)^2}{5s^4+5s^2+1}$$

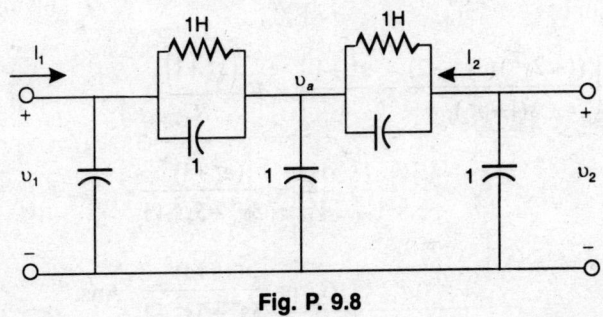

Fig. P. 9.8

Solution:

$$\frac{V_2 - V_a}{\dfrac{s}{1+s^2}} + \frac{V_2}{1/s} = 0$$

or $\quad \left(\dfrac{1+s^2}{s}\right)(V_2 - V_a) + sV_2 = 0$

or $\quad (1+s^2)(V_2 - V_a) + s^2 V_2 = 0$

or $\quad (1+2s^2)V_2 = (1+s^2)V_a$

or $\quad V_a = \dfrac{(1+2s^2)V_2}{1+s^2}$

$$\frac{V_1 - V_a}{\dfrac{s}{1+s^2}} + sV_a = I_1$$

$$\frac{(V_a - V_2)}{\dfrac{s}{1+s^2}} + sV_a + \frac{(V_a - V_1)}{\dfrac{s}{1+s^2}} = 0$$

or $$\frac{(1+s^2)(V_a - V_2)}{s} + sV_a + \frac{(1+s^2)}{s}(V_a - V_1) = 0$$

or $$V_a\left(\frac{3s^2 + 2}{s}\right) = (V_1 + V_2)\left(\frac{s^2 + 1}{s}\right)$$

or $$\frac{(1+2s^2)V_2}{1+s^2}\frac{(3s^2 + 2)}{s} = (V_1 + V_2)\left(\frac{s^2 + 1}{s}\right)$$

or $$V_2\left(\frac{(1+2s^2)(3s^2 + 2)}{s(1+s^2)} - \frac{s^2 + 1}{s}\right) = V_1\frac{(s^2 + 1)}{S}$$

or $$\frac{V_2}{V_1} = \frac{(s^2 + 1)^2}{5s^4 + 5s^2 + 1}$$

$\therefore$ $$G_{12} = \frac{(s^2 + 1)^2}{5s^4 + 5s^2 + 1} \quad \textbf{Ans.}$$

Problem 9.9. For each of the networks shown in Fig. P. 9.9 (a) – (g), connect a voltage source V_1 to port 1 and designate polarity references for V_2 at post 2. For each network, determine $G_{12} = V_2/V_1$.

Fig. P. 9.9 (a), (b), (c), (d)

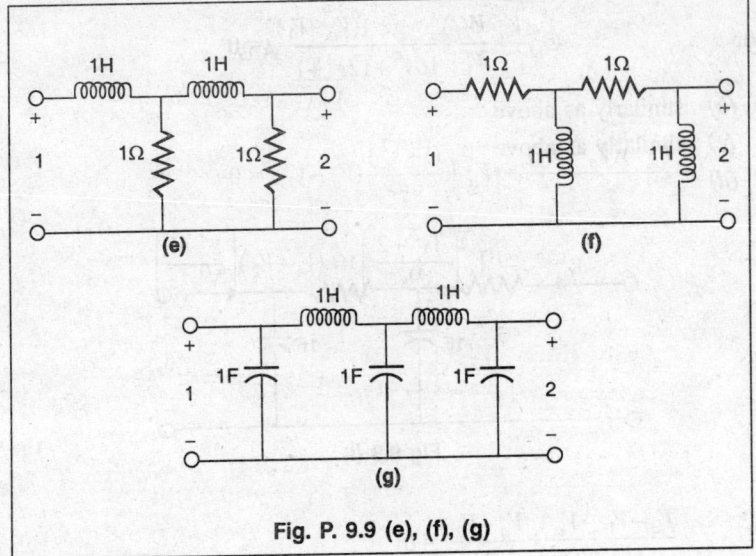

Fig. P. 9.9 (e), (f), (g)

Solution:

(a)

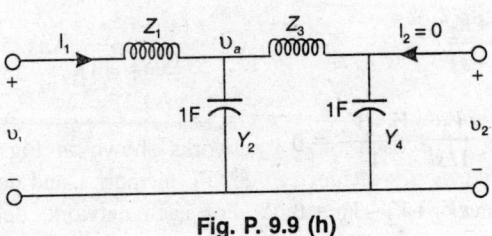

Fig. P. 9.9 (h)

$$I_b = Y_4 V_2 = 2s V_2$$

$$V_a = V_2 + I_b Z_3 = V_2 + 2s V_2 \times 2s = V_2(1+4s^2)$$

$$\therefore \quad V_a = V_2(1+4s^2)$$

$$I_1 = I_b + Y_2 V_a = 2s V_2 + 2s(V_2(1+4s^2))$$
$$= V_2(2s + 2s(1+4s^2)) = 2s V_2(1+1+4s^2)$$

or $\quad I_1 = 2s V_2(2+4s^2) = 4s V_2(1+2s^2)$

$$\therefore \quad V_1 = V_a + Z_1 I_1 = V_2(1+4s^2) + 2s(4s V_2(1+2s^2))$$

$$V_1 = V_2(1+4s^2 + 8s^2(1+2s^2))$$

or
$$G_{12} = \frac{V_2}{V_1} = \frac{1}{16s^4 + 12s^2 + 1} \quad \text{Ans.}$$

(b) similarly as above

(c) similarly as above

(d)

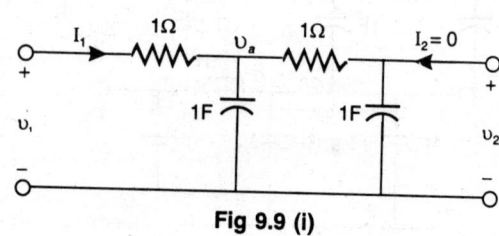

Fig 9.9 (i)

$$\frac{V_a - V_1}{1} + \frac{V_a}{1/s} + \frac{V_a - V_2}{1} = 0$$

or
$$(s+2)V_a - V_1 - V_2 = 0$$

or
$$V_a = \frac{V_1 + V_2}{(2+s)} \qquad (1)$$

$$\frac{V_2}{1/s} + \frac{V_2 - V_a}{1} = 0$$

or
$$sV_2 + V_2 - V_a = 0$$

or
$$V_2(1+s) = V_a = \frac{V_1 + V_2}{(2+s)}$$

or
$$(2+s)(1+s)V_2 - V_2 = V_1$$

or
$$V_2((2+s)(1+s) - 1) = V_1$$

or
$$\frac{V_2}{V_1} = \frac{1}{(2+s)(1+s) - 1} = \frac{1}{s^2 + 3s + 2 - 1}$$

or
$$\frac{V_2}{V_1} = \frac{1}{s^2 + 3s + 1}$$

or
$$G_{12} = \frac{1}{s^2 + 3s + 1} \quad \textbf{Ans.}$$

(e) similarly as above any methods

(f) similarly as above any methods.

(g)

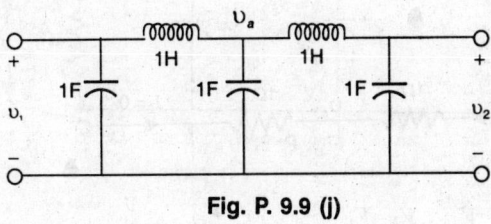

Fig. P. 9.9 (j)

$$\frac{V_1}{1/s} + \frac{V_1 - V_a}{s} = 0$$

or

$$\left(\frac{s^2 + 1}{s}\right)V_1 = \frac{V_a}{s}$$

or

$$(s^2 + 1)V_1 = V_a \qquad (1)$$

$$\frac{V_a - V_1}{s} + \frac{V_a}{1/s} + \frac{V_a - V_2}{s} = 0$$

or

$$\left(\frac{s^2 + 2}{s}\right)V_a - \frac{V_1}{s} - \frac{V_2}{s} = 0$$

or

$$(s^2 + 2)V_a = V_1 + V_2 \qquad (2)$$

or

$$(s^2 + 2)(s^2 + 1)V_1 = V_1 + V_2$$

or

$$V_1(s^4 + 3s^2 + 2 - 1) = V_2$$

or

$$\frac{V_2}{V_1} = \frac{1}{s^4 + 3s^2 + 1}$$

or

$$G_{12} = \frac{V_2}{V_1} = \frac{1}{s^4 + 3s^2 + 1} \textbf{ Ans.}$$

Problem 9.10. For the network given in Fig. P. 9.9 (a), terminate port 2 in a 1-Ω resistor and connect a voltage source at port 1. Let I_1 be the current in the voltage source and I_2 be the current in the 1-Ω load. Assign reference directions for each. For this network, compute $G_{12} = V_2/V_1$ and $\alpha_{12} = I_2/I_1$.

Solution:

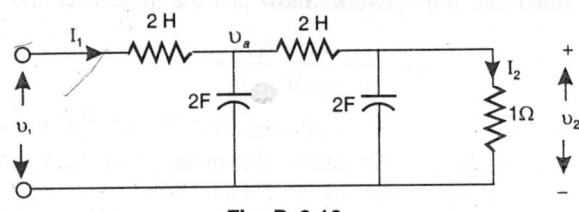

Fig. P. 9.10

$$\frac{V_a - V_1}{2s} + \frac{V_a}{1/2s} + \frac{V_a - V_2}{2s} = 0$$

$$(2s^2 + 1)V_a = \frac{V_1 + V_2}{2} \qquad (1)$$

$$\frac{V_2 - V_a}{2s} + \frac{V_2}{1/2s} + \frac{V_2}{1} = 0$$

or $$V_2\left(1 + 2s + \frac{1}{2s}\right) = \frac{V_a}{2s}$$

or $$V_2\left(\frac{2s + 4s^2 + 1}{2s}\right) = \frac{V_a}{2s}$$

or $$V_2\left(4s^2 + 2s + 1\right) = V_a = \frac{V_1 + V_2}{2(2s^2 + 1)}$$

$$\therefore \quad G_{12} = \frac{V_2}{V_1} = \frac{1}{4s^2\left(4s^2 + 2s + 3 + \dfrac{1}{s} + \dfrac{1}{4s^2}\right)}$$

$$I_1 = \frac{V_a}{1/2s} + \frac{V_a - V_2}{2s} = 2sV_a + \frac{V_a}{2s} - \frac{V_2}{2s}$$

$$= V_a\left(\frac{4s^2 + 1}{2s}\right) - \frac{V_2}{2s} = \frac{V_1 + V_2}{2(2s^2 + 1)}\frac{(4s^2 + 1)}{2s} - \frac{V_2}{2s}$$

$$I_2 = (8s^3 + 4s^2 + 4s + 1)V_2$$

$$\therefore \quad V_2 = I_2 \cdot 1 = I_2$$

$$\therefore \quad I_1 = (8s^3 + 4s^2 + 4s + 1)I_2$$

or $$\frac{I_2}{I_1} = \frac{1}{(8s^3 + 4s^2 + 4s + 1)} \quad \textbf{Ans.}$$

Problem 9.11. The network shown in (a) of Fig. P. 9.11 is known as a *shunt peaking* network. Show that the impedance has the form

$$Z(s) = \frac{K(s-z_1)}{(s-p_1)(s-p_2)}$$

and determine z_1, p_1 and p_2 in terms of R, L and C. If the poles and zeros of $Z(s)$ have the locations shown in (b) of the figure with $Z(jo) = 1$, find the values for R, L and C.

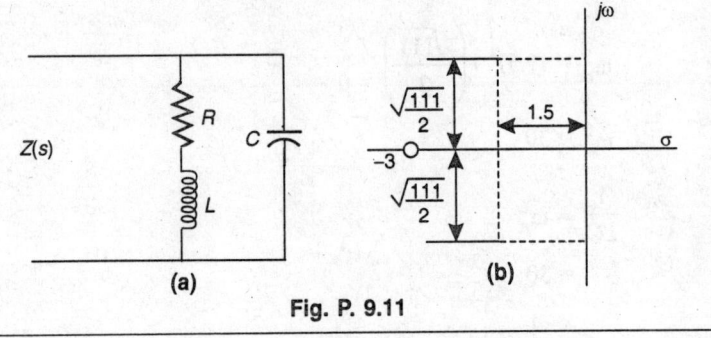

Fig. P. 9.11

Solution:

$$Z(s) = \frac{(sL+R)}{s^2 CL + sRC + 1}$$

$$= \frac{L}{LC} \frac{(s+R/L)}{\left(s^2 + s\dfrac{R}{L} + \dfrac{1}{LC}\right)}$$

$$Z(s) = \frac{1}{C} \frac{(s+R/L)}{\left(s^2 + s\dfrac{R}{L} + \dfrac{1}{LC}\right)}$$

$$\therefore \qquad Z(s) = \frac{1}{C} \frac{(s+R/L)}{(s-p_1)(s-p_2)} \qquad Z_1 = -R/L$$

$$p_1 = -\frac{R}{2L} + \sqrt{\left(\frac{R}{2L}\right)^2 - \frac{1}{LC}}$$

$$p_2 = -\frac{R}{2L} - \sqrt{\left(\frac{R}{2L}\right)^2 - \frac{1}{LC}}$$

$$\because \qquad Z_1 = -3$$

$$\therefore \qquad -\frac{R}{L} = -3$$

or $\qquad R = 3L$

$\because \qquad Z(j0) = 1 = R$

$$\therefore \qquad L = \frac{1}{3}H$$

$$\because \qquad \omega_n^2 = (1.5)^2 + \left(\frac{\sqrt{111}}{2}\right)^2$$

$$\therefore \qquad \omega_n^2 = 30$$

$$\therefore \qquad \frac{1}{LC} = \omega_n^2$$

$$= 30$$

or $\qquad C = \frac{1}{10}F$ **Ans.**

Problem 9.12. A system has a transfer function with a pole at $s = -3$ and a zero which may be adjusted in position at $s = -a$. The response of this system to a step input has a term of the form $K_1 e^{-3t}$. Plot the value of K_1 as a function of a for values of a between 0 and 5.

Solution:

$$Z(s) = \frac{(s+a)}{(s+3)}$$

$$R(t) = K_1 e^{-3t} u(t)$$

$$(R/s) = \frac{K_1}{s(s+3)}$$

$$K_1 = \frac{+(3-a)H}{3}$$

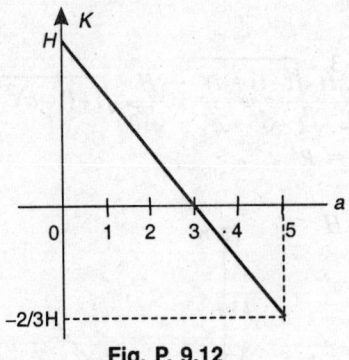

Fig. P. 9.12

or $\qquad K_1 = H\left(1 - \dfrac{a}{3}\right)$

for $\qquad a = 0$

$\qquad K_1 = H$

for $\qquad a = 5$

$\qquad K_1 = \dfrac{-2}{3}H$ **Ans.**

Problem 9.13. A system has a transfer function with poles at $s = -1 \pm j1$ and at $s = -3$, and a zero which may be adjusted in position at $s = -a$. One term of the response of this system to a step input is of the form. $K_3 e^{-t} \sin(t + \phi)$. Plot the value of K_3 as a function of a for values of a between 0 and 5.

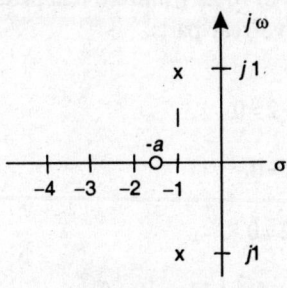

Fig. P. 9.13

Solution:

Similarly, as above

$$K_3 = \frac{2H\sqrt{1+(1-a)^2}}{\sqrt{2}\times\sqrt{5}\times 2} = \frac{H}{\sqrt{10}}\sqrt{1+(1-a)^2}$$

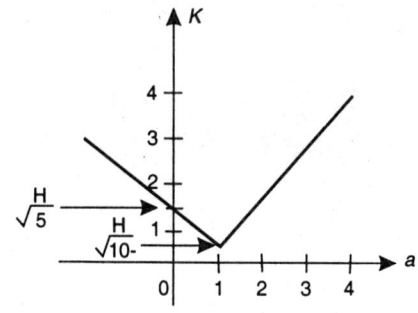

Fig. P. 9.13 (a)

For $a = 0$

$$K_3 = \frac{H}{\sqrt{5}}$$

For $a = 5$

$$K_3 = H\sqrt{\frac{17}{10}} \quad \textbf{Ans.}$$

Problem 9.14. Apply the Routh-Hurwitz criterion to the following equations and determine: (a) the number of roots with positive real parts, (b) the number of roots with zero real parts, and (c) the number of roots with negative real parts.

(a) $4s^3 + 7s^2 + 7s + 2 = 0$

(b) $s^3 + 3s^2 + 4s + 1 = 0$

(c) $5s^3 + s^2 + 6s + 2 = 0$

(d) $s^5 + 2s^4 + 2s^3 + 4s^2 + 11s + 10 = 0$

Solution:

(A) $4s^3 + 7s^2 + 7s + 2 = 0$

$$
\begin{array}{c|cc}
s^3 & 4 & 7 \\
s^2 & 7 & 2 \\
s^1 & \dfrac{7^2 - 8}{7} & \\
s^0 & 2 &
\end{array}
$$

No sign is changing

(a) There is no root with positive real parts.

(b) There is no root with zero real parts.

(c) There are three roots with negative real parts.

(B) $s^3 + 3s^2 + 4s + 1 = 0$

Similarly as above.

(C) $5s^3 + s^2 + 6s + 2 = 0$

$$
\begin{array}{c|cc}
s^3 & 5 & 6 \\
s^2 & 1 & 2 \\
s^1 & -4 & \\
s^0 & 2 &
\end{array}
$$

There is no change in sign.

So   (a) There are two roots with + ve real parts.

(b) There is no root with zero real parts.

(c) There is 1 root with – ve real parts.

(D) $s^5 + 2s^4 + 2s^3 + 4s^2 + 11s + 10 = 0$

$$
\begin{array}{c|cc}
s^5 & 1 & 21 \\
s^4 & 2 & 410 \\
s^3 & \epsilon & 6 \\
s^2 & 4 - \dfrac{12}{\epsilon} & 10 \\
s^1 & 6 & \\
s^0 & 10 &
\end{array}
$$

There is no sign change. But there is zero ϵ So

(a) There is no root with + ve real parts.

(b) There is no root with zero real parts.

(c) There are five root with – ve real parts.

Problem 9.15. Given the equation $s^3 + 5s^2 + Ks + 1 = 0$ (a) For what range of values of K will the roots of the equation have negative real parts? (b) Determine the value of K such that the real part vanishes.

Solution:

$$
\begin{array}{c|cc}
s^3 & 1 & K \\
s^2 & 5 & 1 \\
s^1 & \dfrac{5K-1}{5} & \\
s^0 & 1 &
\end{array}
$$

(a) $\dfrac{5K-1}{5} > 0$ or $5K > 1$

   or $K > \dfrac{1}{5}$, so that roots has – ve real parts.

(b) $5K \geq 1$ or $K \geq \dfrac{1}{5}$

   or $K = \dfrac{1}{5}$ **Ans.**

Problem 9.16. Repeat the tests of Prob. 9.14 for the following equations:

(a) $720s^5 + 144s^4 + 214s^3 + 38s^2 + 10s + 1 = 0$

(b) $25s^5 + 105s^4 + 120s^3 + 120s^2 + 20s + 1 = 0$

(c) $s^5 + 5.5s^4 + 14.5s^3 + 8s^2 - 19s - 10 = 0$

(d) $s^5 - s^4 - 2s^3 + 2s^2 - 8s + 8 = 0$

(e) $s^6 + 1 = 0$

Solution:

(A)

$$
\begin{array}{c|ccc}
s^5 & 720 & 214 & 10 \\
s^4 & 144 & 38 & 1 \\
s^3 & 24 & 4.368 & \\
s^2 & 11.78 & 1 & \\
s^1 & 2.334 & & \\
s^0 & 1 & &
\end{array}
$$

There is two change in sign. So

(a) No. of roots with + ve real parts is zero.
(b) No. of roots with zero real parts is zero.
(c) No. of roots with – ve real parts is 5.

(B) Similarly as above.

(C)

$$
\begin{array}{c|ccc}
s^5 & 1 & 14.5 & -19 \\
s^4 & 5.5 & 8 & -10 \\
s^3 & 13.05 & -17.18 & \\
s^2 & 15.26 & -10 & \\
s^1 & -8.63 & & \\
s^0 & -10 & &
\end{array}
$$

There is two change in sign. So

(a) No. of roots with + ve real parts is 1.
(b) No. of roots with zero real parts is 0.
(c) No. of roots with – ve real parts is 4.

(D)

$$
\begin{array}{c|ccc}
s^5 & 1 & -2 & -8 \\
s^4 & -1 & 2 & 8 \\
s^3 & 0 & 0 &
\end{array}
$$

So, differentiating the above equation

$$\frac{d}{ds}(-s^4 + 2s^2 + 8) = -4s^3 + 4s + 0$$

$$
\therefore \quad
\begin{array}{c|ccc}
s^5 & 1 & -1 & -8 \\
s^4 & -1 & 2 & 8 \\
s^3 & -4 & 4 & \\
s^2 & 1 & 8 & \\
s^1 & 36 & & \\
s^0 & 8 & &
\end{array}
$$

There is two changes in sign and there is one row zero. So

    (a) No. of roots with + ve real parts is 2.

    (b) No. of roots with – ve real parts is 2.

    (c) No. of roots with zero real parts is 1.

(E) $\qquad\qquad s^6 + 1 = 0$

$\because \qquad\qquad a^3 + b^3 = (a+b)(a^2 + b^2 - ab)$

Let $\qquad\qquad a = s^2, b = 1^2$

$\therefore \qquad (s^2)^3 + (1^2)^3 = (s^2 + 1)\left((s^2)^2 + (1^2)^2 - s^2\right) = 0$

or $\qquad\qquad s^2 + 1 = 0$

$\therefore \qquad\qquad\qquad s = \pm j1 \quad \therefore s_1 = +j1, \ s_2 = -j$

$\qquad\qquad s^2(s^2 - 1) + 1 = 0$

or $\qquad\qquad s^2(s^2 - 1) = -1$

or $\qquad\qquad s^4 - s^2 + 1 = 0$

$$
\begin{array}{c|cc}
s^4 & 1 & -\ 1 \\
s^3 & 0 & 0 \\
s^3 & 4 & -2 \\
s^2 & -1 & 1 \\
s^1 & \in \\
s^0 & 1 \\
\end{array}
$$

There is two changes in sign. So

Hence,

    (a) No. of roots with + ve real parts is 2.

    (b) No. of roots with zero real parts is 2.

    (c) No. of roots with – ve real part is 2.

Problem 9.17. For the following polynomials, (1) determine the number of zeros in the right half of the s plane, (2) determine the number of zeros on the imaginary axis of the s plane. Show method.

(a) $2s^6 + 2s^5 + 3s^4 + 2s^3 + 4s^2 + 3s + 2 = p_1(s)$

(b) $s^6 + 2s^5 + 6s^4 + 10s^3 + 11s^2 + 12s + 6 = p_2(s)$

(c) $2s^6 + 2s^5 + 4s^4 + 3s^3 + 5s^2 + 4s + 1 = p_3(s)$

Solution:

(A) $2s^6 + 2s^5 + 3s^4 + 2s^3 + 4s^2 + 3s + 2 = p_1(s)$

s^6	2	3	4	2
s^5	2	2	3	
s^4	1	1	2	
s^3	$\in$	-1		
s^2	1	2		
s^1	-1			
s^0	2			

(1) *There is two change in sign,* so No. of zero on the right half of the s plane is 2.

∴   (a) No. of zero in the right half of s-plane = 2.

     (b) No. of zero on the imaginary axis = 0.

(B) $s^6 + 2s^5 + 6s^4 + 10s^3 + 11s^2 + 12s + 6 = p_2(s)$

s^6	1	6	11	6
s^5	2	10	12	
s^4	1	5	6	
s^3	0	-10		
s^3	4	10		
s^2	2.5	6		
s^1	1/2.5			
s^0	6			

$$\frac{d}{ds}(s^4 + 5s^2 + 6) = 4s^2 + 10s$$

There is no sign change and $s^4 + 5s^2 + 6$ is a factor of the polynomial

$$s^4 + 5s^2 + 6 = (s^2 + 2)(s^2 + 3), \text{ gives}$$

   (a) No. of zero in the right half of s-plane = 0.

   (b) No. of zero on the Imaginary axis of s-plane = 4.

(C) $2s^6 + 2s^5 + 4s^4 + 3s^3 + 5s^2 + 4s + 1 = p_3(s)$

s^6	2	4	5	1
s^5	2	3	4	
s^4	1	1	1	
s^3	1	2		
s^2	-1	1		
s^1	2	6		
s^0	1			

(a) No. of zero on the right half of s-plane = 2.

(b) No. of zero on the imaginary axis of s-plane = 0. **Ans.**

Problem 9.18. Consider the equation

$$a_0 s^4 + a_1 s^3 + a_2 s^2 + a_3 s + a_4 = 0$$

Use the Routh-Hurwitz criterion to determine a set of conditions necessary in order that all roots of the equation have negative real parts. Assume that all coefficients in the equation are positive.

Solution:

s^4	a_0	a_2	a_4
s^3	a_1	a_3	
s^2	$\dfrac{a_1 a_2 - a_0 a_3}{a_1}$	a_4	
s^1	$\dfrac{(a_1 a_2 - a_0 a_3) a_3 - a_1^2 a_4}{(a_1 a_2 - a_0 a_3)}$		
s^0	a_4		

For all roots have – ve real parts, the necessary condition is

(1) a_0 to be + ve

(2) a_1 to be + ve

(3) $(a_1 a_2 - a_0 a_3) > 0$, i.e. $a_1 a_2 > a_0 a_3$

(4) $\dfrac{(a_1 a_2 - a_0 a_3) a_3 - a_1^2 a_4}{(a_1 a_2 - a_0 a_3)} > 0$ i.e., $(a_1 a_2 - a_0 a_3) a_3 > a_1^2 a_4$

(5) $a_4 > 0$ **Ans.**

Problem 9.19. For the network of Fig. P. 9.19,, let $R_1 = R_2 = 1\,\Omega$, $C_1 = 1F$ and $C_2 = 2\ F$. For what values of k will the network be stable? In other words, for what values of k will the roots of the characteristic equation have real parts in the left half of the s-plane?

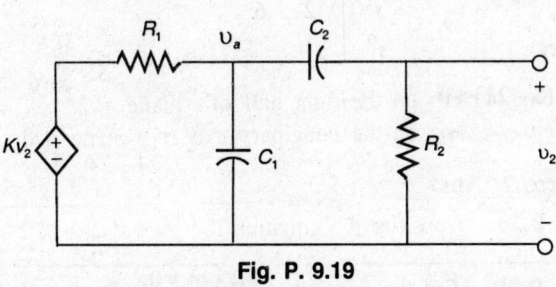

Fig. P. 9.19

Solution:

$$\frac{V_a - kV_2}{R_1} + \frac{V_a}{1/C_1 s} + \frac{V_a - V_2}{1/C_2 s} = 0$$

or

$$\frac{V_a - kV_2}{1} + \frac{V_a}{1/s} + \frac{V_a - V_2}{1/2s} = 0$$

or

$$V_2(2s + k) = V_a(1 + 3s) \qquad (1)$$

or

$$\frac{V_2}{R_2} + \frac{V_2 - V_a}{1/C_2 s} = 0$$

or

$$\frac{V_2}{1} + \frac{V_2 - V_a}{1/2s} = 0$$

or

$$V_2(1 + 2s) = 2sV_a \qquad (2)$$

Dividing equation (1) by equation (2), we get

$$\frac{(2s + k)}{(1 + 2s)} = \frac{(1 + 3s)}{2s}$$

or

$$2s(2s + k) = (1 + 3s)(1 + 2s)$$

or

$$4s^2 + 2sk = 1 + 5s + 6s^2$$

or $2s^2 + s(5-2k)+1=0$

$$
\begin{array}{ccc}
s^2 & 2 & 1 \\
s^1 & (5-2\dot{k}) & \\
s^0 & 1 &
\end{array}
$$

For, stability

$$(5-2k)>0$$

or $5 > 2k$

or $k<5/2$ **Ans.**

Problem 9.20.   For the network of Prob. 9.18, let $k = 2$, $C = 1$ F and $R_2 = 1\,\Omega$. Determine the relationship that must exist between R_1 and C_2 for the system to oscillate, that is, for the roots of the characteristic equation to be conjugate and have zero real parts.

Solution:

$$V_a\left(\frac{1}{R_1}+C_1s+C_2s\right) = V_2\left(\frac{2}{R_1}+C_2s\right)$$

or $$V_a\left(\frac{1}{R_1}+s+C_2s\right) = V_2\left(\frac{2}{R_1}+C_2s\right) \qquad (1)$$

$$\frac{V_2-V_a}{1/C_2s}+\frac{V_2}{R_2} = 0$$

or $$(V_2-V_a)C_2s+\frac{V_2}{1} = 0$$

or $$V_2(1+C_2s) = C_2sV_a \qquad (2)$$

Dividing equation (1) by equation (2) we get

$$s^2R_1C_2 +(R_1C_2 + R_1 -C_2)s+1 = 0$$

For the roots to be imaginary and have zero real parts coefficient of s is to be zero.

i.e., $$\left(R_1C_2 +R_1 -C_2)\right) = 0$$

or $\qquad R_1(1+C_2) = C_2$

or $\qquad R_1 = \dfrac{C_2}{1+C_2}$ **Ans.**

Problem 9.21. The network of the Fig. P. 21 represents a *phase-shift oscillator*. (a) Show that the condition necessary for oscillation is $g_m R_L \geq 29$. (b) Show that the frequency of oscillation when $g_m R_L = 29$ is $\omega_0 = 1/\sqrt{6}RC$.

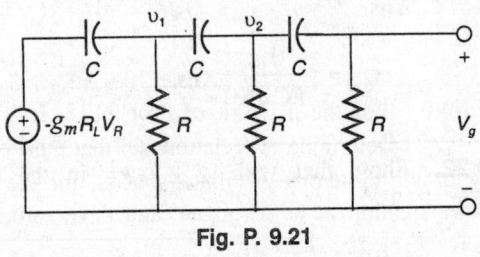

Fig. P. 9.21

Solution:

$$V_1(2sC+G) - V_2\,sC + g_m\,R_L\,sCV_g = 0 \tag{1}$$

$$V_2(2sC+G) - V_1 sc - V_g sC = 0 \tag{2}$$

$$V_g(sC+G) - V_2 sC = 0 \tag{3}$$

Here $G = 1/R\,\Delta$

$$
= \begin{vmatrix}
(2scC+G) & -sc & -g_m R_L sc \\
-sC & (2sC+G) & -sC \\
0 & -sC & sc+G
\end{vmatrix}
$$

$$= s^3(1C^2 - g_m R_L C^3) + s^2/6C^2 G + 5G^2 Cs + G^3 = 0$$

$$
\therefore \quad
\begin{array}{c|cc}
s^3 & \left(1C^2 - g_m R_L C^3\right) & 5G^2 c \\[4pt]
s^2 & \left(6C^2 G\right) & G^3 \\[4pt]
s^1 & \dfrac{(29 - g_m R_L)G^2}{6} & \\[8pt]
s^0 & G^3 &
\end{array}
$$

So, (a) circuit will oscillatory it g_m is assumed + ve so that

$$29 - g_m R_L \leq 0$$

or $\quad g_m R_L \geq 29$

(b) If $\quad g_m R_L = 29$

$$6GC^2 s^2 + G^3 = 0$$

$$6(sC)^2 + G^2 = 0$$

or $\quad 6(sC)^2 = -G^2, s = \pm \dfrac{G}{j\sqrt{6}}$

$\therefore \qquad \omega_o = \dfrac{1}{RC\sqrt{6}}$ **Ans.**

Problem 9.22. Show that with $Z_a Z_b = R_0^2$ in the bridged-T network of Fig. P. 9.22.

$$\frac{V_2}{V_1} = \frac{1}{1 + Z_a / R_0}$$

and the input impedance at port 1 is $Z_{in} = R_0$

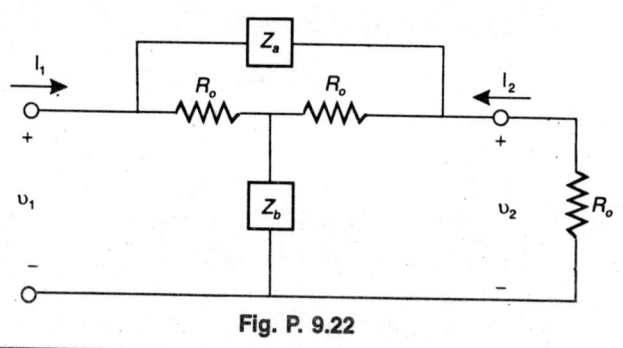

Fig. P. 9.22

Solution:

Given $\quad \dfrac{V_2}{V_1} = \dfrac{1}{1 + Z_a / R_0}$

or $\quad \dfrac{V_2}{V_1} = \dfrac{R_0}{R_0 + Z_a}$

$$Z_{in} = \frac{(R_0 + Z_a)(R_0 + Z_b)}{Z_a + Z_b + 2R_0} = \frac{R_0^2 + R_0(Z_a + Z_b) + Z_a Z_b}{Z_a + Z_b + 2R_0}$$

$\because \quad Z_a Z_b = R_0^2$, So

$$Z_{in} = \frac{R_0^2 + R_0(Z_a + Z_b) + R_0^2}{Z_a + Z_b + 2R_0} = \frac{R_0(2R_0 + Z_a + Z_b)}{(Z_a + Z_b + 2R_o)}$$

$$Z_{in} = R_o \quad \textbf{Ans.}$$

Problem 9.23. An active network is described by the characteristic equation $s^2 + (3 + 6K_1)s + 6K_2 = 0$

It is required that the network be stable and that no component of its response decay more rapidly than $K_1 e^{-3t}$. Show that these conditions are satisfied if $K_2 > 0$, $|K_1| < \dfrac{1}{2}$, and $K_2 > 3K_1$. Crosshatch the area of permitted value of K_1 and K_2 in the $K_1 - K_2$ plane.

Solution:

$$s^2 + (3 + 6K_1)s + 6K_2 = 0$$

or

$$
\begin{array}{lll}
s^2 & 1 & 6K_2 \\
s^1 & (3 + 6K_1) & \\
s^0 & 6K_2 &
\end{array}
$$

For stability: $6K_2 > 0$ and $3 + 6K_1 > 0$

i.e. $K_2 > 0$ and $K_1 > -\dfrac{1}{2}$

or $|K_1| < \dfrac{1}{2}$

When the function is shifted by 3 to the right

$$(s-3)^2 + (3 + 6K_1)(s-3) + 6K_2 = 0$$

$$s^2 - 6s + 9 + 3s + 6K_1 s - 9 - 18K_1 + 6K_2 = 0$$

or

$$s^2 + (6K_1 - 3)s + (6K_2 - 18K_1) = 0$$

$$\begin{array}{c|cc}
s^2 & 1 & 6K_2 - 18K_1 \\
s^1 & 6K_1 - 3 \\
s^0 & 6K_2 - 18K_1
\end{array}$$

For both roots in the left half of s plane.

$$6K_1 - 3 \geq 0$$

or $\quad 6K_1 \geq 3$

or $\quad K_1 \geq 1/2$

and $\quad 6K_2 - 18K_1 \geq 0$

or $\quad 6K_2 \geq 18K_1$

or $\quad K_2 \geq 3K_1$ **Ans.**

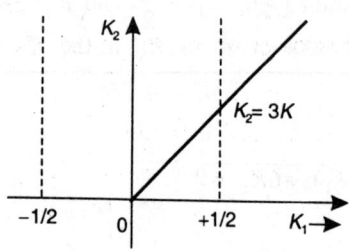

Fig. P. 9.22 (a)

Two Port Parameters

Problem 10.1. The Z-parameters of the two port network in Fig. P. 10.1 (a) are $Z_{11} = 4\,s$, $Z_{12} = Z_{21} = 3\,s$ and $Z_{22} = 9\,s$. Find the input current i_1 for $v_s = \cos 1000\,t\,(\text{V})$ by using the open circuit impedance terminal characteristic equations of N, together with KCL equations at nodes A, B and C.

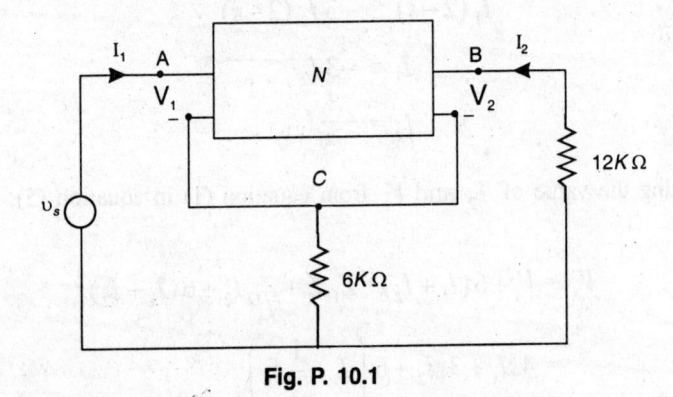

Fig. P. 10.1

Solution:

From the two port network N.

$$V_1 = Z_{11}I_1 + Z_{12}I_2 \tag{1}$$

$$V_2 = Z_{21}I_1 + Z_{22}I_2 \tag{2}$$

At node C $\qquad \dfrac{V_c}{6} = I_1 + I_2$

$\therefore \qquad V_c = 6(I_1 + I_2) \tag{3}$

At node B
$$\frac{V_2 - V_c}{12} = I_2$$

or
$$\frac{V_2 - 6(I_1 + I_2)}{12} = I_2$$

$$V_2 = 12I_2 + 6(I_1 + I_2)$$

$$V_2 = v6I_1 + 18I_2 \qquad (4)$$

At node A
$$I_1 = \frac{V_s - V_1}{6} - I_2$$

or
$$6(I_1 + I_2) + V_1 = V_s \qquad (5)$$

Since equations (2) and (4) are same:

$$6I_1 + 18I_2 = Z_{21}I_1 + Z_{22}I_2 = 3sI_2 + 9sI_2$$

or
$$I_1(2-s) = I_2(3s-6)$$

$$I_1(2-s) = I_2 3(s-2)$$

$$I_1(2-s) = -3I_2(2-s)$$

$$I_1 = -3I_2$$

$$I_2 = -\frac{1}{3}I_1$$

Putting the value of I_2 and V_1 from equation (1) in equation (5), we get

$$V_s = V_1 + 6(I_1 + I_2) = Z_{11}I_1 + Z_{12}I_2 + 6(I_1 + I_2)$$

$$= 4sI_1 + 3sI_2 + 6\left(I_1 - \frac{1}{3}I_1\right)$$

$$= 4sI_1 + 3s\left(\frac{-I_1}{3}\right) + 6\left(\frac{2I_1}{3}\right) = 3sI_1 + 4I_1$$

$$V_s = (3s+4)I_1$$

For
$$V_s = \cos 1000t = \cos 10^3 t$$

$$Z' = 3s + 4 = (3j + 4)\,k\Omega$$

$\therefore$
$$1\angle 0° = 5\angle 36.9°\,I_1$$

or
$$I_1 = \frac{1\angle 0°}{5\angle 36.9°}$$

$$I_1 = 0.2\angle -36.9°$$

$\therefore$
$$I_1 = 0.2\cos(1000t - 36.9°)\ \text{Ans.}$$

Problem 10.2. Express the reciprocity criteria in terms of h, g and T parameters.

Solution:

For h-parameters:
$$V_1 = h_{11}I_1 + h_{12}V_2$$
$$I_2 = h_{21}I_1 + h_{22}V_2$$

With output terminals shorted, $V_2 = 0$

$\therefore$ $\qquad\qquad\qquad V_1 = h_{11}I_1$ $\qquad\qquad\qquad$ (1)

and $\qquad\qquad\qquad I_2 = h_{21}I_1 + h_{22}V_2$

for $\qquad\qquad\qquad V_2 = 0,\ -I_2 = h_{21}I_1$ $\qquad\qquad$ (2)

$\qquad\qquad\qquad\qquad$ [direction of I_2 is reversed]

Dividing eqn (2) by (1) $\quad \dfrac{I_2}{V_1} = \dfrac{-h_{21}}{h_{11}}$ $\qquad\qquad$ (3)

Now, the excitation and response are interchanged such that voltage V_2 is applied at output terminal while input terminals are short circuited. The direction of I_1 is now reversed. Thus

$$0 = -h_{11}I_1 + h_{12}V_2$$

or $\qquad\qquad\qquad \dfrac{I_1}{V_2} = \dfrac{h_{12}}{h_{11}}$ $\qquad\qquad\qquad$ (4)

Assuming $V_1 = V_2$, following the principle of reciprocity, the left hand sides of equations (3) and (4) will be same. Thus $h_{21} = -h_{12}$

exhibits the condition of reciprocity.

For ABCD parameters: $V_1 = AV_2 - BI_2$

$$I_1 = CV_2 - DI_2$$

with excitation V_1 and shorting the output port, i.e. $V_2 = 0$ gives

$$V_1 = -BI_2$$

or $$\frac{I_2}{V_1} = -\frac{1}{B} \qquad\qquad (5)$$

Now, with interchange of excitation and response, the voltage source V_2 is at the output port while short circuit current I_1 is obtained at input port. Thus

$$0 = AV_2 - BI_2 \text{ and } I_1 = CV_2 - DI_2$$

These equations give

$$I_2 = \frac{AV_2}{B} \text{ and } I_1 = CV_2 - \frac{AD}{B}V_2$$

or $$I_1 = V_2\left(\frac{BC - AD}{B}\right)$$

or $$\frac{I_1}{V_2} = \frac{BC - AD}{B} \qquad\qquad (6)$$

From eqns (5) and (6) the reciprocity condition follows when

$$\frac{-1}{B} = \frac{BC - AD}{B} \left[i.e., \frac{I_2}{V_1} = \frac{I_1.}{V_2}\right]$$

$\therefore$ $$\frac{-1}{B} = \frac{BC - AD}{B}$$

or $AD - BC = 1$

Thus $AD - BC = 1$ exhibits the condition of reciprocity in ABCD parameters.

For g-parameters: $I_1 = g_{11}V_1 + g_{12}I_2$

$$V_2 = g_{21}V_1 + g_{22}I_2$$

Similarly the reciprocity relation in term of g-parameters is obtained by putting $V_1 = 0$, gives.

$$\frac{I_1}{I_2} = g_{12} \text{ and } \frac{V_2}{-I_2} = g_{22}$$

or $$\frac{I_1}{V_2} = \frac{-g_{12}}{g_{22}} \tag{7}$$

Now interchanging the excitation and response, with $V_2 = 0$. The direction of I_2 get reversed.

Thus $$0 = g_{21}V_1 - g_{22}I_2$$

or $$g_{22}I_2 = g_{21}V_1$$

$\therefore$ $$\frac{I_2}{V_1} = \frac{g_{21}}{g_{22}} \tag{8}$$

For the condition of reciprocity

$$\frac{I_1}{V_2} = \frac{I_2}{V_1}$$

$\therefore$ $$\frac{-g_{12}}{g_{22}} = \frac{g_{21}}{g_{22}}$$

or $$g_{12} = -g_{21}$$

Thus, $g_{12} = -g_{21}$ exhibits the condition of reciprocity in g parameters.

Thus, for reciprocity:

(1) $h_{12} + h_{21} = 0$ for h-parameter.

(2) $AD - BC = 1$ for ABCD parameter.

(3) $g_{12} + g_{21} = 0$ for g-parameter. Ans.

Problem 10.3. Find the T-parameters of a two-port device whose Z-parameters are $Z_{11} = s$, $Z_{12} = Z_{21} = 10s$ and $Z_{22} = 100s$

Solution:

$$Z_{11} = s, \ Z_{12} = Z_{21} = 10s \text{ and } Z_{22} = 100s$$

Z-parameters in terms of T-parameters is given as:

$$A = \frac{Z_{11}}{Z_{21}}, B = \frac{D_{22}}{Z_{21}}, C = \frac{1}{Z_{21}}$$

and $$D = \frac{Z_{22}}{Z_{21}}$$

where $$D_{22} = Z_{11}Z_{22} - Z_{12}Z_{21}$$

$$\therefore \qquad A = \frac{Z_{11}}{Z_{21}} = \frac{s}{10s} = \frac{1}{10} = 0.1$$

$$B = \frac{Z_{11}Z_{22} - Z_{12}Z_{21}}{Z_{21}} = \frac{100s^2 - 10s \times 10s}{10s} = 0$$

$$C = \frac{1}{Z_{21}} = \frac{1}{10s} = 10^{-1}/s$$

$$D = \frac{Z_{22}}{Z_{21}} = \frac{100s}{10s} = 10$$

$$\therefore \quad A = 0.1, B = 0, C = 1/10s, D = 10 \text{ Ans.}$$

Problem 10.4. Find the T-parameters of a two-port device whose Z-parameter are $Z_{11} = 10^6 s, Z_{12} = Z_{21} = 10^7 s$ and $Z_{22} = 10^8 s$.

Solution:

$$A = \frac{Z_{11}}{Z_{21}} = \frac{10^6 s}{10^7 s} = 0.1$$

$$B = \frac{Z_{11}Z_{22} - Z_{12}Z_{21}}{Z_{21}} = \frac{10^{14} s - 10^{14} s}{10^7 s} = 0$$

$$C = \frac{1}{Z_{21}} = \frac{1}{10^7 s}$$

$$D = \frac{Z_{22}}{Z_{21}} = \frac{10^8 s}{10^7 s} = 10$$

$$\therefore \quad A=0.1, B=0, C=\frac{1}{10^7 s}, D=10 \text{ Ans.}$$

Problem 10.5. The Z-parameters of a two port device N are $Z_{11}=Ks, Z_{12}=Z_{21}=10\,ks$ and $Z_{22}=100\,ks$. A 1Ω resistor is connected across the output port (Fig. P. 10.5). (a) Find the input impedance $Z_{in}=V_1/I_1$ and construct its equivalent circuit. (b) Give the values of the element for $k=1$ and 10^6.

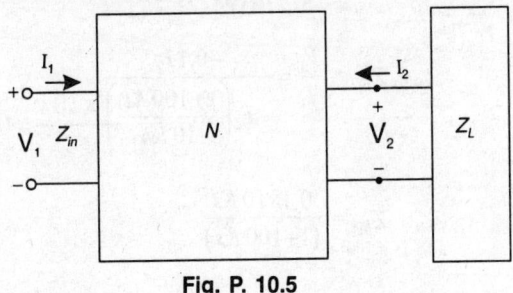

Fig. P. 10.5

Solution:

$$Z_{11} = Ks, Z_{12}=Z_{21}=10\,Ks, Z_{22}=100\,Ks$$

$$\therefore \quad A = \frac{Z_{11}}{Z_{21}}=\frac{Ks}{10\,Ks}=0.1$$

$$B = \frac{Z_{11}\,Z_{22}-Z_{12}\,Z_{21}}{Z_{21}}=\frac{100\,K^2\,s^2-100\,K^2\,s^2}{10\,Ks}=0$$

$$C = \frac{1}{Z_{21}}=\frac{1}{10\,Ks}$$

$$D = \frac{Z_{22}}{Z_{21}}=\frac{100\,Ks}{10\,Ks}=10$$

$$\therefore \quad A = 0.1, \ B = 0, \ C = 1/10\ Ks, \ D = 10$$

T-paramoters are defined by:

$$V_1 = AV_2-B\,I_2$$

$$I_1 = CV_2-D\,I_2$$

$$\therefore \quad V_1 = 0.1V_2$$

and $\qquad I_1 = \dfrac{1}{10\,ks}V_2 - 10\,I_2$

Since $\qquad V_2 = -Z_L\,I_2 = -I_2 \qquad\qquad [\text{as } R = Z_L^{\cdot} = 1\,\Omega]$

$\therefore \qquad\qquad V_1 = -0.1\,I_2 \qquad\qquad\qquad\qquad (1)$

$$I_1 = \dfrac{1}{10\,Ks}(-I_2) - 10\,I_2$$

$$I_1 = -I_2\left(\dfrac{1 + 100\,Ks}{10\,Ks}\right) \qquad\qquad (2)$$

(a) $\therefore \qquad\qquad Z_{in} = \dfrac{V_1}{I_1} = \dfrac{-0.1\,I_2}{-I_2\left(\dfrac{1 + 100\,Ks}{10\,Ks}\right)}$

or, $\qquad\qquad Z_{in} = \dfrac{0.1 \times 10\,Ks}{(1 + 100\,Ks)}$

or, $\qquad\qquad Z_{in} = \dfrac{Ks}{1 + 100\,Ks}$

or, $\qquad\qquad Z_{in} = \dfrac{1}{100 + 1/ks}$

Its equivalent circuit is:

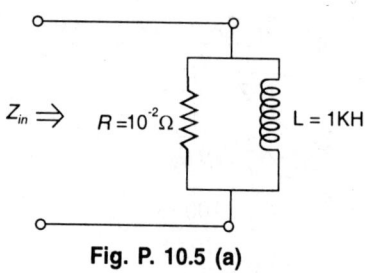

$Z_{in} \Rightarrow \qquad R = 10^{-2}\,\Omega \qquad L = 1\,KH$

Fig. P. 10.5 (a)

(b) For, $\qquad\qquad K = 1,\ R = \dfrac{1}{100}\,\Omega,\ L = 1\,\text{H}$

For $\qquad\qquad K = 10^6,\ R = \dfrac{1}{100}\,\Omega,\ L = 10^6\,\text{H} \quad$ **Ans.**

Problem 10.6. The device N in Fig. P. 10.6. is specified by its following Z-parameters: $Z_{22} = N^2 Z_{11}$ and $Z_{12} = Z_{21} = \sqrt{Z_{11} Z_{22}}$ $= N Z_{11}$. Find $Z_{in} = V_1 / I_1$ when a load Z_L is connected to the output terminal. Show that if $Z_{11} \gg Z_L / N^2$, we have impedance scaling such that $Z_{in} = Z_L / N^2$.

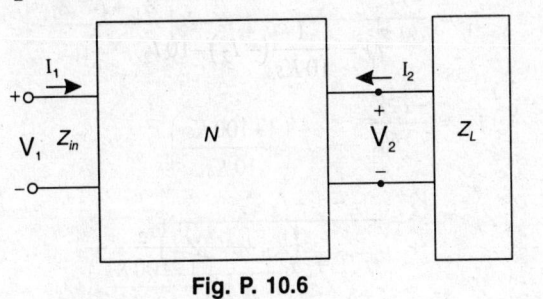

Fig. P. 10.6

Solution:

The T-parameters are defined by

$$V_1 = AV_2 - BI_2 \tag{1}$$

$$I_1 = CV_2 - DI_2 \tag{2}$$

$$A = \frac{Z_{11}}{Z_{21}} = \frac{Z_{11}}{N Z_{11}} = \frac{1}{N}$$

$$B = \frac{Z_{11} Z_{22} - Z_{12} Z_{21}}{Z_{21}} = \frac{N^2 Z_{11}^2 - N^2 Z_{11}}{N Z_{11}}$$

$$\therefore \quad B = 0$$

$$C = \frac{1}{Z_{21}} = \frac{1}{N Z_{11}}$$

$$D = \frac{Z_{22}}{Z_{21}} = \frac{N^2 Z_{11}}{N Z_{11}} = N$$

$$\therefore \quad A = \frac{1}{N}, B = 0, C = \frac{1}{N Z_{11}}, D = N$$

$$\therefore \quad V_1 = \frac{1}{N} V_2 - 0 \qquad \text{[From 1]}$$

$$\therefore \quad V_1 = \frac{V_2}{N}$$

$$I_1 = \frac{1}{N Z_{11}} V_2 - N I_2 \qquad \text{[From 2]}$$

$\because \qquad V_2 = -I_2 Z_L$

$\therefore \qquad I_1 = \frac{-I_2 Z_L}{N Z_{11}} - N I_2, \quad I_1 = -I_2 \left(\frac{Z_L + N^2 Z_{11}}{N Z_{11}} \right)$

and $\qquad V_1 = \frac{-I_2 Z_L}{N}$

$\therefore \qquad Z_{in} = \frac{V_1}{I_1} = \dfrac{-I_2 Z_L}{N \times -I_2 \dfrac{\left(Z_L + N^2 Z_{11} \right)}{N Z_{11}}}$

or, $\qquad Z_{in} = \dfrac{Z_L Z_{11}}{Z_L + N^2 Z_{11}} \quad$ or, $\quad Z_{in} = \dfrac{Z_L}{N^2 + Z_L / Z_{11}}$

When $Z_{11} \gg Z_L / N^2$, then Z_L / Z_{11} is very small and can be neglected.

$\therefore \qquad Z_{in} = \dfrac{Z_L}{N^2 + 0}$

$\therefore \qquad Z_{in} = \dfrac{Z_L}{N^2} \quad$ **Ans.**

Problem 10.7. Find the Z-parameters in the circuit of Fig. P. 10.7. (*Hint:* Use the series connection rule.)

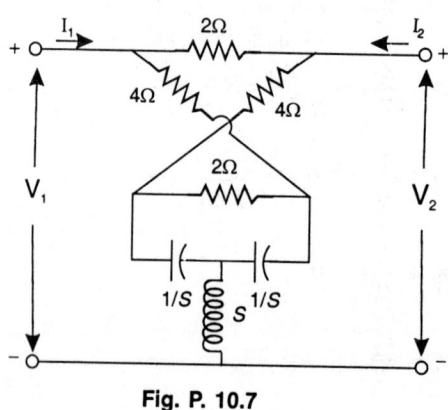

Fig. P. 10.7

Solution:

In the above circuit there are two circuits connected in series. The first circuit is:

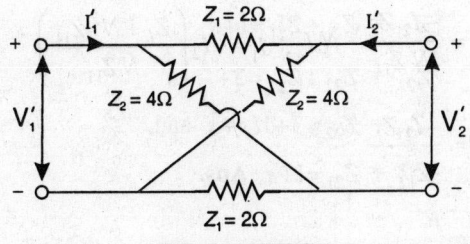

Fig. P. 10.7 (a)

Its Z-parameter is given as

$$Z_{11}' = Z_{22}' = \frac{Z_1 + Z_2}{2} = \frac{2+4}{2} = 3\,\Omega$$

$$Z_{12}' = Z_{21}' = \frac{Z_2 - Z_1}{2} = \frac{4-2}{2} = 1\,\Omega$$

And other circuit is:

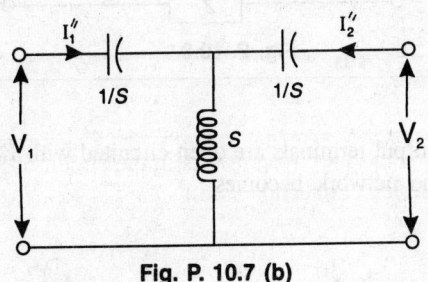

Fig. P. 10.7 (b)

Its Z-parameter is given as:

$$Z_{11}'' = Z_a + Z_c = \frac{1}{s} + s$$

$$Z_{12}'' = Z_{21}'' = Z_c = s$$

$$Z_{22}'' = Z_b + Z_c = \frac{1}{s} + s$$

when two Z-parameter circuits are connected in series. Its equivalent Z-parameter is given as:

$$Z_{11} = Z'_{11} + Z''_{11} = 3 + \frac{1}{s} + s$$

$$Z_{12} = Z'_{12} + Z''_{12} = 1 + s$$

$$Z_{21} = Z'_{21} + Z''_{21} = 1 + s$$

$$Z_{22} = Z'_{22} + Z''_{22} = 3 + \frac{1}{s} + s$$

$\therefore$ $Z_{11} = Z_{22} = 3 + 1/s + s$ and.

$$Z_{12} = Z_{21} = 1 + s \quad \textbf{Ans.}$$

Problem 10.8. Find Z-parameters of the two-port circuit of Fig. P. 10.8.

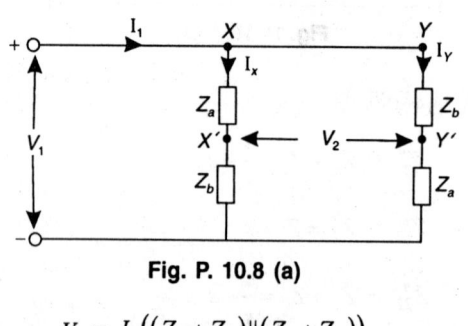

Fig. P. 10.8

Solution:

In the first step, output terminals are open circuited with V_1 at the input terminal. Now, the network becomes

Fig. P. 10.8 (a)

Hence $V_1 = I_1 \left((Z_a + Z_b) \| (Z_b + Z_a) \right)$

or, $V_1 = I_1 \left(\dfrac{Z_a + Z_b}{2} \right)$

or, $\qquad \dfrac{V_1}{I_1} = \dfrac{Z_a + Z_b}{2}$ at $I_2 = 0$

Since $\qquad \left. \dfrac{V_1}{I_1} \right|_{I_2=0} = Z_{11}$

Hence $\qquad Z_{11} = \dfrac{Z_a + Z_b}{2}$

Similarly $\qquad \left. \dfrac{V_2}{I_1} \right|_{I_2=0} = Z_{21}$

Now $\qquad V_2 = V_{x'} - V_{y'} = \left(V_1 - I_x Z_a \right) - \left(V_1 - I_y Z_b \right) = I_y Z_b - I_x Z_a$

Since $\qquad I_x = \dfrac{I_1 \left(Z_a + Z_b \right)}{\left(Z_a + Z_b \right) + \left(Z_a + Z_b \right)} = \dfrac{I_1}{2}$

and $\qquad I_y = \dfrac{I_1}{2}$

hence $\qquad V_2 = \dfrac{I_1}{2} Z_b - \dfrac{I_1}{2} Z_a.$

or, $\qquad \dfrac{V_2}{I_1} = \dfrac{Z_b}{2} - \dfrac{Z_a}{2} = \dfrac{Z_b - Z_a}{2}$

or, $\qquad \left. \dfrac{V_2}{I_1} \right|_{I_2=0} = Z_{21} = \dfrac{Z_b - Z_a}{2}$

Since the given network is symmetrical, hence $Z_{12} = Z_{21}$ and $Z_{11} = Z_{22}$

$\therefore \qquad Z_{11} = Z_{22} = \dfrac{Z_a + Z_b}{2}$

and $\qquad Z_{12} = Z_{21} = \dfrac{Z_b - Z_a}{2}$ **Ans.**

Problem 10.9. Compute the short-circuit admittance parameters of the circuit in Fig. P. 10.9.

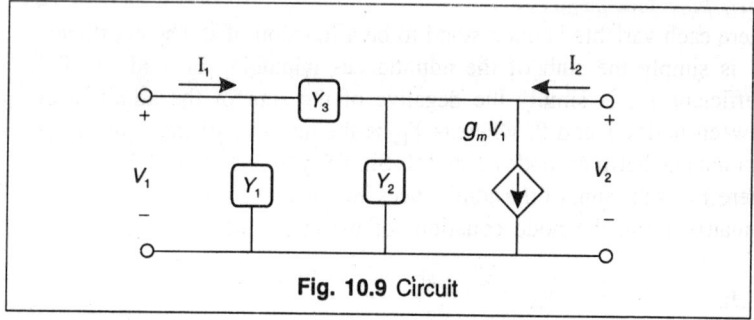

Fig. 10.9 Circuit

Solution:

From the above figure, using KCL, we have

$$I_1 = Y_1V_1 + Y_3(V_1 - V_2) = (Y_1 + Y_3)V_1 - Y_3V_2 \tag{1}$$

and $I_2 = g_mV_1 + Y_2V_2 + Y_3(V_2 - V_1) = (g_m - Y_3)V_1 + (Y_2 + Y_3)V_2 \tag{2}$

In matrix form, these equations can be expressed as

$$\begin{bmatrix} I_1 \\ I_2 \end{bmatrix} = \begin{bmatrix} Y_1 + Y_3 & -Y_3 \\ g_m - Y_3 & Y_2 + Y_3 \end{bmatrix} \begin{bmatrix} V_1 \\ V_2 \end{bmatrix}$$

Hence, short-circuit admittance parameters, we have,

$$y_{11} = Y_1 + Y_3, y_{12} = -Y_3, y_{21} = g_m - Y_3 \text{ and } y_{22} = Y_2 + Y_3$$

Problem 10.10. Compute the *y*-parameters for the circuit of Fig. P. 10.10

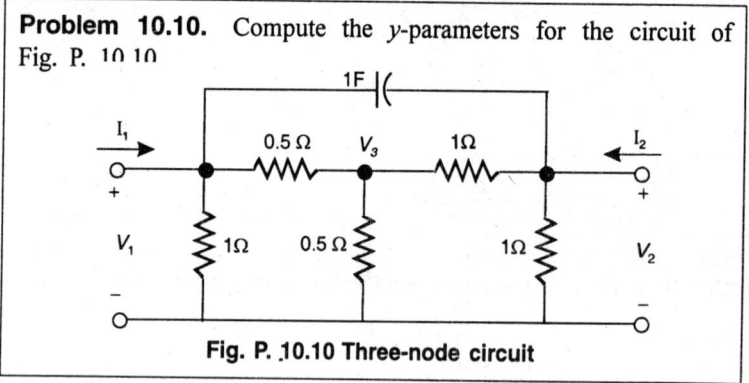

Fig. P. 10.10 Three-node circuit

Solution:

Consider, for example, node 1, in which the current I_1 must be determined. I_1 is of the form

$$I_1 = Y_{11}V_1 + Y_{12}V_2 + Y_{13}V_3$$

where each variable is understood to be a function of s. The coefficient Y_{11} is simply the sum of the admittances impinging on node 1. The coefficient Y_{12} is simply the negative of the sum of the admittances between nodes 1 and 2, whereas Y_{13} is the negative of the sum of the admittances between nodes 1 and 3. Similarly, $I_2 = Y_{21}V_1 + Y_{22}V_2 + Y_{23}V_3$ where Y_{22} is the sum of the admittances impinging on node 2, etc. Hence, in matrix form, the node equations of the circuit of Fig. P. 10.10 are

$$\begin{bmatrix} I_1 \\ I_2 \\ \hline 0 \end{bmatrix} = \begin{bmatrix} s+3 & -s & | & -2 \\ -s & s+2 & | & -1 \\ - & - & - & - \\ -2 & -1 & | & 5 \end{bmatrix} \begin{bmatrix} V_1 \\ V_2 \\ \hline V_3 \end{bmatrix} \equiv \begin{bmatrix} W_{11} & | & W_{12} \\ - & - & - \\ W_{21} & | & W_{21} \end{bmatrix} \begin{bmatrix} V_1 \\ V_2 \\ \hline V_3 \end{bmatrix}$$

(1)

This nodal equation matrix is symmetric because the *RLC* network contains no dependent sources. Using the method of matrix partitioning induced by equation (1) yields the *y*-parameter equations,

$$\begin{bmatrix} I_1 \\ I_2 \end{bmatrix} = \begin{bmatrix} W_{11} - W_{12}W_{22}^{-1}W_{21} \end{bmatrix} \begin{bmatrix} V_1 \\ V_2 \end{bmatrix}$$

$$= \left(\begin{bmatrix} s+3 & -s \\ -s & s+2 \end{bmatrix} - \frac{1}{5} \begin{bmatrix} 2 \\ 1 \end{bmatrix} [2 \quad 1] \right) [s+3][s+3] \begin{bmatrix} V_1 \\ V_2 \end{bmatrix}$$

$$\begin{bmatrix} I_1 \\ I_2 \end{bmatrix} = \begin{bmatrix} s+2.2 & -(s+0.4) \\ -(s+0.4) & s+1.8 \end{bmatrix} \begin{bmatrix} V_1 \\ V_2 \end{bmatrix}$$

Hence the *y*-parameter areas

$$y_{11} = s+2.2, \ Y_{12} = Y_{21} = -(s+0.4)$$
$$y_{22} = s+1.8$$

Problem 10.11. Compute the *y*-parameters of the circuit in Fig. P. 10.11

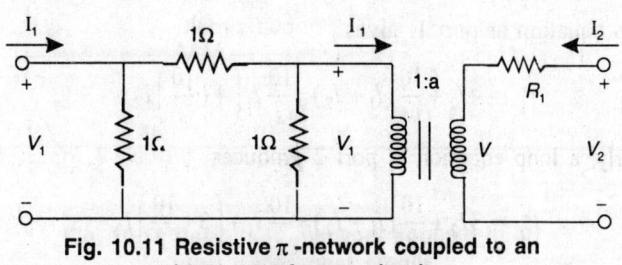

Fig. 10.11 Resistive π-network coupled to an ideal transformer circuit

Solution:

First, at port 1,

$$I_1 = V_1 + (V_1 - \hat{V}_1) = 2V_1 - \hat{V}_1 = 2V_1 - \frac{1}{a}V_2 \tag{1}$$

Now considering that $aI_2 = -\hat{I}_1$, a node equation at the primary of the transformer, gives

$$I_2 = -\frac{1}{a}\hat{I}_1 = -\frac{1}{a}[-\hat{V}_1 + (V_1 - \hat{V}_1)] = -\frac{1}{a}V_1 + \frac{2}{a^2}V_2 \tag{2}$$

where the last equality uses the relationship $a\hat{V}_1 = V_2$. Putting equations (1) and (2) in matrix form we have, y-parameter relationship expressed as,

$$\begin{bmatrix} I_1 \\ I_2 \end{bmatrix} = \begin{bmatrix} 2 & -\dfrac{1}{a} \\ -\dfrac{1}{a} & \dfrac{2}{a^2} \end{bmatrix} \begin{bmatrix} V_1 \\ V_2 \end{bmatrix}$$

Problem 10.12 Compute the z-parameters for the circuit of Fig. P. 10.12

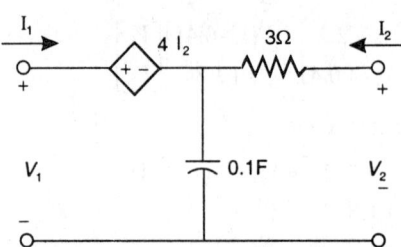

Fig. 10.12 SimpleT-circuit for computing z-parameters

Solution:

A loop equation at port 1, gives

$$V_1 = 4I_2 + \frac{10}{s}(I_1 + I_2) = \frac{10}{s}I_1\left(4 + \frac{10}{s}\right)I_2 \tag{1}$$

Similarly, a loop equation at port 2 produces

$$V_2 = 3I_2 + \frac{10}{s}(I_1 + I_2) = \frac{10}{s}I_1 + \left(3 + \frac{10}{s}\right)I_2 \tag{2}$$

Using equations (1) and equation (2), the z-paramters are

$$Z = \begin{bmatrix} z_{11} & z_{12} \\ z_{21} & z_{22} \end{bmatrix} = \begin{bmatrix} \dfrac{10}{s} & \dfrac{4s+10}{s} \\ \dfrac{10}{s} & \dfrac{3s+10}{s} \end{bmatrix} \Omega$$

Problem 10.13. Compute the z-parameters of the π-network of Fig. P. 10.13.

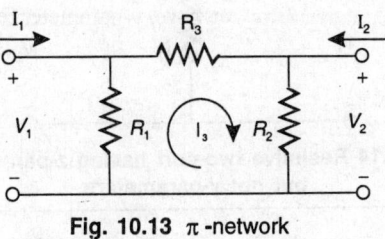

Fig. 10.13 π-network

Solution:

Three loop equations are:

$$V_1 = R_1 I_1 - R_1 I_3 \tag{1}$$

$$V_2 = R_2 I_2 - R_2 I_3 \tag{2}$$

and

$$0 = -R_1 I_1 + R_2 I_2 + (R_1 + R_2 + R_3) I_3 \tag{3}$$

Putting equations (1), (2) and (3) in matrix form and partitioning the matrix appropriately, we have

$$\begin{bmatrix} V_1 \\ V_2 \\ \hline 0 \end{bmatrix} = \begin{bmatrix} R_1 & 0 & | & -R_1 \\ 0 & R_2 & | & R_2 \\ \hline -R_1 & R_2 & | & R_1 + R_2 + R_3 \end{bmatrix} \begin{bmatrix} I_1 \\ I_2 \\ \hline I_3 \end{bmatrix}$$

Hence, using the matrix partitioning formula, we obtain

$$\begin{bmatrix} z_{11} & z_{12} \\ z_{21} & z_{22} \end{bmatrix} = \begin{bmatrix} R_1 & 0 \\ 0 & R_2 \end{bmatrix} - \frac{1}{R_1 + R_2 + R_3} \begin{bmatrix} -R_1 \\ R_2 \end{bmatrix} \begin{bmatrix} -R_1 & R_2 \end{bmatrix}$$

$$= \begin{bmatrix} R_1 & 0 \\ 0 & R_2 \end{bmatrix} - \frac{1}{R_1 + R_2 + R_3} \begin{bmatrix} R_1^2 & -R_1 R_2 \\ -R_1 R_2 & R_2^2 \end{bmatrix}$$

$$= \frac{1}{R_1 + R_2 + R_3} \begin{bmatrix} R_1(R_2 + R_3) & R_1 R_2 \\ R_1 R_2 & R_2(R_1 + R_3) \end{bmatrix}$$

Problem 10.14. Compute the z-parameters of the circuit of Fig. P. 10.14 and determine whether there are y-parameters.

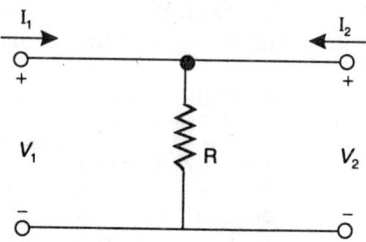

Fig. 10.14 Resistive two-port having z-parameters, but not y-parameters.

Solution:

By inspection, the z-parameter two-port equations are easily computed as

$$\begin{bmatrix} V_1 \\ V_2 \end{bmatrix} = \begin{bmatrix} R & R \\ R & R \end{bmatrix} \begin{bmatrix} I_1 \\ I_2 \end{bmatrix}$$

The z-parameter matrix Z is singular since det $[Z] = R^2 - R^2 = 0$. Because the Z-matrix does not have an inverse, the circuit fails to have y-parameters. Also consider Fig. P. 10.14 (a). Because $V_2 = 0$, there is also a short circuit across V_1, i.e. $V_1 = 0$, making the ratio

$$y_{11} = \frac{I_1}{V_1}\bigg|_{V_2 = 0} \quad \text{undefined.}$$

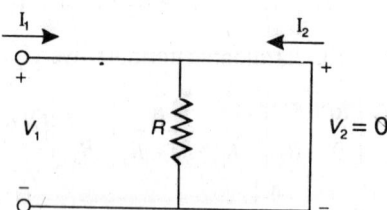

Fig. 10.14 (a) Equivalent circuit for computing y_{11} in which port 2 is short-circuit, so that $V_2 = 0$.

Problem 10.15. Find the *h*-parameters of the two-port in Fig. P. 10.15

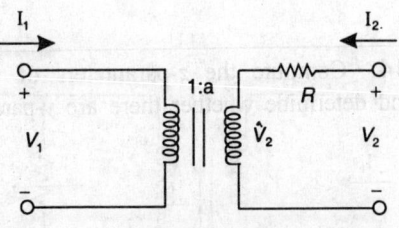

Fig. P. 10.15 Ideal transformer circuit

Solution:

From the primary and secondary voltage relationship of an ideal transformer, we have

$$V_1 = \frac{1}{a}\hat{V}_2$$

By KVL at port 2, $\hat{V}_2 = V_2 - RI_2$ then,

$$V_1 + \frac{R}{a}I_2 = \frac{1}{a}V_2 \qquad (1)$$

From the primary and secondary current relationship of an ideal transformer, we have

$$I_2 = \frac{-1}{a}I_1 \qquad (2)$$

In matrix form, equations (1) and (2) can be written as

$$\begin{bmatrix} 1 & \dfrac{R}{a} \\ 0 & 1 \end{bmatrix} \begin{bmatrix} V_1 \\ I_2 \end{bmatrix} = \begin{bmatrix} 0 & \dfrac{1}{a} \\ -\dfrac{1}{a} & 0 \end{bmatrix} \begin{bmatrix} I_1 \\ V_2 \end{bmatrix} \qquad (3)$$

Solving for the vector $\begin{bmatrix} V_1 & I_2 \end{bmatrix}^T$ produces the *h*-parameter equation

$$\begin{bmatrix} V_1 \\ I_2 \end{bmatrix} = \begin{bmatrix} 1 & \dfrac{R}{a} \\ 0 & 1 \end{bmatrix}^{-1} \begin{bmatrix} 0 & \dfrac{1}{a} \\ -\dfrac{1}{a} & 0 \end{bmatrix} \begin{bmatrix} I_1 \\ V_2 \end{bmatrix} = \begin{bmatrix} \dfrac{R}{a^2} & \dfrac{1}{a} \\ -\dfrac{1}{a} & 0 \end{bmatrix} \begin{bmatrix} I_1 \\ V_2 \end{bmatrix}$$

Problem 10.16. Find the *h*-parameters of the linear transformer circuit of Fig. P. 10.16

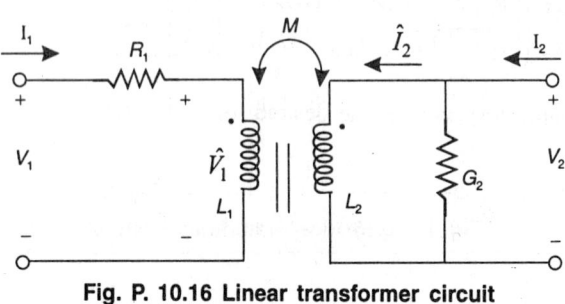

Fig. P. 10.16 Linear transformer circuit

Solution:

Writing a node equation at the primary leads to the relationship

$$\hat{I}_2 = I_2 - G_3 V_3 \tag{1}$$

From the defining equation for a linear transformer (coupled inductors), the voltage V_2 across port 2 satisfies

$$V_2 = L_2 s \hat{I}_2 + M s I_1 \tag{2}$$

Since $\hat{I}_2 = I_2 - G_2 V_2$ then

$$V_2 = L_2 s I_2 - G_2 L_2 s V_2 + M s I_1$$

Rearranging gives, $L_2 s I_2 = (1 + G_2 L_2 s) V_2 - M s I_1 \tag{3}$

A loop equation at port 1 implies that $V_1 = R_1 I_1 + \hat{V}_1$ or equivalently,

$$\hat{V}_1 = V_1 - R_1 I_1$$

From the coupled inductor relationship

$$\hat{V}_1 = L_1 s I_1 + M s \hat{I}_2$$

and since $\hat{I}_2 = I_2 - G_2 V_2$ we have

$$V_1 - R_1 I_1 = L_1 s I_1 + M s (I_2 - G_2 V_2)$$

Grouping like terms with V_1 and I_2 on the left of the equal sign and I_1 and V_2 on the right leads to

$$V_1 - M s I_2 = (L_1 s + R_1) I_1 - G_2 M s V_2 \tag{4}$$

Writing equations (3) and (4) in matrix form and solving for $\begin{bmatrix} V_1 & I_2 \end{bmatrix}^T$ we get,

$$\begin{bmatrix} 0 & L_2s \\ 1 & -Ms \end{bmatrix}\begin{bmatrix} V_1 \\ I_2 \end{bmatrix} = \begin{bmatrix} -Ms & G_2L_2s+1 \\ L_1s+R_1 & -G_2Ms \end{bmatrix}\begin{bmatrix} I_1 \\ V_2 \end{bmatrix} \tag{5}$$

Solving equation (5), we get the desired h-parameter equation

$$\begin{bmatrix} V_1 \\ I_2 \end{bmatrix} = \begin{bmatrix} 0 & L_2s \\ 1 & -Ms \end{bmatrix}^{-1}\begin{bmatrix} -Ms & G_2L_2s+1 \\ L_1s+R_1 & -G_2Ms \end{bmatrix}^{-1}\begin{bmatrix} I_1 \\ V_2 \end{bmatrix}$$

$$= \frac{1}{L_2s}\begin{bmatrix} Ms & L_2s \\ 1 & 0 \end{bmatrix}\begin{bmatrix} -Ms & G_2L_2s+1 \\ L_1s+R_1 & -G_2Ms \end{bmatrix}\begin{bmatrix} I_1 \\ V_2 \end{bmatrix}$$

$$\begin{bmatrix} V_1 \\ I_2 \end{bmatrix} = \frac{1}{L_2s}\begin{bmatrix} (L_1L_2-M^2)s^2+R_1L_2s & Ms \\ -Ms & G_2L_2s+1 \end{bmatrix}\begin{bmatrix} I_1 \\ V_2 \end{bmatrix}$$

Hence the h-parameters are

$$h_{11} = \frac{(L_1L_2-M^2)s^2+R_1L_2s}{L_2s}$$

$$h_{12} = \frac{M}{L_2}, \; h_{21} = \frac{-M}{L_2}$$

$$h_{22} = \frac{G_2L_2s+1}{L_2s}$$

Problem 10.17. For the network shown in Fig. P. 10.17, find the Z and Y parameters if they exist.

Fig. P. 10.17 (a), (b)

Solution:

For Z parameter

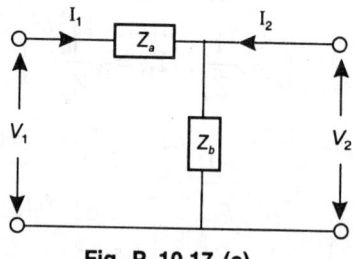

Fig. P. 10.17 (c)

Now, $V_1 = Z_a I_1 + Z_b (I_1 + I_2) = (Z_a + Z_b) I_1 + Z_b + I_2$

Also $V_2 = Z_b (I_1 + I_2) = Z_b I_1 + Z_b I_2$

Comparing with, $V_1 = Z_{11} I_1 + Z_{12} I_2$

$V_2 = Z_{21} I_1 + Z_{22} I_2$

We get $Z_{11} = (Z_a + Z_b)$

$Z_{12} = Z_b$

$Z_{21} = Z_b$

$Z_{22} = Z_b$

For Y parameters $I_1 = \dfrac{V_1 - V_2}{Z_a} = \dfrac{V_1}{V_a} - \dfrac{V_2}{Z_a}$

Now $(I_1 + I_2) Z_b = V_2$

$I_1 + I_2 = \dfrac{V_2}{Z_b}$

$I_2 = \dfrac{V_2}{Z_b} - I_1 = \dfrac{V_2}{Z_b} - \dfrac{(V_1 - V_2)}{Z_a}$

$= \left(-\dfrac{1}{Z_a}\right) V_1 + \left(\dfrac{1}{Z_a} + \dfrac{1}{Z_b}\right) V_2$

Comparing with $\quad I_1 = Y_{11}V_1 + Y_{12}V_2$

$$I_2 = Y_{21}V_1 + Y_{22}V_2$$

We get, $\qquad Y_{11} = \dfrac{1}{Z_a}, \quad Y_{21} = -\dfrac{1}{Z_a}$

$$Y_{12} = \dfrac{-1}{Z_a}, \quad Y_{22} = \left(\dfrac{1}{Z_a} + \dfrac{1}{Z_b}\right)$$

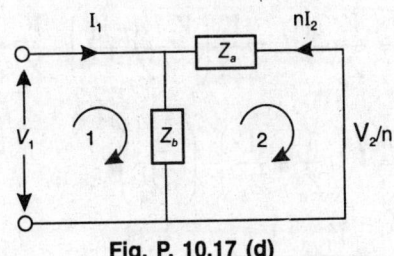

Fig. P. 10.17 (d)

(b) For Z parameter

KVL for loop (1) $\quad V_1 = Z_b (I_1 + nI_2) = Z_b I_1 + nZ_b I_2$

For loop (2) $\qquad \dfrac{V_2}{n} = nI_2 Z_a + Z_b (I_1 + nI_2)$

$$\dfrac{V_2}{n} = Z_b I_1 + (Z_a + Z_b) nI_2$$

$$V_2 = nZ_b I_1 + (Z_a + Z_b) n^2 I_2$$

Comparing V_1 and V_2 with equation representing Z parameters

$$V_1 = \cdot Z_{11} I_1 + Z_{12} I_2$$

$$V_2 = Z_{21} I_1 + Z_{22} I_2$$

We get $\qquad Z_{11} = Z_b \quad Z_{12} = nZ_b$

$$Z_{21} = nZ_b \quad Z_{22} = n^2 (Z_a + Z_b)$$

For Y parameter

By KVL, we get
$$\begin{bmatrix} Z_b & nZ_b \\ nZ_b & n^2(Z_a+Z_b) \end{bmatrix} \begin{bmatrix} I_1 \\ I_2 \end{bmatrix} = \begin{bmatrix} V_1 \\ V_2 \end{bmatrix}$$

$$I_1 + nI_2 = \frac{V_1}{Z_b}$$

$$nI_2 = \frac{V_2/n - V_1}{Z_a}$$

Hence,
$$I_1 = \frac{V_1}{Z_b} - nI_2 = \frac{V_1}{Z_b} - \left(\frac{V_2/n - V_1}{Z_a}\right) = \frac{V_1}{Z_b} + \frac{V_1}{Z_a} - \frac{V_2}{nZ_a}$$

$$= V_1\left(\frac{1}{Z_a} + \frac{1}{Z_b}\right) - \left(\frac{1}{nZ_a}\right)V_2 \qquad (1)$$

$$I_2 = \frac{-V_1}{nZ_a} + \frac{V_2}{n^2Z_a} \qquad (2)$$

Comparing equations (1) and (2) with the Y parameter equation we get,

$$Y_{11} = \frac{1}{Z_a + Z_b} \qquad Y_{12} = \left(\frac{1}{nZ_a}\right)$$

$$Y_{21} = \left(\frac{-1}{nZ_a}\right) \qquad Y_{22} = \frac{1}{n^2Z_a} \quad \textbf{Ans.}$$

Problem 10.18. Find the y and z parameters for the resistive network of Fig. P. 10.18.

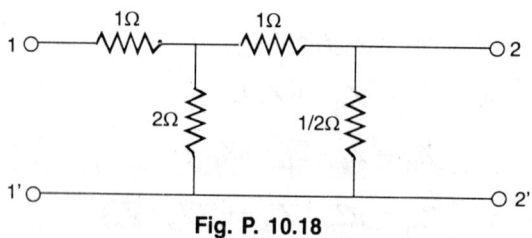

Fig. P. 10.18

Solution:

For Z parameter

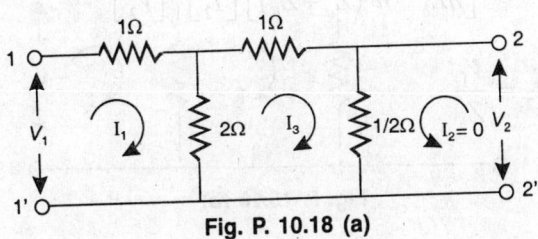

Fig. P. 10.18 (a)

Open circuit port 2 and apply V_1

Writing loop equation we get,

$$V_1 = 3I_1 - 2I_3$$

$$-2I_1 + (2+1+\frac{1}{2})I_3 = 0$$

$$-2I_1 + \frac{7}{2}I_3 = 0, \quad I_1 = \frac{7}{4}I_3$$

$$V_1 = 3I_1 - 2 \times \frac{4}{7}I_1 = \frac{13}{7}I_1 \tag{1}$$

$$V_2 = I_3 \times \frac{1}{2} = \frac{4}{7} \times \frac{I_1}{2} = \frac{2}{7}I_1 \tag{2}$$

From (1) $\qquad Z_{11} = \dfrac{V_1}{I_1}\Big|_{I_2=0} = \dfrac{13}{7}$

From (2) $\qquad Z_{21} = \dfrac{V_2}{I_1}\Big|_{I_2=0} = \dfrac{2}{7}$

Again, by definition, $\qquad Z_{12} = \dfrac{V_1}{I_2}\Big|_{I_1=0}$

$$Z_{22} = \dfrac{V_2}{I_2}\Big|_{I_1=0}$$

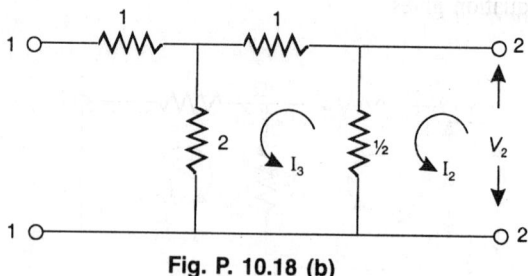

Fig. P. 10.18 (b)

Open circuit. Port 1 and apply V_2 to Port 2
Writing loop equation

$$V_2 = \frac{1}{2}(I_2 - I_3)$$

$$I_3\left(2+1+\frac{1}{2}\right) - \frac{1}{2}I_2 = 0$$

$$\frac{1}{2}I_2 = \frac{7}{3}I_3, I_3 = \frac{1}{7}I_2$$

$$V_2 = \frac{1}{2}\left(I_2 - \frac{1}{7}I_2\right) = \frac{3}{7}I_2 \tag{1}$$

Now
$$V_1 = 2I_3 = 2\times\frac{1}{7}I_2 = \frac{2}{7}I_2 \tag{2}$$

From (1) and (2) we get,

$$Z_{12} = \frac{V_1}{I_2} = \frac{2}{7}$$

$$Z_{22} = \frac{V_2}{I_2} = \frac{3}{7}$$

Hence
$$Z_{11} = \frac{13}{7} \quad Z_{12} = \frac{2}{7}$$

$$Z_{21} = \frac{2}{7} \quad Z_{22} = \frac{3}{7}$$

For Y parameter: Short circuit Port 2 and apply V_1 to Port 1. By KVL, the loop equation gives

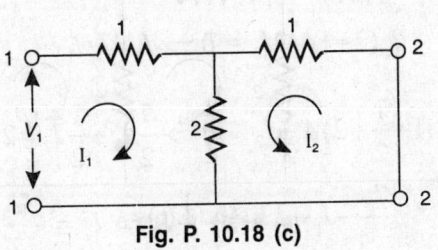

Fig. P. 10.18 (c)

$$\begin{bmatrix} 3 & 2 \\ 2 & 3 \end{bmatrix} \begin{bmatrix} I_1 \\ I_2 \end{bmatrix} = \begin{bmatrix} V_1 \\ 0 \end{bmatrix}$$

Now

$$3I_1 + 2I_2 = V_1$$

$$2I_1 + 3I_2 = 0$$

$$2I_1 = -3I_2, \quad I_2 = -\frac{2}{3} I_1$$

$$V_1 = 3I_1 + 2 \times \frac{-2}{3} I_1, \quad V_1 = \frac{5}{3} I_1$$

$$Y_{11} = \frac{I_1}{V_1} \bigg|_{v_2 = 0} = \frac{3}{5}$$

$$Y_{21} = \frac{I_2}{V_1} \bigg|_{v_2 = 0} = \frac{\dfrac{-2}{3} I_1}{5/3\, I_1} = \frac{-2}{5}$$

Again, short circuit Port 1 and apply V_2 to Port 2

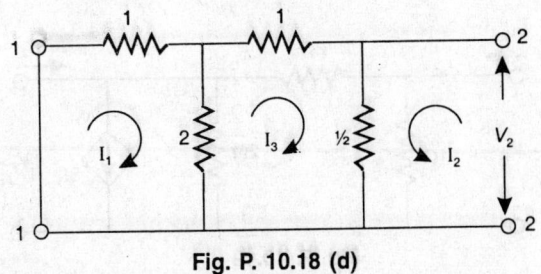

Fig. P. 10.18 (d)

$$V_2 = \frac{1}{2}(I_2 + I_3)$$

$$I_1(2+1) - 2I_3 = 0 \Rightarrow I_3 = \frac{3}{2}I_1$$

$$-2I_1 + I_3\left(1 + \frac{1}{2} + 2\right) + \frac{1}{2}I_2 = 0 \Rightarrow \frac{7}{2}I_3 + \frac{1}{2}I_2 = 2I_1$$

$$\frac{7 \times 3}{4}I_1 - 2I_1 = -\frac{1}{2}I_2 \Rightarrow I_2 = -\frac{13}{2}I_1$$

$\therefore$
$$V_2 = \frac{1}{2}\left(\frac{-13}{2}I_1 + I_3\right)$$

$$V_2 = \frac{1}{2}\left(\frac{-13}{2}I_1 + \frac{3}{2}I_1\right) = \frac{1}{2}\left(-\frac{10}{2}I_1\right) = \frac{-5}{2}I_1$$

$\therefore$
$$Y_{12} = \frac{I_1}{V_2}\bigg|_{V_1=0} = -2/5$$

$$Y_{22} = \frac{I_2}{V_2}\bigg|_{V_1=0} = \frac{\dfrac{-13}{2}I_1}{-5/2\,I_1} = \frac{13}{5}I_1$$

Hence
$$Y_{11} = \frac{3}{5} \quad Y_{12} = \frac{-2}{5}$$

$$Y_{21} = \frac{-2}{5} \quad Y_{22} = \frac{13}{5}$$

Problem 10.19. The network of Fig. P. 10.19 contains a current-controlled current source. For this network, find the y and z parameters.

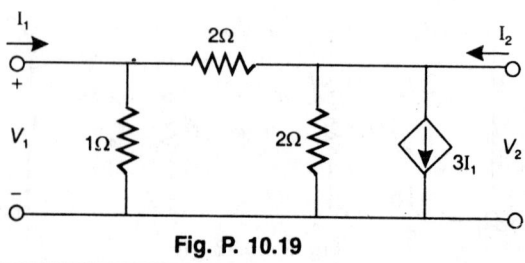

Fig. P. 10.19

Solution:

For Z parameters

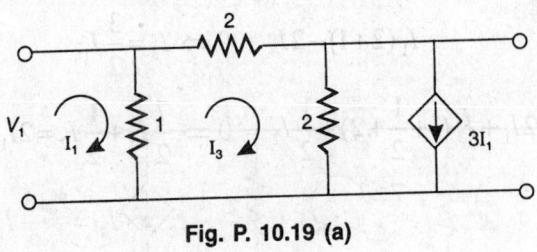

Fig. P. 10.19 (a)

$$Z_{11} = \frac{V_1}{I_1}\bigg|_{I_2 = 0}$$

$$Z_{21} = \frac{V_2}{I_1}\bigg|_{I_2 = 0}$$

Open circuit Port 2 and apply V_1 to Port 1.

$$V_1 = I_1 - I_3$$

$$5I_3 - I_1 - 6I_1 = 0$$

$$5I_3 - 7I_1 = 0, \quad I_3 = \frac{7}{5}I_1$$

$$V_1 = I_1 - \frac{7}{5}I_1 = \frac{-2}{5}I_1$$

$$Z_{11} = \frac{V_1}{I_1} = \frac{-2}{5}$$

$$Z_{21} = \frac{V_2}{I_1} = \frac{2I_3 - 2 \times 3I_1}{I_1} = \frac{2 \times \frac{7}{5}I_1 - 6I_1}{I_1} = \frac{(14 - 30)I_1}{5I_1} = \frac{\frac{-16}{5}I_1}{I_1}$$

$$Z_{21} = \frac{-16}{5}$$

Again
$$Z_{12} = \frac{V_1}{I_2}\bigg|_{I_1=0}$$

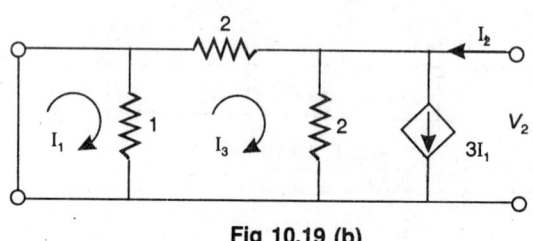

Fig 10.19 (b)

$$Z_{22} = \frac{V_2}{I_2}\bigg|_{I_1=0}$$

Open circuit Port 1 and apply V_2 to Port 2

$$V_2 = 2I_2 - 6I_1 - 2I_3 = 2I_2 - 2I_3 \quad (\because I_1 = 0)$$

also,
$$5I_3 - 2I_2 = 0$$

$$I_3 = \frac{2}{5}I_2$$

$$V_2 = 2I_2 - \frac{4}{5}I_2 = \frac{6}{5}I_2$$

$$Z_{22} = \frac{V_2}{I_2} = 6/5$$

$$Z_{12} = \frac{V_1}{I_2} = \frac{1 \times I_3}{I_3} = \frac{2}{5}$$

$$Z_{11} = -2/5 \quad Z_{12} = 2/5$$

$$Z_{21} = \frac{-16}{5} \quad Z_{22} = 6/5$$

For Y Parameters

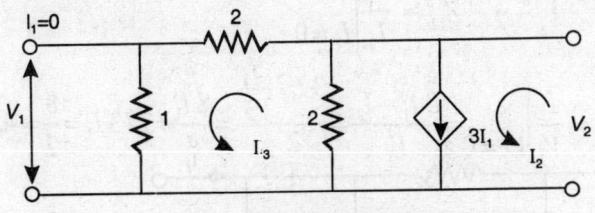

Fig. P. 10.19 (c)

Short in circuit Port 2 and apply V_1 to Port 1

Then,

$$V_1 = I_1 - I_3 \tag{1}$$

$$5I_3 - I_1 + 2I_2 - 6I_1 = 0$$

$$5I_3 - 7I_1 + 2I_2 = 0 \tag{2}$$

again

$$2(I_2 - 3I_1) + 2I_3 = 0$$

$$I_2 - 3I_1 + I_3 = 0$$

$$I_2 = 3I_1 - I_3 \tag{3}$$

$$Y_{11} = 3 \quad Y_{12} = -2/3$$

Putting equation (3) in equation (2) we get

$$5I_3 - 7I_1 + 2(3I_1 - I_3) = 0$$

$$5I_3 - 7I_1 + 6I_1 - 2I_3 = 0$$

$$5I_3 - I_1 - 2I_3 = 0$$

$$3I_3 - I_1 = 0$$

$$I_3 = \frac{I_1}{3} \tag{4}$$

Putting equation (4) in equation (1) we have

$$V_1 = I_1 - \frac{I_1}{3} = 2\frac{I_1}{3}$$

$$\therefore \qquad Y_{11} = \frac{I_1}{V_1} = \frac{3}{2}$$

$$Y_{21} = \frac{I_2}{V_1}\bigg|_{V_2=0} = \frac{3I_1-I_3}{V_1} = \frac{3I_1-\dfrac{I_1}{3}}{\dfrac{2}{3}I_1} = \frac{8I_1}{3}/2/3I_1 = \frac{8}{2} = 4$$

Again short circuit Port 1 and apply V_2 to Port 2

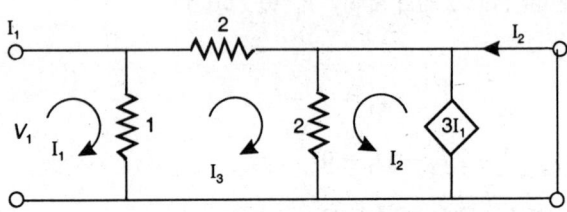

Fig. P. 10.19 (d)

$$V_2 = 2(I_2-3I_1)+2I_3$$

$$I_1-I_3 \doteq 0 \Rightarrow I_1 = I_3$$

$$4I_3+2I_2-6I_1 = 0$$

$$2I_3+I_2-3I_1 = 0$$

$$I_2+2I_3 = 3I_1$$

$$I_2+2I_1 = 3I_1$$

$$I_2 = I_1$$

$$2(I_2-3I_1)+2I_3 = V_2$$

$$2(I_1-3I_1)+2I_1 = V_2$$

$$2(-2I_1)+2I_1 = V_2$$

$$V_2 = -2I_1$$

$$Y_{12} = \frac{I_1}{V_2}\bigg|_{V_1=0} = \frac{-1}{2}$$

$$Y_{22} = \frac{I_2}{V_2}\bigg|_{V_1=0} = \frac{I_1}{-2I_1} = \frac{-1}{2}$$

$$Y_{21} = 0 \quad Y_{22} = 8/3$$

Z parameters in terms of Y–parameters can be obtained as $[Z]=[Y]^{-1}$

Since

$$\begin{bmatrix} Z_{11} & Z_{12} \\ Z_{21} & Z_{22} \end{bmatrix} = \begin{bmatrix} Y_{11} & Y_{12} \\ Y_{21} & Y_{22} \end{bmatrix}^{-1}$$

$$\Delta Y = 3 \times 8/3 = 8$$

$$Z_{11} = \frac{Y_{22}}{\Delta Y} = \frac{8}{3 \times 8} = \frac{1}{3}$$

$$Z_{22} = \frac{Y_{11}}{\Delta Y} = \frac{3}{8}$$

$$Z_{12} = \frac{-Y_{12}}{\Delta Y} = \frac{2/3}{8} = \frac{1}{12}$$

$$Z_{21} = \frac{-Y_{21}}{\Delta Y} = 0$$

$$Z = \begin{bmatrix} 1/3 & 1/12 \\ 0 & \dfrac{3}{8} \end{bmatrix}$$

Problem 10.20. Figure P. 10.20 shows a resistive network containing a single controlled source. For this network, find the y and z parameters.

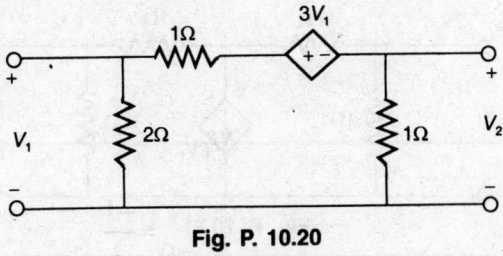

Fig. P. 10.20

Solution:

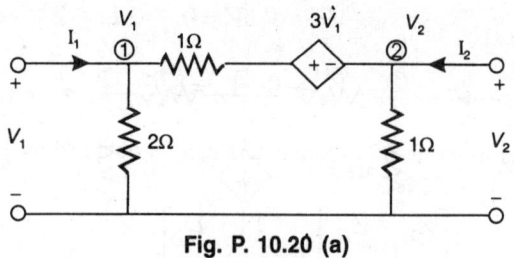

Fig. P. 10.20 (a)

Using nodal analysis, we get on node (1)

$$I_1 - V_1(1 + 1/2) + V_2 + 3V_1 = 0$$

or
$$I_1 = \frac{3}{2}V_1 - V_2 - 3V_1 \qquad (1)$$

on node (2)

$$I_2 - V_2(1 + 1) + V_1 - 3V_1 = 0$$

$$I_2 = 2V_2 - 2V_1 \qquad (2)$$

Hence,
$$Y_{11} = 3/2 \quad Y_{21} = 4$$

$$Y_{12} = -1/2 \quad Y_{22} = -1/2$$

Problem 10.21. Find the y and z parameters for the resistive network containing a controlled source as shown in Fig. P. 10.21.

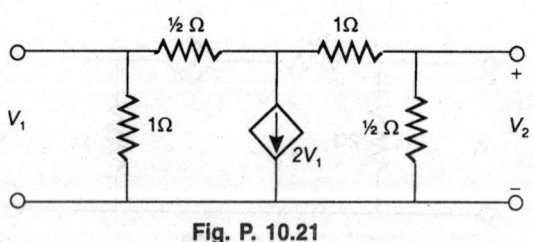

Fig. P. 10.21

Solution:

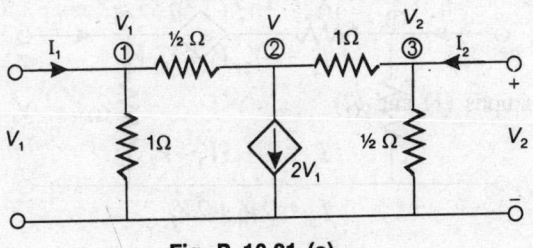

Fig. P. 10.21 (a)

Applying nodal analysis:

At node (1)

$$I_1 - V_1\left(1+\frac{1}{1/2}\right)+V\frac{1}{1/2} = 0$$

$$I_1 = 3V_1 - 2V \qquad (1)$$

At node (3)

$$I_2 - 2V_2 - (V_2 - V)1 = 0$$

$$I_2 = 2V_2 + V_2 - V = 3V_2 - V \qquad (2)$$

At mode (2)

$$-2V_1 + (V_1 - V)2 - (V - V_2) = 0$$

$$-2V_1 + 2V_1 - 2V_1 - V + V_2 = 0$$

$$V_2 = 3V \qquad (3)$$

From equations (3) and (1), we get

$$I_1 = 3V_1 - \frac{2}{3}V_2 \qquad (4)$$

Again from equation, (3) and (2), we get

$$I_2 = 3V_2 - \frac{V_2}{3} = \frac{8}{3}V_2$$

$$I_2 = 0V_1 + \frac{8}{3}V_2 \qquad (5)$$

Comparing equations (4) and (5) with Y parameter equation

$$I_1 = Y_{11} V_1 + Y_{12} V_2$$

$$I_2 = Y_{21} V_1 + Y_{22} V_2$$

From equations (1) and (2)

$$I_1 = -1.5 V_1 - V_2$$

$$I_2 = 2 V_1 + 2 V_2$$

Comparing with

$$I_1 = Y_{11} V_1 + Y_{12} V_2$$

$$I_2 = Y_{21} V_1 + Y_{22} V_2$$

We get,

$$Y_{11} = -1.5 \quad Y_{12} = -1$$

$$Y_{21} = 2 \quad Y_{22} = 2$$

Z parameters in terms of Y parameter is given by

$$[Z] = [Y]^{-1} = \begin{bmatrix} -1.5 & -1 \\ 2 & 2 \end{bmatrix}^{-1}$$

$$\Delta Y = -3 + 2 = -1$$

$$\begin{bmatrix} Z_{11} & Z_{12} \\ Z_{21} & Z_{22} \end{bmatrix} = \frac{\begin{bmatrix} Y_{22} & -Y_{12} \\ -Y_{21} & Y_{11} \end{bmatrix}}{\Delta Y} = \frac{\begin{bmatrix} 2 & -(-1) \\ -2 & -1.5 \end{bmatrix}}{-1} = \begin{bmatrix} -2 & -1 \\ 2 & 1.5 \end{bmatrix}$$

Problem 10.22. The network of the following figure contains both a dependent current source and a dependent voltage source. For the element values given, determine the y and z parameters.

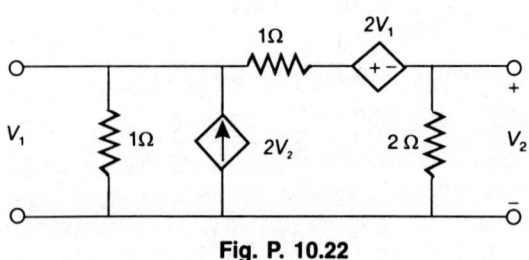

Fig. P. 10.22

Solution:

For Y parameters short circuit port (2) and apply voltage V_1 to port (1), we get

$$I_1 = V_1 + 3V_1 = 4V_1$$

Now $$Y_{11} = \frac{I_1}{V_1}\bigg|_{V_2=0} = \frac{1}{4}$$

$$I_2 = -3V_1$$

$$Y_{21} = \frac{I_2}{V_1} = -3$$

When port (1) is short circuited and V_2 is applied to port (2) we get,

$$Y_{12} = \frac{I_1}{V_2}\bigg|_{V_1=0}$$

$$Y_{22} = \frac{I_2}{V_2}\bigg|_{V_1=0}$$

Now $$I_1 = -3V_2$$

$\therefore$ $$Y_{12} = -3$$

Also $$I_2 = (1+1/2)V_2 = 3/2V_2$$

$$Y_{22} = \frac{I_2}{V_2} = \frac{\frac{3}{2}V_2}{V_2} = \frac{3}{2}$$

Again $$[Z] = [Y]^{-1} = \begin{bmatrix} 4 & -3 \\ -3 & 3/2 \end{bmatrix}^{-1}$$

$$\Delta Y = 4 \times 3/2 - 9 = 6 - 9 = -3$$

$\therefore$ $$[Z] = \frac{\begin{bmatrix} 3/2 & 3 \\ 3 & 4 \end{bmatrix}}{-3} = \begin{bmatrix} -1/2 & -1 \\ -1 & -4/3 \end{bmatrix}$$

$$Z_{11} = -1/2 \quad Z_{12} = -1$$

$$Z_{21} = -1 \quad Z_{22} = -4/3$$

Problem 10.23. Find the y and z partameters for the RC ladder network of Fig. P. 10.23.

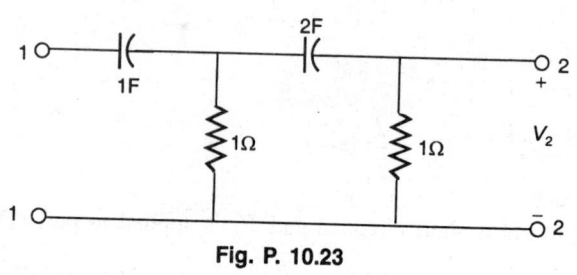

Fig. P. 10.23

Solution:

For Z parameter

When Port 2 is open, i.e. $I_2 = 0$

Then
$$Z_{11} = \frac{2s^2 + 5s + 1}{s(4s+1)}$$

Also
$$Z_{21} = \frac{V_2}{I_1}\bigg|_{I_2 = 0} = \frac{2s}{4s+1}$$

Again when port 1 is open

$$Z_{12} = \frac{V_1}{I_2}\bigg|_{I_1 = 0} = \frac{2s}{4s+1}$$

$$Z_{22} = \frac{V_2}{I_2}\bigg|_{I_1 = 0} = \frac{2s}{4s+1}$$

Problem 10.24. The network of Fig. P. 10.24 is a bridged-T RC network. For the given network find the y and z parameters.

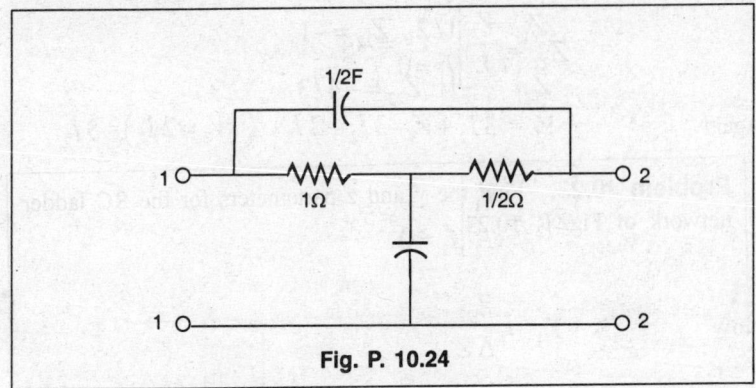

Fig. P. 10.24

Solution:

Y parameters can be obtained as

$$Y_{11} = \frac{I_1}{V_1}\bigg|_{V_2=0} = \frac{s^2+8s+8}{2(s+6)}$$

and

$$Y_{21} = \frac{I_2}{V_1}\bigg|_{V_2=0} = \frac{-(s^2+6s+4)}{2(s+6)}$$

Calculation for Z parameter

When $\quad I_2 = 0, \quad V_1 = 3I_1$

$$Z_{11} = \frac{V_1}{I_1}\bigg|_{I_2=0} = \frac{3I_1}{I_1} = 3$$

$$Z_{21} = \frac{V_2}{I_1}\bigg|_{I_2=0}$$

Now $\quad V_2 + 4V_3 - V_3 = 0, \quad V_2 = -3V_3$

From loop, $\quad V_3 = 2I_1$

$\therefore \qquad V_2 = -3V_3 = -6I_1, \quad Z_{21} = \dfrac{V_2}{I_1} = -6$

Again when $\quad I_1 = 0, \quad V_3 = 2I_2$

$$V_2 = 2(I_2 - 2V_3) + V_3 = 2I_2 - 4V_3 + V_3$$

$$= 2I_2 - 3V_3 = 2I_2 - 3 \times 2I_2 = -4I_2$$

$$Z_{22} = \frac{V_2}{I_2}\bigg|_{I_1=0} = -4$$

Again

$$V_1 = 3I_2 + V_3 = 3I_2 + 2I_2 \quad (\because V_3 = 2I_2) = 5I_2$$

$$Z_{12} = \frac{V_1}{I_2}\bigg|_{I_1=0} = 5$$

Now

$$Y_{11} = \frac{Z_{22}}{\Delta Z}$$

here

$$\Delta Z = (Z_{11}Z_{22} - Z_{21}Z_{12}) = (3 \times (-4) - (-6) \times 5)$$

$$= -12 + 30 = 18$$

$\therefore$

$$Y_{11} = \frac{-4}{18} = \frac{-2}{9}$$

$$Y_{12} = \frac{-Z_{12}}{\Delta Z} = \frac{-5}{18}$$

$$Y_{21} = \frac{-Z_{21}}{\Delta Z} = \frac{6}{18} = \frac{1}{3}$$

$$Y_{22} = \frac{Z_{11}}{\Delta Z} = \frac{3}{18} = \frac{1}{6}$$

For *Z* parameters

$$Z_{11} = \frac{V_1}{I_1}\bigg|_{I_2=0} = \frac{0.5s^2 + 5s + 4}{s(1.5s+2)}$$

$$Z_{21} = \frac{V_2}{I_1}\bigg|_{I_2=0} = \frac{0.5s^2 + 3s + 4}{s(1.5s+2)}$$

$$Z_{22} = \frac{0.5s^2 + 3s + 4}{s(1.5s+2)}$$

$$Z_{12} = \frac{V_1}{I_2}\bigg|_{I_1=0} = \frac{0.5s^2 + 3s + 4}{s(1.5s+2)}$$

Problem 10.25. Determine the ABCD (transmission) parameters for the network.

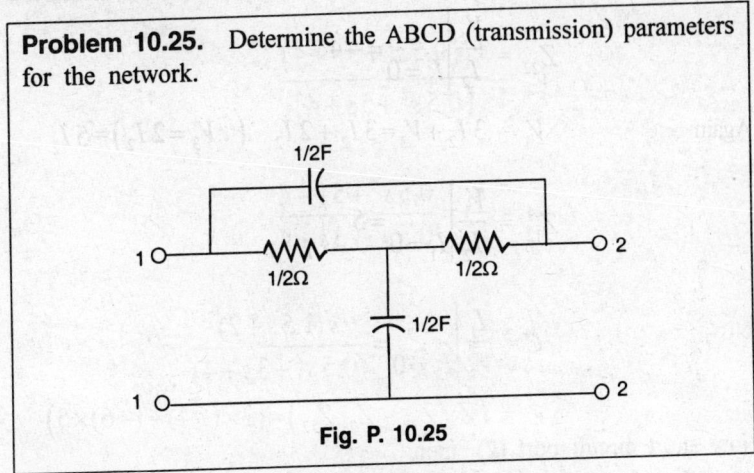

Fig. P. 10.25

Solution:

We know that transmission parameters are given by

$$V_1 = AV_2 + BI_2$$
$$I_1 = CV_2 + DI_2$$

Now, when $I_2 = 0$

i.e.

$$A = \frac{V_1}{V_2}\bigg|_{I_2 = 0}$$

$$V_1 = \left(1 + \frac{2}{s}\right)I_1 - I_a$$

$$0 = -I_1 + \left(1.5 + \frac{2}{s}\right)I_a$$

$$I_1 = \frac{V_1\left(1.5 + \frac{2}{s}\right)}{0.5 + \frac{5}{s} + \frac{4}{s^2}}$$

$$I_a = \frac{V_1}{0.5 + \frac{5}{s} + \frac{4}{s^2}} = \frac{s^2 V_1}{0.5s^2 + 5s + 4}$$

$$V_2 = \frac{V_1\left(0.5s^2+3s+4\right)}{\left(0.5s^2+5s+4\right)}$$

$$A = \frac{V_1}{V_2} = \frac{0.5s^2+5s+4}{0.5s^2+3s+4}$$

$$C = \left.\frac{I_1}{V_2}\right|_{I_2=0} = \frac{s(1.5s+2)}{(0.5s^2+3s+4)}$$

Now short circuit port (2). then

$$V_2 = 0 \quad \text{and} \quad V_a = \frac{V_1}{\dfrac{s}{2}+3} = \frac{2V_1}{s+6}$$

$$I_2 = V_1\left(s/2+\frac{4}{s+6}\right) = \frac{V_1(s^2+6s+8)}{2(s+6)}$$

$$I_1 = V_1\left(\frac{s}{2}+1-\frac{2}{s+6}\right) = \frac{V_1(s^2+8s+8)}{2(s+6)}$$

Now $$B = \frac{V_1}{I_2} \quad \text{when} \quad V_2=0$$

$$B = \frac{2(s+6)}{(s^2+6s+8)}$$

Also $$D = \frac{I_1}{I_2} \quad \text{when} \quad V_2=0$$

$$D = \left(\frac{s^2+8s+8}{s^2+6s+8}\right)$$

Problem 10.26. The following figure shows a network with passive elements and two ideal transformers having 1 : 1 turns-ratios. For the element values specified, determine the Z parameters.

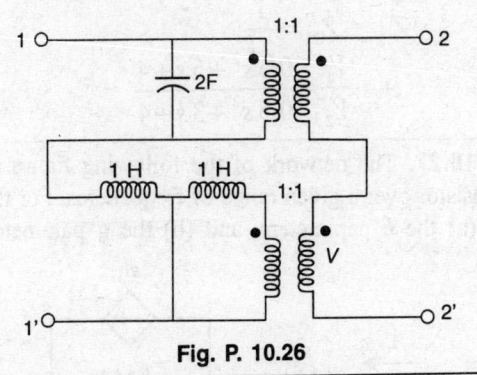

Fig. P. 10.26

Solution:

For the upper part of the given circuit diagram

$$V_{1x} = \frac{1}{2s} I_1 + \frac{1}{2s} I_2$$

$$V_{2y} = \frac{1}{2s} I_1 + \frac{1}{2s} I_2$$

Again consider bottom part of circuit.

$$V_{1y} = sI_1 + 0I_2$$

$$V_{2x} = 0I_1 + sI_2$$

$$V_1 = V_{1x} + V_{1y} = \left(s + \frac{1}{2s}\right) I_1 + \frac{1}{2s} I_2$$

$$V_2 = V_{2x} + V_{2y} = \frac{1}{2s} I_1 + \left(s + \frac{1}{2s}\right) I_2$$

Comparing with $V_1 = Z_{11} I_1 + Z_{12} I_2$

$$V_2 = Z_{21} I_1 + Z_{22} I_2$$

We have $\qquad Z_{11} = Z_{22} = s + \dfrac{1}{2s}$

and $\qquad\qquad Z_{12} = Z_{21} = \dfrac{1}{2s}$

Problem 10.27. The network of the following figure represents a certain transistor over a given range of frequencies. For this network, determine (a) the h parameters, and (b) the g parameters.

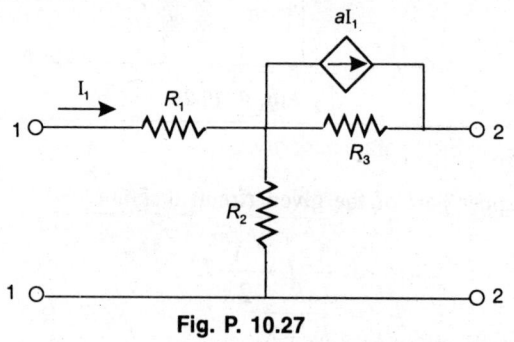

Fig. P. 10.27

Solution:

(1) The hybrid parameter of two port is given by

$$V_1 = h_{11} I_1 + h_{12} V_2$$

$$V_2 = h_{21} I_1 + h_{22} V_2$$

When we make $V_2 = 0$

$$V_x = \frac{(1-a) I_1 R_2 R_3}{(R_2 + R_3)}$$

$$h_{11} = \frac{V_1}{I_1}\bigg|_{V_2=0} = R_1 + V_x$$

$$= R_1 + \frac{(1-a) I_1 R_2 R_3}{(R_2 + R_3)} = \frac{R_1 R_2 + R_1 R_3 + (1-a) R_2 R_3}{R_2 + R_3}$$

$$h_{21} = \left. \frac{I_2}{I_1} \right|_{V_2=0} = -a - \frac{(1-a)R_2}{R_2+R_3} = \frac{-(aR_3+R_2)}{(R_2+R_3)}$$

Hence, $\quad h_{21} = \dfrac{-(aR_3+R_2)}{(R_2+R_3)}$

Now we open circuit port (1), then, $I_1 = 0$

Hence, $\qquad h_{12} = \dfrac{V_1}{V_2} = \dfrac{R_2}{R_2+R_3}$

$$h_{22} = \frac{I_2}{V_2} = \frac{1}{R_2+R_3}$$

Therefore, hybrid parameters are

$$h_{11} = \frac{R_1R_2 + R_1R_3 + (1-a)R_2R_3}{R_2+R_3}$$

$$h_{12} = \frac{R_2}{R_2+R_3}$$

$$h_{21} = \frac{-(aR_2+R_2)}{(R_2+R_3)}$$

$$h_{22} = \frac{1}{(R_2+R_3)}$$

(2) Now g parameters are given by the equations

$$I_1 = g_{11}V_1 + g_{12}I_2$$

$$V_2 = g_{21}V_1 + g_{22}I_2$$

When $I_2 = 0$, i.e. we open circuit port (2), then

$$g_{11} = \frac{I_1}{V_1} = \frac{1}{R_1+R_2}$$

$$g_{21} = \frac{V_2}{V_1} = \frac{R_2+aR_3}{R_1+R_2}$$

Now, short circuit port (1) then

$$g_{12} = \frac{I_1}{I_2} = \frac{-R_2}{R_1 + R_2}$$

$$g_{22} = \frac{V_2}{I_2} = \left(\frac{-a R_2}{R_1 + R_2} + 1\right) R_3 + \left(\frac{-R_2}{R_1 + R_2} + 1\right) R_2$$

$$g_{22} = \frac{R_1 R_2 + R_1 R_3 + (1-a) R_2 R_3}{R_1 + R_2}$$

Problem 10.28. The network of the following figure represents the transistor of Prob. 10.27 over a different range of frequencies. For this network, determine (a) the h parameters, and (b) the g parameters.

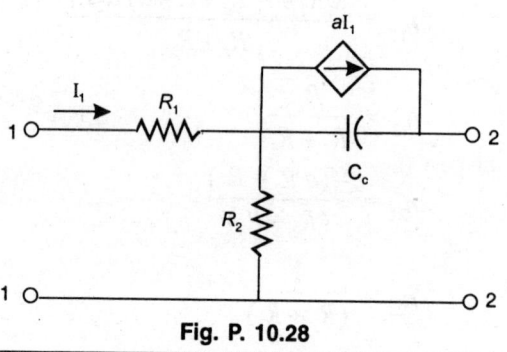

Fig. P. 10.28

Solution:

(1) h-parameters are found from equation

$$V_1 = h_{11} I_1 + h_{12} V_2$$

$$I_2 = h_{21} I_1 + h_{22} V_2$$

$h_{11} = \dfrac{V_1}{I_1} \bigg|_{V_2 = 0}.$ We short circuit port (2)

then $h_{11} = \dfrac{sC_c R_1 R_2 + R_1 + (1-a) R_2}{sC_c R_2 + 1}$

$$h_{21} = \frac{I_2}{I_1}\bigg|_{V_2=0} = \frac{sC_cR_2+a}{sC_cR_2+1}$$

Now we open circuit port (1), then

$$h_{12} = \frac{V_1}{V_2}\bigg|_{I_1=0}, h_{22} = \frac{V_1}{V_2}\bigg|_{I_1=0}$$

$$h_{12} = \frac{sC_cR_2}{sC_cR_2+1}, h_{22} = \frac{sC_c}{sC_cR_2+1}$$

(2) g parameters are given by

$$I_1 = g_{11}V_1 + g_{12}I_2$$

$$V_2 = g_{21}V_1 + g_{22}I_2$$

For
$$g_{11} = \frac{I_1}{V_1} \quad g_{21} = \frac{V_2}{V_1}$$

We open circuit port (2)

$$g_{11} = \frac{I_1}{V_1}\bigg|_{I_2=0} \Rightarrow g_{11} = \frac{1}{R_1+R_2}$$

and
$$g_{21} = \frac{sC_cR_2+a}{sC_c(R_1+R_2)}$$

For obtaining g_{12} and g_{22} short circuit port (1) then

$$g_{12} = \frac{I_1}{I_2}\bigg|_{V_1=0}, g_{22} = \frac{V_2}{I_2}\bigg|_{V_1=0}$$

$$g_{12} = \frac{-R_2}{R_1+R_2}$$

and
$$g_{22} = \frac{V_2}{I_2}\bigg|_{V_1=0} = \frac{sC_cR_1R_2+R_1+(1-a)R_2}{sC_c(R_1+R_2)}$$

Problem 10.29. Show that the standard T section representation of a two-port network may be expressed in terms of the h parameters by the equations shown in the following figure.

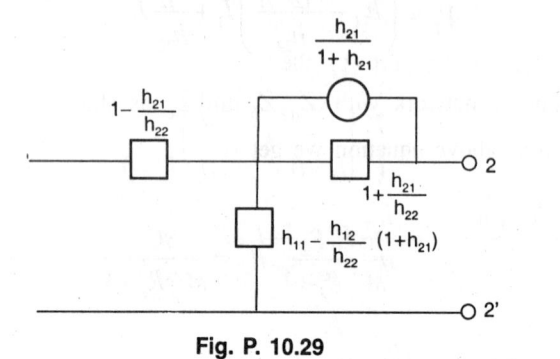

Fig. P. 10.29

Solution:

The h-parameter is given by equation

$$V_1 = h_{11}I_1 + h_{12}V_2 \qquad (1)$$

$$I_2 = h_{21}I_1 + h_{22}V_2 \qquad (2)$$

From the figure we can write

$$V_1 = (Z_a + Z_c)I_1 + Z_c I_2 \qquad (3)$$

and

$$V_2 = Z_cI_1 + (Z_b + Z_c)I_2 \qquad (4)$$

From equation (2) we can write

$$I_2 = h_{21}I_1 + h_{22}V_2$$

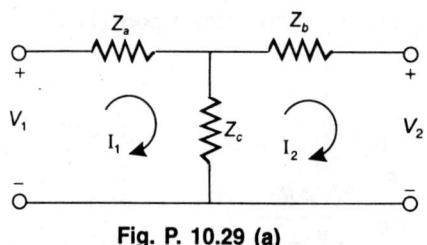

Fig. P. 10.29 (a)

or

$$V_2 = \frac{-h_{21}}{h_{22}}I_1 + \frac{1}{h_{22}}I_2$$

Put value of V_2 in equation (1) we get

$$V_1 = \left(h_{11} - \frac{h_{12} h_{21}}{h_{22}} \right) I_1 + \frac{h_{12}}{h_{22}} I_2$$

Now consider T network with Z_a, Z_b and Z_c as shown.

Now arranging above equation we get

$$V_2 + \frac{h_{21} + h_{12}}{h_{22}} I_1 = \frac{h_{12}}{h_{22}} I_1 - \frac{1}{h_{22}} I_2, Z_c = \frac{h_{12}}{h_{22}}$$

$$Z_a = h_{11} - \frac{h_{12}}{h_{22}} (1 + h_{21})$$

$$Z_b = \frac{1 - h_{12}}{h_{22}}$$

The above circuit modifies to

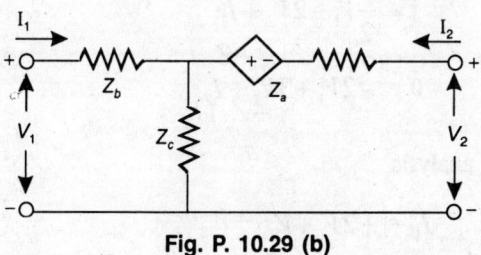

Fig. P. 10.29 (b)

again

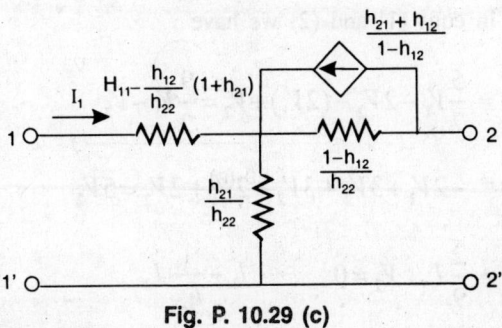

Fig. P. 10.29 (c)

Problem 10.30. The network of the figure may be considered as a two-port network embedded in another resistive network. The resistive network is described by the following short-circuit admittances: $y_{11} = y_{22} = 2b$, $y_{21} = 2J$, and $y_{12} = 1J$. I_a is a constant equal to 1 amp, find the voltage and the two ports of the network N, V_1 and V_2.

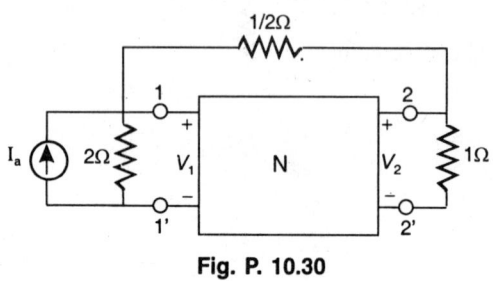

Fig. P. 10.30

Solution:

If network contains feedback which is dependent on some voltage V_2 then two port parameters cannot provide solution. Hence we assume $V_2 = 0$. Then writing node equation, we get

$$I = \frac{5}{2}V_1 - 2V_2 + I_1 \tag{1}$$

$$0 = -2V_1 + 3V_2 + I_2 \tag{2}$$

From two port analysis

$$I_1 = +2V_1 + V_2$$

$$I_2 = 2V_1 + 2V_2$$

Putting value in eqns (1) and (2) we have

$$I = \frac{5}{2}V_1 - 2V_2 + (2V_1) + V_2 = \frac{9}{2}V_1 - V_2$$

and $\qquad 0 = -2V_1 + 3V_2 + 3V_2 + 2V_1 + 2V_2 = 5V_2$

$\Rightarrow \qquad V_1 = \frac{2}{9}I, \quad V_2 = 0$

Problem 10.31. The network shown in Fig. P. 10.31 consists of a resistive T-and a resistive π-network connected in parallel. For the element values given, determine the Y parameters.

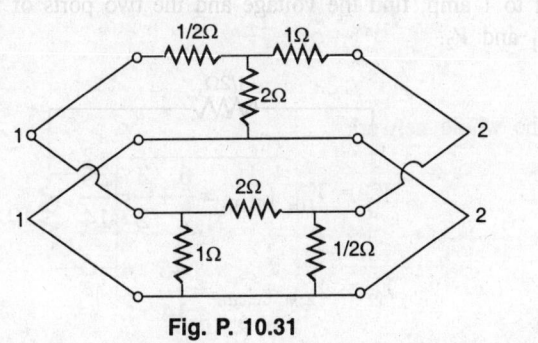

Fig. P. 10.31

Solution:

We consider network 'm' to be

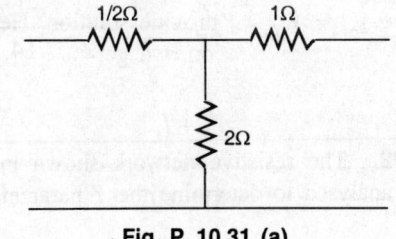

Fig. P. 10.31 (a)

Now apply star-delta conversion:

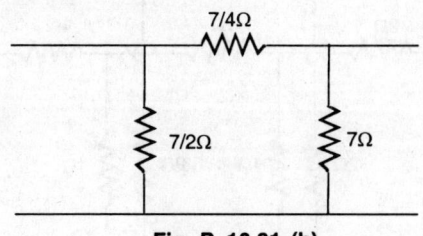

Fig. P. 10.31 (b)

It gives,

$$Y_{12m} = Y_{21m} = -\frac{4}{7}$$

and

$$Y_{11m} = \frac{6}{7}, \quad Y_{22m} = \frac{5}{7}$$

Now we consider network $'n'$ which on simplification gives,

$$Y_{12n} = Y_{21n} = \frac{-1}{2}$$

$$Y_{11n} = \frac{3}{2}, \quad Y_{22n} = 5$$

For the whole network

$$Y_{11} = Y_{11m} + Y_{11n} = \frac{6}{7} + \frac{3}{2} = \frac{33}{14}$$

$$Y_{22} = Y_{22m} + Y_{22n} = \frac{45}{14}$$

$$Y_{12} = Y_{12n} + Y_{12m} = \frac{-4}{7} + \frac{-1}{2} = \frac{-15}{14}$$

$$Y_{21} = Y_{21n} + Y_{21m} = \frac{-4}{7} + \frac{-1}{2} = \frac{-15}{14}$$

Problem 10.32. The resistive network shown in the following figure is to be analysed to determine the Y parameters.

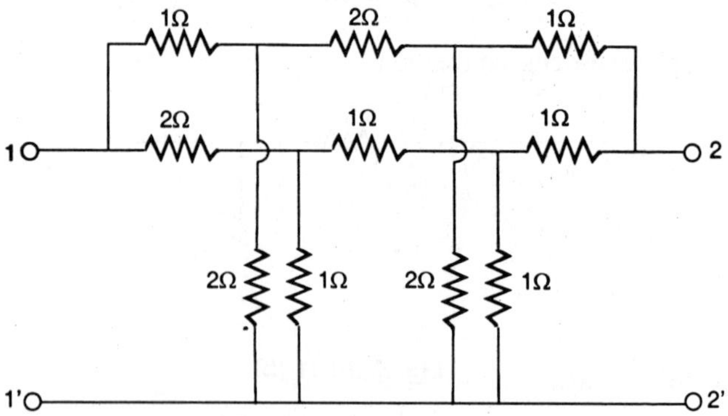

Fig. P. 10.32

Solution:

Let us consider one of the network to be 'm', then m network is

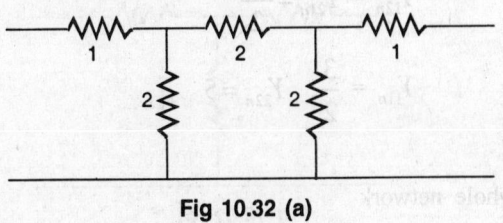

Fig 10.32 (a)

By star-Delta conversion. We have

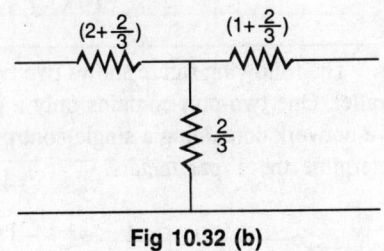

Fig 10.32 (b)

Which gives

$$Y_{11m} = Y_{22m} = \frac{7}{15}$$

$$Y_{12m} = Y_{21m} = \frac{-2}{15}$$

By star-Delta conversion network 'n' gives

$$Y_{11n} = \frac{5}{13}, Y_{12n} = Y_{21n} = \frac{-1}{13},$$

$$Y_{22n} = \frac{8}{13}$$

Hence for the whole network, we have

$$Y_{11} = Y_{11m} + Y_{11n} = \frac{7}{15} + \frac{5}{13} = \frac{166}{195}$$

$$Y_{22} = Y_{22m} + Y_{22n} = \frac{7}{15} + \frac{8}{13} = \frac{211}{195}$$

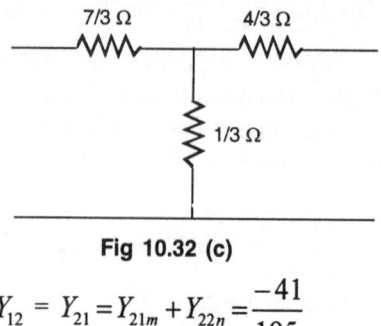

Fig 10.32 (c)

$$Y_{12} = Y_{21} = Y_{21m} + Y_{22n} = \frac{-41}{195}$$

Problem 10.33. The following figure shows two two-port network connected in parallel. One two-port contains only a gyrator, and the other is a resistive network containing a single controlled source. For this network, determine the Y parameters.

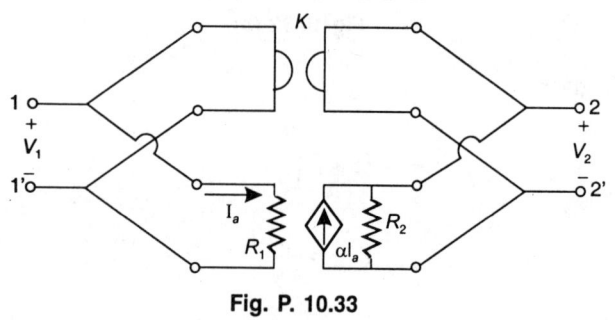

Fig. P. 10.33

Solution:

We first take gyrator network

Here
$$V_1 = KI_2 \quad Y_{11m} = 0 \quad Y_{12m} = \frac{-1}{K}$$

$$V_2 = -KI_1 \quad Y_{21m} = \frac{1}{K} \quad Y_{22m} = 0$$

Considering the second remaining network

$$Y_{11n} = \frac{1}{R_1}, \quad Y_{12n} = 0, \quad Y_{21n} = \frac{-\alpha}{R_1}, \quad Y_{22n} = \frac{1}{R_2}$$

$\therefore$ For the whole network

$$Y_{11} = Y_{11m} + Y_{11n} = 0 + \frac{1}{R_1} = \frac{1}{R_1}, \quad Y_{12m} + Y_{12n} = Y_{12} = \frac{-1}{K}$$

$$Y_{21} = Y_{21m} + Y_{21n} = \frac{1}{k} - \frac{-\alpha}{R_1}, Y_{22} = Y_{22m} + Y_{22n} = \frac{1}{R_2} + 0 = \frac{1}{R_2}$$

Problem 10.34. The network of the figure is of the type used for the so-called "notch filter." For the element values that are given, determine the Y parameters.

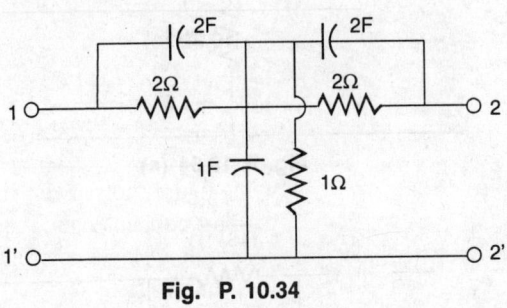

Fig. P. 10.34

Solution:

From the given figure we consider first the network shown below
From star-delta conversion

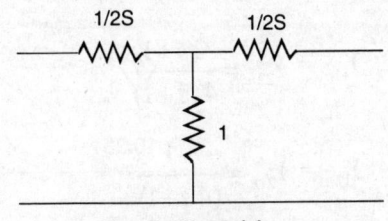

Fig. P. 10.34 (a)

Now
$$Y_{11m} = Y_{22m} = \frac{2s(2s+1)}{4s+1}$$

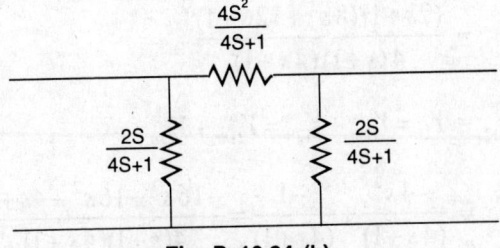

Fig. P. 10.34 (b)

$$Y_{12m} = Y_{21m} = \frac{-4s^2}{4s+1}$$

Next network is given below
Delta conversion

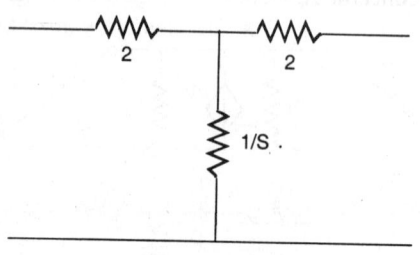

Fig. P. 10.34 (c)

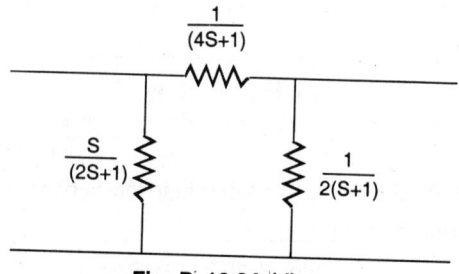

Fig. P. 10.34 (d)

$$Y_{11n} = Y_{22n} = \frac{2s+1}{4(s+1)}$$

$$Y_{12n} = Y_{21n} = \frac{-1}{(4s+1)}$$

Hence $\quad Y_{11} = Y_{11m} + Y_{11n} = \frac{2s(2s+1)}{(4s+1)} + \frac{(2s+1)}{4(s+1)}$

$$= \frac{(2s+1)(8s^2+12s+1)}{4(s+1)(4s+1)}$$

$$Y_{12} = Y_{21} = Y_{12m} + Y_{12n} = Y_{21m} + Y_{21n}$$

$$= \frac{-4s^2}{(4s+1)} + \frac{-1}{(4s+1)} = \frac{-16s^3+16s^2+4s+1}{4(s+1)(4s+1)}$$

Problem 10.35. The following figure shows two networks as (a) and (b). It is asserted that one is the equivalent of the other. Is this assertion correct? Show reasoning. If it is, might one network has an advantage over the other as far as the calculation of network parameters is concerned?

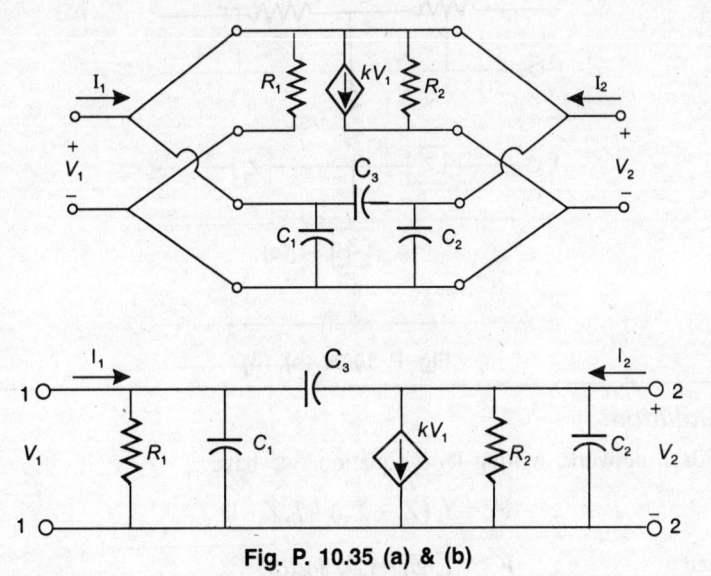

Fig. P. 10.35 (a) & (b)

Solution:

Two network are equivalent only if short circuit admittance parameters of them are equal. Here the two networks have the same short circuit admittance parameter and hence are equivalent. As far as calculation of network parameters is concerned both network involve the same amount of calculation, however network of figure (a) indicates the low frequency behaviour of the network.

Problem 10.36. Two two-port networks are said to be equivalent if they have identical Y or Z parameters (or other of the characterizing parameters). In this problem, we wish to study the conditions under which the π-network of (a) is equivalent to the T-network of (b). Show that the two networks are equivalent if

$$Y_a = \frac{Z_2}{D}, \quad Y_b = \frac{Z_3}{D}, \quad \text{and} \quad Y_c = \frac{Z_1}{D}$$

where $$D = Z_1 Z_2 + Z_2 Z_3 + Z_3 Z_1$$

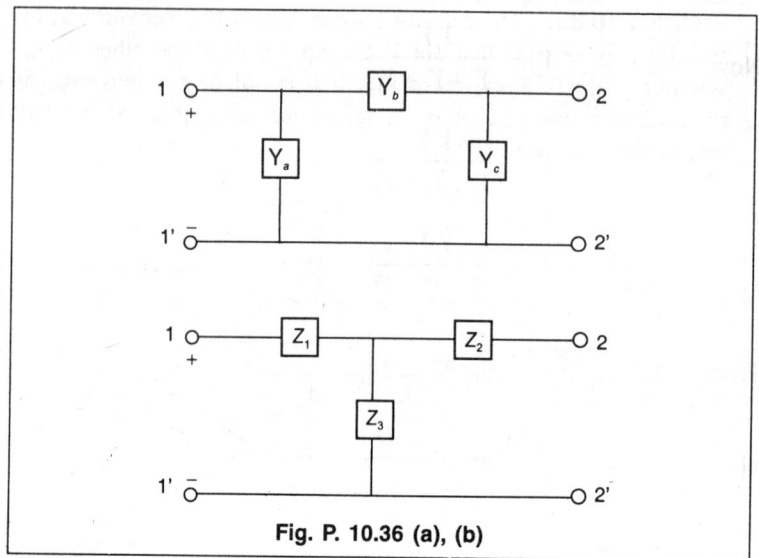

Fig. P. 10.36 (a), (b)

Solution:

For T network, writing loop equation, we have

$$V_1 = I_1(Z_1 + Z_3) + I_2 Z_3$$

and

$$V_2 = I_3 Z_3 + (Z_2 + Z_3) I_2$$

Solving for I_1 and I_2 we get

$$I_1 = \frac{V_1(Z_2 + Z_3) - V_2 Z_3}{Z_1 Z_2 + Z_1 Z_3 + Z_2 Z_3}$$

and

$$I_2 = \frac{V_1 Z_3 + V_2(Z_1 + Z_3)}{Z_1 Z_2 + Z_3(Z_1 + Z_2)}$$

Now for the π-network, the γ-parameters equation are:

$$I_1 = (Y_a + Y_b)V_1 - Y_b V_2$$

and

$$I_2 = -Y_b V_1 + (Y_b + Y_c)V_2$$

$$Y_a = \frac{Z_2}{D} \quad \text{where}$$

$$D = Z_1 Z_2 + Z_2 Z_3 + Z_3 Z_1$$

$$Y_b = \frac{Z_3}{D}, \quad Y_c = \frac{Z_1}{D}$$

Now $$Y_a = \frac{Y_1 Y_3}{Y_1 + Y_2 + Y_3}$$

$$Y_b = \frac{Y_2 Y_1}{Y_1 + Y_2 + Y_3}$$

$$Y_c = \frac{Y_2 Y_3}{Y_1 + Y_2 + Y_3}$$

From above $$Y_a = \frac{Z_2}{D} \Rightarrow \frac{1}{Z_a} = \frac{Z_2}{D} \Rightarrow Z_a = \frac{D}{Z_2}$$

also $$Z_b = \frac{D}{Z_3}, Z_c = \frac{D}{Z_1}$$

Problem 10.37. Apply the T-π transformation to obtain an equivalent (a) T-network and (b) π-network for the capacitive network given in Fig. P. 10.37.

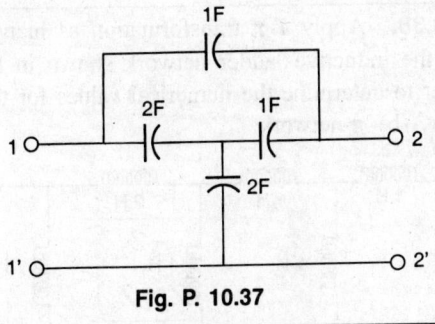

Fig. P. 10.37

Solution:

We have the given network

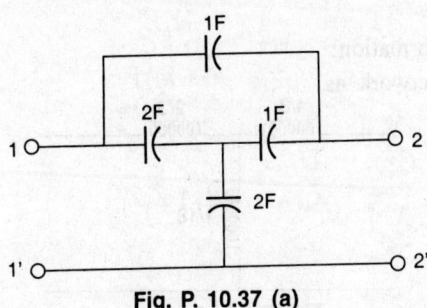

Fig. P. 10.37 (a)

Applying T-π transformation we have
(a) π-network is

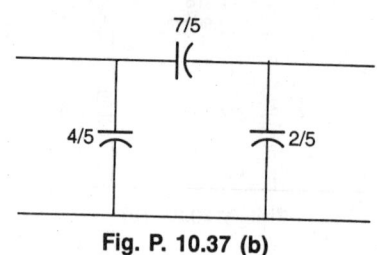

Fig. P. 10.37 (b)

(b) T-network is

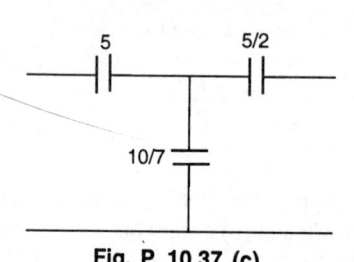

Fig. P. 10.37 (c)

Problem 10.38. Apply T-π transformation as many times as is necessary to the inductive ladder network shown in the following figure in order to determine the numerical values for the equivalent (a) T-network, (b) π-network.

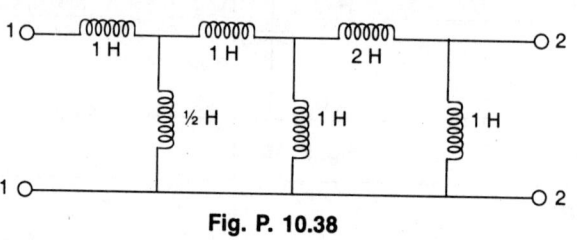

Fig. P. 10.38

Solution:

Using T-π transformation:
We have (a) T-network as

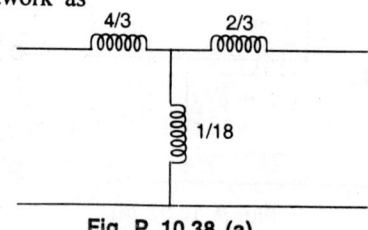

Fig. P. 10.38 (a)

and, (b) π -network

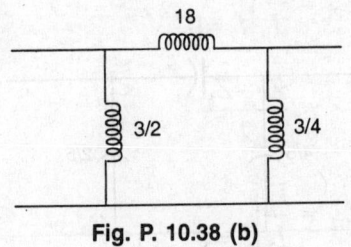

Fig. P. 10.38 (b)

Problem 10.39. The network given in Fig. P. 10.39 is known as a lattice network; this lattice is symmetrical in the sense that two arms of the lattice have impedance Z_a and two have impedance Z_b. For this network, (a) determine the Z parameters, and (b) express Z_a and Z_b in terms of Z parameters.

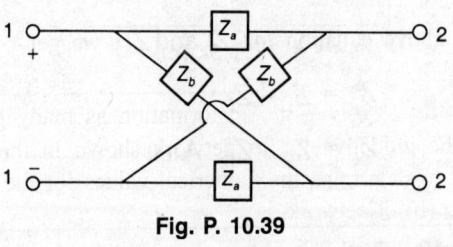

Fig. P. 10.39

Solution:

(a) For the given lattice network, we can write Z-parameters equation as

$$V_1 = Z_{11}I_1 + Z_{12}I_2$$

$$V_2 = Z_{21}I_1 + Z_{22}I_2$$

where

$$Z_{11} = Z_{22} = \frac{Z_a + Z_b}{2}$$

Now

$$V_2 = \frac{V_1(Z_b - Z_a)}{(Z_a + Z_b)}$$

$$I_1 = \frac{2V_1}{Z_a + Z_b}$$

$$Z_{21} = \frac{V_2}{I_1}\bigg|_{I_2=0}$$

$$= \frac{Z_b - Z_a}{2}$$

$$Z_{12} = \frac{V_1}{I_2}\bigg|_{I_1=0}$$

$$= \frac{Z_b - Z_a}{2}$$

(b) For the lattice network, using Z parameters, we have

$$2Z_{11} = Z_a + Z_b$$

$$2Z_{12} = Z_b - Z_a$$

Solving above equation for Z_a and Z_b, we get

$$Z_a = Z_{11} - Z_{12}$$

and $$Z_b = Z_{11} + Z_{12} \text{ **Ans.**}$$

Problem 10.40. In this problem, we consider two-port network having a symmetry property illustrated in (a) of the figure: If the network is divided at the dashed line, the two half networks have mirror symmetry with respect to the dashed line. The two half networks are connected by any number of wires as shown, and we will consider only the cases in which these wires do not cross. If a network meeting these specifications is bisected at the dashed line, then with the connecting wires open, the input impedance at either port is $Z_{1/2oc}$ as shown in (b). Similarly, with the connecting wires shorted, the impedance at either port is $Z_{1/2sc}$ as shown in (c). A theorem due to Bartlett states that these impedances are related to those given for the arms of the lattice in Prob. 10.39 by the equations.

$$Z_a = Z_{1/2sc}, \quad Z_b = Z_{1/2oc}$$

This is known as Bartlett's bisection theorem, and permits an equivalent lattice network to be found for any symmetrical network. Prove the theorem.

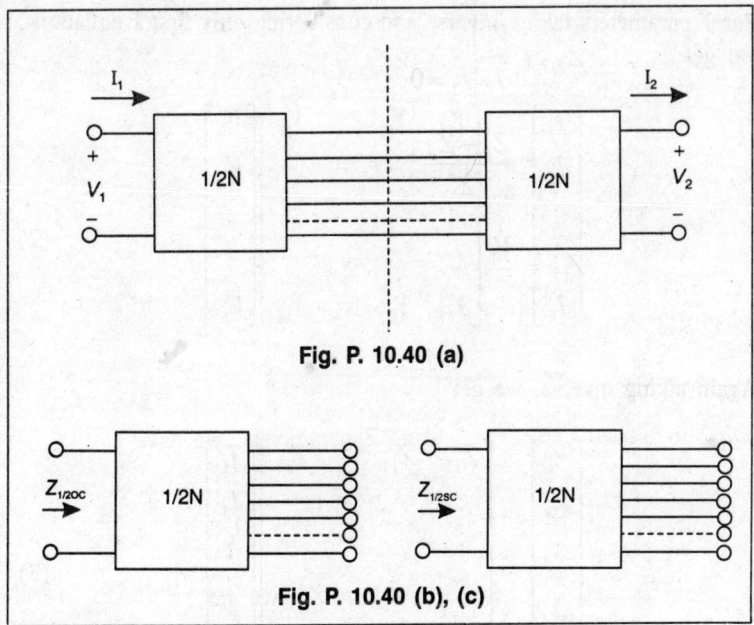

Fig. P. 10.40 (a)

Fig. P. 10.40 (b), (c)

Solution:

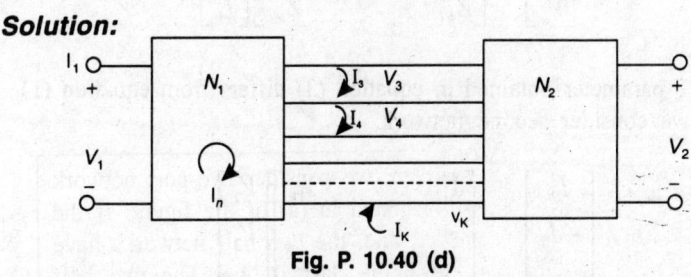

Fig. P. 10.40 (d)

Here $I_3, I_4 - - - I_k$ are the mesh currents on the axis of symmetry.

$I_{k+1} - - - I_n$ are mesh currents intermal of O/P terminal

N_1 and N_2 are two identical linear networks.

$$
\begin{bmatrix} V_1 \\ V_3 \\ \vdots \\ \vdots \\ \vdots \\ \vdots \\ V_k \end{bmatrix} = \begin{bmatrix} Z_{11} & Z_{13}\cdots & Z_{1k}\cdots & Z_{1n} \\ - & - & - & - \\ - & - & - & - \\ Z_{k1} & & Z_{kk} & Z_{kn} \\ - & & - & - \\ - & & - & - \\ Z_{n1} & & Z_{nk} & Z_{nn} \end{bmatrix} \begin{bmatrix} I_1 \\ I_3 \\ \vdots \\ I_k \\ \vdots \\ \vdots \\ I_n \end{bmatrix} \tag{1}
$$

For Y parameters taking inverse and considering only first k equations, we get

$$
\begin{bmatrix} I_1 \\ I_3 \\ : \\ : \\ I_k \end{bmatrix} = \begin{bmatrix} Y_{11} & Y_{12} & & Y_{1k} \\ : & - & - & - \\ : & & & \\ : & & & \\ Y_{k1} & Y_{k2} & & Y_{kk} \end{bmatrix} \begin{bmatrix} V_1 \\ V_3 \\ : \\ : \\ V_k \end{bmatrix}
\tag{2}
$$

Again taking inverse, we get

$$
\begin{bmatrix} V_1 \\ V_3 \\ : \\ : \\ V_k \end{bmatrix} = \begin{bmatrix} Z_{11} & Z_{13} & & Z_{1k} \\ : & - & - & - \\ : & & & \\ : & & & \\ Z_{k1} & & & Z_{kk} \end{bmatrix} \begin{bmatrix} I_1 \\ I_3 \\ : \\ : \\ I_k \end{bmatrix}
\tag{3}
$$

Then Z parameter obtained in equation (3) differs from equation (1) Now we consider second network.

$$
\begin{bmatrix} I_2 \\ -I_3 \\ : \\ : \\ I_k \end{bmatrix} = \begin{bmatrix} Y_{11} & & & Y_{1k} \\ : & - & - & - \\ : & & & \\ : & & & \\ Y_{k1} & & & Y_{kk} \end{bmatrix} \begin{bmatrix} V_2 \\ V_3 \\ : \\ : \\ V_k \end{bmatrix}
\tag{4}
$$

and on transformation we have

$$
\begin{bmatrix} V_2 \\ V_3 \\ : \\ : \\ V_k \end{bmatrix} = \begin{bmatrix} Z_{11} & & & Z_{1k} \\ : & - & - & - \\ : & & & \\ : & & & \\ Z_{k1} & & & Z_{kk} \end{bmatrix} \begin{bmatrix} I_2 \\ -I_3 \\ : \\ : \\ -I_k \end{bmatrix}
\tag{5}
$$

If we substract the first equation of (4) from (2), we get

$$I_1 - I_2 = Y_{11}(V_1 - V_2)$$

Adding equation (3) to (4). We have

$$V_1 + V_2 = Z_{11}(I_1 + I_2)$$

and

$$V_1 - V_2 = \frac{1}{Y_{11}}(I_1 - I_2)$$

By observing equation (2), we have

$$Y_{11} = \frac{I_1}{V_1}\bigg|_{V_3 - V_k = 0}$$

And from equation (3) we get

$$Z_{11} = \frac{V_1}{I_1}\bigg|_{I_3 - I_k = 0}$$

Now

$$V_1 = \frac{Z_{oc} + Z_{sc}}{2}I_1 + \frac{Z_{oc} - Z_{sc}}{2}I_2$$

$$V_2 = \frac{Z_{oc} - Z_{sc}}{2}I_1 - \frac{Z_{oc} + Z_{sc}}{2}I_2$$

We can write from network

$$V_1 = \frac{Z_a + Z_b}{2}I_1 + \frac{Z_a - Z_b}{2}I_2$$

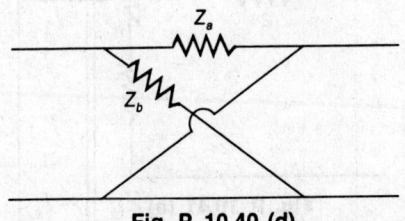

Fig. P. 10.40 (d)

$$V_2 = \frac{Z_b - Z_a}{2} I_1 + \frac{Z_a + Z_b}{2} I_2$$

Hence,

$$Z_a = Z_{sc}$$

and

$$Z_b = Z_{sc}$$

and we get the lattice network which is found from the original symmetrical network.

Problem 10.41. Apply the theorem of Prob. 10.40 to the network given in Prob. 10.41 with the terminating resistor at port 2 removed, and so obtain a lattice equivalent network.

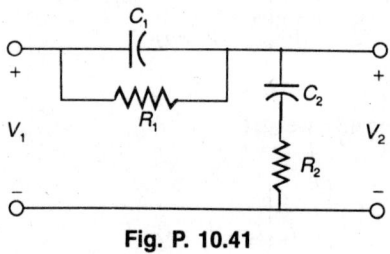

Fig. P. 10.41

Solution:

From the network, we get on simplification

Now

$$Z_s = \frac{1}{2}$$

$$Z_0 = 2$$

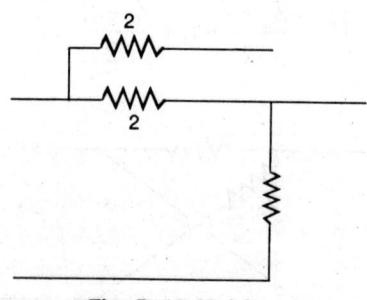

Fig. P. 10.41 (a)

and the lattice network is

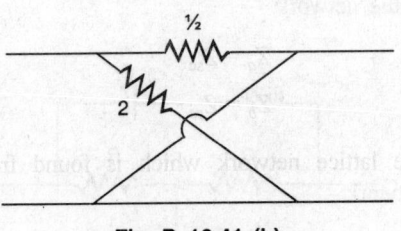

Fig. P. 10.41 (b)

Problem 10.42. Apply the theorem of Prob. 10.40 to the network of Prob. 10.41 with the terminating resistor R_0 removed to find the lattice equivalent of the given network.

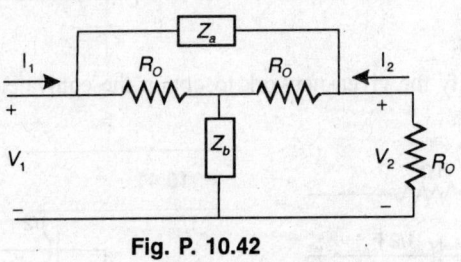

Fig. P. 10.42

Solution:

From the network

$$Z_s = \frac{Z_a R_0 / 2}{\dfrac{Z_0}{2} + R_0} = \frac{Z_a R_0}{Z_a + 2 R_0}$$

and the simplified network is

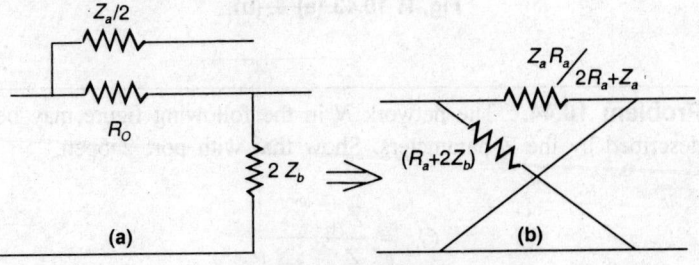

Fig. P. 10.42 (a) & (b)

Problem 10.43. (a) Show that the network of the following figure satisfies the requirements described in Prob. 10.40 (b). Find the lattice equivalent of the network.

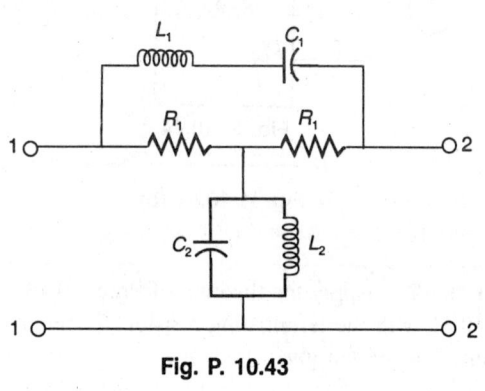

Fig. P. 10.43

Solution:

We simplify the given network to obtain the equivalent lattice network as below

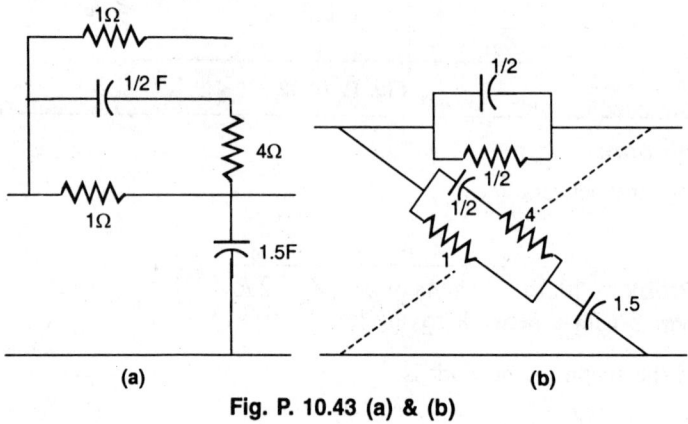

Fig. P. 10.43 (a) & (b)

Problem 10.44. The network N in the following figure may be described by the Z parameters. Show that with port 2 open,

$$G_{12} = \frac{Z_{21}}{Z_{11}}$$

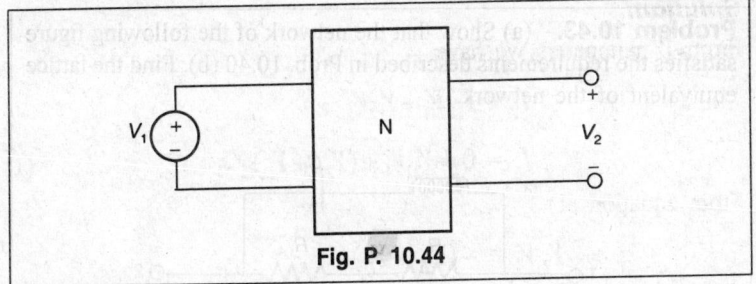

Fig. P. 10.44

Solution:

Now with port (2) open we have

$$V_1 = Z_{11}I_1 \qquad (1)$$

and $\qquad V_2 = Z_{21}I_1 \qquad (2)$

with port (2) open $I_2 = 0$ and hence

$$Z_{11} = \frac{V_1}{I_1}\bigg|_{I_2=0}$$

$$Z_{21} = \frac{V_2}{I_1}\bigg|_{I_2=0}$$

Now dividing eqns (1) and (2), we get

$$\frac{V_2}{V_1} = \frac{Z_{21}}{Z_{11}}$$

Problem 10.45. The network N in Fig. P. 10.45 is terminated at port 2 with a network having impedance $Z_L = 1/Y_L$. Show that

$$G_{12} = \frac{-y_{21}}{y_{22} + Y_L}$$

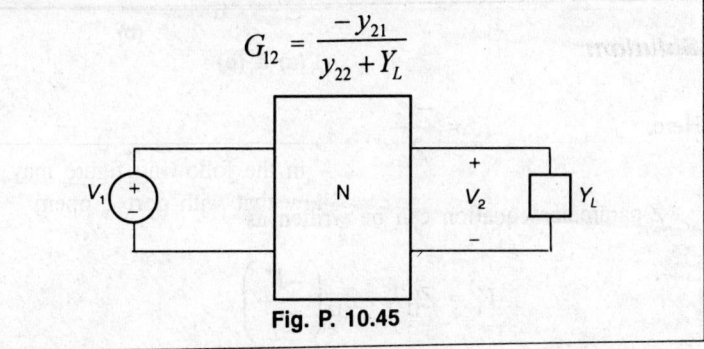

Fig. P. 10.45

Solution:

From Y parameters we have

$$I_1 = Y_{11}V_1 + Y_{12}V_2$$

$$I_2 = 0 = Y_{21}V_1 + (Y_{22} + Y_L)V_2 \tag{1}$$

From equation (1)

$$Y_{21}V_1 = -(Y_{22} + Y_L)V_2$$

or

$$\frac{V_2}{V_1} = \frac{-Y_{21}}{(Y_{22} + Y_L)}$$

Hence

$$G_{12} = \frac{V_2}{V_1} = \frac{-Y_{21}}{(Y_{22} + Y_L)}$$

Problem 10.46. The network N of the figure is terminated at port 2 in impedance $Z_L = 1/Y_L$. Show that the transfer impedance for the combination is

$$Z_{12} = \frac{z_{21}Z_L}{z_{22} + Z_L}$$

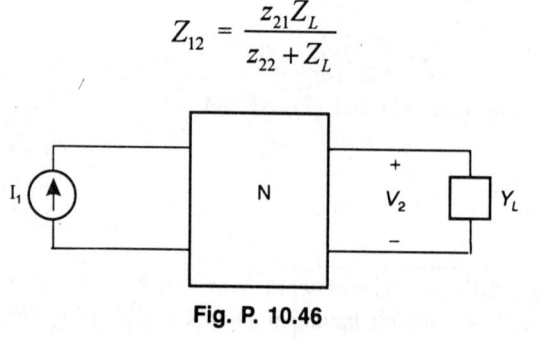

Fig. P. 10.46

Solution:

Here,

$$I_2 = \frac{-V_2}{Z_L}$$

∴ Z-parameter equation can be written as

$$V_1 = Z_{11}I_1 + Z_{12}\left(\frac{-V_2}{Z_L}\right) \tag{1}$$

$$0 = Z_{21}I_1 + (Z_{22} + Z_L)\left(\frac{-V_2}{Z_L}\right) \quad (2)$$

$$\Rightarrow \quad (Z_{22} + Z_L)\frac{V_2}{Z_L} = Z_{21}I_1$$

$$I_1 = \frac{(Z_{22} + Z_L)V_2}{Z_{21}Z_L} \quad (3)$$

Hence

$$\frac{V_1}{I_1} = Z_{12} = \frac{Z_{21}Z_L}{Z_{22} + Z_L}$$

Problem 10.47. Figure P. 11.47 shows two two-port network connected in cascade. The two networks are distinguished by the subscripts a and b. Show that the combined network may be described by the equations.

$$Z_{12} = \frac{Z_{12a}Z_{12b}}{Z_{11b} + Z_{22a}}$$

and

$$Y_{12} = \frac{-y_{12a}y_{12b}}{y_{11b}' + y_{22a}}$$

for the transfer functions.

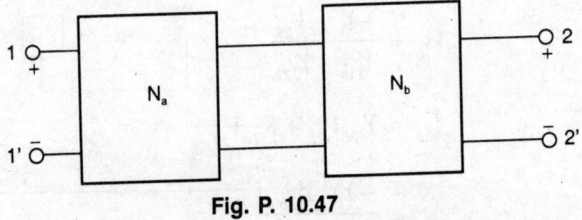

Fig. P. 10.47

Solution:

Writing the Z parameter equation.

$$V_1 = Z_{11}I_1 + Z_{12}I_2$$

$$V_2 = Z_{21}I_1 + Z_{22}I_2$$

Now we know that $\quad Z_{12} = \frac{V_1}{I_2}\bigg|\text{when } I_1 = 0 = \frac{V_{1a}}{I_{2b}}\bigg|_{I_{1a}=0}$

If $I_1 = 0$ than we can have from network 'b' load as Z_{22a} again

$$V_{1b} = -Z_{12a}I_1 = Z_{11b}I_{1b} + Z_{12b}I_{2b}$$

hence

$$I_{1b} = \frac{-Z_{22b} \cdot I_{2b}}{Z_{11b} + Z_{22a}}$$

Also

$$V_{1a} = Z_{12a} \cdot I_{2a} = -Z_{12a} \cdot I_{1b} = \frac{Z_{12a} \cdot Z_{12b}}{Z_{11b} + Z_{22a}} I_{2b}$$

$$\Rightarrow \qquad Z_{12} = \frac{Z_{12a} \cdot Z_{12b}}{Z_{11b} + Z_{22a}}$$

Again from Y-parameter equation

$$I_1 = Y_{11}V_1 + Y_{12}V_2$$

$$I_2 = Y_{21}V_1 + Y_{22}V_2$$

We know by short circuiting port (1) we have

$$Y_{12} = \frac{I_1}{V_2}\bigg|_{V_1=0} = \frac{I_{1a}}{V_{2b}}\bigg|_{V_{1a}=0}$$

Now when $V_2 = 0$, then the load on network 'b' is equal to Y_{22a}

hence $I_{1b} = -Y_{12a}V_{1b} = Y_{11b} \cdot V_{1b} + Y_{12b} \cdot V_{2b}$

$$\Rightarrow \qquad V_{1b} = \frac{-Y_{12b} \cdot V_{2b}}{Y_{11b} + Y_{22b}}$$

Also

$$I_{2b} = Y_{12b}V_{2b} + Y_{12a}V_{1b}$$

$$= \frac{-Y_{12a}Y_{12b} \cdot V_{2b}}{Y_{11b} + Y_{22a}}$$

$$\Rightarrow \qquad Y_{12} = \frac{-Y_{12a}Y_{12b}}{Y_{11b} + Y_{22a}}$$

11

Sinusoidal Steady State Analysis

Problem 11.1. For the sinusoidal waveform of the following figure, write an equation for using numerical values for the magnitude, phase, and frequency.

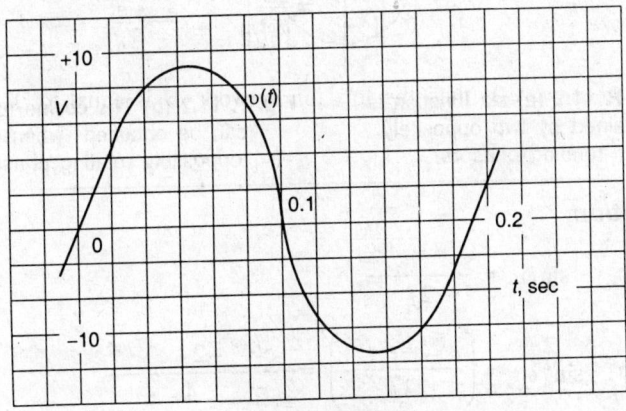

Fig. P. 11.1

Solution:

Equation for sinusoidal waveform is

$$V = A \sin(\omega t + \phi)$$

Where
- A = Magnitude
- ω = Frequency
- ϕ = Phase

Using numerical value of $A = \dfrac{70}{8}, \omega = 2\pi f = \dfrac{2\pi}{0.2}, \phi = \dfrac{180}{5} = 36$

Then equation is $V = \dfrac{70}{8}\sin\left(\dfrac{2\pi}{0.2}t + 36°\right) = \dfrac{70}{8}\sin(10\pi t + 36°)$

Problem 11.2.   Starting with the rotating phasors, $e^{\pm j\omega t}$, show by a construction similar to that illustrated in Figs P. 11.2 (a) and 11.2 (b) that $\sin^2 \omega t + \cos^2 \omega t = 1$

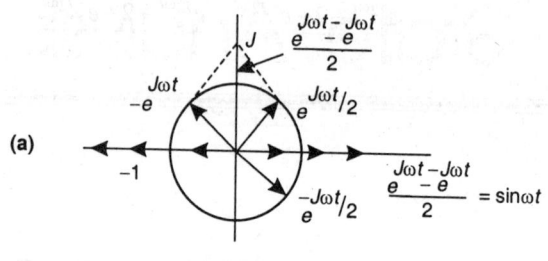

(a)

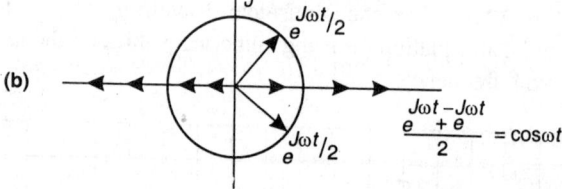

(b)

Fig. P. 11.2 (a) sin function is obtained by two oppositely rotating phasors.

Fig. P. 11.2 (b) The cosine function can be obtained when two oppositely rotating phasors

---

Solution:

Now $\sin\omega t = \dfrac{e^{j\omega t} - e^{j\omega t}}{2j}$

Hence $\sin^2 \omega t = \left(\dfrac{e^{j\omega t} - e^{-j\omega t}}{2j}\right)^2 = -\dfrac{e^{j2\omega t} - 2 + e^{-2j\omega t}}{4}$

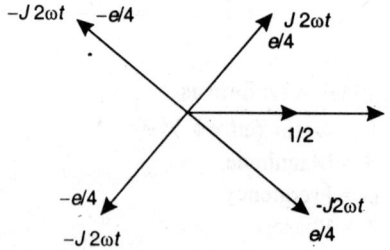

Fig. P. 11.2 (c)

Similarly, we know that

$$\cos \omega t = \frac{e^{j\omega t} + e^{-j\omega t}}{2}$$

Hence

$$\cos^2 \omega t = \left(\frac{e^{j\omega t} + e^{-j\omega t}}{2}\right)^2 = \frac{e^{j2\omega t} + 2 + e^{-j2\omega t}}{4}$$

Now

$$\sin^2 \omega t + \cos^2 \omega t = -\frac{e^{j2\omega t} - 2 + e^{-j2\omega t}}{4} + \frac{e^{j2\omega t} + 2 + e^{-j2\omega t}}{4}$$

$$= \frac{(e^{j2\omega t} - e^{j2\omega t}) + (2+2) + (e^{-j2\omega t} - e^{-j2\omega t})}{4} = \frac{4}{4} = 1$$

Hence $\sin^2 \omega t + \cos^2 \omega t = 1$

Problem 11.3. Given the equation

$$\sin 377t + 3\sqrt{2} \sin\left(377t + \frac{\pi}{4}\right) = A\cos(377t + \theta)$$

Determine A and θ

Solution:

Given $\sin 377t + 3\sqrt{2} \sin\left(377t + \frac{\pi}{4}\right) = A\cos(377t + \theta)$

Now $\qquad\qquad \sin 377t = I_m e^{j377t}$ $\qquad\qquad$ (1)

and $\quad 3\sqrt{2} \sin(377t + \pi/4) = I_m\left[3\sqrt{2} e^{j(377t + \pi/4)}\right]$

$$= I_m\left[3\sqrt{2} e^{j\pi/4} \cdot e^{j377t}\right]$$

$$\left(\because e^{j(377t + \pi/4)} = e^{j377t} \cdot e^{j\pi/4}\right)$$

$$= I_m\left[3\sqrt{2}(\cos\pi/4 + j\sin\pi/4) \cdot e^{j377t}\right]$$

$$= I_m\left[3\sqrt{2}\left(\frac{1}{\sqrt{2}} + j\frac{1}{\sqrt{2}}\right) e^{j377t}\right]$$

$$= I_m\left[3(1+j) e^{j377t}\right] \qquad\qquad (2)$$

$$= I_m\left[(3 + j3) e^{j377t}\right]$$

Adding equations (1) and (2), we have

$$\sin 377t + 3\sqrt{2} \sin\left(377t + \frac{\pi}{4}\right) = I_m\left[(3+j3)+1\right]e^{j377t}$$

$$= I_m\left[(4+j3)\right]e^{j377t} = \left[5e^{j36.87}\,e^{j377t}\right]I_m = 5\,I_m\left[e^{j(36.87+377t)}\right]$$

$$\doteq 5I_m\left[\sin(377t+36.87) + \cos(377t+90-36.87)\right]$$

$$= 5I_m\left[\sin(377t+36.89) + \cos(377t-53.13)\right]$$

$$= 5\,I_m\sin(377t+36.87°) + 5I_m\cos(377t-53.13)$$

Comparing with $A\cos(377t+\theta)$ we get $A = 5$ and $\theta = -53.13°$

Problem 11.4.   Show that $\displaystyle\sum_{k=1}^{n} A_k \sin(\omega_1 t - \phi_k) = C\sin(\omega_1 t + \theta)$

In other words, show that the sum of any number of sinusoids of arbitrary amplitude and phase angle but all of the same frequency is a sinusoid of the same frequency.

Solution:

Given $\displaystyle\sum_{k=1}^{n} A_k \sin(\omega_1 t + \phi_k) = C\sin(\omega_1 t + \theta)$

Now $\displaystyle\sum_{k=1}^{n} A_k \sin(\omega_1 t + \phi_k) = \sum_{k=1}^{n} A_k(\sin\omega_1 t\cos\phi_k + \cos\omega_1 t + \sin\phi_k)$

Let $\displaystyle\sum_{k=1}^{n} A_k\cos\phi_k = A$ (1)

and $\displaystyle\sum_{k=1}^{n} A_k\sin\phi_k = B$ (2)

Then $\displaystyle\sum_{k=1}^{n} A_k\sin(\omega_1 t + \phi_k) = A\sin\omega_1 t + B\cos\omega_1 t$ (3)

Now divide and multiply eqn (3) by $(A^2+B^2)^{\frac{1}{2}}$ then equation (3) becomes

$$A\sin\omega_1 t + B\cos\omega_1 t = \left(A^2+B^2\right)^{1/2}\left[\frac{A\sin\omega_1 t}{\left(A^2+B^2\right)^{1/2}}+\frac{B\cos\omega_1 t}{\left(A^2+B^2\right)^{1/2}}\right]$$

$$= C\left[\cos\theta\sin\omega_1 t + \sin\theta\cos\omega_1 t\right] \qquad (4)$$

Where $\quad C = \left(A^2+B^2\right)^{1/2}$

and $\quad \cos\theta = \dfrac{A}{(A^2+B^2)^{1/2}}$

$\qquad \sin\theta = \dfrac{B}{(A^2+B^2)^{1/2}}$

$\qquad \tan\theta = \dfrac{B}{A} \Rightarrow \theta = \tan^{-1}(B/A)$

Fig. P. 11.4

Hence equation (4) becomes

$$C\left[\cos\theta\sin\omega_1 t + \sin\theta\cos\omega_1 t\right] = C\sin(\omega_1 t+\theta)$$

Therefore $\quad \displaystyle\sum_{k=1}^{n} A_k\sin(\omega_1 t+\phi_k) = C\sin(\omega_1 t+\theta)$

Problem 11.5. Using the equation of Problem 11.4 with $n = 2$, determine C and θ in terms of A_1, A_2, ϕ_1, and ϕ_2.

Solution:

When $n = 2$, then

$$\sum_{k=1}^{2} A_k\sin(\omega_1 t+\phi_k) = A_1\sin(\omega_1 t+\phi_1) + A_2\sin(\omega_1 t+\phi_2)$$

and $\quad A_1\sin(\omega_1 t+\phi_1) + A_2\sin(\omega_1 t+\phi_2) = (A_1\cos\phi_1 + A_2\cos\phi_2)\sin\omega_1 t$

$$+ (A_1\sin\phi_1 + A_2\sin\phi_2)\cos\omega_1 t$$

Now $\quad C = \left[\left(A_1\cos\phi_1 + A_2\cos\phi_2\right)^2 + \left(A_1\sin\phi_1 + A_2\sin\phi_2\right)^2\right]^{1/2}$

$$= [A_1^2\cos^2\phi_1 + A_1^2\sin^2\phi_1 + A_2^2\cos^2\phi_2 + A_2^2\sin^2\phi_2 +$$

$$2A_1 A_2\cos\phi_1\cos\phi_2 + 2A_1 A_2\sin\phi_1\sin\phi_2]^{1/2}$$

$$= \left[A_1^2 + A_2^2 + 2 A_1 A_2 \left(\cos \phi_1 \cos \phi_2 + \sin \phi_1 \sin \phi_2 \right) \right]^{1/2}$$

$$= \left[A_1^2 + A_2^2 + 2 A_1 A_2 \cos (\phi_1 - \phi_2) \right]^{1/2}$$

Now we write $\quad \cos \theta = \dfrac{A_1 \cos \phi_1 + A_2 \cos \phi_2}{C}$

and $\quad\quad\quad\quad \sin \theta = \dfrac{A_1 \sin \phi_1 + A_2 \sin \phi_2}{C}$

Hence $\quad\quad\quad\quad \tan \theta = \dfrac{A_1 \sin \phi_1 + A_2 \sin \phi_2}{A_1 \cos \phi_1 + A_2 \cos \phi_2}$

$\Rightarrow \quad\quad\quad\quad \theta = \tan^{-1} \left(\dfrac{A_1 \sin \phi_1 + A_2 \sin \phi_2}{A_1 \cos \phi_1 + A_2 \cos \phi_2} \right)$

Problem 11.6. Solve the following differential equations for the steady-state solution:

(a) $\dfrac{di}{dt} + 2i = \sin 2t$ $\quad\quad\quad\quad$ (b) $\dfrac{di}{dt} + i = \cos 3t$

(c) $\dfrac{di}{dt} + 3i = \cos (2t + 45°)$ $\quad\quad$ (d) $\dfrac{d^2 i}{dt^2} + 2 \dfrac{di}{dt} + i = 5 \sin (2t + 30°)$

(e) $\dfrac{d^2 i}{dt^2} + i = 2 \sin t$

Solution:

(a) Given $\quad\quad\quad \dfrac{di}{dt} + 2i = \sin 2t$ $\quad\quad\quad\quad\quad\quad$ (1)

Now $\quad\quad\quad\quad \sin 2t = \dfrac{e^{j2t} - e^{-j2t}}{2j}$

Let $\quad\quad\quad\quad I_1(t) = I_1 e^{j2t}$

Hence from equation (1)

$$I_1 \left(2j e^{j2t} + 2 e^{j2t} \right) = \left(\dfrac{e^{j2t} - e^{-j2t}}{2j} \right)$$

Comparing coefficient of e^{j2t}, we get

$$(2j+2)I_1 = \frac{1}{2j} \Rightarrow I_1 = \frac{1}{4(-1-j)} = \frac{-1-j}{8}$$

$$i_1 = -\frac{1}{4}\cos 2t + \frac{1}{2}\sin 2t$$

(b) Given $\dfrac{di}{dt} + i = \cos 3t$ (1)

Put $I_1(t) = I_1 e^{j3t}$ (2)

From eqns (1) and (2), we have

$$(j3+1)I_1 = 1/2 \Rightarrow I_1 = \frac{1}{2(1+j3)} = \frac{1-j3}{20}$$

We use $\cos 3t = \dfrac{e^{j3t} + e^{-j3t}}{2}$

$$i = \frac{1}{10}(\cos 3t + 3\sin 3t)$$

(c) Given $\dfrac{di}{dt} + 3i = \cos(2t + 45°)$

Put $I_1(t) = I_1 e^{j2t}$

Since $\cos(2t+45°) = \dfrac{e^{j(2t+45°)} + e^{-j(2t+45°)}}{2}$ (1)

Now $\dfrac{di}{dt} + 3i = I_1(j2e^{j2t} + 3e^{j2t})$ (2)

Comparing eqns (1) and (2) the coefficient of e^{j2t}, we get

$$I_1(j2+3) = \frac{e^{j45°}}{2}$$

$$\Rightarrow \qquad I_1 = 0.139 e^{j11.3°}$$

$$\Rightarrow \qquad i = 0.139\left[e^{j(2t+11.3°)} + e^{-j(2t+11.3°)}\right]$$

$$= 0.278\cos(2t+11.3°)$$

(d) Given equation is $\dfrac{d^2i}{dt^2} + 2\dfrac{di}{dt} + i = 5\sin(2t+30°)$

Put $\qquad\qquad I(t) = Ie^{j2t}$

Then $\dfrac{d^2i}{dt^2} + \dfrac{2di}{dt} + i = \left[-4 + j4 + 1\right)]Ie^{j2t}$ (1)

$$5\sin(2t+30°) = \dfrac{5\left[e^{j(2t+30°)} - e^{-j(2t+30°)}\right]}{2j}$$

$$= \dfrac{5\left[e^{j2t}\cdot e^{j30°} - e^{-j(2t+30°)}\right]}{2j} \qquad (2)$$

From eqns (1) and (2), equating coefficient of e^{j2t}, we get

$$(-4 + j4 + 1)I = \dfrac{5e^{j30°}}{2j} \Rightarrow I_1 = \dfrac{5e^{j30°}}{2j(-3+j4)} = \dfrac{-5e^{j30°}}{(8+j6)}$$

Hence $\qquad\qquad I = 0.5e^{j173.1°}$

$\therefore \qquad\qquad i = 0.5\left[e^{j(2t+173.1)} + e^{-J(2t+173.1)}\right]$

$$i = \cos(2t+173.1°)$$

(e) Given $\dfrac{d^2i}{dt^2} + i = 2\sin t$

Put $\qquad\qquad I(t) = Ite^{jt}$

Then $\qquad \dfrac{d^2i}{dt^2} + i = Ie^{jt}\left[2j-t\right] + Ie^{jt}$

$\because \qquad\qquad \dfrac{dI}{dt} = \dfrac{d}{dt}\left(I + e^{jt}\right) = I\left[tje^{jt} + e^{jt}\right]$

and $\qquad\qquad \dfrac{d^2I}{dt^2} = I\left[tj^2\,e^{jt} + je^{jt} + je^{jt}\right] = Ie^{jt}\left[2-t\right]$

$$\dfrac{d^2I}{dt^2} + I = Ie^{jt}\left[2-t+t\right] = 2Ie^{jt} \qquad (1)$$

$$2\sin t = 2\left[\frac{e^{jt} - e^{-jt}}{2j}\right] = \frac{e^{jt} - e^{-jt}}{2} \qquad (2)$$

From eqns (1) and (2), comparing coefficient of e^{jt}, we get

$$(j2)I = \frac{1}{j}, I = -\frac{1}{2} \quad \therefore \quad i = -\frac{1}{2}t\left(e^{jt} + e^{-jt}\right) = -t\cos t$$

Problem 11.7. Repeat Prob. 11.6 for the following differential equations, solving only for the steady-state solution:

(a) $\dfrac{d^2i}{dt^2} + 2\dfrac{di}{dt} + 2i = 3\cos(t + 30°)$

(b) $\dfrac{d^2i}{dt^2} + 4i = 3\cos(2t + 45°)$

(c) $\dfrac{di}{dt} + 2i = \sin 2t + \cos t$

Solution:

(a) $\dfrac{d^2i}{dt^2} + 2\dfrac{di}{dt} + 2i = 3\cos(t + 30°)$

$$3\cos(t + 30°) = \frac{3}{2} \times 2\cos(t + 30°)$$

$$= 1.5 \times 2\frac{\left[e^{j(t+30°)} + e^{-j(t+30°)}\right]}{2}$$

$$= 1.5\left[e^{j(t+30°)} + e^{-j(t+30°)}\right]$$

$$= 1.5\left[e^{jt} \cdot e^{j30°} + e^{-jt} \cdot e^{-j30°}\right]$$

Taking only e^{jt} coefficient, we have

$$A_1 = \frac{1.5}{(1 + j2)}e^{j30} = \frac{1.5e^{j30°}}{2.24e^{j63.4°}}$$

$$= 0.67e^{j(30 - 63.4)} = 0.67e^{-j33.4}$$

Hence

$$i = \frac{0.67 \times 2}{2} e^{-j33.4} = 1.34 \frac{e^{-j33.4}}{2} = 1.34\cos(t - 33.4).$$

$$= 1.34\cos(t - 33.4°)$$

(b) $\cdot \dfrac{d^2 i}{dt^2} + 4i = 3\cos(2t + 45°) = 1.5\left[e^{j(2t+45°)} + e^{-j(2t+45°)} \right]$

We take homogenous solution as Ite^{j2t} and assume

$$i = Ite^{j2t} , \frac{di}{dt} = Ie^{j2t} + I2\,jte^{j2t}$$

$$\frac{d^2 i}{dt^2} = I2\,je^{j2t} + I(-4)te^{j2t} + I2\,je^{j2t}$$

$$= (-4t + 4j)\,Ie^{j2t}$$

$$\frac{d^2 i}{dt^2} + 4i = (4\,jIe^{j2t}) = 1.5e^{j45°}$$

$\Rightarrow$ $\qquad\qquad I = \dfrac{1.5e^{j45°}}{4j} = \dfrac{0.75}{2}e^{-j45°}$

Hence $\qquad i = 0.75t\cos(2t - 45°)$

(c) Given equation $\dfrac{di}{dt} + 2i = \sin 2t + \cos t$

Now $\qquad\qquad \sin 2t = \dfrac{e^{j2t} - e^{-j2t}}{2j}$

$$\cos t = \dfrac{e^{jt} + e^{-jt}}{2}$$

We take $\qquad I_1(t) = I_1 e^{j2t}$ then

$$\frac{di}{dt} + 2i = (I_1 e^{j2t})2j + 2Ie^{j2t}$$

$$= I_1 e^{j2t}[2j+1] = \dfrac{e^{j2t} - e^{-j2t}}{2j}$$

Comparing coefficient of e^{j2t}, we get from eqn (1)

$$I_1(2j+1) = \frac{1}{2j} \Rightarrow I_1 = \frac{1}{-4+2j} = 0.1766\,e^{-j135°}$$

Again let $\quad I_2(t) = I_2 e^{jt}$ then

$$\frac{di}{dt} + 2i = (I_2 e^{jt})j + 2I_2 e^{jt} = I_2 e^{jt}\left[j+2\right]$$

$$I_2 e^{jt}\left[j+2\right] = \frac{e^{jt} - e^{-jt}}{2} \tag{2}$$

Comparing coefficient of e^{jt} from eqn (2), we get

$$I_2\left[j+2\right] = \frac{1}{2} \Rightarrow I_2 = \frac{1}{(2j+4)} = 0.2446\,e^{-j26.6°}$$

Hence, $i = 0.353\cos(2t-135°) + 0.448\cos(t-26.6)$

Problem 11.8. The network of Fig. P. 11.8 has a sinusoidal voltage source and is operating in the steady state. Determine the steady-state current $i(t)$ if $v_1 = 2\,\cos 2t$.

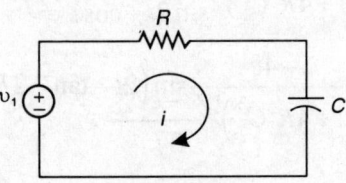

Fig. P. 11.8

Solution:

Using KVL for the given figure:

$$Ri + \frac{1}{C}\int i\,dt = V_1$$

Differentiating both side w.r.t. t, we have

$$R\frac{di}{dt} + \frac{i}{c} = -2(2)\sin 2t \qquad \because\; V_1 = 2\cos 2t$$

$$= -4\sin 2t = -4\left[\frac{e^{j2t}-e^{-j2t}}{2j}\right] = 2j\left[e^{j2t}-e^{-j2t}\right].$$

Let

$$I = I_1 e^{j2t}$$

Then

$$R\frac{d}{dt}(I_1 e^{j2t}) + \frac{I_1 e^{j2t}}{c} = j2e^{j2t}$$

$$R I_1 (2j)e^{j2t} + \frac{I_1 e^{j2t}}{c} = j2e^{j2t}$$

$$\Rightarrow \qquad \left(j2R+\frac{1}{c}\right)I_1 = 2je^{j2t}$$

$$\Rightarrow \qquad I_1 = \frac{2j}{\left(j2R+\dfrac{1}{c}\right)} = \frac{j2c}{(1+j2Rc)}$$

$$= \frac{j2c}{(1+4R^2C^2)^{1/2}}e^{-j\tan^{-1}2RC}$$

$$I = \frac{-4C}{(1+4R^2C^2)^{1/2}}\frac{e^{j\left(2t-\tan 2^{-1}RC\right)}-e^{-j\left(2t-\tan 2^{-1}RC\right)}}{2j}$$

$$= \frac{-4c}{\left(1+4R^2C^2\right)^{1/2}}\sin\left(2t-\tan^{-1}2RC\right)$$

Problem 11.9. In the network of Fig. P. 11.9 $i_1 = 3\cos(t+45°)$ and the network is operating in the steady state. Determine the node-to-datum voltage $v_1(t)$.

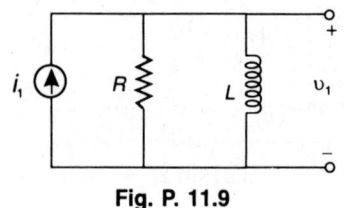

Fig. P. 11.9

Solution:

The differential equation having v_1 and its derivative and having terms including i_1, R and L only

we know $$v_1 = \frac{di_L}{dt}$$

where i_L = steady state current flowing through L

From KCL, we get

$$i_L = i_1 - \frac{V_1}{R} = i_1 - Gv_1 \tag{1}$$

Differentiating equation (1) w.r.t. t, we get

$$\frac{di_L}{dt} = \frac{di_1}{dt} - G\frac{dv_1}{dt}$$

$$\frac{di_1}{dt} = \frac{di_1}{dt} + G\frac{dv_1}{dt} = \frac{v_1}{L} + G\frac{dv_1}{dt}$$

Also

$$\frac{di_1}{dt} = \frac{d}{dt}(3\cos(t+45°)) = -3\sin(t+45°)$$

$$= 3\cos(t+45°+90°)$$

$$= 3\cos(t+135°)$$

$$= 1.5[e^{j(t+135°)} + e^{-j(t+135°)}]$$

$$\therefore \quad G\frac{dv_1}{dt} + \frac{v_1}{L} = 1.5[e^{j(t+135°)} + e^{-j(t+135°)}]$$

Suppose $$v_1 = Ae^{jt}$$

A can be calculated to be

$$A = \frac{1.5e^{j135}}{\left(\frac{1}{L} + jG\right)} = \frac{1.5\,RLe^{j135°}}{(R+jL)} = \frac{1.5\,RL}{(R^2+L^2)^{1/2}}e^{j(135-\tan^{-1}\frac{L}{R})}$$

$$v_1 = \frac{3RL}{(R^2+L^2)^{1/2}}\cos\left(t+135° - \tan^{-1}\frac{L}{R}\right)$$

Problem 11.10. For the given network, find $v_a(t)$ in the steady state if $v_1 = 2\sin 2t$.

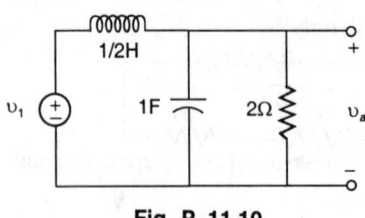

Fig. P. 11.10

Solution:

Given
$$v_1 = 2\sin 2t$$

The node equation for v_a can be written as:

$$\frac{v_a}{2} + 1 \cdot \frac{dv_a}{2} + \frac{1}{1/2}\int (v_a - v_1)dt = 0$$

Differentiating above equation w.r.t. t, we get

$$\frac{1}{2}\frac{dv_a}{dt} + \frac{d^2 v_a}{dt^2} + 2v_a - 2v_1 = 0$$

$$0.5\frac{d^2 v_a}{dt^2} + 0.25\frac{dv_a}{dt} + v_a = v_1$$

$$v_1 = 2\sin 2t = \frac{2(e^{j2wt} - e^{-j2wt})}{2j} = \frac{e^{j2t} - e^{-j2t}}{j}$$

Neglecting terms containing e^{-j2t} we get,

$$0.5\frac{d^2 V_a}{dt^2} + 0.25\frac{dV_a}{dt} + V_a = \frac{e^{j2t}}{j}$$

Let
$$V_a = Be^{j2t}$$

Where $B = \dfrac{-j}{-2 + j0.5 + 1} = \dfrac{j}{1 - j0.5} = 0.895e^{j116.6°}$

Then $v_a = 0.895e^{j(2t + 116.6°)} + 0.895e^{-j(2t + 116.6)}$

$$= 1.799\cos(2t + 116.6°)$$

Problem 11.11. In the resistive network shown in Fig. P. 11.11, $v_1 = 2\sin(2t + 45°)$ for all t. (a) Determine $i_a(t)$. (b) Determine $i_b(t)$.

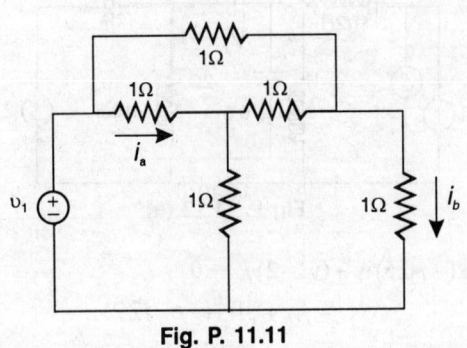

Fig. P. 11.11

Solution:

The network shown in the figure is of a balanced bridge and therefore by observation.

(a) $\qquad i_a(t) = \sin(2t + 45°)$

and (b) $\qquad i_b(t) = \sin(2t + 45°)$

Problem 11.12. The network shown in the following figure is operating in the steady state with sinusoidal voltage sources. If $v_1 = 2\cos 2t$ and $v_2 = 2\sin 2t$, determine the voltage $v_a(t)$.

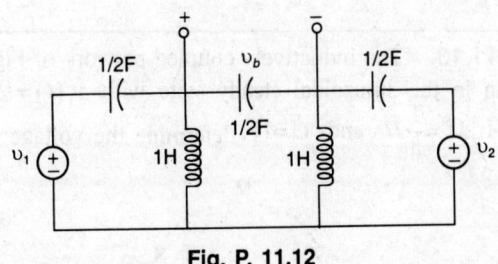

Fig. P. 11.12

Solution:

From the network given, we can write

$$v_1 = j2, \; v_2 = 2, \; Y_1 = -j0.5, \; v_c = j$$

Consider node b, $(v_b - j2)j = v_b(-j0.5) + (v_b - v_c)j = 0$

$$jv_b + 2 - j0.5v_b + jv_b - jv_c = 0$$

$$j1.5v_b - jv_c = -2 \qquad (1)$$

At node C,

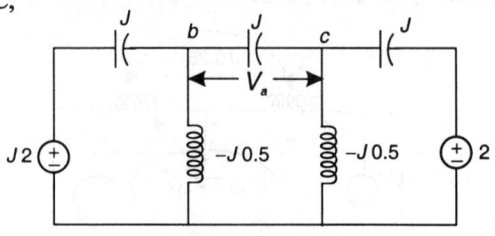

Fig P. 11.12 (a)

$$(v_c - v_b)j + (-j0.5)v_c + (v_c - 2)j = 0$$
$$-jv_b + j1.5v_c = j2 \qquad (2)$$

We can write $\qquad v_b - v_c = v_a$

By Cramer's Rule for eqns (1) and (2), we have

$$v_a = v_b - v_c = \frac{\begin{bmatrix} -2 & -j \\ j2 & j1.5 \end{bmatrix} \begin{bmatrix} j1.5 & -2 \\ -5 & j2 \end{bmatrix}}{\begin{bmatrix} j1.5 & -j \\ -J & j1.5 \end{bmatrix}}$$

Then $\qquad v_a = \dfrac{-j3 - 2 + 3 + j2}{-2.25 + 1} = \dfrac{1.5}{-1.25} = 1.13e^{j135}$

Hence $\qquad v_a = 1.13\cos(2t + 45°)$

Problem 11.13. The inductively coupled network of Fig. P. 11.13 is operating in the sinusoidal steady state with $v_1(t) = 2\cos t.$. If $L_1 = L_2 = 1\text{H}, M = \dfrac{1}{4}H$ and $C = 1\text{F}$, determine the voltage $v_{a(t)}$.

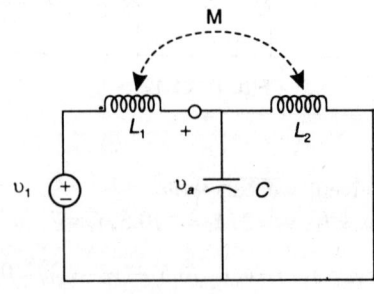

Fig. P. 11.13

Solution:

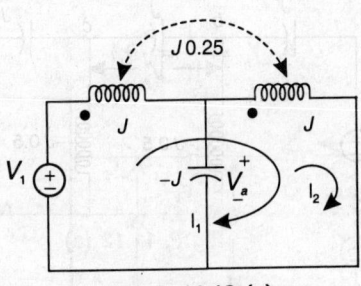

Fig. P. 11.13 (a)

$v_1 = 2$, and $X_1 = X_2 = 1$, mutual inductance $X_m = 0.25$, and $X_c = 1$

$$X_m = \frac{1}{4} = 0.25$$

Writing loop equation, we have

$$2 = j0.25\,I_1 + j1.25\,I_2$$

and $$0 = j1.25\,I_1$$

When $$I_1 = 0, \quad I_2 = -j1.6$$

Hence $$V_a = 1.6\,\cos t$$

Problem 11.14. The network of the following figure is operating in the sinusoidal steady state. In the network, it is determined that $v_a = 10\sin(1000t + 60)$ and $v_b = 5\sin(1000t - 45°)$. The magnitude of the impedance of the capacitor is 10 Ω. Determine the impedance at the input terminals of the network N.

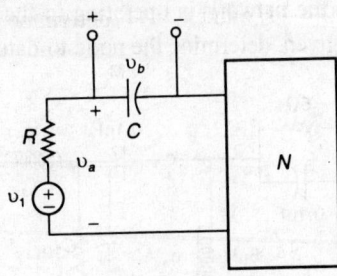

Fig. P. 11.14

Solution:

For the network given in the problem, we have

$$v_a = 10e^{j60°} \text{ and } v_b = 5e^{-j45°}$$

$$Y_b = \frac{1}{10} = j0.1$$

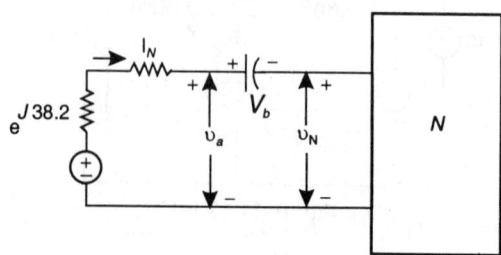

Fig. P. 11.14 (a)

Now $$v_N = v_a - v_b$$

and $$I_N = v_b Y_b$$

Impedance at the input terminal =

$$Z_N = \frac{v_N}{I_N} = \frac{v_a - v_b}{v_a Y_b}$$

$$= \frac{10e^{j60} - 5e^{-j45°}}{0.5e^{j45°}} = 24.5e^{j38.2°}$$

Problem 11.15. In the network shown, $v_1 = 10\sin 10^6 t$ and $i_1 = 10\cos 10^6 t$ and the network is operating in the steady state. For the element values given, determine the node-to-datum voltage $v_a(t)$.

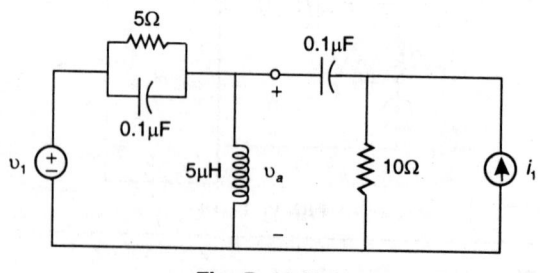

Fig. P. 11.15

Solution:

The given network can be transformed as:

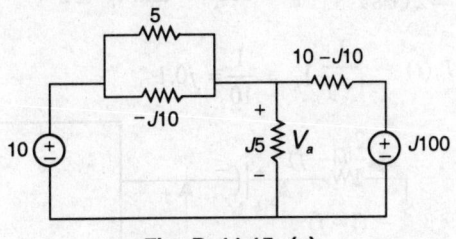

Fig. P. 11.15. (a)

put
$$V_1 = 10, \quad I_1 = j10$$
$$Z_L = -j10, \quad Z_L = j5$$

Now
$$\frac{1}{10-j10} = \frac{10+j10}{100+100} = \frac{10\,j10}{200} = 0.05 + j0.05$$

$$v_a = \frac{2+j-j+j5}{0.2+j0.1-j0.2+j0.05} = \frac{-3+j6}{0.25-j0.05} = \frac{6.7e^{j116.6°}}{0.255e^{-j11.3°}}$$

Hence
$$v_a = 26.3\,e^{j(116.6+11.3)} = 26.3\,e^{j(127.9)}$$
$$= 26.3\sin(10^6 t + 127.9°)$$

or,
$$v_a = 26.3\sin(10^6 t + 90° + 37.9°) = 26.3\cos(10^6 t + 37.9°)$$

Problem 11.16. For the bridged-T network of the following figure $v_1 = 2\cos t$ and the system is in the steady state. For this network, (a) determine $i_a(t)$ and (b) determine $i_b(t)$.

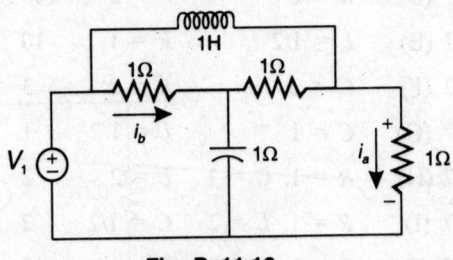

Fig. P. 11.16

Solution:

The given bridged T network is a balanced bridge for all frequencies.
Now given $v_1 = 2\cos t$

(a) $I_a(t) = \dfrac{2}{1+j}$

$= \dfrac{2}{2}(1-j)$

$= (1-j)$

$= 1.414 e^{-j45°}$

$= 1.414 \cos(t-45°)$

(b) $I_b(t) = \dfrac{2}{1-j}$

$= \dfrac{2}{2}(1+j)$

$= (1 + j)$

$= 1.414 e^{j45°}$

$= 1.414 \sin(t+45°)$

Problem 11.17					
	Network 1	*Network 2*	V_m	ω	ϕ
11.17 (A)	$R = 1$	$C = 2$	2	1	$-30°$
11.17 (B)	$R = 2$	$C = 1$	10	2	$45°$
11.17 (C)	$R = 20$	$C = 1/2$	1	0.1	$0°$
11.17 (D)	$R = 2$	$L = 2$	100	1/2	$30°$
11.17 (E)	$L = 1/2$	$R = 1$	10	1/2	$0°$
11.17 (F)	$C = 2$	$R = 2$	3	1	$45°$
11.17 (G)	$C = 1$	$L = 1/2$	1	2	$0°$
11.17 (H)	$R = 1, C = 1$	$L = 2$	2	1	$30°$
11.17 (I)	$R = 1, L = 2$	$C = 1/2$	2	1	$45°$
11.17 (J)	$L = 1, C = 2$	$R = 1$	10	1/2	$0°$

Solution: 11.17 (A)

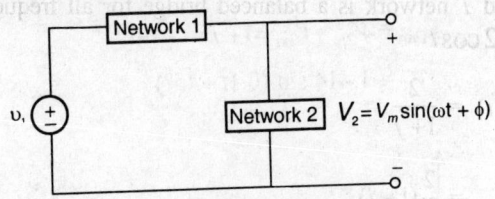

Fig. P. 11.17 (a)

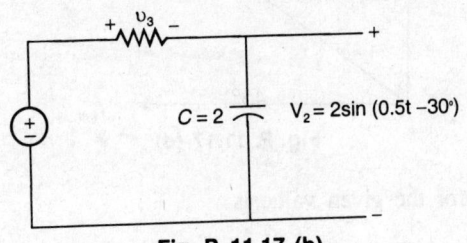

Fig. P. 11.17 (b)

We can write $\quad V_2 = \sin(0.5t - 30°)$

$$V_{2m} = 2e^{-j30°}, \; Y_2 = 2e^{-j60°}, \; Z_1 = 1$$

$$V_{3m} = I_{cm}Z_1 = 2e^{j60} \quad (\because I_{cm} = V_{2m})$$

Now $\quad V_{1m} = V_{2m} + V_{3m} = 2.732 + j0.732 = 2.82e^{j15}$

Hence $\quad V_1 = 2.82 \sin(0.5t + 15°)$

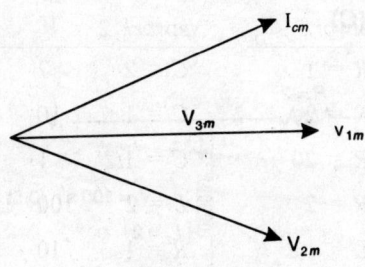

Fig. P. 11.17 (c)

Solution: 11.17 (B)

We can write from the given data

$$V_2 = \sin(0.1t)$$
$$V_{2m} = 1, \; Y_2 = j0.05$$
$$I_{c2m} = V_{2m}Y_2 = j0.05, \; Z_1 = 20$$

$$V_{3m} = I_{c2m}, \ Z_1 = j$$

Now $\qquad V_{1m} = V_{2m} + V_{3m} = 1 + j = 1.414 \, e^{j45°}$

$\therefore \qquad\qquad V_1 = 1.414 \sin (0.1t + 45°)$

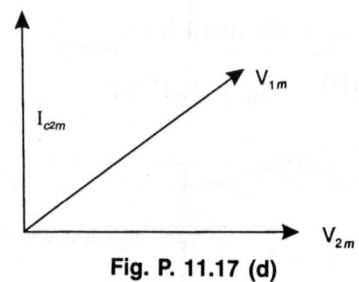

Fig. P. 11.17 (d)

The network for the given value is

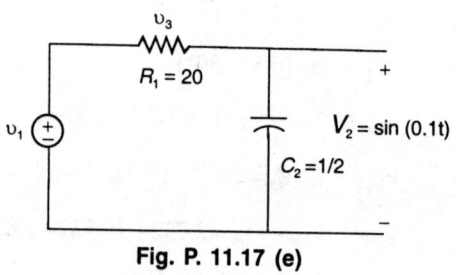

Fig. P. 11.17 (e)

Solution: 11.17 (C)

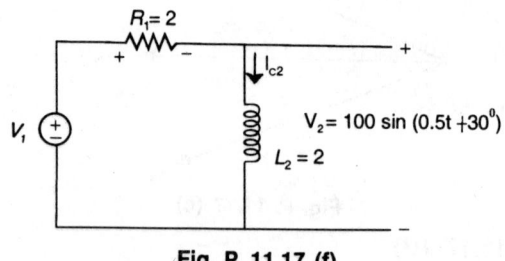

Fig. P. 11.17 (f)

Given $\qquad V_{2m} = 100 \, e^{j30°} \quad Y_2 = -j$

Now $\qquad I_{c2m} = V_{2m} Y_2 = 100 \, e^{-j60°}, \quad Z_1 = 2$

$$V_{3m} = I_{L2m}Z_1 = 200\,e^{-j60°}$$

Hence
$$V_{1m} = V_{2m} + V_{3m} = 100\,e^{j30} + 200\,e^{-j60°}$$
$$= (186.6 - j123.2) = 224\,e^{-J33.5°}$$
$$V_1 = 224 \sin (0.5t - 33.5°)$$

Solution: 11.17 (D)

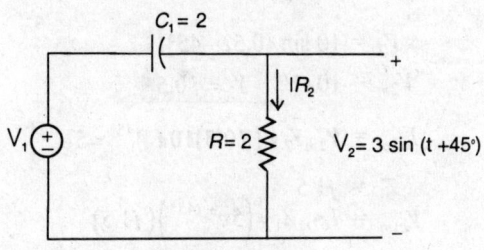

Fig. p. 11.17 (g)

Now
$$V_2 = 3 \sin (t + 45°)$$
$$V_{2m} = 3\,e^{j45°}, \; Y_2 = 0.5$$

Again
$$I_{R2m} = V_{2m}Y_2 = 1.5\,e^{j45°}$$
$$Z_1 = j0.5$$
$$V_c I_m = I_{R2m}Z_1 = 0.75\,e^{-j45°}$$
$$V_{1m} = I_{R2m}Z_1 = 0.75\,e^{-j45°}$$
$$V_{1m} = V_{c1m} + V_{2m} = 0.75\,e^{-j45°} + 3\,e^{j45°} = 3.09\,e^{j31°}$$
$$V_1 = 3.09 \sin (t + 31°)$$

Solution: 11.17 (E)

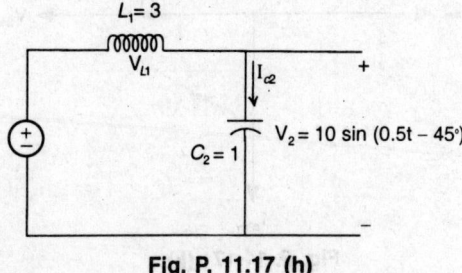

Fig. P. 11.17 (h)

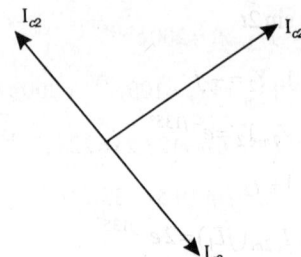

Fig. P. 11.17 (i)

$$V_2 = 10 \sin (0.5t - 45°)$$
$$V_{2m} = 10e^{-j45°}\ Y_2 = j0.5$$

Hence $\qquad I_{c2m} = V_{2m}Y_2 = (j0.5)10\,e^{-j45°} = 5\,e^{-j45°}$

and $\qquad Z_1 = j1.5$
$$V_{c1m} = I_{c2m}Z_1 = \left(5e^{+j45°}\right)(j1.5)$$
$$= (7.5)e^{+j(45°+90)} = 7.5\,e^{j(135°)}$$

Now $\qquad V_{1m} = V_{L1m} + V_{2m} = 3.09\,e^{j31°}$

and $\qquad V_1 = 2.5 \sin (0.5t - 45°)$

Solution: **11.17 (F)**

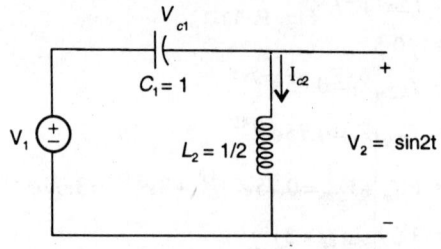

Fig. P. 11.17 (j)

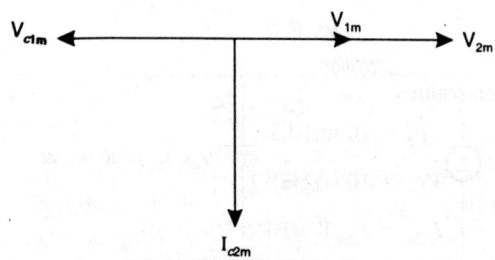

Fig. P. 11.17 (k)

From the figure $V_2 = \sin 2t$

and

$$V_{2m} = 1,\ Y_2 = -j$$
$$I_{L2m} = V_{2m}Y_2 = e^{-j135°}$$
$$Z_1 = 1 + j2$$
$$V_{L1m} = I_{c2m}(jL_1) = 2e^{-j135}$$
$$V_{R1m} = I_{c2m}\ R_1 = e^{j135°}$$
$$V_{1m} = V_{R1m} + V_{L1m} + V_{2m} = \frac{-1+j}{\sqrt{2}}$$
$$= e^{j135°} \text{ and } v_1 = \sin(t + 135°)$$

Solution: **11.17(G)**

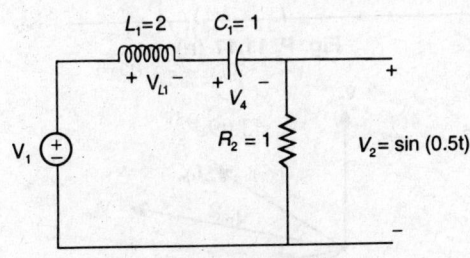

Fig. P. 11.17 (l)

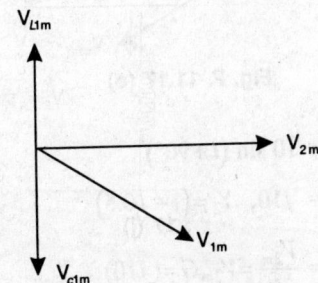

Fig. P. 11.17 (m)

From the given values

$$V_2 = 10\sin(0.5t)$$
$$V_{2m} = 10,\ Y_2 = 1$$
$$I_{R2m} = V_{2m}Y_2 = 10$$
$$Z_1 = (j0.5 - j) = -j0.5$$

$$V_{c1m} = (I_{R2m})(j \times X_L) = j0.5$$

$$I_{c1m} = (I_{R2m})(-jX_L) = -j10$$

Again $$V_{1m} = V_{L1m} + V_{c1m} + V_{2m} = 11.2 e^{-j26.6°}$$

$$V = 11.2 \sin (0.5t + 26.6°)$$

Solution: 11.17 (H)

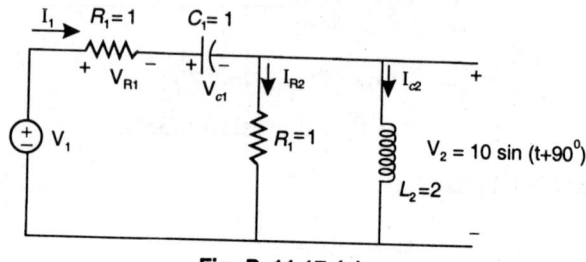

Fig. P. 11.17 (n)

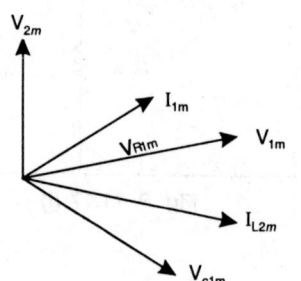

Fig. P. 11.17 (o)

From the values $$V_2 = 10 \sin (t + 90°)$$

$$V_{2m} = j10, \quad Y_2 = (1 - j0.5)$$

$$I_{R2m} = \frac{V_{2m}}{R} = V_{2m}G = (j10)$$

$$I_{L2m} = V_{2m}(j2) = 2$$

$$I_{1m} = 5 + j10$$

$$V_{R1m} = (I_{1m}R_1) = (5 + j10)1 = (5 + j10)$$

$$V_{c1m} = (I_{1m})(j \times X_c) = (5 + j10)(-j) = (-5 + j10)$$

Therefore $$V_{1m} = V_{R1m} + V_{c1m} + V_{2m} = (5 + j10) + (10 - j5) + j10$$

$$= (15 + j15) = 21.2 e^{j45°}$$

$$V_1 = 21.2\sin(t+45°)$$

Solution: 11.17 (I)

From the given values we form the network as below

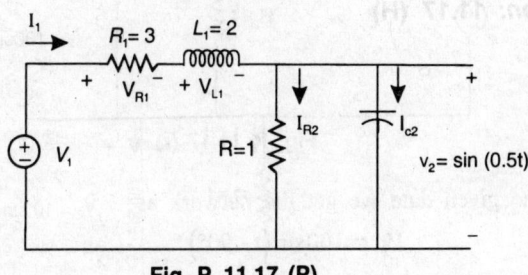

Fig. P. 11.17 (P)

Now

$$V_2 = \sin 0.5t$$

$$V_{2m} = 1, \ Y_2 = (1+j0.25), \ Z_1 = (3+j)$$

$$I_{R2m} = V_{2m}G = 1$$

$$I_{c2m} = V_{2m}(j2) = j0.25$$

$$I_{1m} = I_{R2m} + I_{c2m} = (1+j0.25)$$

$$V_{L1m} = (I_{1m})(jX_L) = (0.25+j)$$

Then

$$V_{1m} = 3.75 + j1.75$$

$$V_{1m} = 41.2 e^{j25°}$$

Fig. P. 11.17 (q)

Hence

$$V_1 = 4.12[\sin(0.5t+25°)]$$

Solution: 11.17 (J)

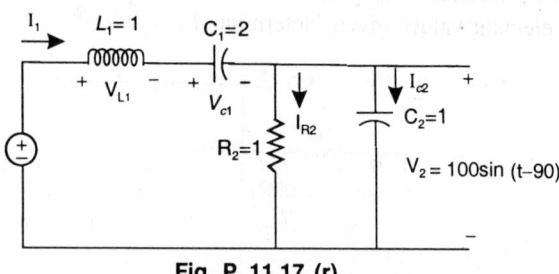

Fig. P. 11.17 (r)

From the given data we get the network as

$$V_2 = 100\sin(t - 90°)$$

$$V_{2m} = j100, \quad Y_2 = (1 + j), Z_1 = j - j0.5 = j0.5$$

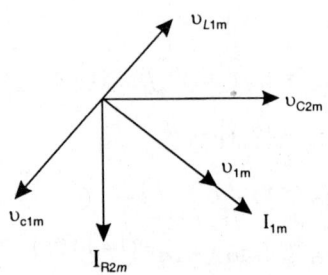

Fig. P. 11.17 (s)

$$I_{R2m} = V_{2m}G = -j100,$$

$$I_{c2m} = V_{2m}(jX_c) = 100$$

$$I_{1m} = I_{c1m} - I_{R2m} = (100 - j100)$$

$$V_{c1m} = I_{1m}(jX) = -50 - j50$$

$$V_{L1m} = I_{1m}(jX) = 100 + j100$$

$$V_{1m} = V_{L1m} + V_{c1m} + V_{Lm}$$

$$= (100 + j100) + (-50 - j50) + (j100)$$

$$= (50 - j150) = 70.7\,e^{-j45°}$$

$$V_1 = 70.7\sin(t - 45°)$$

Problem 11.18. The network of the following figure is operating in the sinusoidal steady state and it is known that $v_3 = 2\sin 2t$. For the element values given, determine $V_2/V_1 = Ae^{j\phi}$

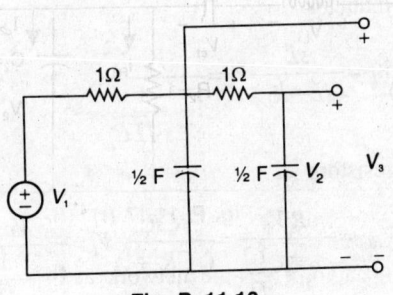

Fig. P. 11.18

Solution:

$$V_3 = 2\sin 2t$$

$$V_{30} = 2 \quad Z_c = \frac{-j}{\omega c} = -j$$

$$V_{30} = \frac{-jV_3}{(1-j)} = (1-j) = \sqrt{2}\,e^{-j45°}$$

Then

$$I_1 = jV_3 + V_3 - V_2 = j2 + 2 - 1 + j = 1 + j3$$

$$V_{30} = 1 + j3 + 2 = 3 + j3 = 3\sqrt{2}\,e^{j45°}$$

Problem 11.19. The network of the following figure is adjusted so that $R_L = R_C = \sqrt{L/C}$. (a) Draw a complete phasor diagram showing all voltages and currents (and their relationships to each other) for the condition $|I_L| = |I_C|$. (b) Let the frequency for the condition of part (a) be ω_1. Draw a phasor diagram for a frequency $\omega_2 > \omega_1$. (c) Repeat part (b) for a frequency $\omega_3 < \omega_1$.

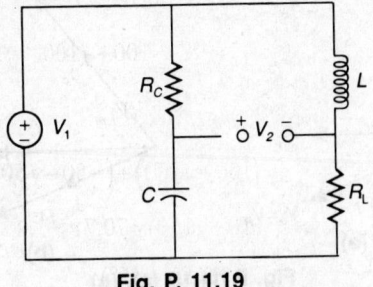

Fig. P. 11.19

Solution:

Given
$$R_L = R_C = \sqrt{\frac{L}{C}}$$

Voltage across inductor

$$L = V_L = \frac{sL}{(sL + R_L)}V_1 = \frac{V_1}{1 + \dfrac{1}{s\sqrt{LC}}} \tag{1}$$

Voltage across resistors

$$R_C = V_{RC} = \frac{R_c V_1}{(R_c + \dfrac{1}{sC})} = \frac{sCR_c}{(sCR_c + 1)}V_1 = \frac{V_1}{1 + \dfrac{1}{s\sqrt{LC}}} \tag{2}$$

From eqns (1) and (2) we observe, $V_L = R_c$ and this is true for all frequencies

$$V_c = V_{RL} = V_1\left[1 - \frac{1}{1 + \dfrac{1}{s\sqrt{LC}}}\right] = \frac{1}{1 + s\sqrt{LC}}$$

For all frequencies $I_L = \left(\dfrac{C}{L}\right)^{1/2}$

$$V_{RL} = \left(\frac{C}{L}\right)^{1/2}\frac{1}{1 + s\sqrt{LC}}$$

$$I_c = \left(\frac{C}{L}\right)^{1/2}V_{RC} = \left(\frac{C}{L}\right)^{1/2}\frac{V_1}{1 + \dfrac{1}{s\sqrt{LC}}}$$

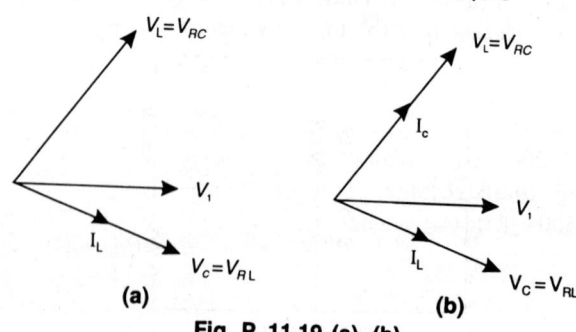

(a) (b)

Fig. P. 11.19 (a), (b)

given
$$|I_L| = |I_c| \text{ when } \omega_1 = \frac{1}{\sqrt{LC}}$$

$$V_L(\omega t) = V_{Rc}(\omega_1) = 0.707 V_1 e^{j45°}$$

$$I_L = -0.707 \left(\frac{C}{L}\right)^{1/2} V_1 e^{-j45°}$$

$$V_L(\omega_1) = V_{RL}(\omega_1) = 0.707 V_1 e^{j45°}$$

$$I_c = 0.707 \left(\frac{C}{L}\right)^{1/2} V_1 e^{j45°}$$

Now
$$\omega = 2\omega_1$$

$$V_L = V_{Rc} = \frac{V_1}{1 - j0.5} = 0.895 V_1 e^{j26.6°}$$

$$V_c = V_{RL} = \frac{V_1}{(1 + j2)} = 0.447 V_1 e^{-j63.4°}$$

Then current
$$I_L = 0.447 \left(\frac{C}{L}\right)^{1/2} e^{-j63.4°}$$

$$I_c = 0.895 \sqrt{\frac{C}{L}} e^{j26.6°}$$

For
$$\omega = \frac{\omega_1}{2}, V_L = V_{Rc} = \frac{V_1}{1 - j2} = 0.447 e^{j63.4°}$$

$$V_c = V_{RL} = \frac{V_1}{1 + j0.5} = 0.895 e^{-j26.6°}$$

∴
$$I_c = 0.447 \frac{C}{L} e^{j63.4°}$$

Problem 11.20. Two circuit elements in a series connection have current and total voltage, $i = 13.42 \sin(500t - 53.4°)$ (A) and $v = 150 \sin(500 + 10°)$ (V). Identify the two elements.

Solution:

Since i lag v by 63.4°, hence element, i.e. R and L

$$Z = \frac{150}{13.42} = 11.17\,\Omega$$

$$Z = \sqrt{R^2 + (\omega L)^2}$$

$$\tan\theta = \tan(63.4°) = \frac{\omega L}{R} = \frac{500L}{R} = 1.99638$$

or

$$L = \frac{1.99638R}{500}$$

∴

$$Z = \sqrt{R^2 + (1.99638R)^2}$$

$$11.17 = R\sqrt{1 + 3.98707} = R(2.23)$$

∴

$$R = \frac{11.17}{2.23} = 5.01\,\Omega$$

and

$$L = \frac{1.996 \times 5.01}{500}$$

$$L = 20\ \text{mH}\ \textbf{Ans.}$$

Problem 11.21. Two circuit elements in a series connection have current and total voltage

$$i = 4.0\cos(2000t + 13.2°)\,\text{A}$$

$$v = 200\sin(2000t + 50.0°)\,\text{V}$$

Identify the two elements

Solution:

Since $\quad \sin(90° + \theta) = \cos\theta$

∴

$$i = 4.0\cos(2000t + 13.2°)$$

$$= 4.0\sin(2000t + 103.8°)$$

$$v = 200\sin(2000t + 50.0°)$$

Since *i* leads *v* by an angle 53.8°, hence circuit contain *R* and *C*.

∴

$$Z = \sqrt{R^2 + \left(\frac{1}{\omega C}\right)^2} = \frac{200}{4} = 50$$

and

$$\theta = \tan^{-1}(1/\omega RC)$$

∴

$$1/\omega RC = \tan\theta = 1.366$$

or $$\frac{1}{\omega C} = 1.366\,R$$

∴ $$Z = \sqrt{R^2 + \left(1.366R\right)^2} = 50$$

$$1.69\ R = 50$$

$$R = \frac{50}{1.69} = 30\,\Omega$$

∴ $$R = 30\,\Omega$$

$$\frac{1}{\omega C} = 1.366\ R = 1.366 \times 30$$

or $$C = \frac{1}{2000 \times 1.366 \times 30}$$

or $$C = 1.24\,\mu F \quad \textbf{Ans.}$$

Problem 11.22. A series RC circuit with $R = 27.5\,\Omega$ and $C = 66.7\,\mu F$, has sinusoidal voltages and current, with angular frequency 1500 rad/sec. Find the angle by which the current leads the voltage.

Solution:

$$Z = \sqrt{R^2 + \left(\frac{1}{\omega c}\right)^2} = \sqrt{(27.5)^2 + \left(\frac{1}{1500 \times 66.7 \times 10^{-6}}\right)^2}$$

$$= \sqrt{(27.5)^2 + (10)^2}$$

$$Z = 29.26$$

$$\tan\theta = \frac{1}{\omega R C} = \frac{1}{1500 \times 27.5 \times 66.7 \times 10^{-6}}$$

$$\tan\theta = 0.36$$

∴ $$\theta = 20°$$

Hence current leads voltage by 20° angle. **Ans.**

Problem 11.23. A series RLC circuit, with $R = 15\,\Omega, L = 80\,mH$ and $C = 30\,\mu F$, has a sinusoidal current at angular frequency 500 rad/s. Determine the phase angle and whether the current leads or lags the total voltage.

Solution:

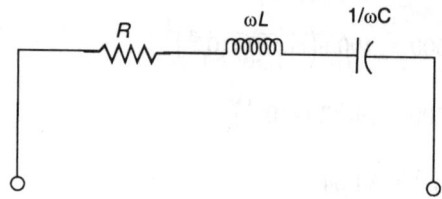

Fig. P. 11.23

$$Z = R + J\left(\omega L - \frac{1}{\omega C}\right)$$

$$\tan\theta = \frac{X_L - X_c}{R} = \frac{\omega L - \dfrac{1}{\omega C}}{R} = \frac{\omega^2 LC - 1}{R} = \frac{-4}{2.25}$$

$$\therefore \qquad \theta = -60.6°$$

since $\quad \theta = -$ ve, hence current leads the voltage.

Problem 11.24. A capacitance $C = 35\,\mu F$ is in parallel with a certain element. Identify the element, given that the voltage and total current are $v = 150 \sin 3000t$, $i_T = 16.5 \sin(3000t + 72.4°)$

Solution:

For parallel RC $\qquad\qquad \tan\theta = \omega RC$

Here $\qquad\qquad\qquad \theta = 72.4°$ and $\omega = 3000$

$\therefore \qquad\qquad\qquad \tan 72.4° = \omega RC$

$$3.152 = 3000\,R \times 35 \times 10^{-6}$$

$\therefore \qquad\qquad\qquad R = 30.1\,\Omega$ **Ans.**

Problem 11.25. A two-element series circuit, with $R = 20\,\Omega$, and $L = 20$ mH, has an impedance $40.0\angle\theta\,\Omega$. Determine the angle θ and the frequency.

Solution:

$\because \qquad\qquad\qquad Z = \sqrt{R^2 + (\omega L)^2}$

$$\therefore \quad 40 = \sqrt{(20)^2 + \left(\omega \times 20 \times 10^{-3}\right)^2}$$

or
$$1600 = 400 + \left(\omega \times 20 \times 10^{-3}\right)^2$$

or
$$1200 = \left(\omega \times 20 \times 10^{-3}\right)^2$$

or
$$\omega \times 20 \times 10^{-3} = 34.64$$

or
$$\omega = 1732 \text{ rad/sec}$$

$$\therefore \quad \tan\theta = \frac{\omega L}{R}$$

or
$$\tan\theta = \frac{34.64}{20} = 1.732$$

$$\therefore \quad \theta = \tan^{-1} 1.732$$

$$\theta = 60°$$

$$\because \quad 2\pi f = \omega = 1732$$

$$\therefore \quad f = \frac{1732}{2\pi}$$

$$\therefore \quad f = 275.6 \text{ Hz } \textbf{Ans.}$$

Problem 11.26. Determine the impedance of the series RL circuit with $R = 25\,\Omega$, $L = 10$ mH at *(a)* 100 Hz *(b)* 500 Hz *(c)* 1000 Hz.

Solution:

(a) For frequency 100 Hz

$$\omega = 2\pi \times 100 = 628.31 \text{ rad/sec.}$$

$$\therefore \quad \omega L = 628.31 \times 10 \times 10^{-3} = 6.28$$

$$\therefore \quad |Z| = \sqrt{R^2 + (\omega L)^2} \quad \text{For series } RL$$

$$\therefore \quad |Z| = \sqrt{(25)^2 + (6.28)^2}$$

$$\therefore \quad |Z| = \sqrt{664.47}$$

$$|Z| = 25.8$$

$$\theta = \tan^{-1}\left(\frac{\omega L}{R}\right) = \tan^{-1}\frac{6.28}{25}$$

$\therefore$ $\qquad\qquad\qquad\theta = 14.1°$

$\therefore$ $\qquad\qquad\qquad Z = 25.8 \angle 14.1°$

(b) at 500 Hz $\qquad\omega = 2\pi \times 500 = 3141.6\,\text{rad/sec.}$

$\therefore$ $\qquad\qquad\qquad\omega L = 31.416\,\Omega$

$\therefore$ $\qquad\qquad\qquad |Z| = \sqrt{25^2 + 31.41^2} = \sqrt{1611.96}$

$\therefore$ $\qquad\qquad\qquad |Z| = 40.14\,\Omega$

$$\theta = \tan^{-1}\left(\frac{\omega L}{R}\right) = \tan^{-1}\left(\frac{31.41}{25}\right)$$

$\therefore$ $\qquad\qquad\qquad\theta = 51.5°$

$\therefore$ $\qquad\qquad\qquad Z = 40.14 \angle 51.5°$

(c) at 1000 Hz $\qquad\omega = 2\pi \times 1000 = 6283.18\,\text{rad/sec.}$

$\therefore$ $\qquad\qquad\qquad\omega L = 62.83\,\Omega$

$\therefore$ $\qquad\qquad\qquad |Z| = \sqrt{25Z^2 + 62.83^2} = 67.62$

and $\qquad\qquad\qquad\theta = \tan^{-1}\frac{62.83}{25} = 68.3°$

$\therefore$ $\qquad\qquad\qquad Z = 67.62 \angle 68.3°$ **Ans.**

Problem 11.27. Determine the circuit constants of a two-element circuit if the applied voltage $v = 150 \sin(5000\,t + 45°)$ results in a current $i = 3.0 \sin(5000\,t - 15°)$.

Solution:

Since current lags voltage by angle 60°. Hence circuit is *RL* circuit.

$\therefore$ $\qquad\qquad R = Z \cos\theta = \dfrac{150}{3}\cos 60° = 50 \times 1/2 = 25\,\Omega$

and $\qquad\qquad\omega L = Z \sin\theta$

$\qquad\qquad\qquad 5000\,L = 50 \sin 60°$

or $\qquad\qquad\qquad L = \dfrac{50 \times 0.866}{5000}$

or $\qquad\qquad\qquad L = 8.66\ \text{mH}$ **Ans.**

Problem 11.28. A series circuit of $R = 10\Omega$ and $C = 40\mu F$ has an applied voltage $v = 500 \cos(2500\,t - 20°)$ V. Find the resulting current i.

Solution:

$$\omega = 2500$$

$$\tan\theta = \frac{1}{\omega RC} = \frac{1}{2500 \times 10 \times 40 \times 10^{-6}}$$

or

$$\theta = \tan^{-1}(1)$$

$$\therefore \qquad \theta = 45°$$

Since circuit is RC, hence current leads the voltage by an angle $45°$, so current angle

$$\theta' = 45° - 20° = 25°$$

$$\therefore \qquad Z = \sqrt{R^2 + \left(\frac{1}{\omega c}\right)^2} = \sqrt{10^2 + (10)^2}$$

$$Z = 10\sqrt{2}$$

current

$$i = A\cos(\omega t + \theta')$$

$$A = \frac{500}{10\sqrt{2}} = 25\sqrt{2}$$

$$\therefore \qquad i_2 = 25\sqrt{2}\cos(2500t + 25°) \quad \textbf{Ans.}$$

Problem 11.29. Three impedances are in series: $Z_1 = 30\angle 45°$, $Z_2 = 10\sqrt{2}\angle 45°$ and $Z_3 = 5.0\angle -90°\,\Omega$. Find the applied voltage V_1 if the voltage across Z_1 is $27.0\angle -10°$ V.

Solution:

Z_1, Z_2 and Z_3 are in series

$$\therefore \qquad Z_{eq} = Z_1 + Z_2 + Z_3$$

$$= 3.0\angle 45° + 10\sqrt{2}\angle 45° + 5.0\angle -90°$$

$$= 2.12 + j2.12 + 10\,j10 + 0 - j5$$

$$= 12.12 + j7.12$$

$$Z_{eq} = 14\angle 30°$$

$\because$ $V_1 = 27.0\angle-10°$

$\therefore$ $I_1 = \dfrac{V_1}{Z_1} = \dfrac{27.0\angle-10°}{3.0\angle45°} = 9\angle-55°$

$\therefore$ Total applied voltage $= Z_{eq}I_1$

$\therefore$ $V = 14\angle30°\times9\angle-55°$

$= 126\angle30°-55°$

$V = 126\angle-25°$ **Ans.**

Problem 11.30. For the three-element series circuit in Fig. P. 11.30 (a) Find the current I; (b) find the voltage across each impedance and construct the voltage phasor diagram which shows that $V_1 + V_2 + V_3 = 100\angle0°V$.

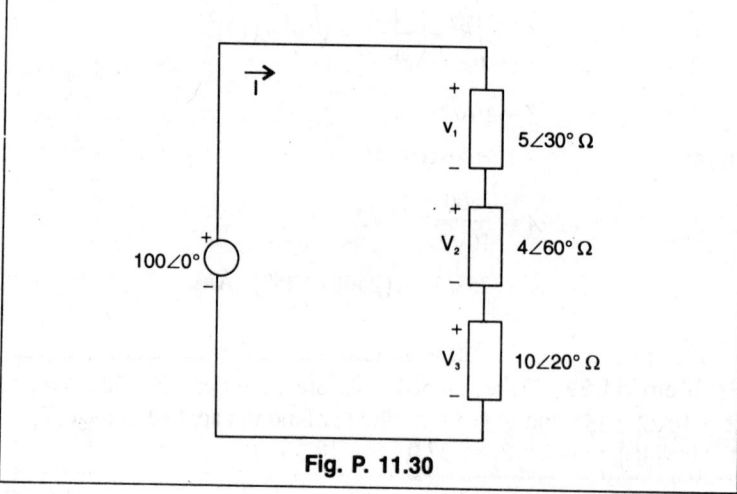

Fig. P. 11.30

Solution:

Since Z_1, Z_2 and Z_3 are in series

$\therefore$ $Z_{eq} = Z_1 + Z_2 + Z_3 = 5\angle30° + 4\angle60° + 10\angle-20°$

$= 4.33 + j2.5 + 2 + j3.46 + 9.39 - j3.42$

$= 15.72 + j2.54 = 16\angle9.17°$

(a)   Current $I = \dfrac{100\angle0°}{16\angle9.17°}$

$$I = 6.25\angle-9.17°$$

(b) $V_1 = Z_1 I = 5\angle30°\times6.25\angle-9.17°$

$V_1 = 31.25\angle20.83°$

$V_2 = Z_2 I = 4\angle60°\times6.25\angle-9.17°$

$V_2 = 25\angle50.83°$

$V_3 = 10\angle-20°\times6.25\angle-9.17°$

$V_3 = 62.5\angle-29.17°$

$V_{eq} = V_1+V_2+V_3$

$= 31.25\angle20.83+25\angle50.83+6.25\angle-29.17°$

$= 29.2+j11.11+15.79+j19.38+54.57-j30.46$

$= 99.56+j0=99.56\angle0°+V_{eq}\approx100\angle0°$

Its voltage phasor can be expressed as

Taking $100\angle0°$ as reference line

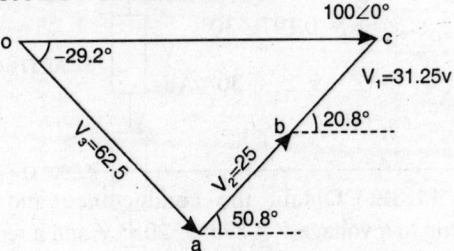

Fig. P. 11.30 (a)

Problem 11.31. Find Z in the parallel circuit of Fig. P. 11.31, if $V=50.0\angle30.0°$V and $I=27.9\angle57.8°$A

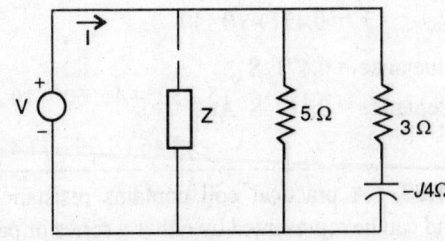

Fig. P. 11.31

Solution:

$$\frac{I}{V} = \frac{1}{Z} + \frac{1}{50} + \frac{1}{3 - jJ4}$$

or $\qquad \dfrac{27.9\angle 57.8°}{50\angle 30.0} = \dfrac{1}{Z} + 0.2 + \dfrac{1}{5\angle -53.13°}$

$$0.558\angle 27.8° = \frac{1}{Z} + 0.2 + 0.2\angle 53.13°$$

$$0.493 + j0.26 = \frac{1}{Z} + 0.2 + 0.12 + j0.16$$

or $\qquad \dfrac{1}{Z} = 0.493 - 0.32 + j0.26 - j0.16$

or $\qquad \dfrac{1}{Z} = 0.17 + j0.1$

or $\qquad \dfrac{1}{Z} = 0.197\angle 30°$

or $\qquad Z = 5\angle -30°$ **Ans.**

Problem 11.32. Obtain the conductance and susceptance corresponding to a voltage $V = 85.0 \angle 205°$ V and a resulting current $I = 41.2 \angle -141.0°$ A.

Solution:

$$Y = \frac{I}{V} = \frac{41.2\angle -141.0°}{85.0\angle 205°} = 0.48\angle -346°$$

or $\qquad Y = 0.471 + j0.117$.

$\because \qquad$ Conductance $= 0.471$ S

and $\qquad$ Susceptance $= 0.117$ S **Ans.**

Problem 11.33. A practical coil contains resistance as well as inductance and can be represented by either a series or parallel circuit, as suggested in Fig. P. 11.33. Obtain R_p and L_p in terms of R_s and L_s.

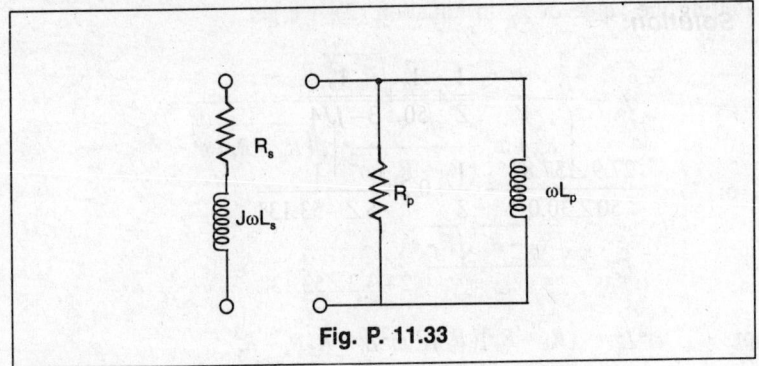

Fig. P. 11.33

Solution:

For series R_s, L_s, $Z_s = R_s + j\omega L_s$

For parallel R_p, L_p

$$Z_p = \frac{j\omega R_p L_p}{R_p + j\omega L_p} = \frac{j\omega R_p L_p \left(R_p - j\omega L_p\right)}{R_p^2 + (\omega L_p)^2}$$

$$Z_p = \frac{\omega^2 R_p L_p^2 + j\omega R_p^2 L_p}{R_p^2 + (\omega L_p)^2}$$

$\because$ $\qquad Z_p = Z_s$

$\because$ $\qquad R_s = \dfrac{\omega^2 R_p L_p^2}{R_p^2 + (\omega L_p)^2}$ $\hfill (1)$

and $\qquad L_s = \dfrac{\omega R_p^2 L_p}{R_p^2 + (\omega L_p)^2}$ $\hfill (2)$

from equation (1)

$$\omega^2 R_p L_p^2 = R_s R_p^2 + R_s (\omega L_p)^2$$

or $\qquad \omega^2 L_p^2 \left(R_p - R_s\right) = R_s R_p^2$

or $\qquad L_p^2 = \dfrac{R_s R_p^2}{(R_p - R_s)\omega^2}$

Putting the value of L_p in equation (2), we get

$$L_s = \frac{R_p^2 \sqrt{R_s R_p^2}}{\left(R_p^2 + \omega^2 \dfrac{R_s R_p^2}{\left(R_p - R_s\right)\omega^2}\right)\sqrt{\left(R_p - R_s\right)\omega^2}}$$

$$L_s = \frac{\sqrt{R_p - R_s}\ \sqrt{R_s}}{\omega}$$

or $\qquad \omega^2 L_s^2 = \left(R_p - R_s\right)\left(R_s\right), \quad \omega^2 L_s^2 = R_p R_s - R_s^2$

or $\qquad R_p = \dfrac{R_s^2 + \omega^2 L_s^2}{R_s} \quad$ or $\quad R_p = R_s + \dfrac{\omega^2 L_s^2}{R_s}$

and $\quad \because \ L_p = \dfrac{R_p \sqrt{R_s}}{\sqrt{R_p - R_s}\ \omega}$

$$\therefore \qquad L_p = \frac{\left(R_s + \dfrac{\omega^2 L_s^2}{R_s}\right)\sqrt{R_s}}{\omega \sqrt{\left(R_s + \dfrac{\omega^2 L_s^2}{R_s} - R_s\right)}}$$

or $\qquad L_p = L_s + \dfrac{R_s^2}{\omega^2 L_s} \quad$ **Ans.**

Problem 11.34. In the network shown in Fig. P. 11.34 the 60 Hz current magnitudes are known to be $I_T = 29.9$ A, $I_1 = 22.3$ A, and $I_2 = 8.0$ A. Obtain the circuit constant R and L.

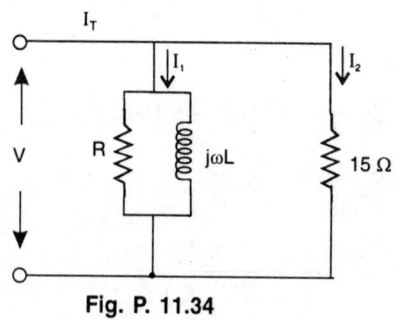

Fig. P. 11.34

Solution:

$$V = I_2 \times 15 = 8 \times 15 = 120\,\text{V}$$

$$Y_1 = \frac{I_1}{V} = \frac{22.3}{120} = 0.185\,s$$

$$Y_{eq} = \frac{I_T}{V} = \frac{29.9}{120} = 0.249\,s$$

$$Y_2 = \frac{1}{15} = 0.0665$$

Let
$$G = 1/R, \quad B = \frac{1}{\omega L}$$

$$\therefore \qquad (0.249\,s)^2 = (0.066\,s + G)^2 + B^2 \tag{1}$$

and
$$(0.185)^2 = G^2 + B^2 \tag{2}$$

Substracting eqn (2) from (1)

$$(0.249)^2 - (0.185)^2 = (0.066)^2 + 2 \times 0.66\,G$$

or
$$G = 1/5.8\,\Omega$$

$$\therefore \qquad R = 5.8\,\Omega$$

$$B = 0.067$$

$$\therefore \qquad \omega L = 14.9\,\Omega$$

or
$$2\pi \times 50\,L = 14.9$$

or
$$L = 38.5\ \text{mH}$$

Problem 11.35. Obtain the magnitude of the voltage V_{AB} in the two-branch parallel network of Fig. P. 11.35, if X_L is (a) $5\,\Omega$ (b) $15\,\Omega$ (c) $0\,\Omega$.

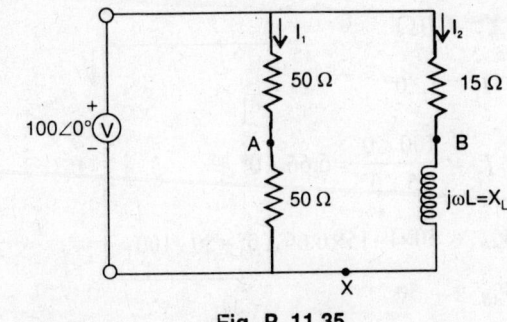

Fig. P. 11.35

Solution:

(a) When $X_L = 5\,\Omega$

$$I_1 = \frac{100\angle 0°}{100\angle 0°} = 1\angle 0°$$

$$I_2 = \frac{100\angle 0°}{15+j5} = \frac{100\angle 0°}{15.8\angle 18.43°}$$

$$= 6.329\angle -18.43°$$

$$V_{AB} = 50\times I_1 - 15 I_2 \;\left(i.e.\; V_A - V_B\right)$$

$$= 50\times 1 - 15\times 6.329\angle -18.43°$$

$$= 50\times 1 - 94.93\angle -18.43°$$

$$= 50-(90-j30)=50-90+j30=-40+j30$$

$$V_{AB} = 50\angle 143.13°$$

(b) When $X_L = 15\,\Omega$

$$I_1 = 1\angle 0°$$

$$I_2 = \frac{100\angle 0°}{15+j15} = \frac{100\angle 0°}{21.21\angle 45°} = 4.71\angle -45°$$

$\therefore$ $$V_{AB} = 50 I_1 - 15\times 4.71\angle -45°$$

$$= 50\times 1 - 70.71\angle -45°$$

$$= 50-(50-j50)=j50$$

$\therefore$ $$V_{AB} = 50\angle 90°$$

(c) When $X_L = 0\,\Omega$

$$I_1 = 1\angle 0°$$

$$I_2 = \frac{100\angle 0°}{15\angle 0°} = 6.66\angle 0°$$

$\therefore$ $$V_{AB} = 50\times 1 - 15\times 6.66\angle 0° = 50-100$$

$$V_{AB} = -50$$

$$\therefore \qquad V_{AB} = 50 \angle 180°$$

Hence in all three case

$$V_{AB} = 50 \text{ volt.}$$

Problem 11.36. In the network shown in Fig. P. 11.36, $v_{AB} = 36.1°$ $\angle 3.18°$. Find the source voltage V.

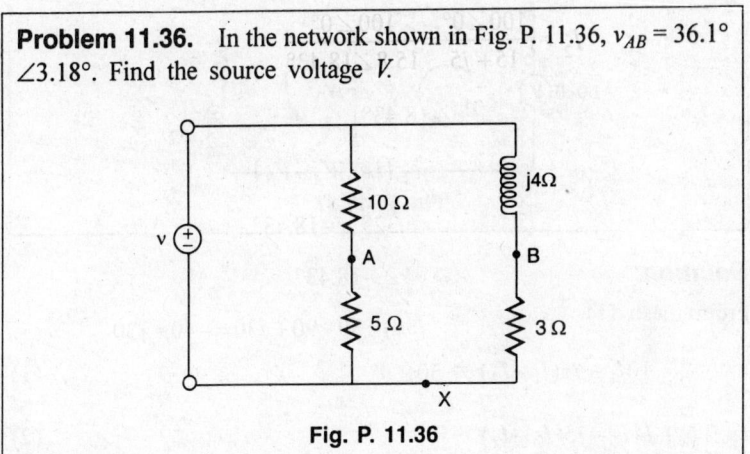

Fig. P. 11.36

Solution:

$$V_{AX} = \frac{V \times 5}{10 + 5} = \frac{V}{3}$$

$$V_{BX} = \frac{V \times 3}{3 + j4} = \frac{3V}{3 + j4}$$

$$V_{AB} = V_{AX} - V_{BX} = \frac{V}{3} - \frac{3V}{3 + j4} = \frac{(3 + j4)V - 9V}{3(3 + j4)}$$

$$36.1 \angle 3.18° = \frac{(-6 + j4)V}{9 + j12}$$

$$36.1 \angle 3.18° = \frac{7.21 \angle 146.3}{15 \angle 53.13} V$$

$$36.1 \angle 3.18° = 0.48 \angle 93.18° V$$

$$\therefore \qquad V = \frac{36.1 \angle 3.18°}{0.48 \angle 93.18°}$$

$$V = 75 \angle -90° \quad \textbf{Ans.}$$

Problem 11.37. For the network of Fig. P. 11.37 assign two different sets of mesh currents and show that for each $\Delta_z = 55.9\angle-26.57°\,\Omega^2$. For each choice, calculate the phasor voltage V. Obtain the phasor voltage across the $3+j4\,\Omega$ impedance and compare with V.

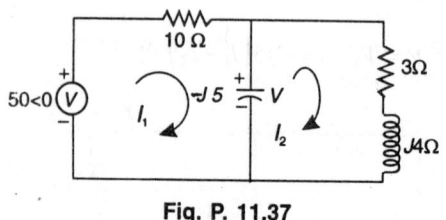

Fig. P. 11.37

Solution:

From mesh (1)

$$10I_1 - j5(I_1 - I_2) = 50\angle0° \tag{1}$$

$$3I_2 + j4I_2 - j5(I_2 - I_2) = 0 \tag{2}$$

or $\begin{bmatrix} 10-j5 & j5 \\ j5 & 3-j_1 \end{bmatrix}\begin{bmatrix} I_1 \\ I_2 \end{bmatrix} = \begin{bmatrix} 50\angle0° \\ 0 \end{bmatrix}$

$\therefore$ $\quad\quad\quad \Delta_z = \begin{vmatrix} 10-j5 & j5 \\ j5 & 3-j1 \end{vmatrix}$

$\quad\quad\quad\quad\quad = (10-j5)(3-j1) - j5(j5)$

$\quad\quad\quad\quad\quad = (11.18\angle-26.56)(3.162\angle-18.43) + 25$

$\quad\quad\quad\quad\quad = 35.35\angle-45° + 25 = 25 - j25 + 25 = 50 - j25$

$\Delta_z = 55.9\angle-26.57°\,\Omega$ **Ans.**

$$I_1 = \frac{\begin{vmatrix} 50\angle0° & j5 \\ 0 & 3-j1 \end{vmatrix}}{\Delta_2} = \frac{50(3.16\angle-18.43)}{55.9\angle-26.57°}$$

$$I_1 = 2.82 \angle 8.14°$$

$$I_2 = \frac{\begin{vmatrix} 11.18\angle{-26.56} & 50\angle0° \\ 5\angle{+90°} & 0 \end{vmatrix}}{\Delta_z} = \frac{-50\angle0°\times5\angle90°}{55.9\angle{-26.57°}}$$

$$I_2 = -4.47\angle116.57°$$

$$V = V_{3+j4} = -j5(I_1 - I_2)$$

$$= -j5(2.8 + j0.4 - 2 + j4)$$

$$= -j5(0.8 + j4.4) = 22 - j4$$

$$V = V_{3+j4} = 22.36\angle{-10.30°} \quad \textbf{Ans.}$$

Problem 11.38. Find the network of Fig. P. 11.38 use the mesh current method to find the current in the $2 + j3\Omega$ impedance due to each of the source V_1 and V_2.

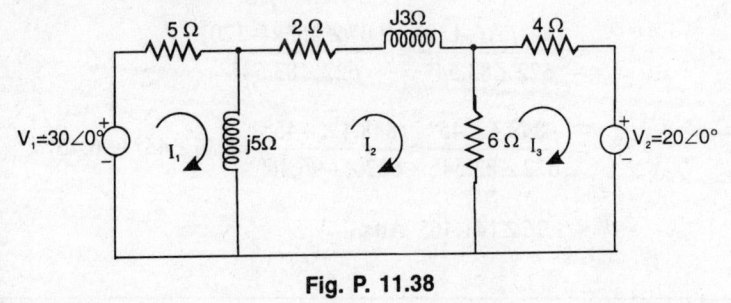

Fig. P. 11.38

Solution:

$$5I_1 + j5(I_1 - I_2) = 30\angle0° \tag{1}$$

$$2I_2 + j3I_2 + 6(I_2 - I_3) + j5(I_2 - I_1) = 0 \tag{2}$$

$$4I_3 + 6(I_3 - I_2) = 20\angle0° \tag{3}$$

or

$$\begin{bmatrix} (5+j5) & -j5 & 0 \\ -j5 & (8+j8) & -6 \\ 0 & -6 & 10 \end{bmatrix} \begin{bmatrix} I_1 \\ I_2 \\ I_3 \end{bmatrix} = \begin{bmatrix} 30\angle0° \\ 0 \\ 20\angle0° \end{bmatrix}$$

I_2 (when $V_2 = 0$), which flows in $(2 + j3)\Omega$

$$I_2 = \frac{\begin{vmatrix} 5+j5 & 30 & 0 \\ -j5 & 0 & -6 \\ 0 & 0 & 10 \end{vmatrix}}{\begin{vmatrix} 5+j5 & -j5 & 0 \\ -j5 & 8+j8 & -6 \\ 0 & -6 & 10 \end{vmatrix}} = \frac{j1500}{622\angle 83.54°} = \frac{1500\angle 90°}{622\angle 83.54°}$$

$$I_2 = 2.41\angle 6.45°$$

Similarly I_2 when $V_1 = 0$ is given as:

$$I_2' = \frac{\begin{vmatrix} 5+j5 & 0 & 0 \\ -j5 & 0 & -6 \\ 0 & 20 & 10 \end{vmatrix}}{\begin{vmatrix} 5+j5 & -j5 & 0 \\ -j & 8+j8 & -6 \\ 0 & -6 & 10 \end{vmatrix}}$$

$$= \frac{(5+j5)(-120)}{622\angle 83.54°} = \frac{7.07\angle 45° \times (-120)}{622\angle 83.54°}$$

$$= \frac{-848.4\angle 45°}{622\angle 83.54°} = \frac{848.4\angle -45°}{622\angle -96.46°} = 1.36\angle 45° + 96.46°$$

$$I_2' = 1.36\angle 141.46° \textbf{ Ans.}$$

Problem 11.39. For the network of Fig. P. 11.39, obtain the current ration I_1/I_3.

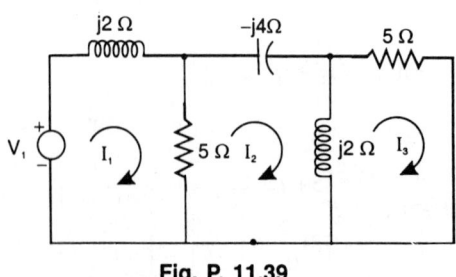

Fig. P. 11.39

Solution:

$$j2I_1 + 5(I_1 - I_2) = V_1 \tag{1}$$

$$-j4I_2 + j2(I_2 - I_3) + 5(I_2 - I_1) = 0 \tag{2}$$

$$5I_3 + j2(I_3 - I_2) = 0 \tag{3}$$

$$\begin{bmatrix} 5+j2 & -0 & 0 \\ -5 & 5-j2 & -j2 \\ 0 & -j2 & 5+j2 \end{bmatrix} \begin{bmatrix} I_1 \\ I_2 \\ I_3 \end{bmatrix} = \begin{bmatrix} V_1 \\ 0 \\ 0 \end{bmatrix}$$

$$I_1 = \begin{bmatrix} V_1 & -5 & 0 \\ 0 & 5-j2 & -j2 \\ 0 & -j2 & 5+j2 \end{bmatrix} \div \Delta 2$$

$$I_3 = \begin{bmatrix} 5+j2 & -5 & V_1 \\ -5 & 5-j2 & 0 \\ 0 & -j2 & 0 \end{bmatrix} \div \Delta 2$$

or

$$\frac{I_1}{I_3} = \frac{V_1((5-j2)(5+j2)+4)}{V_1(j10)} = \frac{(25+4+4)}{j10}$$

$$\frac{I_1}{I_3} = \frac{33}{j10} = 3.3 \angle -90° \text{ **Ans.**}$$

Problem 11.40. For the network of Fig. P. 11.38, obtain $z_{input\ 7}$ and $z_{transfer\ 13}$. Show that $z_{transfer\ 31} = z_{transfer\ 13}$.

Solution:

$$z_{input1} = \frac{V_1}{I_1} = \frac{V_1}{\dfrac{V_1(33)}{\Delta z}} = \frac{\Delta z}{33}$$

$$\Delta z = \begin{vmatrix} 5+j2 & -5 & 0 \\ -5 & 5-j2 & -j2 \\ 0 & -j2 & 5+j2 \end{vmatrix}$$

$(5 + j2) ((5 - j2) (5 + j2) + 4) + 5 (- 5 (5 + j2)$

$$= (5 + j2) (33) - 25(5 + j2)$$

$$= (5 + j2) (8)$$

$$= 43.08\angle21.8$$

$\therefore$ $$Z_{input\,1} = \frac{43.08\angle21.8°}{33}$$

$$Z_{input\,1} = 1.30\angle21.8° \textbf{ Ans.}$$

$$Z_{transfer\,13} = \frac{V_1}{I_3} = \frac{V_1 \times \Delta z}{j10V_1} = \frac{43.08\angle21.18°}{10\angle90°}$$

$$Z_{transfer\,3,1} = \frac{\Delta_z}{\Delta_{31}} = \frac{43.08\angle21.8°}{j10} = 4.308\angle-68.2°$$

$\therefore$ $$Z_{transfer\,1,3} = Z_{transfer\,3,1} \text{ Proved}$$

Problem 11.41. The network of Fig. P. 11.41, obtain the voltage ratio V_1/V_2 by application of the node voltage method.

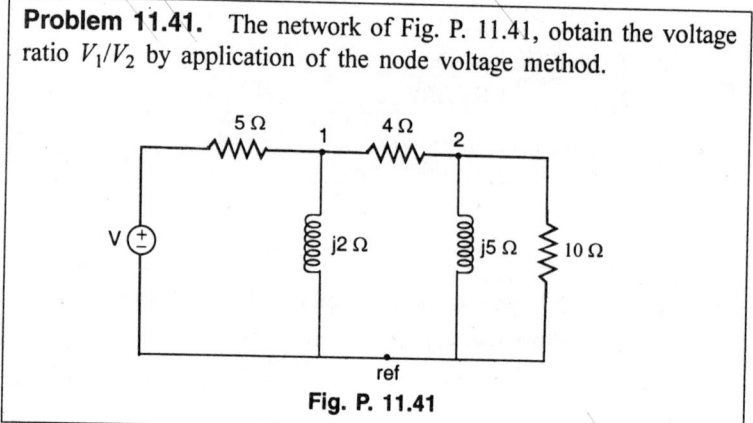

Fig. P. 11.41

Solution:

Applying node voltage method:

$$\frac{V_1-V}{5}+\frac{V_1}{j2}+\frac{V_1-V_2}{4} = 0 \qquad (1)$$

$$\frac{V_2-V_1}{4}+\frac{V_2}{j5}+\frac{V_2}{10} = 0 \qquad (2)$$

$$\left[\begin{array}{cc} \left(\dfrac{1}{5}+\dfrac{1}{4}+\dfrac{1}{j2}\right) & \dfrac{-1}{4} \\[3mm] \dfrac{-1}{4} & \left(\dfrac{1}{4}+\dfrac{1}{10}+\dfrac{1}{j5}\right) \end{array}\right]\left[\begin{array}{c} V_1 \\[2mm] V_2 \end{array}\right]=\left[\begin{array}{c} \dfrac{V}{5} \\[2mm] 0 \end{array}\right]$$

$\therefore \qquad V_1 = \begin{vmatrix} \dfrac{V}{5} & \dfrac{-1}{4} \\[3mm] 0 & \left(\dfrac{1}{4}+\dfrac{1}{10}+\dfrac{1}{j5}\right) \end{vmatrix} \div \Delta Z$

$\qquad\quad V_2 = \begin{vmatrix} \left(\dfrac{1}{5}+\dfrac{1}{4}+\dfrac{1}{j2}\right) & \dfrac{V}{s} \\[3mm] -\dfrac{1}{4} & 0 \end{vmatrix} \div \Delta Z$

$\therefore \qquad \dfrac{V_1}{V_2} = \dfrac{\dfrac{V}{5}\left(\dfrac{1}{4}+\dfrac{1}{10}+\dfrac{1}{j5}\right)}{\dfrac{V}{5}\left(\dfrac{1}{4}\right)}$

$\qquad \dfrac{V_1}{V_2} = \dfrac{(0.25+0.1-j0.2)}{0.25} = \dfrac{(0.35-j0.2)}{0.25} = \dfrac{0.40}{0.25}\angle -29.74°$

$\qquad \dfrac{V_1}{V_2} = 1.612 \angle -29.74°$ **Ans.**

12

Passive Network

Problem 12.1. Test whether the following function is a partial fraction.

$$f(s) = \frac{2s^4 + 7s^3 + 11s^2 + 12s + 4}{s^2 + 5s^3 + 9s^2 + 11s + 6}$$

Solution:

Suppose $P_1(s)$, $P_2(s)$ and $Q_1(s)$, $Q_2(s)$ are the even and odd parts of the numerator and denominator.

Hence

$$P_1(s) = 2s^4 + 11s^2 + 4$$
$$Q_1(s) = 7s^3 + 12s$$
$$P_2(s) = s^4 + 9s^2 + 6$$
$$Q_2(s) = 5s^3 + 11s$$

On the $j\omega$-axis $P_2^2(s)$ and Q_2^2 (s) being positive, for all ω if

$$\beta(\omega^2) = (P_1(s)P_2(s) - Q_1(s)Q_2(s))/_{s=j\omega} \geq 0$$

∴

$$\beta(\omega^2) = (P_1(s)P_2(s) - Q_1(s)Q_2(s))$$

$$= 2s^8 - 6s^6 - 22s^4 - 30s^2 + 24$$

$$= 2\omega^8 + 6\omega^6 - 22\omega^4 + 30\omega^2 + 24$$

Suppose $x = \omega^2$

$$\beta_0(x) = 2x^4 + 6x^3 - 22x^2 + 30x + 24$$

$$\beta_1(x) = 8x^3 + 18x^2 - 44x + 30$$

or

$$\frac{\beta_0(x)}{\beta_1(x)} = (0.25x + 0.1875) - \frac{14.375x^2 - 30.75x - 22.94}{\beta_1(x)}$$

$$\beta_2(x) = 14.375x^2 - 30.75x - 22.94$$

$$\frac{\beta_1(x)}{\beta_2(x)} = (0.556x + 2.44) - \frac{-43.73x - 85.97}{\beta_2(x)}$$

$$\beta_3(x) = -43.73x - 85.97$$

$$\frac{\beta_2(x)}{\beta_3(x)} = (-0.328x + 1.16) - \frac{-123}{A_3(x)}$$

$$\beta_4(x) = -123$$

Hence Sturm's function is expressed by

$$\beta_0(x) = 2x^4 + 6x^3 - 22x^2 + 30x + 24$$

$$\beta_1(x) = 8x^3 + 18x^2 - 44x + 30$$

$$\beta_2(x) = 14.375x^2 - 30.75x - 22.94$$

$$\beta_3(x) = -43.73x - 85.97$$

$$\beta_4(x) = -123$$

	A_0	A_1	A_2	A_3	A_4	No. of changes of sign
$x = 0$	+	+	−	−	−	$S_0 = 1$
$x = \infty$	+	+	+	−	−	$S_\infty = 1$

The no. of zeros in the interval $0 < x < \infty$ is

$$s_\infty - s_0 = 0$$

Therefore, Re. $(f(j\omega)) \geq 0$ for all ω

$$F(s) = \frac{m(s)}{n(s)}$$

The continued fraction of the ratio $\dfrac{P_1(s)}{Q_1(s)}$ and $\dfrac{P_2(s)}{Q_2(s)}$

$$\frac{P_1(S)}{Q_1(S)} = \frac{2s^4 + 11s^2 + 4}{7s^3 + 12s} = 0.285s + \cfrac{1}{0.924s + \cfrac{1}{0.912s + \cfrac{1}{2s}}}$$

$$\frac{P_2(s)}{Q_2(s)} = \frac{s^4+9s^2+6}{5s^3+11s} = 0.2s + \cfrac{1}{0.735s+\cfrac{1}{1.032s+\cfrac{1}{1.098s}}}$$

$m(s)$ and $n(s)$ are Hurwitz polynomial and $F(s)$ is a partial fraction.

Problem 12.2 If, $Y(s) = \dfrac{s(s+2)}{(s+1)(s+3)}$

Then find Foster's form of $R\text{-}C$ network

Solution:

At first we expand $Y(s)/s$ instead of $Y(s)$ otherwise signs of the residues of $Y(s)$ at its poles are negative hence

$$\frac{Y(s)}{s} = \frac{(s+2)}{(s+1)(s+3)} = \frac{1/2}{s+1} + \frac{1/2}{s+3}$$

$\therefore$ $$Y(s) = \frac{s/2}{s+1} + \frac{s/2}{s+3} = \frac{1}{2+\dfrac{2}{s}} + \frac{1}{2+\dfrac{6}{s}}$$

So network is realized by the figure below.

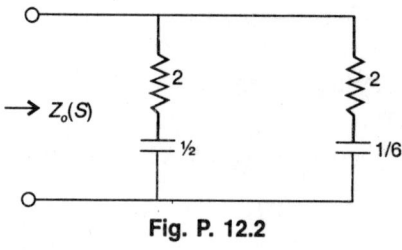

Fig. P. 12.2

Problem 12.3. Determine the Foster and Cauer form of realization of the given driving-point impedance function

$$Z(s) = \frac{4(s^2+1)(s^2+9)}{s(s^2+4)}$$

Solution:

We have
$$Z(s) = \frac{4(s^2+1)(s^2+9)}{s(s^2+4)}$$

After partial fraction, we have
$$Z(s) = 4s + \frac{9}{s} + \frac{15s}{s^2+4}$$

In the above term, $4s$ is recognised by impedance of four-unit inductor, $9/s$ is recognised by impedance of 1/9 unit capacitor.

For parallel LC branch, the impedance is
$$Z = \frac{s/C}{s^2+1/LC}$$

By direct comparison, $C = 1/1sF$, $L = \frac{15}{4}$ henry.

The network is realized as given below

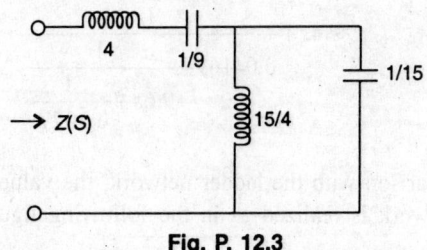

Fig. P. 12.3

Foster second form of realization can be obtained from the driving point admittance function as

$$\frac{Y(s)}{s} = \frac{(s^2+4)}{4(s^2+1)(s^2+9)} = \frac{3/32}{s^2+1} + \frac{5/32}{s^2+9}$$

So
$$Y(s) = \frac{(3/3s)s}{s^2+1} + \frac{(5/32)s}{s^2+9}$$

The above admittance function consists of two terms, each of which is realized by an inductor and capacitor in series. The admittance of such series tuned circuit is $Y = \dfrac{s/L}{s^2+1/LC}$

Foster second form is realized as given in the following network

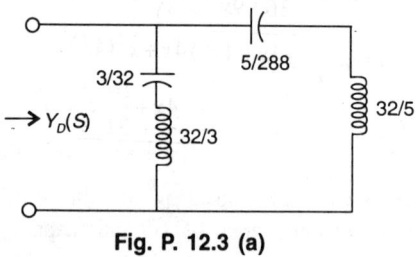

Fig. P. 12.3 (a)

For Cauer first form of realization from the driving point impedance function.

Continued fraction expansion

$$Z(s) = \frac{4s^2 + 40s^2 + 36}{s^3 + 4s}$$

$$= 4s + \cfrac{1}{0.0416s + \cfrac{1}{9.6s + \cfrac{1}{0.3077s}}}$$

By direct comparison with the ladder network, the value of L and C is found. The network is realized as in the following figure.

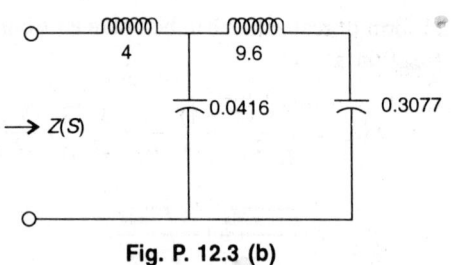

Fig. P. 12.3 (b)

The second form of ladder network realization is obtained by fraction expansion of Z/s otherwise function is exhausted. Putting $s^2 = x$

$$\therefore \qquad \frac{Z}{s} = \frac{4(x+1)(x+9)}{x(x+4)} = \frac{36 + 40x + 4x^2}{4x + x^2}$$

Passive Network

Continued fraction is

$$4x+x^2)\overline{36+40x+4x^2}(9/x$$
$$\underline{36+9x}$$
$$31x+4x^2)\overline{4x+x^2}(4/31$$
$$\underline{4x+\frac{16}{31}x^2}$$
$$\frac{15}{31}x^2)\overline{31x+4x^2}\left(\frac{961}{15x}\right.$$
$$\underline{31x}$$
$$4x^2)\frac{15}{31}x^2(\frac{15}{124}$$
$$\underline{\frac{15}{31}x^2}$$

Hence

$$\frac{Z}{s} = \frac{9}{s}+\cfrac{1}{\cfrac{4}{31}+\cfrac{1}{\cfrac{961}{15x}+\cfrac{1}{15/124}}}$$

utting back $x=s^2$ again

$$Z(s) = \frac{9}{s}+\cfrac{1}{\cfrac{4}{31s}+\cfrac{1}{\cfrac{961}{159}+\cfrac{1}{\cfrac{15}{124s}}}}$$

ie network is realized by

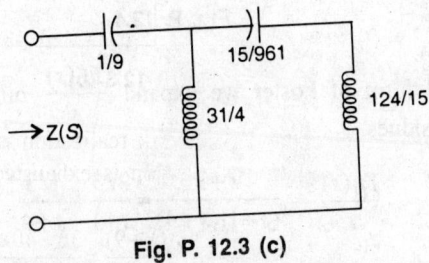

Fig. P. 12.3 (c)

Problem 12.4. Realize the two canonical foster networks from the RC driving point impedance function

$$Z_D(s) = \frac{(s+1)(s+4)}{s(s+3)}$$

Solution:

Two canonical Foster are first and second form of Foster.

First form of Foster can be realized by partial fraction expansion.

$$Z_D(s) = \frac{s^2+5s+4}{s^2+3s} = 1 + \frac{2s+4}{s^2+3s} = 1 + \frac{A_1}{s} + \frac{A_2}{s+3}$$

$$A_1 = \left.\frac{2s+4}{s+3}\right|_{s=0} = \frac{4}{3}$$

$$A_2 = \left.\frac{2s+4}{s}\right|_{s=-3} = \frac{2}{3}$$

Since residues are positive hence network can be realized

$$Z_D(s) = 1 + \frac{1}{\left(\frac{3}{4}\right)s} + \frac{2/3}{s+3}$$

The realized network is below

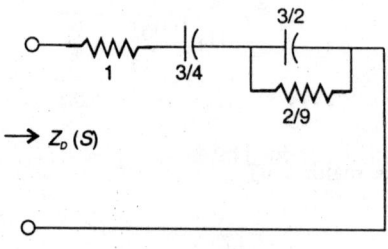

Fig. P. 12.4

For second form of Foster we expand $\dfrac{Y_D(s)}{s}$ otherwise $Y_D(s)$ has negative residues.

$$\therefore \qquad \frac{Y_D(s)}{s} = \frac{s+3}{(s+1)(s+4)} = \frac{2/3}{s+1} + \frac{1/3}{s+4}$$

$$Y_D(s) = \frac{(2/3)s}{s+1} + \frac{(1/3)s}{s+4}$$

Since the driving point impedance function has three numbers of internal poles and zeros, so networks realized contains four elements and hence are canonic.

By direct comparison

$$\rightarrow R_1 = \frac{3}{2}\Omega, C_1 = \frac{2}{3}F, R_2 = 3\Omega, C_2 = \frac{1}{12}F$$

The realized network is below.

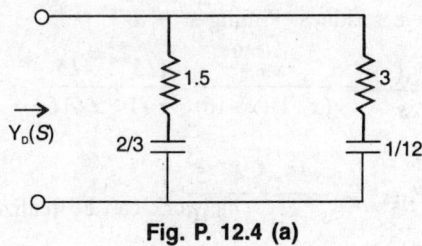

Fig. P. 12.4 (a)

Problem 12.5. Synthesize first and second Foster and Cauer forms of L-C driving-point impedance function

$$Z_D(s) = \frac{(s^2+1)(s^2+16)}{s(s^2+4)}$$

Solution:

Foster first form can be obtained by expanding $\dfrac{Z_D(s)}{s}$ by partial fraction expansion. Let's put $s^2 = x$

$$\frac{Z_D(s)}{S} = \frac{(x+1)(x+16)}{x(x+4)}$$

$$= 1 + \frac{4.25}{x} + \frac{8.75}{x+4}$$

Putting back $x = s^2$

$$Z_D(s) = s + \frac{4.25}{s} + \frac{8.75s}{s^2+4}$$

Realized network is below

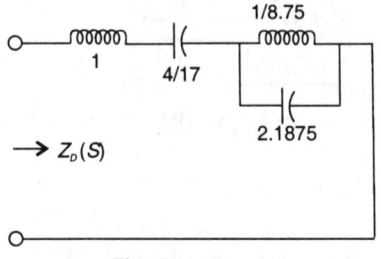

Fig. P. 12.5

Foster second form can be obtained by expanding $\dfrac{Y_D(s)}{s}$ otherwise $Y_D(s)$ has negative residues. Putting $s^2 = x$

$$\frac{Y_D(s)}{s} = \frac{x+4}{(x+1)(x+16)} = \frac{1/5}{x+1} + \frac{4/5}{x+16}$$

$$Y_D(s) = \frac{s/5}{s^2+1} + \frac{4s/5}{s^2+16}$$

After putting back $x = s^2$

By direct comparison

$$L_1 = 5H, C_1 = \frac{1}{5}F, L_2 = \frac{5}{4}H, C_2 = \frac{1}{20}F$$

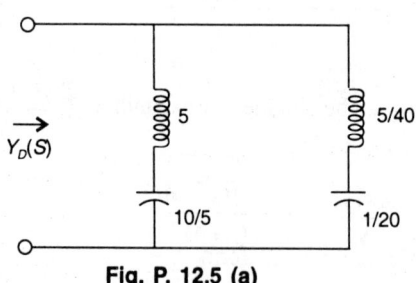

Fig. P. 12.5 (a)

Cauer first form can be obtained by expanding $Z_D(s)$

$$Z_D(s) = \frac{s^4 + 17s^2 + 16}{s^3 + 4s}$$

$$\begin{array}{r} s^3 + 4s \overline{\smash{\big)}\, s^4 + 17s^2 + 16} \ \underline{} \, s \to z \\ s^4 + 4s^2 \end{array}$$

$$\begin{array}{r} 13s^2 + 16 \overline{\smash{\big)}\, s^3 + 4s} \ \dfrac{s}{13} \to y \\ s^3 + \dfrac{16s}{13} \end{array}$$

$$\begin{array}{r} \dfrac{36s}{13} \overline{\smash{\big)}\, 13s^2 + 16} \ \dfrac{169s}{36} \to z \\ 13s^2 \end{array}$$

$$\begin{array}{r} 16 \overline{\smash{\big)}\, \dfrac{36s}{13}} \ \dfrac{36s}{208} \to Y \\ \dfrac{36s}{13} \end{array}$$

Hence,

$$Z_D(s) = s + \cfrac{1}{\cfrac{s}{13} + \cfrac{1}{\cfrac{169s}{36} + \cfrac{1}{\cfrac{36s}{208}}}}$$

The realized Cauer first form of LC network is given below

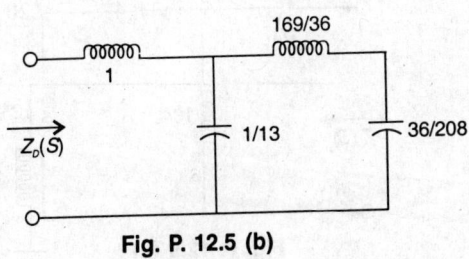

Fig. P. 12.5 (b)

For obtaining cauer second form we expand $Z_D(s)$ by fraction till all the terms are exhausting

$$Z_D(s) = \frac{16 + 17s^2 + s^4}{4s + s^3}$$

$$4s + s^3 \overline{\left| 16 + 17s^2 + s^4 \right.} \; 4/s \rightarrow z$$
$$\underline{16 + 4s^3}$$
$$13s^2 + s^4 \overline{\left| 4s + s^3 \right.} \; \dfrac{4}{13s} \rightarrow y$$
$$\underline{4s + \dfrac{4s^3}{13}}$$
$$\dfrac{9s^3}{13} \overline{\left| 13s^2 + s^4 \right.} \; \dfrac{169}{9s} \rightarrow z$$
$$\underline{13s^2}$$
$$s^4 \overline{\left| \dfrac{9s^3}{13} \right.} \; \dfrac{9}{13s} \rightarrow Y$$
$$\underline{\dfrac{9}{13} s^3}$$

Hence,

$$Z_D = \frac{4}{5} + \cfrac{1}{\cfrac{4}{13s} + \cfrac{1}{\cfrac{169}{9s} + \cfrac{1}{\cfrac{9}{13s}}}}$$

The realized network are given below $\rightarrow$

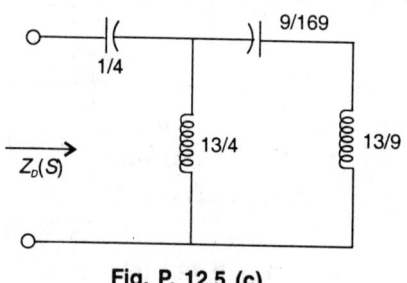

Fig. P. 12.5 (c)

Problem 12.6. The RC driving-point impedance function is given as

$$Z_D(s) = \frac{H(s+1)(s+4)}{s(s+3)}$$

Realize the impedance function in ladder form given $Z_D(-2) = 1$

Solution:

$$Z_D(-2) = 1 \leftarrow \text{given in question}$$

$$1 = H\left[\frac{(s+1)(s+4)}{s(s+3)}\right]_{s=-2} = H$$

$$\therefore \qquad H = 1$$

Hence the impedance function will be

$$Z_D(s) = \frac{(s+1)(s+4)}{s(s+3)}$$

The Cauer first form can be obtained by fraction expansion of $Z_D(s)$ till all the terms are exhausted

$$Z_D(s) = \frac{s^2+5s+4}{s^2+3s} = 1 + \frac{2s+4}{s^2+3s}$$

$$= 1 + \frac{1}{\dfrac{s^2+3s}{2s+4}} = 1 + \frac{1}{0.5s + \dfrac{s}{2s+4}}$$

$$= 1 + \frac{1}{0.5s + \dfrac{1}{\dfrac{2s+4}{s}}} = 1 + \frac{1}{0.5s + \dfrac{1}{2 + \dfrac{1}{0.25s}}}$$

Hence, the network is

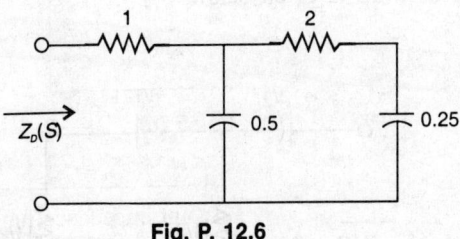

Fig. P. 12.6

We expand $Z_D(s)$ by fraction for second form of Cauer realization.

$$Z_D(s) = \frac{4+5s+s^2}{3s+s^2}$$

$$3s + s^2 \overline{\left| 4 + 55 + s^2 + \frac{4}{3s} \right.} \to \frac{1}{C_1 s}$$

$$\frac{4 + \dfrac{4s}{3}}{}$$

$$\frac{11s}{3} + s^2 \overline{\left| 3s + s^2 \right.} \dfrac{9}{11} \to \frac{1}{R_2}$$

$$\frac{3s + \dfrac{9}{11}s^2}{}$$

$$\frac{2}{11}s^2 \overline{\left| \dfrac{11}{3}s + s^2 \right.} \dfrac{121}{6s} \to \frac{1}{C_3 s}$$

$$\frac{\dfrac{11}{3}s}{}$$

$$s^2 \overline{\left| \dfrac{2s^2}{11} \right.} \dfrac{2}{11} \to \frac{1}{R_4}$$

$$\frac{\dfrac{2s^2}{11}}{}$$

$$Z_D(-s) = \frac{4}{3s} + \cfrac{1}{\cfrac{9}{11} + \cfrac{1}{\cfrac{121}{6s} + \cfrac{1}{\cfrac{2}{11}}}}$$

The resulting network is given below

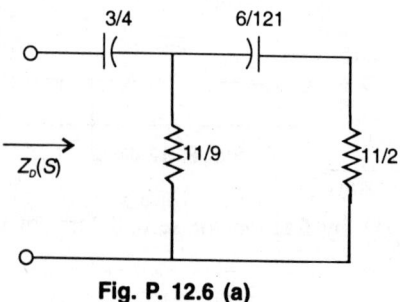

Fig. P. 12.6 (a)

Problem 12.7. Determine the Foster first form and Cauer second form after sythesizing the RL driving-point impedance function.

$$Z(s) = \frac{2(s+1)(s+3)}{(s+2)(s+4)}$$

Solution:

For Foster first form, we expand $\dfrac{Z(s)}{s}$ by partial fraction and then multiply by s. We do not expand $Z(s)$ by partial fraction because this will result negative residues at $s = -2$, and $s = -4$.

$$\frac{Z(s)}{s} = \frac{2(s+1)(s+3)}{s(s+2)(s+4)} = \frac{A_1}{s} + \frac{A_2}{s+2} + \frac{A_3}{s+4}$$

$$A_1 = \frac{2(s+1)(s+3)}{(s+2)(s+4)}\bigg|_{s=0} = \frac{3}{4}$$

$$A_2 = \frac{2(s+1)(s+3)}{s(s+4)}\bigg|_{s=-2} = \frac{1}{2}$$

$$A_3 = \frac{2(s+1)(s+3)}{s(s+2)}\bigg|_{s=-4} = \frac{3}{4}$$

After multiplying by s and putting the value

$$Z(s) = \frac{3}{4} + \frac{(1/2)s}{s+2} + \frac{(3/4)s}{s+4}$$

By direct comparison, we get values

$$R = \frac{3}{4}\Omega, R_1 = \frac{1}{2}\Omega, L_1 = \frac{1}{4}H, R_2 = \frac{3}{4}\Omega, L_2 = \frac{3}{16}H$$

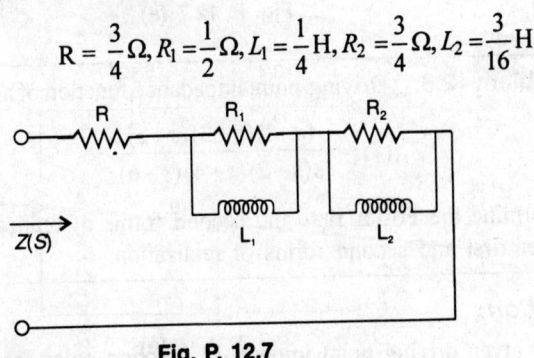

Fig. P. 12.7

Cauer second form of realization is obtained by repeated removal of poles at origin.

We continued fraction expansion

$$Z(s) = \frac{2(s+1)(s+3)}{(s+2)(s+4)} = \frac{6+8s+2s^2}{8+6s+s^2}$$

$$= 0.75 + \frac{3.4s+1.25s^2}{8+6s+s^2} = 0.75 + \cfrac{1}{\cfrac{8+6s+s^2}{3.5s+1.25s^2}}$$

$$= 0.75 + \cfrac{1}{\cfrac{2.286}{s} + \cfrac{3.143s+s^2}{3.5s+1.25s^2}}$$

Hence, $Z(s) = 0.75 + \cfrac{1}{\cfrac{1}{0.4375s} + 1.1135 + \cfrac{1}{\cfrac{1}{0.434s} + \cfrac{1}{0.1365}}}$

So, the Cauer second form of RL network is realized is given below

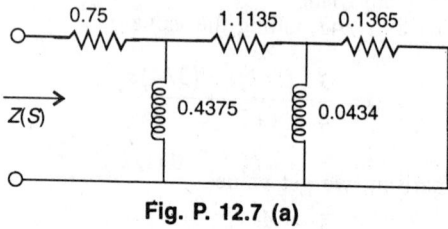

Fig. P. 12.7 (a)

Problem 12.8.   Driving-point impedance function of an RC network

$$Z_D(s) = \frac{(s+1)(s+3)(s+5)}{s(s+2)(s+4)(s+6)}$$

Determine the Foster first and second forms of realization and the Cauer first and second forms of realization.

Solution:

In the given driving point impedance function poles are real, negative and simple.

(i) First form of Foster realization can be obtained by the partial fraction expansion of Z_D

$$Z_D(s) = \frac{A_o}{s} + \frac{A_1}{s+2} + \frac{A_2}{s+4} + \frac{A_3}{s+6}$$

$$A_o = sZ_D(s)\Big|_{s=0} = 0.3125$$

$$A_1 = (s+2)Z_D(s)\Big|_{s=2} = 0.1875$$

$$A_2 = (s+4)Z_D(s)\Big|_{s=-4} = 0.1875$$

$$A_3 = (s+6)Z_D(s)\Big|_{s=-6} = 0.3125$$

All residues are positive. Putting the value of A_0, A_1, A_2, A_3. Therefore

$$Z_D(s) = \frac{0.3125}{s} + \frac{0.1875}{s+2} + \frac{0.1875}{s+4} + \frac{0.3125}{s+6}$$

With direct comparison

$$C_0 = 1/A_0 = 3.2\,\text{F}$$

$$C_1 = 1/A_1 = 5.33\,\text{F}$$

$$R_1 = -s_1/C_1 = 0.0937\,\Omega$$

$$C_2 = 1/A_2 = 5.33\,\text{F}$$

$$R_2 = -s_2/C_2 = 0.0469\,\Omega$$

$$C_3 = 1/A_3 = 3.2\,\Omega$$

$$R_3 = \frac{-s_3}{C_3} = 0.052\,\Omega$$

Since the total number of internal poles and zeros are 6 so the realized network contains 7 elements.

The realized network is given below

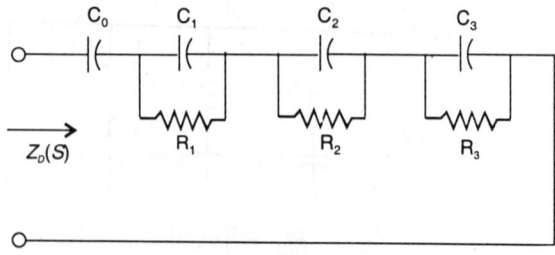

Fig. P. 12.8

(ii) Foster second form can be obtained by expanding $\dfrac{Y_D(s)}{s}$ otherwise $Y_D(s)$ has negative residues.

$$\therefore \qquad \frac{Y_D(s)}{s} = \frac{(s+2)(s+4)(s+6)}{(s+1)(s+3)(s+5)}$$

$$= \beta_0 + \frac{\beta_1}{s+1} + \frac{\beta_2}{s+3} + \frac{\beta_3}{s+5}$$

$$\beta_1 = \left[(s+1)Y_D(s)/s\right]_{s=-1} = 2.5$$

$$\beta_2 = \left[(s+3)Y_D(s)/s\right]_{s=-3} = 0.75$$

$$\beta_3 = \left[S+6 Y_D(s)/s\right]_{s=-5} = 0.375$$

$$\beta_0 = s \to \infty \left[\frac{Y_D(s)}{s}\right] = 1 = C_0$$

$$\therefore \qquad Y_D(s) = \beta_0 s + \frac{\beta_1 s}{s+1} + \frac{\beta_2 s}{s+3} + \frac{\beta_3 s}{s+5}$$

By direct comparison, we get

$$R_1 = 0.4\,\Omega \qquad C_1 = 2.5\text{F}$$

$$R_2 = 1.33\,\Omega \qquad C_2 = 0.25\text{F}$$

$$R_3 = 2.66\,\Omega \qquad C_3 = 0.075\text{F}$$

The realized network is

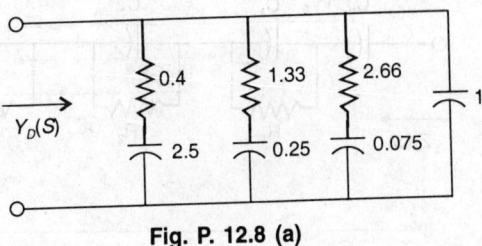

Fig. P. 12.8 (a)

(iii) Cauer first form of realization is obtained by continued fraction expansion

$$Z_D(s) = \cfrac{1}{s + \cfrac{1}{0.33 + \cfrac{1}{1.5s + \cfrac{1}{0.67 + \cfrac{1}{0.6s + \cfrac{1}{3.33 + \cfrac{1}{0.1s}}}}}}}$$

The Cauer first form of network realization is given below

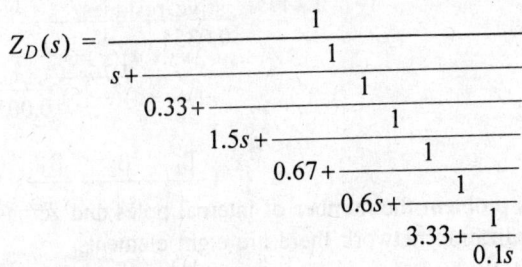

Fig. P. 12.8 (b)

Cauer second form is obtained by expanding the given driving-point function into continued fraction expansion about the origin.

In the given impedance function has a pole at the origin and is removed by the continued fraction expansion of $Z_D(s)$ dealing with zero freq. behaviour.

$$Z_D(s) = \frac{15 + 23s + 9s^2 + s^3}{48s + 44s^2 + 12s^3 + s^4}$$

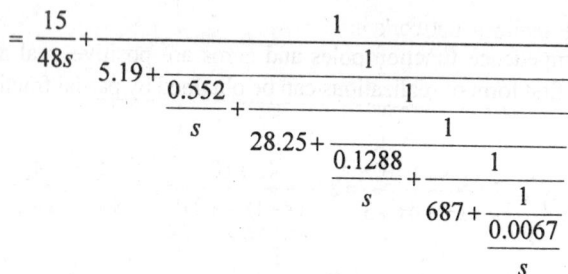

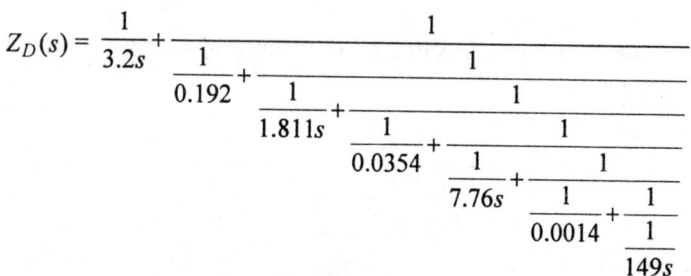

In this problem the number of internal poles and zeros is six, hence in the realization network there are eight elements.

The realization network is given below.

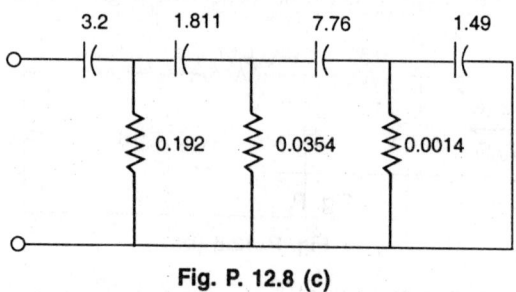

Fig. P. 12.8 (c)

Problem 12.9. Synthesize the Foster I and II forms of realization of the R–C driving point function

$$Z_D(s) = \frac{2s^2 + 12s + 16}{s^2 + 4s + 3}$$

Solution:

In the given impedence function poles and zeros are positive, real and simple, Foster first form of realizations can be obtained by partial fraction expansion

$$Z_D(s) = 2 + \frac{4s+10}{s^2+4s+3} = 2 + \frac{4s+10}{(s+1)(s+3)} = 2 + \frac{A_1}{S+1} + \frac{A_2}{s+3}$$

$$A_1 = \frac{4s+10}{s+3}\Big|_{s=-1} = 3$$

$$A_2 = \frac{4s+10}{s+1}\Big|_{s=-3} = 1$$

All residues are positive, hence, putting the value of A_1 and A_2

$$Z_D(s) = 2 + \frac{3}{s+1} + \frac{1}{s+3}$$

By direct comparison,

$$C_1 = \frac{1}{3}F, R_1 = 3\Omega, C_2 = 1F, R_2 = \frac{1}{3}\Omega \text{ and } R = 2\Omega$$

The realized network is below.

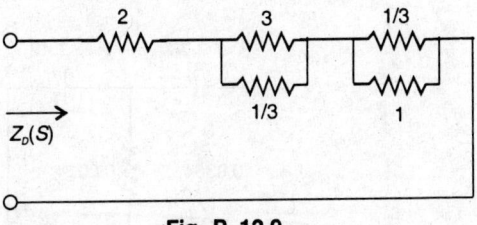

Fig. P. 12.9

Foster record form can be realized from pansion of $\dfrac{Y_D(s)}{s}$ by partial fraction expansion

$$\frac{Y_D(s)}{s} = \frac{s^2+4s+3}{2s^2+12s+16} = \left(\frac{s^2+4s+3}{2s^2+12s+16} - \frac{3}{16}\right) + \frac{3}{16}$$

$$Y_D(s) = \frac{s(5s+14)}{16(s+2)(s+4)} + \frac{3}{16}$$

$$\frac{Y_D(s)}{s} = \frac{5s+14}{16(s+2)(s+4)} + \frac{3}{16}$$

$$= \frac{1/8}{s+2} + \frac{3/16}{s+4} + \frac{3}{16s}$$

Since, $$Y_D(s) = \frac{(1/8)s}{s+2} + \frac{(3/16)s}{s+4} + \frac{3}{16}$$

The realized network is given below

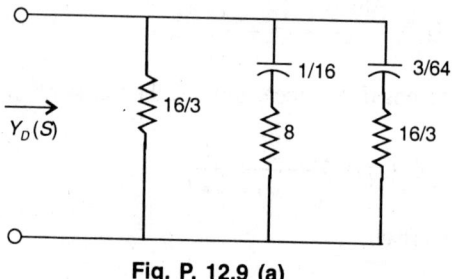

Fig. P. 12.9 (a)

Reader's Notes

Reader's Notes

Reader's Notes

Reader's Notes

Reader's Notes

Reader's Notes